Singular Limits of Dispersive Waves

NATO ASI Series

Advanced Science Institutes Series

A series presenting the results of activities sponsored by the NATO Science Committee, which aims at the dissemination of advanced scientific and technological knowledge, with a view to strengthening links between scientific communities.

The series is published by an international board of publishers in conjunction with the NATO Scientific Affairs Division

A	Life Sciences	Plenum Publishing Corporation
B	Physics	New York and London
C	Mathematical and Physical Sciences	Kluwer Academic Publishers
D	Behavioral and Social Sciences	Dordrecht, Boston, and London
E	Applied Sciences	
F	Computer and Systems Sciences	Springer-Verlag
G	Ecological Sciences	Berlin, Heidelberg, New York, London,
H	Cell Biology	Paris, Tokyo, Hong Kong, and Barcelona
I	Global Environmental Change	

Recent Volumes in this Series

Volume 317 —Solid State Lasers: New Developments and Applications
edited by Massimo Inguscio and Richard Wallenstein

Volume 318 —Relativistic and Electron Correlation Effects in Molecules and Solids
edited by G. L. Malli

Volume 319 —Statics and Dynamics of Alloy Phase Transformations
edited by Patrice E. A. Turchi and Antonios Gonis

Volume 320 —Singular Limits of Dispersive Waves
edited by N. M. Ercolani, I. R. Gabitov, C. D. Levermore, and
D. Serre

Volume 321 —Topics in Atomic and Nuclear Collisions
edited by B. Remaud, A. Calboreanu, and V. Zoran

Volume 322 —Techniques and Concepts of High-Energy Physics VII
edited by Thomas Ferbel

Volume 323 —Soft Order in Physical Systems
edited by Y. Rabin and R. Bruinsma

Volume 324 —On Three Levels: Micro-, Meso-, and Macro-Approaches in Physics
edited by Mark Fannes, Christian Maes, and André Verbeure

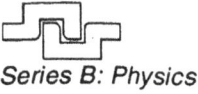

Series B: Physics

Singular Limits of Dispersive Waves

Edited by

N. M. Ercolani
University of Arizona
Tucson, Arizona

I. R. Gabitov
Landau Institute for Theoretical Physics
Moscow, Russia

C. D. Levermore
University of Arizona
Tucson, Arizona

and

D. Serre
Ecole Normale Superieure de Lyon
Lyon, France

Springer Science+Business Media, LLC

Proceedings of a NATO Advanced Research Workshop,
and of a Chaos, Order, and Patterns Panel sponsored workshop on
Singular Limits of Dispersive Waves,
held July 8–12, 1991,
in Lyons, France

NATO-PCO-DATA BASE

The electronic index to the NATO ASI Series provides full bibliographical references (with
keywords and/or abstracts) to more than 30,000 contributions from international scientists
published in all sections of the NATO ASI Series. Access to the NATO-PCO-DATA BASE is
possible in two ways:

—via online FILE 128 (NATO-PCO-DATA BASE) hosted by ESRIN, Via Galileo Galilei,
I-00044 Frascati, Italy

—via CD-ROM "NATO Science and Technology Disk" with user-friendly retrieval software in
English, French, and German (©WTV GmbH and DATAWARE Technologies, Inc. 1989). The
CD-ROM also contains the AGARD Aerospace Database.

The CD-ROM can be ordered through any member of the Board of Publishers or through
NATO-PCO, Overijse, Belgium.

Library of Congress Cataloging-in-Publication Data

Singular limits of dispersive waves / edited by N.M. Ercolani ... [et
al.].
        p.   cm. -- (NATO ASI Series. Series B, Physics ; 320)
    "Proceedings of a NATO Advanced Research Workshop, and of a Chaos,
Order, and Patterns Panel sponsored workshop on Singular Limits of
Dispersive Waves, held July 8-12, 1991, in Lyons, France"--T.p.
verso.
    Includes bibliographical references and index.
    ISBN 978-0-306-44628-3   ISBN 978-1-4615-2474-8 (eBook)
    DOI 10.1007/978-1-4615-2474-8
    1. Wave equation--Congresses.   2. Mathematical physics--Asymptotic
theory--Congresses.   I. Ercolani, Nicolas Michael.   II. Series.
QC174.26.W28S52   1994
530.1'24--dc20   94-891
                                                                    CIP

Additional material to this book can be downloaded from http://extra.springer.com.

ISBN 978-0-306-44628-3

©1994 Springer Science+Business Media New York

Originally published by Plenum Press in1994

SPECIAL PROGRAM ON CHAOS, ORDER, AND PATTERNS

This book contains the proceedings of a NATO Advanced Research Workshop held within the program of activities of the NATO Special Program on Chaos, Order, and Patterns.

Volume 208 — MEASURES OF COMPLEXITY AND CHAOS
edited by Neal B. Abraham, Alfonso M. Albano,
Anthony Passamante, and Paul E. Rapp

Volume 225 — NONLINEAR EVOLUTION OF SPATIO-TEMPORAL STRUCTURES
IN DISSIPATIVE CONTINUOUS SYSTEMS
edited by F. H. Busse and L. Kramer

Volume 235 — DISORDER AND FRACTURE
edited by J. C. Charmet, S. Roux, and E. Guyon

Volume 236 — MICROSCOPIC SIMULATIONS OF COMPLEX FLOWS
edited by Michel Mareschal

Volume 240 — GLOBAL CLIMATE AND ECOSYSTEM CHANGE
edited by Gordon J. MacDonald and Luigi Sertorio

Volume 243 — DAVYDOV´S SOLITON REVISITED: Self-Trapping of Vibrational
Energy in Protein
edited by Peter L. Christiansen and Alwyn C. Scott

Volume 244 — NONLINEAR WAVE PROCESSES IN EXCITABLE MEDIA
edited by Arun V. Holden, Mario Markus, and Hans G. Othmer

Volume 245 — DIFFERENTIAL GEOMETRIC METHODS IN THEORETICAL
PHYSICS: Physics and Geometry
edited by Ling-Lie Chau and Werner Nahm

Volume 256 — INFORMATION DYNAMICS
edited by Harald Atmanspacher and Herbert Scheingraber

Volume 260 — SELF-ORGANIZATION, EMERGING PROPERTIES, AND
LEARNING
edited by Agnessa Babloyantz

Volume 263 — BIOLOGICALLY INSPIRED PHYSICS
edited by L. Peliti

Volume 264 — MICROSCOPIC ASPECTS OF NONLINEARITY IN CONDENSED
MATTER
edited by A. R. Bishop, V. L. Pokrovsky, and V. Tognetti

Volume 268 — THE GLOBAL GEOMETRY OF TURBULENCE: Impact of Nonlinear
Dynamics
edited by Javier Jiménez

SPECIAL PROGRAM ON CHAOS, ORDER, AND PATTERNS

Volume 270 — COMPLEXITY, CHAOS, AND BIOLOGICAL EVOLUTION
edited by Erik Mosekilde and Lis Mosekilde

Volume 272 — PREDICTABILITY, STABILITY, AND CHAOS IN N-BODY
DYNAMICAL SYSTEMS
edited by Archie E. Roy

Volume 276 — GROWTH AND FORM: Nonlinear Aspects
edited by M. Ben Amar, P. Pelcé, and P. Tabeling

Volume 278 — PAINLEVÉ TRANSCENDENTS: Their Asymptotics and Physical
Applications
edited by Decio Levi and Pavel Winternitz

Volume 280 — CHAOS, ORDER, AND PATTERNS
edited by Roberto Artuso, Predrag Cvitanović, and Giulio Casati

Volume 284 — ASYMPTOTICS BEYOND ALL ORDERS
edited by Harvey Segur, Saleh Tanveer, and Herbert Levine

Volume 291 — PROTEIN TRANSFER IN HYDROGEN-BONDED SYSTEMS
edited by T. Bountis

Volume 292 — MICROSCOPIC SIMULATIONS OF COMPLEX HYDRODYNAMIC
PHENOMENA
edited by Michel Mareschal and Brad Lee Holian

Volume 298 — CHAOTIC DYNAMICS: Theory and Practice
edited by T. Bountis

Volume 312 — FUTURE DIRECTIONS OF NONLINEAR DYNAMICS IN PHYSICAL
AND BIOLOGICAL SYSTEMS
edited by P. L. Christiansen, J. C. Eilbeck, and R. D. Parmentier

Volume 320 — SINGULAR LIMITS OF DISPERSIVE WAVES
edited by N. M. Ercolani, I. R. Gabitov, C. D. Levermore, and
D. Serre

PREFACE

The subject of "Singular Limits of Dispersive Waves" had its modern origins in the 1960's when Whitham introduced the first systematic approach to the asymptotic analysis of nonlinear wavepackets. Initially developed through a variational principle applied to the modulation of families of traveling wave solutions, he soon realized that an efficient derivation of *modulation equations* could be accomplished by averaging local conservation laws. He carried out this analysis for a wide variety of dispersive nonlinear wave equations including the nonlinear Klein Gordon, KdV, and NLS equations.

The seminal work of Gardner, Greene, Kruskal and Miura led to the discovery of partial differential equations which are completely integrable through inverse spectral transforms. This provided a larger framework in which to develop modulation theory. In particular, one could consider the local modulation of families of quasiperiodic solutions with an arbitrary number of phases, extending the single phase traveling waves treated by Whitham. The first to extend Whitham's ideas to the multiphase setting were Flaschka, Forest and McLaughlin, who derived N-phase modulation equations for the KdV equation. By using geometric techniques from the theory of Riemann surfaces they presented these equations in Riemann invariant form and demonstrated their hyperbolicity.

By analyzing the limit of the inverse scattering transform, Lax and Levermore independently derived these same equations and obtained a global characterization of the zero dispersion limit for the Korteweg-deVries equation. More specifically, the global limit partitions space-time into regions in which the limit is locally governed by these multiphase modulation equations. This raised the fascinating possibility of phase transitions between regions corresponding to different modulated families.

These differing paths of investigation met in the work of Venakides on second order Lax-Levermore theory. By introducing a "quantization" of the KdV dispersive limit, he was able to argue convincingly that to second order in perturbation theory his limit matched the generalized Whitham ansatz, thus providing a firmer analytical model for modulation theory.

Parallel to these developments in the west, mathematicians in the Soviet Union were exploring the Hamiltonian structure of the modulation equations. Notably, Dubrovin and Novikov introduced a theory of "Hamiltonian systems of hydrodynamic type" which described families of Poisson brackets on the space of N-phase Riemann invariants with respect to which the modulation equations are Hamiltonian. An important recent breakthrough has been Tsarev's discovery that, with respect to these hydrodynamic Poisson structures, the modulation equations are formally completely integrable (in the sense that locally the characteristics of the system may be implicitly defined).

The welcome end of the cold war made possible the timely meeting between researchers in the east and west under the auspices of a NATO Advanced Research Workshop (the first time NATO had funded the participation of so many Soviet scientists). During the week 8-12 July 1991, in the pleasant environs of the Ecole Normale Superieure Lyon, ideas were exchanged and new friendships formed. All participants extend their gratitude to these institutions for the role they played in making this historic meeting possible.

This volume presents a partial record of this workshop. The papers in this collection have all been refereed, which we hope will make them more useful to the larger mathematical community. The editors would like to thank all those involved in this process for the extra time and effort taken to improve the quality of these proceedings.

The papers here fall into three general areas:

1. Whitham Modulation Equations and their Exact Solutions;
2. Asymptotics and Limits;
3. Existence and Regularity of Dispersive Waves.

WHITHAM MODULATION EQUATIONS AND THEIR EXACT SOLUTIONS: Averaging the local conserved densities of KdV provides some local invariants of the modulation equations; however, there are independent integrals arising from other symmetries of the averaged systems. The mechanism by which these invariants are generated stems from a deep relation between the explicit Poisson properties of the brackets and an associated flat metric on the manifold of Riemann invariants. The latter has classical antecedents in Darboux's theory of orthogonal coordinate systems. Tsarev's article in this volume contains further details on this beautiful geometric theory and references.

Tian has applied Tsarev's theory to explicitly work out the transition between zero and one phase representations of the Lax-Levermore weak limit. In particular he constructs an elegant integral representation based on Darboux's equation which permits one to solve, for fairly general local initial data, the "Cauchy problem" for the zero dispersion limit of KdV up through one phase transitions. Continuation through higher phases remains a fascinating open problem. Wright also employs Tsarev's approach to construct the minimizer of the Lax-Levermore variational problem for a restricted (analytic) class of initial data. The minimizer provides an explicit formula for the Young measure associated to the weak limit. This formula can in turn be used to calculate a part of the quantized phases in Venakides' tau function for higher order Lax-Levermore theory.

In regard to this description of phase transitions we also mention the pioneering work of Gurevich and Pitaevsky who numerically investigated self-similar solutions of Whitham's equations in order to model dispersive shocks. They and their collaborators have continued these studies for other modulation problems such as averaged NLS (defocussing). The paper of El', Gurevich and Krylov describes this ongoing research.

Bloch and Kodama explore the problem of regularizing shocks in the modulation equations themselves by formally matching to modulation equations with a larger number of phases. They carry out this study for the Toda shock problem and relate it to the asymptotic and numerical studies of Deift, Venakides and their collaborators.

For focusing NLS and Sine-Gordon the corresponding averaged conservation laws give modulation equations with complex speeds. It is difficult to know what, if any, interpretation may be given to these equations. At this stage most attempts to understand the relation of elliptic modulation equations to the behaviour of oscillations has been numerical. Bronski and McLaughlin introduce a scaled, weakly nonlinear ansätz to describe the saturation of modulational instabilities up to the onset of wavebreaking. Forest and Sinha numerically solve modulation equations adapted to a periodic Sine-Gordon equation in order to resolve the structure of attractors in a numerically observed pre-chaotic regime of a damped driven Sine-Gordon equation.

An interesting variant on Tsarev's method is the analysis of so-called dispersionless hierarchies. This refers to simply setting the dispersive terms to zero in a given soliton equation hierarchy. For KdV, such a truncation does in fact represent a zero dispersion limit prior to wave-breaking. Geogjaev examines dispersionless hierarchies for the case of KdV and of the Benney equations. In particular, he finds a Cauchy integral representation for the characteristics valid up until the time that the first shock forms. Gibbons and Kodama extend this method to solve the initial value problem for any dispersionless Lax equation. Zakharov devlops a general scheme for constructing 2+1 dispersionless soliton equations. His method can reproduce dispersionless KP as well as a generalization of the Benney system to 2+1 dimensions.

In another direction, Novokshenov poses the question of what ideas from Whitham modulation theory may be incorporated into averaging techniques for classical mechanical systems. He makes a particular study of this for the Neumann system.

Finally, there are two papers in this volume which suggest potential applications of modulation theory to current problems in algebraic geometry and conformal field theory. Through the Darboux transformation, Previato examines KP flows associated to stationary Lax equations which represent deformations of vector bundles, rather than just line bundles, on a Riemann surface. These would entail flows on spaces more complicated than tori and could enlarge the scope of modulation theory. Grinevich describes symmetries of Whitham's equations which do not come from isospectral symmetries of KdV. Rather, they form a non-commutative algebra isomorphic to polynomial vector fields on the plane.

ASYMPTOTICS AND LIMITS: In contrast with those mentioned above who studied solutions of limiting equations, a number of papers studied limits of solutions.

The paper of Jin, Levermore and McLaughlin presents a numerical study of the semiclassical limit of the linear and the nonlinear Schrödinger equation in both the defocusing and focusing cases, along with an analytical study of the semiclassical limit of the defocusing case. Not suprisingly, the analytical study parallels the Lax-Levermore approach in carefully establishing the limit of the N-soliton formula. It contains a new convexity result for the N-soliton formula that applies to the KdV case too and had heretofore been overlooked.

The oscillatory nature of modulated solutions of the KdV equation was reviewed by both Bikbaev and Dobrokhotov. Bikbaev treats the long-time behavior of modulated genus g solutions with possibly differing asymptotic boundary conditions as $x \to \pm\infty$. He describes the long time self-similar matching between the two asymptotic boundary behaviours. Dobrokhotov on the other hand studies the solutions of the variational equations for the modulated 2-phase solutions of KdV. This is a natural extension of Whitham's original analysis of one phase modulations. However, this higher phase problem is much more complicated due to the presence of dense resonance points. As a consequence the first corrections in perturbation theory cannot be calculated from a linearized equation and so one must directly study a nonlinear equation.

The so-called Toda shock problem was examined by Kamvissis for the cases of data with critical and subcritical velocities, extending the earlier work of Venakides, Deift and Oba on the supercritical case. Liu and Levermore present some preliminary results from their experimental study of a nonintegrable conservative differencing of the Hopf (inviscid Burgers) equation. They exploit the fact that it is much easier to carry out high-resolution numerical experiments on spatially discrete systems and uncover phenomena that had not been found in integrable cases, identifying a number of ways in which modulated solutions can break down.

Deift and Zhou examine the long-time asymptotics for the autocorrelation function for the spin $-\frac{1}{2}$ XY model at the critical magnetic field, a problem that had a somewhat controversial history in the physics community. They transformed the problem into an oscillatory Riemann-Hilbert problem which could then be solved using methods they had developed in their study of long-time limits of solutions of integrable partial differential equations through the inverse scattering transform.

Finally, using PDE methods, Golse analyzes the "hydrodynamical limit" for systems of Lorentz-like noninteracting billiards that reflect off scatterers. For hard spheres he identifies three asymptotic regimes: the diffusion limit, the kinetic limit, and the streaming limit. He derives an expression for the diffusion coefficient that clearly exhibits its dependence on the geometry of the scatter array.

EXISTENCE AND REGULARITY OF DISPERSIVE WAVES: The rich structure of solitary wave equations has provided an impetus to develop a general theory of dispersive waves on a par with the well established dissipative theory. An important case of this is recent work on local smoothing. The paper by Kenig, Ponce and Vega, three of the principal experts in this field, analyzes the well-posedness of equations in a generalized KdV hierarchy:

$$\partial_t u + \partial_x^{2j+1} u + Q(u, \partial_x u, \ldots, \partial_x^{2j} u) = 0,$$

where Q is a general polynomial without constant or linear terms. The standard PDE techniques which work for dissipative equations cannot handle these. However by extending Kato's smoothing results for linear dispersive equations, questions of existence, uniqueness, and continuous dependence on initial data can be successfully treated.

Integrable PDE's have also encouraged the extension to infinite dimensions of problems and methods from Hamiltonian systems and symplectic geometry such as KAM and normal form theory. The paper by Craig and Wayne for instance studies the problem of existence of periodic solutions of the wave equation

$$u_{tt} - u_{xx} + g(x, u) = 0,$$

on $0 \leq x \leq \pi$ with periodic or Dirichlet boundary conditions in x. The method is perturbative near $u = 0$. The infinite dimensional setting introduces a small divisor problem. They indicate how this can be overcome at least in the case of a finite number of linear frequencies in resonance.

Alber and Marsden also examine hamiltonian structures but in the vicinity of hyperbolic critical points of the Arnold-Liouville foliation of finite phase solutions of integrable pdes. Examples of this are solitons riding over radiation or umbilic solutions of the Neumann problem associated to KdV. They investigate the monodromy of this foliation around the critical value.

Chernykh, Gabitov and Kuznetsov study the physically interesting problem of the propagation of defects in a 1D version of the Newell-Whitehead equation. Although this equation is dissipative rather than dispersive, their geometric approach to the evolution of coherent structures offers several points of analogy to dispersive counterparts such as NLS.

Finally, Prasad presents a novel approach to tracking a single shock front for a system of conservation laws.

The Editors

CONTENTS

WHITHAM MODULATION EQUATIONS AND THEIR EXACT SOLUTIONS

The Whitham Equation and Shocks in the Toda Lattice 1
    A.M. Bloch and Y. Kodama

Semiclassical Behavior in the NLS Equation: Optical Shocks-Focusing
    Instabilities 21
    J.C. Bronski and D.W. McLaughlin

A Numerical Study of Nearly Integrable Modulation Equations 39
    M.G. Forest and A. Sinha

The Quasiclassical Limit of the Inverse Scattering Method 53
    V.V. Geogjaev

Solving Dispersionless Lax Equations 61
    J. Gibbons and Y. Kodama

Nonisospectral Symmetries of the KdV Equations and the Corresponding
    Symmetries of the Whitham Equations 67
    P.G. Grinevich

Breaking Problem in Dispersive Hydrodynamics 89
    G.A. El', A.V. Gurevich, and A.L. Krylov

Whitham Deformations of Two-Dimensional Liouville Tori 105
    V.Y. Novokshenov

Higher Rank Darboux Transformations 117
    G. Latham and E. Previato

On the Initial Value Problem of the Whitham Averaged System 135
    F.R. Tian

On the Integrability of the Averaged KdV and Benney Equations.................. 143
    S.P. Tsarev

Explicit Construction of the Lax-Levermore Minimizer for the KdV Zero
     Dispersion Limit 157
  O.C. Wright

Dispersionless Limit of Integrable Systems in 2+1 Dimensions 165
  V.E. Zakharov

ASYMPTOTICS AND LIMITS

KdV Equation with Nontrivial Boundary Conditions at $x \to \pm\infty$ 175
  R.F. Bikbaev

Long-time Asymptotics for the Autocorrelation Function of the Transverse
     Ising Chain at the Critical Magnetic Field 183
  P. Deift and X. Zhou

Resonances in Multifrequency Averaging Theory 203
  S.Y. Dobrokhotov

Billiards Systems and the Transport Equation 219
  F. Golse

The Behavior of Solutions of the NLS Equation in the Semiclassical Limit 235
  S. Jin, C.D. Levermore and D.W. McLaughlin

Critical and Subcritical Cases of the Toda Shock Problem 257
  S. Kamvissis

EXISTENCE AND REGULARITY OF DISPERSIVE WAVES

Geometric Phases and Monodromy at Singularities 273
  M.S. Alber and J.E. Marsden

Nonlinear Waves and the 1:1:2 Resonance 297
  W. Craig and E. Wayne

Defects of One-Dimensional Vortex Lattices...................................... 315
  A.I. Chernykh, I.R. Gabitov and E.A. Kuznetsov

Oscillations Arising in Numerical Experiments 329
  C.D. Levermore and J.G. Liu

On the Hierarchy of the Generalized KdV Equations 347
  C.E. Kenig, G. Ponce and L. Vega

A New Theory of Shock Dynamics.. 357
  P. Prasad

Index 365

THE WHITHAM EQUATION AND SHOCKS IN THE TODA LATTICE

A.M. Bloch[1] and Y. Kodama[2]

Department of Mathematics
The Ohio State University
Columbus, OH 43210

Abstract: In this paper we present some results on the Whitham equations for the Toda lattice. In particular we show how one can regularize solutions with step initial data by choosing an appropriate Riemann surface on which the equations are defined, and we compare these results with the standard results for the KdV equation.

1. Introduction

In this paper we describe briefly some results on the Whitham equations for the Toda lattice and compare them with analogous results for the KdV equation. Details of our Toda lattice results will appear in the paper Bloch and Kodama (1991).

Our interest is in the asymptotic behavior of the infinite Toda lattice subject to a step-like initial condition. In Holian, Flaschka and McLaughlin (1981) the shock waves arising from such an initial condition were studied from a numerical point of view. Here we show how to characterize some of their observations in terms of the slow modulations of mutiphase wavetrains. An analytic explanation of the Holian et.

---

[1]Supported in part by NSF Grant DMS-90-02136 and NSF PYI Grant DMS-9157556.
[2]Supported in part by the NSF Grant DMS-9109041.

al. results has also recently been obtained independently by Venakides, Deift and Oba (1990). Their approach is quite different however, and relies on a remarkable explicit asymptotic solution of the original Toda lattice equations. We note also that our approach, in contrast to that of Venakides et. al., does not describe the behavior immediately around the shock point, but only as the system moves away from it. However the Whitham equations give a clear geometric picture of the macroscopic behavior of the lattice away from the shock point and are relatively easy to compute with.

Our approach is adopted from the study of the weak dispersion limit of the KdV equation. This problem has been studied in detail over the past two decades by using modulation theory (see Flaschka, Forest and McLaughlin (1980) and the earlier work of Whitham (1974)), and by using asymptotics and inverse scattering (see Lax and Levermore (1983) and Venakides (1985) and also Gurevich and Pitaevskii (1974)). These approaches link the averaged dynamics of g-phase wavetrains with the evolution of Abelian differentials on a Riemann surface of genus g. See also Bikbaev (1989), and the review paper Dubrovin and Novikov (1989) and the references therein.

Our interest in this problem arose in studying the classical zero dispersion limit of Toda (corresponding to a Riemann surface of genus 0) (see Brockett and Bloch (1990) and Kodama (1990)). In Brockett and Bloch (1990) we showed how shocks could develop in finite time from smooth initial data for this system.

To illustrate the idea of dispersive regularization of the Whitham equation we begin by analyzing a simple and well known KdV-Whitham problem. We recall that the presence of shocks in the Whitham equations corresponds to the development of oscillations in the original equations. The essential idea here is to choose the mininal genus such that the Whitham equation does not exhibit a shock for the given initial data. The genus g of the solution then indicates the expected physical behavior of the original system – the presence of g-phase wave trains. One of our key points in this paper is to highlight a subtle difference between the Toda and KdV case. For generic step initial data which leads to shock formation in the genus zero solution, in the KdV case regularization is achieved by going to genus one. In the Toda case, while regularization is achieved by going to genus one, in order to accurately reflect the behavior of the lattice, one needs to go to genus two. This observation was originally made in the analysis of Venakides et. al.(1990) and must be reflected in the Whitham approach.

We recall that the Whitham equations describe solutions of the given system which are slowly modulating g-phase wave trains. The equations are obtained by averaging out the fast oscillations. The spatial variable in the Whitham equations describes the motion of the modulating envelope of the system. Thus, even though our Toda system is discrete, the Whitham equations for it, like those for KdV, depend on a continuous spatial variable.

We begin here by directly writing down the Whitham equation for KdV. The derivation of the equation may be found for example in Flaschka, Forest and McLaughlin (1980). In the next section we outline briefly how to obtain the Whitham equations for Toda.

The Whitham equation for KdV is given by

$$\frac{\partial \omega_1}{\partial T} = \frac{\partial \omega_3}{\partial X} \,. \tag{1.1}$$

where ω_1 and ω_3 are Abelian differentials on the Riemann surface of genus g given by $y^2 = R_g(\lambda)$ where

$$R_g(\lambda) = \prod_{i=0}^{2g} (\lambda - \lambda_i) \,, \tag{1.2}$$

and the branch points λ_i are real and are assumed to satisfy $\lambda_0 < \lambda_2 < \cdots < \lambda_{2g}$.

Explicitly,

$$\omega_1(\lambda) = \frac{\prod_{i=1}^{g}(\lambda - \alpha_i)}{\sqrt{R_g(\lambda)}} d\lambda \,,$$

$$\sim \frac{d\lambda}{\sqrt{\lambda}} + (holomorphic), \quad \text{as } \lambda \to \infty \,, \tag{1.3}$$

$$\omega_3(\lambda) = \frac{\lambda^{g+1} - \frac{1}{2}\sigma_1 \lambda^g + \beta_1 \lambda^{g-1} + \cdots + \beta_g}{\sqrt{R_g(\lambda)}} d\lambda \,,$$

$$\sim \sqrt{\lambda} d\lambda + (holomorphic), \quad \text{as } \lambda \to \infty \,. \tag{1.4}$$

Here the coefficients $\{\alpha_j, \beta_j\}$ are determined by

$$\oint_{a_j} \omega_1 = \oint_{a_j} \omega_3 = 0, \quad j = 1, \ldots, g \,, \tag{1.5}$$

the vanishing of the contour integral along the canonical a_j-cycle and $\sigma_1 = \sum_{i=0}^{2g} \lambda_i$.

Then the averaged solution of the KdV equation is given by

$$\overline{u}(x,t) = \frac{1}{2} \sum_{i=0}^{2g} \lambda_i - \sum_{j=0}^{g} \alpha_j \, . \tag{1.6}$$

Note that in the case $g = 0$ equation (1.1) reduces to the dispersionless KdV (Hopf-Burgers) equation with $\lambda_0 = 2\overline{u}$ and $\lambda_1 = \cdots = \lambda_{2g} = \alpha_1 = \cdots = \alpha_g = 0$, i.e.,

$$\frac{\partial \overline{u}}{\partial T} = \overline{u} \frac{\partial \overline{u}}{\partial X} \, . \tag{1.7}$$

Let us now consider the initial value problem for (1.1) with the step initial condition,

$$\overline{u}(X,0) = \begin{cases} a, & X < 0, \\ b, & X > 0. \end{cases} \tag{1.8}$$

Now our goal in solving the initial value problem is to determine the minimal genus in (1.1) such that we have a global solution in a strong sense. Consider firstly then the case $g = 0$. The characteristic velocity of (1.7) is given by

$$\frac{dX}{dT} = -\overline{u} \, , \tag{1.9}$$

and the general solution is given in the hodograph (implicit) form,

$$X + \overline{u}T = f(\overline{u}), \tag{1.10}$$

where $f(\overline{u})$ is determined from the initial data $\overline{u}(X,0) = f^{-1}(X)$ in the region where it may be inverted. Now note that from (1.8) we have, for $b > a$, $\overline{u}^+ = \lim_{X \to 0+} \overline{u}(X,0) = b > \overline{u}^- = \lim_{X \to 0-} \overline{u}(X,0) = a$. Thus from the characteristic velocity equation (1.9) we can see that equation (1.7) with the initial data (1.8) (an increasing step function) leads to a shock wave solution. On the other hand we obtain a rarefaction wave solution for $b < a$.

For $b > a$ the solution becomes multivalued in the $X - T$ plane in the region between $X^- = -aT$ and $X^+ = -bT$ (one gets three solutions for each X). The theory of the weak dispersion limit of KdV (see Lax and Levermore [1983] and also Bikbaev and Novokshenov [1988]) tells us that the generation of a shock wave corresponds to the generation of quasi-periodic solutions in the original KdV equation, i.e. to changing the genus of the Riemann surface to one of higher genus. Thus after shock formation we have to consider the Whitham equation (1.1) with nonzero genus in order to obtain a global solution.

Consider therefore the case $g = 1$. Then the initial data (1.8) with $b > a$ gives

$$
\begin{aligned}
\lambda_0(X,0) &= 2a, & \forall X, \\
\lambda_1(X,0) &= 2b, & X < 0, \\
&= 2a, & X > 0, \\
\lambda_2(X,0) &= 2b, & \forall X.
\end{aligned}
\tag{1.11}
$$

Note that the initial data $\lambda_1(X,0)$ is a decreasing function of X, while $\overline{u}(X,0)$ is increasing. With this initial data, the Whitham equation with $g = 1$ has a global solution which is obtained as follows. Equation (1.1) with initial data (1.11) can be reduced to a single equation for $\lambda_1(X,T)$:

$$
\frac{\partial \lambda_1}{\partial T} = s(\lambda_1)\frac{\partial \lambda_1}{\partial X},
\tag{1.12}
$$

with $\lambda_0 = 2a$ and $\lambda_2 = 2b$ for all $T \geq 0$. Then computing $s(\lambda_1)$ explicitly for $X \to 0^{\pm}_-$, we obtain $s_1(\lambda_1^+) = \lim_{X \to 0^+} s(\lambda_1) = (2a + b)/3 < s_1(\lambda_1^-) = \lim_{X \to 0^-} s(\lambda_1) = 2b - a$, where $s(\lambda_1)$ is given by

$$
s(\lambda_1) = \frac{\omega_3(\lambda_1)}{\omega_1(\lambda_1)} = \frac{\lambda^2 - \frac{1}{2}\sigma_1\lambda + \beta_1}{\lambda - \alpha_1}.
\tag{1.13}
$$

The characteristic velocity of $\lambda_1(X,T)$ is given by

$$
\frac{dX}{dT} = -s(\lambda_1).
\tag{1.14}
$$

Note that in contrast to the Burgers case where we had $\overline{u}^+ > \overline{u}^-$ and increasing step initial data for $b > a$, here $s_1(\lambda_1^+) < s_1(\lambda_1^-)$ and $\lambda_1(X,0)$ is a decreasing step function. Hence from the characteristic velocity equation (1.14) we see we get a global solution, i.e. a rarefaction wave. (See Levermore (1988), Bikbaev and Novokshenov (1988).)

2. The averaged (Whitham) equations for the Toda lattice

To obtain the Whitham equations which are given by the averaged conservation laws, we first briefly summarize the iso-spectral theory of the Toda lattice equation, and derive the conservation laws.

Recall that in (essentially) Flaschka's form (Flaschka (1976), Moser (1974)) the Toda lattice equations may be written

$$\dot{a}_k = b_{k+1} - b_k,$$
$$\dot{b}_k = b_k(a_k - a_{k-1}), \quad k \in \mathbb{Z}. \tag{2.1}$$

This asymmetrical form of the Toda lattice equations (see Kupershmidt (1985)) is more convenient for our purposes than the original symmetrical form of Flaschka.

The equations (2.1) can be derived from the compatibility conditions for the spectral equations,

$$L\psi = \lambda\psi,$$
$$\psi_t = B\psi, \tag{2.2}$$

where the operators L and B, called the Lax pair, are given by

$$L = \Delta + u_0 + u_1\Delta^{-1},$$
$$B = \Delta + u_0. \tag{2.3}$$

Here Δ is the unit shift operator $\exp(\partial/\partial k)$, ie. $\Delta f(k,t) = f(k+1,t)$, $f : \mathbb{Z} \times \mathbb{R} \to \mathbb{R}$, and $u_0(k,t) = a_k(t)$, $u_1(k,t) = b_k(t)$. In terms of these variables, (2.1) becomes

$$\frac{\partial u_0}{\partial t} = (\Delta - 1)u_1,$$
$$\frac{\partial u_1}{\partial t} = u_1(1 - \Delta^{-1})u_0. \tag{2.4}$$

Note that $B = (L)_+ = L - (L)_-$ where $(\ )_+$ denotes the polynomial part in Δ of L. Let Δ and $(L)_-$ be formal series in L,

$$\Delta = L - \sum_{i=0}^{\infty} G_i L^{-i}, \tag{2.5}$$

$$(L)_- = \sum_{i=1}^{\infty} F_i L^{-i}, \tag{2.6}$$

where $F_i = F_i(u_0, u_1)$ and $G_i = G_i(u_0, u_1)$. Then we have:

Lemma 2.1. *(Kupershmidt (1985))*

The conservation laws for the system (2.4) are given by

$$\frac{\partial}{\partial t} \ln\left(\frac{\Delta\psi}{\psi}\right) = \frac{\partial}{\partial t} \ln(\lambda - \sum_{i=0}^{\infty} G_i \lambda^{-i})$$

$$= (\Delta - 1)\left(\frac{\psi_t}{\psi}\right) = (\Delta - 1)(\lambda - \sum_{n=1}^{\infty} F_i \lambda^{-i}). \tag{2.7}$$

Now let $u^{(-k)} = \Delta^{-k} u$. Setting $\ln\left(\frac{\Delta\psi}{\psi}\right) = \ln\lambda - \frac{P_1}{\lambda} - \frac{P_2}{\lambda_2} \dots$ and solving for the G_i and F_i in terms of the u_i we obtain the conserved densities for the Toda Lattice equation (2.4),

$$P_1 = u_0 \,,$$
$$P_2 = u_1 + \frac{1}{2}u_0^2 \,,$$
$$P_3 = u_1 u_0^{(-1)} + u_0 u_1 + \frac{1}{3}u_0^3 \,, \tag{2.8}$$
$$\vdots$$

and the conservation laws are given by

$$\frac{\partial P_i}{\partial t} = (\Delta - 1)F_i \,. \tag{2.9}$$

where

$$F_1 = u_1 \,,$$
$$F_2 = u_1 u_0^{(-1)} \,,$$
$$F_3 = u_1 u_1^{(-1)} + u_1 (u_0^{(-2)})^2 \,. \tag{2.10}$$
$$\vdots$$

Now the Whitham equations are obtained by averaging. The solutions of the Toda equations are assumed to be slowly modulating g-phase wavetrains. Since the oscillations of the wavetrain are fast compared to the modulations, the idea is to average out these fast oscillations. The procedure for the $g = 1$ case of the KdV equation is given in Whitham (1974) and the general case is analysed in Flaschka, Forest and McLaughlin (1980).

To obtain the Whitham equations one introduces two time scales and then averages the flux and density over the fast variables in the conservation laws.

Suppose F is a flux or density. We denote by $\langle F \rangle$ the spatial average over the fast variables of F (keeping the slow variables fixed). For the discrete Toda system:

$$\langle F \rangle := \lim_{N \to \infty} \frac{1}{2N} \sum_{n=-N}^{N} F(n,t). \tag{2.11}$$

Since these systems are integrable, there exist infinitely many commuting flows. For the case of g-phase wavetrains these flows may be viewed as translations on the g-torus, and each flow will cover the surface of the torus ergodically (if the characteristic frequencies are incommensurate). The integral in the torus variables may then be replaced by integral over the cycles in an associated Riemann surface, and one can thus show that the averaged quantities are abelian differentials on a Riemann surface.

Now to obtain the averaged (Whitham) equations we break up the dynamics into slow and fast scales. Let t_0 and n_0 denote the fast time and spatial variables and let $T = \epsilon t$ and $X = \epsilon n$ be the slow variables.

Then the shift operator Δ and the time derivative in (2.4) may be given, respectively, by

$$\Delta = e^{\frac{\partial}{\partial n}} = e^{\frac{\partial}{\partial n_0} + \varepsilon \frac{\partial}{\partial X}} \sim e^{\frac{\partial}{\partial n_0}} \left(1 + \varepsilon \frac{\partial}{\partial X}\right), \tag{2.12}$$

and

$$\frac{\partial}{\partial t} = \frac{\partial}{\partial t_0} + \varepsilon \frac{\partial}{\partial T}. \tag{2.13}$$

Substituting into (2.12) and averaging over the fast variables, from the terms of order ε we obtain the Whitham equations,

$$\frac{\partial}{\partial T} \langle P_i \rangle = \frac{\partial}{\partial X} \langle F_i \rangle. \tag{2.14}$$

We now wish to write the Whitham equations in terms of the Abelian differentials on a Riemann surface of genus g (see Flaschka, Forest and McLaughlin (1980) for the KdV equation and Krichever (1988) for the KP equation).

Let $y^2 = R_g(\lambda)$ be the Riemann surface of genus g,

$$R_g(\lambda) = \prod_{i=1}^{2g+2} (\lambda - \lambda_i), \tag{2.15}$$

where the branch points λ_i and real and are assumed to satisfy $\lambda_1 < \lambda_2 < \cdots < \lambda_{2g+2}$.

Krichever (1978) showed that a solution of (2.2) was given by the Baker-Akhiezer function,

$$\psi(n, t; \lambda) = r_0 \exp(it \int_{\lambda_0}^{\lambda} \omega_1 + in \int_{\lambda_0}^{\lambda} \omega_0) F(\theta_1, \ldots, \theta_g), \qquad (2.16)$$

where ω_0 and ω_1 are the normalized Abelian differentials of the third and second kinds respectively with singularity at infinity

$$\omega_0(\lambda) = \frac{\prod_{i=1}^{g}(\lambda - \alpha_i)}{\sqrt{R_g(\lambda)}} d\lambda,$$

$$\sim \frac{d\lambda}{\lambda} + O\left(\frac{d\lambda}{\lambda^2}\right), \qquad \text{as } \lambda \to \infty, \qquad (2.17)$$

$$\omega_1(\lambda) = \frac{1}{2} d\lambda + \frac{\lambda^{g+1} - \frac{1}{2}\sigma_1 \lambda^g + \gamma_1 \lambda^{g-1} + \cdots + \gamma_g}{2\sqrt{R_g(\lambda)}} d\lambda,$$

$$\sim d\lambda + O\left(\frac{d\lambda}{\lambda^2}\right), \qquad \text{as } \lambda \to \infty, \qquad (2.18)$$

with the normalization

$$\oint_{a_j} \omega_0 = \oint_{a_j} \omega_1 = 0, \qquad j = 1, \ldots, g, \qquad (2.19)$$

the contour integral along the a_j-cycle, the canonical cycle over the region $[\lambda_{2j}, \lambda_{2j+1}]$. The coefficients α_i and γ_i in (2.17) and (2.18) are uniquely determined by (2.19). Here the coefficient σ_1 in (2.22), and, similarly, σ_2 and σ_3, to be used later, are defined respectively by

$$\sigma_1 = \sum_{i=1}^{2g+2} \lambda_i,$$

$$\sigma_2 = \sum_{i<j} \lambda_i \lambda_j,$$

$$\sigma_3 = \sum_{i<j<k} \lambda_i \lambda_j \lambda_k. \qquad (2.20)$$

$F(\theta_1, \ldots, \theta_g)$ is a quasiperiodic function given by the Riemann Theta function on the Riemann surface. Note that in fact (2.16) is the solution to the symmetric form of the equations (2.2), but all the following calculations are the same for either form.

Taking the differential of the log of (2.16) on the Riemann surface and averaging, using the prescription given above, gives the Whitham equation as:

$$\frac{\partial \omega_0}{\partial T} = \frac{\partial \omega_1}{\partial X}. \qquad (2.21)$$

Note that (2.21) is an averaged form of the conservative law (2.7).

Thus expanding (2.21) for λ large we have from our previous expressions for the averaged quantities:

Theorem 2.2. *The averaged quanties $\langle P_i \rangle$ and $\langle F_i \rangle$ may be characterized in terms of the Riemann surface by*

$$\omega_0(\lambda) = \frac{d\lambda}{\lambda}(1 + \frac{\langle P_1 \rangle}{\lambda} + \frac{2\langle P_2 \rangle}{\lambda^2} + \dots), \tag{2.22}$$

$$\omega_1(\lambda) = d\lambda(1 + \frac{\langle F_1 \rangle}{\lambda^2} + \frac{2\langle F_2 \rangle}{\lambda^3} + \dots), \tag{2.23}$$

where

$$\langle P_1 \rangle = \langle u_0 \rangle = \frac{1}{2}\sigma_1 - \mu_1,$$

$$\langle P_2 \rangle = \langle u_1 + \frac{1}{2}u_0^2 \rangle = \frac{3}{16}\sigma_1^2 - \frac{1}{4}\sigma_2 - \frac{1}{4}\sigma_1\mu_1 + \frac{1}{2}\mu_2,$$

$$\vdots \tag{2.24}$$

$$\langle F_1 \rangle = \langle u_1 \rangle = \frac{1}{16}\sigma_1^2 - \frac{1}{4}\sigma_2 + \frac{1}{2}\gamma_1,$$

$$\langle F_2 \rangle = \langle u_0 u_1 \rangle = \frac{1}{16}\sigma_1^3 - \frac{1}{4}\sigma_1\sigma_2 + \frac{1}{4}\sigma_3 + \frac{1}{4}\sigma_1\gamma_1 + \frac{1}{4}\gamma_2,$$

$$\vdots \tag{2.25}$$

Here σ_1, σ_2 and σ_3 are defined in (2.20), and μ_1 and μ_2 are given by

$$\mu_1 = \sum_{k=1}^{g} \alpha_k, \quad \text{and} \quad \mu_2 = \sum_{k<\ell} \alpha_k \alpha_\ell. \tag{2.26}$$

3. Hyperbolic structure of the Whitham equations

Bikbaev and Novokshenov (1988) and Levermore (1988) showed that the rich structure of the KdV equation enabled one to analyze in the detail the behavior of the solution of the Whitham equation We can adapt their analysis to the Toda equations.

We first note that the $2g + 2$ branch points of the Riemann surface $y^2 = R_g$ of (2.19), $\lambda_1, \lambda_2, \dots \lambda_{2g+2}$, give the Riemann invariants for the Whitham equation. We can see this as follows.

Define

$$s(\lambda) = \frac{\lambda^{g+1} - \frac{1}{2}\sigma_1\lambda^g + \gamma_1\lambda^{g-1} + \dots + \gamma_g}{2\prod_{i=1}^{g}(\lambda - \alpha_i)}, \tag{3.1}$$

where the coefficients are defined by (2.21) and(2.22), i.e. $s(\lambda) = \frac{\tilde{\omega}_1(\lambda)}{\omega_0(\lambda)}$ where

$$\tilde{\omega}_1(\lambda) = \frac{\lambda^{g+1} - \frac{1}{2}\sigma_1\lambda^g + \gamma_1\lambda^{g-1} + \dots + \gamma_g}{2\sqrt{R_g(\lambda)}}. \tag{3.2}$$

Then we have:

Lemma 3.1.

The branch points satisfy the Riemann invariant equation

$$\frac{\partial \lambda_k}{\partial T} = s(\lambda_k)\frac{\partial \lambda_k}{\partial X}, \tag{3.3}$$

where the characteristic velocity $s(\lambda_k)$ is given by (3.1) at $\lambda = \lambda_k$,

$$s(\lambda_k) = \frac{\tilde{\omega}_1(\lambda_k)}{\omega_0(\lambda_k)}. \tag{3.4}$$

We also have:

Lemma 3.2.

$$s(\lambda_j) > s(\lambda_k), \qquad \text{for } \lambda_j > \lambda_k, \tag{3.5}$$

and

$$\frac{\partial s(\lambda_k)}{\partial \lambda_k} > 0, \quad \forall k. \tag{3.6}$$

The proof is similar to that in Bikbaev and Novokshenov (1988) and Levermore (1988).

Corollary 3.3.

For monotonically decreasing smooth initial data the λ_k will not develop shocks.

We of course have discontinuous initial data, but the result still holds. We prove existence and nonexistence of shocks by explicit computation of velocities, using Corollary 3.3 as a guide.

4. Regularization of Shock Waves

In this section we show how to regularize shock waves by going to a higher genus solution.

We consider a canonical example where the initial data is

$$u_0(X,0) = \begin{cases} -2a, & X < 0, \\ 0, & X = 0, \\ 2a, & X > 0, \end{cases}$$
$$u_1(X,0) = 1, \ \forall X, \tag{4.1}$$

which was suggested originally by Holian, Flaschka and McLaughlin (1981). In terms of the original Toda lattice variables, the position of the nth particle, this corresponds to taking the velocity of the nth particle to be a when $n < 0$, $-a$ when $n > 0$, and 0 when $n = 0$, while the displacement of each particle is initially zero.

The basic idea is then to show, by studying the behavior of the Riemann invariants that the $g = 0$ solution may develop a shock while the higher genus solution does not. Other initial conditions with step profile may be treated in an analogous fashion.

The nature of the solution is quite different for the parameter values $a > 1$, $0 < a < 1$, $-1 < a < 0$ and $a < -1$. Note that the sound speed in the lattice is unity. We also note that the solutions for $a > 0$ and $t > 0$ are related to the solutions for $a < 0$ and $t < 0$ because of the symmetry of equations (2.4). We will consider here only the case $a > 1$ The other cases are discussed in Bloch and Kodama [1991].

First we have:

Lemma 4.1. *The genus $g = 0$ equations develop a shock for the initial data (4.1) with $a > 1$.*

Proof.

From equations (2.24) and (2.25) with $g = 0$ we have

$$
\begin{aligned}
u_0 &= \frac{1}{2}(\lambda_1 + \lambda_2)\,, \\
u_1 &= \frac{1}{16}(\lambda_1 - \lambda_2)^2\,.
\end{aligned}
\tag{4.2}
$$

Hence for the given initial conditions λ_1 and λ_2 may be given by

$$
\begin{aligned}
\lambda_1 &= -2a - 2\,, & X &< 0\,, \\
&= 2a - 2\,, & X &> 0\,, \\
\lambda_2 &= -2a + 2\,, & X &< 0\,, \\
&= 2a + 2\,, & X &> 0\,,
\end{aligned}
\tag{4.3}
$$

as illustrated in Figure 1. Now from (3.4) the characteristic velocities of the Riemann invariants λ_1 and λ_2 are given by

$$
\begin{aligned}
s(\lambda_1) &= -\sqrt{u_1} = \frac{1}{4}(\lambda_1 - \lambda_2)\,, \\
s(\lambda_2) &= +\sqrt{u_1} = -\frac{1}{4}(\lambda_1 - \lambda_2)\,.
\end{aligned}
\tag{4.4}
$$

We can now see that λ_1 and λ_2 develop shocks as follows. Firstly, from (4.4) we see that λ_1 moves to the right and λ_2 moves to the left. An explicit calculation of the velocities at the corners of λ_1 and λ_2, gives

$$s(\lambda_2^+) = -s(\lambda_1^-) = 1 + a\,,$$
$$s(\lambda_2^-) = -s(\lambda_1^+) = 1\,, \qquad\qquad (4.5)$$

where $+$ denotes the limiting velocity at 0 from the right, and $-$ that from the left. Now since $s(\lambda_i^+) > s(\lambda_i^-)$, $i = 1, 2$, a shock develops. □

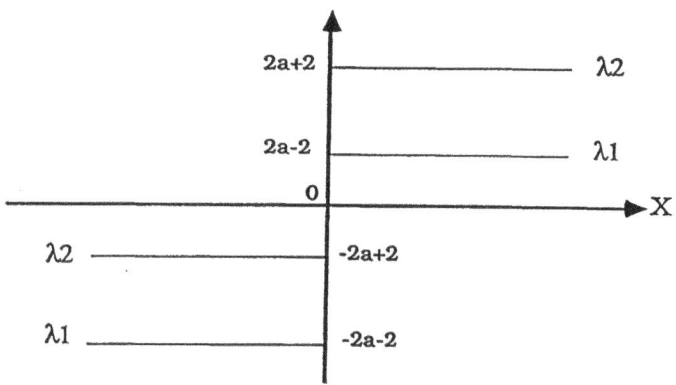

Fig. 1. Genus $g = 0$ initial data for $a > 1$

We now show that regularization of the equations for $a > 1$ requires genus $g = 1$. The basic idea is to note that the initial data (4.1) can also be considered as the degenerate data of the higher genus Whitham equation.

We have:

Theorem 4.2.

For $a > 1$ the genus $g = 1$ Whitham equations with the initial data (4.1) do not develop a shock.

Proof. For the given set of initial data one calculates firstly that λ_i, $i = 1, \ldots 4$, satisfying (2.24) and (2.25) with $g = 1$ are given by

$$\lambda_1 = -2a - 2, \qquad \forall X,$$

$$\lambda_2 = -2a + 2, \qquad X < 0,$$

$$= -2a - 2 = \lambda_1, \qquad X > 0,$$

$$\lambda_3 = 2a + 2 = \lambda_4, \qquad X < 0,$$

$$= 2a - 2, \qquad X > 0,$$

$$\lambda_4 = 2a + 2, \qquad \forall X,$$

(4.6)

as illustrated in Figure 2.

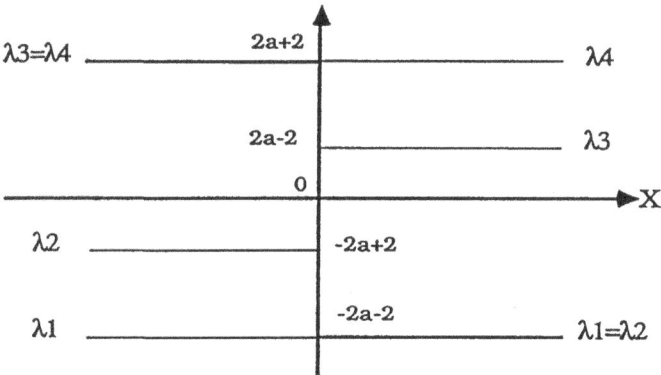

Fig. 2. Degenerate $g = 1$ initial data for $a > 1$

It should be noted that for fixed g the λ_i are determined uniquely since they have to satisfy the relationship between the λ_i and the first four conserved densities $\langle P_1 \rangle, \ldots, \langle P_4 \rangle$. Initially of course, $\langle P_3 \rangle$ and $\langle P_4 \rangle$ are linearly dependent on $\langle P_1 \rangle$ and $\langle P_2 \rangle$ since the $g = 0$ and $g = 1$ solution match at $t = 0$. In fact, we see that $\lambda_3 = \lambda_4$ ($X < 0$) implies $\alpha_1 = \lambda_3 = \lambda_4$ and $\gamma_1 = \frac{\lambda_3}{2}(\lambda_1 + \lambda_2)$, and $\lambda_1 = \lambda_2$ ($X > 0$) implies $\alpha_1 = \lambda_1 = \lambda_2$ and $\gamma_1 = \frac{\lambda_1}{2}(\lambda_3 + \lambda_4)$, and for these values ω_0 and ω_1 reduce to the Abelian differentials for the $g = 0$ case.

Note also that the λ_i's are all monotonic decreasing and from the Levermore theory, the no shock condition for smooth initial data is given by $\frac{\partial s(\lambda_i)}{\partial \lambda_i} > 0$. We now prove the equivalent statement for our nonsmooth initial data by explicit calculation of the velocities.

We can now calculate explicitly the characteristic speeds $s_i = s(\lambda_i)$ of (3.4),

$$s(\lambda) = \frac{\lambda^2 - \frac{1}{2}\sigma_1\lambda + \gamma_1}{2(\lambda - \alpha_1)}, \tag{4.7}$$

where α_1 and γ_1 are determined by the normalization (2.23), i.e.,

$$I_1^1 - \alpha_1 I_1^0 = 0, \tag{4.8}$$

$$I_1^2 - \frac{1}{2}\sigma_1 I_1^1 + \gamma_1 I_1^0 = 0, \tag{4.9}$$

with

$$I_1^\ell(\lambda_1, \ldots, \lambda_4) = \int_{\lambda_2}^{\lambda_3} \frac{\lambda^\ell d\lambda}{\sqrt{(\lambda - \lambda_1)(\lambda - \lambda_2)(\lambda - \lambda_3)(\lambda - \lambda_4)}}. \tag{4.10}$$

Thus (4.7) becomes

$$s(\lambda) = -\frac{1}{4}\sigma_1 + \frac{\lambda^2 - I_1^2/I_1^0}{2(\lambda - I_1^1/I_1^0)}. \tag{4.11}$$

We now need to compute $s(\lambda_3^-)$, the limit of $s(\lambda_3)$ from the left (as $X \to 0^-$), as well as $s(\lambda_3^+)$, $s(\lambda_2^-)$ and $s(\lambda_2^+)$. We omit the details here. We find the characteristic velocity $s(\lambda_3^-)$ is given by

$$s(\lambda_3^-) = 1 + 2a - 2a \frac{\int_0^{\pi/2} \frac{\cos^3\theta}{\sqrt{1+a\sin^2\theta}}\,d\theta}{\int_0^{\pi/2} \frac{\cos\theta}{\sqrt{1+a\sin^2\theta}}\,d\theta}$$

$$= \frac{\sqrt{a}\sqrt{1+a}}{\log|\sqrt{a} + \sqrt{1+a}|} > 1. \tag{4.12}$$

Note that the wave front speed is greater than unity, which is the sound speed in the lattice, reflecting shock formation.

Similarly we can show $s(\lambda_2^+) = -s(\lambda_3^-)$. The velocities $s(\lambda_3^+)$ and $s(\lambda_2^-)$ may be found in terms of elliptic integrals:

$$s(\lambda_2^-) = -s(\lambda_3^+) = (-2a + 2)(1 - \Gamma) < 0, \tag{4.13}$$

where

$$\Gamma = \frac{\int_0^{\pi/2} \frac{\sin^4\theta}{\sqrt{(a\cos^2\theta+\sin^2\theta)(a\sin^2\theta+\cos^2\theta)}}\,d\theta}{\int_0^{\pi/2} \frac{\sin^2\theta}{\sqrt{(a\cos^2\theta+\sin^2\theta)(a\sin^2\theta+\cos^2\theta)}}\,d\theta} < 1.$$

From these velocity calculations we can see that the $g = 1$ Whitham equation has a regular single phase solution and no shock occurs as shown in Figure 3. $\square$

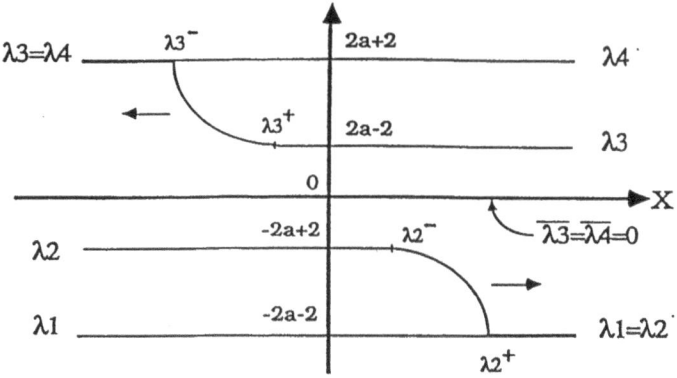

Fig. 3. Degenerate $g = 2$ solution for $a > 1$

Note also that, for example, in the region $s(\lambda_3^+) \leq -\frac{X}{T} \leq s(\lambda_3^-)$ the unique solution λ_3, with all other λ_i fixed, can be given in hodograph form (for the general case see Kodama (1990), Kodama and Gibbons (1989) and Tsarev (1985)),

$$X + s(\lambda_3)T = 0. \tag{4.14}$$

Since λ_3 is monotonic, one can invert (4.17) to find λ_3 as a function of X and T. This solution may be connected to the steady $g = 1$ solution in $0 < -\frac{X}{T} < s(\lambda_3^+)$ and the $g = 0$ solution in $-\frac{X}{T} \geq s(\lambda_3^-)$ (outside the domain of influence). We can also see from (4.17) that for fixed T, $\frac{\partial X}{\partial \lambda_3} \to 0$ as $\lambda_3 \to \lambda_3^-$ and $\frac{\partial X}{\partial \lambda_3} \to \infty$ as $\lambda_3 \to \lambda_3^+$.

We wish to remark that we have chosen the solution here of minimum genus. One can chose a solution of any higher genus with degeneracy in the spectrum since if any two λ_i become degenerate the corresponding factors in the abelian differentials cancel out, as the spectrum is real. Hence the velocity calculations for the higher genus case remain precisely the same. Such a degeneracy is in fact needed to explain the detailed behavior of the original Toda lattice, while the solution of the $g = 1$ Whitham equations describes the averaged behavior of the original lattice. As pointed out to us by S. Venakides, to satisfy the boundary conditions at $n = 0$ in the original lattice one needs a degenerate genus two Riemann surface with two branch points coinciding at $\lambda = 0$. The calculations above are then precisely the same except that now one has the additional characteristics $\overline{\lambda}_3 = \overline{\lambda}_4 = 0$ and the total ordering $\lambda_1 \leq \lambda_2 \leq \overline{\lambda}_3 = \overline{\lambda}_4 \leq \lambda_3 \leq \lambda_4$, as illustrated in Fig. 3. This ensures that the particle at $X = 0$ (i.e. $n = 0$ in the lattice) remains fixed. The solution near $X = 0$ then corresponds to the binary oscillations of the lattice behind the shock as observed in

Holian, Flaschka and McLaughlin (1981) and as predicted in Venakides, Deift and Oba (1990).

We have seen in the case $a > 1$ how regularity of solutions dictates the choice of Whitham equations for the Toda lattice with step initial data. The same is true for other values of a, as shown in Bloch and Kodama [1991].

We remark that the problem with smooth initial data may be solved by the generalized hodograph method (Kodama (1990), Kodama and Gibbons (1989) and Tsarev (1985)), with similarity variables. This method is similar to that used in Gurevich and Pitaevskii (1974) for the KdV case. In a future communication we hope to give a method for connecting the different genus solutions before and after the shock.

We remark finally on the physical significance of our solutions. The physical behavior predicted by these solutions agrees with the numerical studies of Holian, Flaschka and McLaughlin for $a > 0$. Their initial condition may be viewed as being imparted by a "piston" moving with velocity a. They then discuss three regions of the shock profile – the shock front, the rear of the shock profile near the piston and a transition region between front and rear. For a less than and above the critical value 1 (the sound speed in the lattice) one gets quite different behavior.

For $a > 1$ one observes regular periodic behavior in the rear of the shock region corresponding to "binary" oscillation – displacement of particles being equal and opposite and thus oscillating at the highest frequency the lattice can support. This is consistent with the behavior one predicts from the degenerate $g = 2$ solution (with $\overline{\lambda}_3 = \overline{\lambda}_4 = 0$) illustrated in Fig. 3.

For $0 < a < 1$ one observes essentially zero oscillation in the rear of the shock and nonzero oscillation elsewhere. Again this can be predicted from the degenerate $g = 2$ solution we obtain in Bloch and Kodama [1990], as the interior exhibits zero gap.

Note that $a > 0$ corresponds to the lattice contracting, which can generate shock waves and $a < 0$, on the other hand, corresponds to the lattice expanding which can generate rarefaction waves. Again, for the latter, one gets different behavior for $-1 < a < 0$ and $a < -1$. For $-1 < a < 0$, since the velocity of the expansion is less than the sound velocity, no shock occurs and a $g = 0$ solution provides a good description for the lattice behavior. For $a < -1$, the expansion velocity is greater than the sound velocity, and a degenerate $g = 1$ solution is required to describe the behavior. However, the gap never opens and in this case u_0 and u_1 near $X = 0$ never become stationary except at $T = \infty$. Physically, this implies the lattice is continually expanding. The analysis of these cases via the Whitham equation is given in Bloch and Kodama [1991].

Acknowledgements. We would like to thank R. Brockett, P. Deift and S. Venakides for some valuable conversations.

References

R. F. Bikbaev (1989), Structure of a shock wave in the theory of the Korteveg-de Vries equation, Phys. Lett. 147A, 289-293.

R. F. Bikbaev and V. Yu. Novokshenov (1988), Self-similar solutions of the Whitham equations, and KdV equations with f.-g. boundary conditions, in the Proceedings of the III Intern. Workshop, Kiev 1987, Vol. 1, 32-35.

A. M. Bloch and Y. Kodama (1991), Dispersive regulation of the Whitham equation for the Toda lattice, SIAM J. App. Math., to appear.

R. W. Brockett and A. M. Bloch (1990), Sorting with the dispersionless limit of the Toda lattice, in the Proceedings of the CRM Workshop on Hamiltonian Systems, Transformation Groups and Spectral Transform Methods (eds. J. Harnad and J. E. Marsden), Publications CRM, Montreal, 103-112.

B. A. Dubrovin and S. P. Novikov, Hydrodynamics of weakly deformed soliton lattices. Differential geometry and Hamiltonian theory, Russian Math. Surveys 44:6, 35-124.

H. Flaschka (1976), The Toda lattice, Phys. Rev. B9, 1924-1925.

H. Flaschka, M. G. Forest and D. W. McLaughlin (1980), Multiphase averaging and the inverse spectral solution of the Korteweg-de Vries equation, Comm. Pure and Appl. Math 33, 739-784.

A. V. Gurevich and L. D. Pitaevskii (1974), Nonstationary structure of noncolliding shock waves, JETP 38, 291.

B. L. Holian, H. Flaschka and D. W. McLaughlin (1981), Shock waves in the Toda lattice: analysis, Phys. Rev. A 24, 2595-2623.

Y Kodama (1990), Solutions of the dispersionless Toda equation, Phys. Lett. 147A, 477-482.

Y. Kodama and J. Gibbons (1990), Integrability of the Dispersionless KP Hierarchy, Proc. of the Workshop on Nonlinear and Turbulent Processes in Physics, Kiev, (World Scientific, Singapore).

I. M. Krichever (1978), Algebraic curves and nonlinear difference equations, Russian Mathematical Surveys 33, 255-256.

I. M. Krichever (1988) Method of averaging for two-dimensional "integrable" equations, Funct. Anal, and its Appl. 22, 37-52.

B. A. Kupershmidt (1985), Discrete Lax Equations and Differential Difference Calculus, (Asterique 23), Soc. Math. France.

P. D. Lax and C. D. Levermore (1983), The small dispersion limit of the Korteweg-de Vries equation I, II, III, Comm. in Pure and Appl. Math 30, 253-290, 571-593, 809-829.

C. D. Levermore (1988), The hyperbolic nature of the zero dispersion KdV limit, Comm. in Partial Diff. Eqns, 13(4), 495-514.

J. Moser (1974), Finitely many mass points on the line under the influence of an exponential potential, Battelles Recontres, Springer Notes in Physics, 417-497.

S. P. Tsarev (1985), On Poisson brackets and one-dimensional Hamiltonian systems of hydrodynamic type, Soviet Math. Dokl. 31, 488-491.

S. Venakides (1985), The generation of modulated wavetrains in the solution of the Korteweg-de Vries equation, Comm. in Pure and Appl. Math. 38, 833-909.

S. Venakides, P. Deift and R. Oba (1990), The Toda shock problem, preprint.

G. B. Whitham (1974), Linear and Nonlinear Waves, John Wiley, New York.

SEMICLASSICAL BEHAVIOR IN THE NLS EQUATION: OPTICAL SHOCKS - FOCUSING INSTABILITIES

Jared C. Bronski [*]

Program in Applied and Computational Math
Princeton University
Princeton, New Jersey 08540

David W. McLaughlin [†]

Department of Mathematics
Program in Applied and Computational Math
Princeton University
Princeton, New Jersey 08540

Abstract

We consider the semiclassical limit of the non-linear Schroedinger equation in both the defocusing and the focusing cases. We review the theory of the defocusing case, and verify it with numerical experiments. We also point out an application of that theory to the physical problem of optical shocking in fibers. Following that we consider the focusing case. We argue that the the failure of the existing theory is a result of the modulational instability. We consider the weakly nonlinear case and make an ansatz that allows the modulational instability to express itself. We derive from this ansatz a set of equations we claim are valid prebreaking. These equations succesfully treat the modulational instability, describe an interaction between the laboratory scale and the intermediate scale of the instability, and allow for the generation of oscillations on the shortest scale. We begin to test the validity of these equations with a series of numerical experiments.

1 Introduction

In this paper we explore the semiclassical limit of the nonlinear Schroedinger equation for the focusing and defocusing cases. We will begin by reviewing the theory of the defocusing NLS. We consider the defocusing case for two reasons: Our primary purpose is to call attention to an application of the semiclassical theory of defocusing NLS to the physical problem of short pulses and optical shocking in optical fibers. A secondary purpose is to provide a test for the numerical routines we use to integrate the focusing equation. The fact that the defocusing limit is well understood theoretically makes it a useful test case. The remainder of the paper will be concerned with the focusing NLS. We develop a semiclassical theory for focusing NLS with weak nonlinearity and run numerical simulations to support our theory.

---

[*]Funded in part by a graduate fellowship from DOD.
[†]Funded in part by AFOSR - 90-0161 and NSF DMS 8922717 A01.

2 The Semiclassical Limit of Defocusing NLS

2.1 The Integrable Theory

In [1][2] Jin, Levermore and McLaughlin, following the work of Lax and Levermore [3][4][5], consider the $\epsilon \to 0$ limit of the defocusing NLS equation

$$i\epsilon\psi_t \;+\; \epsilon^2\psi_{xx} - |\psi|^2\psi = 0 \tag{2.1}$$
$$\psi(x,0) \;=\; A(x,0)\exp(iS(x,0)/\epsilon). \tag{2.2}$$

They analyze this problem by considering the $\epsilon \to 0$ limit of the associated scattering problem and deriving a WKB-type condition for the eigenvalues. Under some assumptions (the most important being a vanishing reflection coefficient) they derive expressions for $\rho = |\psi|^2$ and $\mu = i\epsilon(\psi_x\psi^* - \psi_x^*\psi)$ in terms of the unique solution of a certain minimization problem. They show that ρ^ϵ and μ^ϵ approach strong limits for $t < t_{break}$, where t_{break} is the time of break of a certain hyperbolic system. In the pre-breaking region the $\epsilon \to 0$ limits of ρ, μ are completely described by two quantities related to a variational problem. These are basically the Riemann invariants we will encounter in the next section. In the post-breaking regime the limits exist only weakly, since $|\psi|$ exhibits rapid oscillations with $O(1)$ amplitude, and the number of quantities required to characterize the limit increases. We will make use of some of the above results at several points in the remainder of the paper, but in general we will take a more general modulation theory approach which does not rely on integrability.

2.2 Strongly Nonlinear Geometric Optics

We begin with the nonlinear Schroedinger equation

$$i\epsilon\psi_t + \epsilon^2\psi_{xx} + \beta|\psi|^2\psi = 0. \tag{2.3}$$

and make the geometric optics ansatz

$$\psi(x,t) = A(x,t)\exp(iS(x,t)/\epsilon). \tag{2.4}$$

If we insert this ansatz into the nonlinear Schroedinger equation and equate powers of ϵ we find that $A(x,t)$ and $S(x,t)$ satisfy the coupled system of equations

$$S_t + S_x^2 - \beta|A|^2 \;=\; 0 \tag{2.5}$$
$$A_t + 2S_xA_x + S_{xx}A \;=\; 0. \tag{2.6}$$

The usual evolution equations of linear geometric optics have an eikonal equation which evolves independently of the transport equation and a transport equation which is slaved to the eikonal equation. Here the fact that the nonlinearity is not small puts the eikonal and transport equations on more equal footing. It is worth noting at this point that the same equations come out of a variety of approaches, including a Whitham type modulation applied to the constant amplitude solution of NLS[6]. In that context the strong coupling of the eikonal and transport equations comes from the fact that the dispersion relation for the constant amplitude solution is a function of k, ω, and $|A|$.

If we introduce new variables $U = 2S_x$ and $\rho = |A|^2$, the local velocity and mass density, we get the following system

$$\begin{pmatrix} U \\ \rho \end{pmatrix}_t = \begin{pmatrix} -U & 2\beta \\ -\rho & -U \end{pmatrix} \begin{pmatrix} U \\ \rho \end{pmatrix}_x . \tag{2.7}$$

System 2.5 is exactly that of one dimensional gas dynamics with gas velocity U, mass density ρ, local sound speed $c^2(\rho) = -2\beta\rho$ and gas law $\mathcal{P} = -\beta\rho^2$. In the defocusing case ($\beta < 0$) the

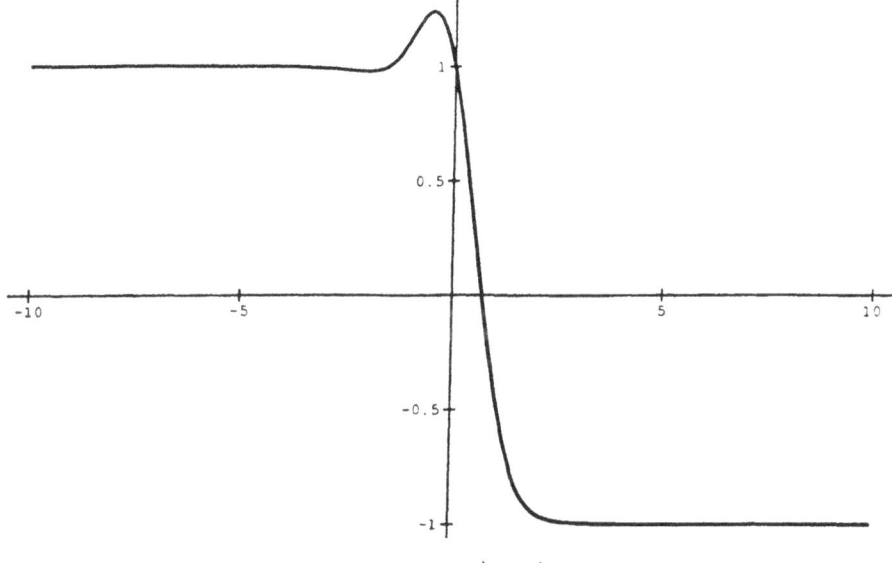

Figure 1. $c^+(x, 0)$

dynamics is that of an ordinary gas i.e. $\frac{\partial p}{\partial \rho} > 0$. In the focusing case $\frac{\partial p}{\partial \rho} < 0$ the pressure *decreases* as the density increases. We have a gas that likes to clump together around local density variations. This is clearly related to the existence of soliton solutions for the focusing NLS–the strange gas law is a manifestation of the modulational instability. We will discuss this in greater detail later in this paper.

The above system can be written in the Riemann invariant form

$$\Gamma_t^+ + c^+ \Gamma_x^+ = 0 \tag{2.8}$$
$$\Gamma_t^- + c^- \Gamma_x^- = 0, \tag{2.9}$$

where $c^\pm$, $\Gamma^\pm$ are given by

$$c^\pm = U \pm \sqrt{-2\beta\rho} \tag{2.10}$$
$$\Gamma^\pm = -\frac{U}{2} \pm \sqrt{-2\beta\rho}. \tag{2.11}$$

In the case $\beta < 0$, the defocusing sign of NLS, the eigenspeeds of this system are real and the system is well-posed as an initial value problem. (We will consider the case $\beta > 0$ in Section 3.) A very general result by Lax guarantees that if the $c^\pm(x, 0)$ are anywhere decreasing, then the above system develops shocks. After shocking, a "caustic region" is crucial to understanding the $\epsilon \to 0$ limit of defocusing NLS. We next describe the shape of this caustic region.

2.3 The Caustic Region

We can understand the shape of the caustic region by considering the initial wavespeeds $c^\pm(x, 0)$ for some particular choice of initial data. The choice we make is the following:

$$A(x, 0) = \exp(-x^2) \tag{2.12}$$
$$S_x(x, 0) = -\tanh(x). \tag{2.13}$$

This choice of initial data corresponds to a distribution of mass centered about the origin. The mass to the left of the origin has momentum directed towards the right, while the mass to the right of the origin has momentum directed towards the left. In figure 1 we plot $c^+(x, 0)$. (Since $c^{\pm}$ satisfy $c^-(x, 0) = -c^+(-x, 0)$, the characteristics associated with c^- are the mirror images of those associated with c^+.)

In the generic case (for symmetric initial data) both c^+ and c^- will shock and we will see the development of two caustic regions symmetrically placed about the origin, and the extent of these regions will grow in time. The integrable theory of Jin, Levermore, and McLaughlin predicts that in the caustic region we should observe oscillations with $O(1)$ height and $O(\epsilon)$ wavelength. We have run some numerical experiments on defocusing NLS to verify this, and to provide a first test for our numerical routines.

2.4   Numerics

We integrated NLS using a standard split-step algorithm, with Richardson extrapolation to improve the convergence. The linear part of the evolution was calculated with an FFT, while the nonlinear step was calculated exactly. The number of Fourier modes employed varied, but it was generally $2^{12} = 4096$. The time step was on the order of .0005. The left-hand set of graphs for Experiment 2 (figure 3) shows the Fourier spectrum on a Log-Linear scale. Obviously the spatial scales are being well resolved. As we can see from these graphs there is very little energy in wavenumbers above $\kappa \approx 15$, and that the amplitudes of these wave-packets are less than one. If we insert these numbers into the dispersion relation, $\omega = \epsilon \kappa^2 + |A|^2 \epsilon^{-1}$, we can see that $\omega \delta t \approx .02$, so the time evolution is extremely well resolved.

2.4.1   Experiment 1–The Development Of Rapid Oscillations

Experiment 1 shows the evolution of a wave packet $\psi_o = A \exp(iS/\epsilon)$, where $A(x)$ is gaussian and $S_x(x) = -\tanh(x)$, for $\epsilon = .05, .025$. Physically we can interpret this initial data as a distribution of particles with momenta directed towards the origin. The phase of this initial data encourages the development of caustics. Notice that the initially smooth ψ steepens up and develops oscillations with characteristic wavelength $O(\epsilon)$, with an envelope which seems to be independent of ϵ, just as the theory predicts. There is another region of rapid oscillations placed symmetrically with respect to the origin. Also notice that the spatial extent of the region of rapid oscillations grows in time, again in accordance with the theory.

2.4.2   Experiment 2 - Gaussian Initial Data Develops Caustics

Experiment 2 shows the evolution of a *real* gaussian wave packet under defocusing NLS and the associated power spectrum. Notice that caustics still develop even when the phase is initially trivial. This phenomenon is *not present* in the semi-classical limit of the linear equation, where the amplitude simply spreads. It is interesting to compare these pictures to the ultrashort optical pulse experiments of Rothenberg and Grischkowski [7].

2.5   Short Pulses in Optical Fibers

Short pulses in optical fibers have been a topic of interest in recent optics literature [7][8][9][10] since they potentially provide a procedure for generating ultrashort ($\approx$ 200 femtoseconds) pulses from short (picosecond) pulses, a procedure which might be technologically important. In all of these papers the authors have observed, numerically and experimentally, that initial pulses tend to 'square up' (flatten and steepen). The leading and trailing edges of the pulse steepen until a train of rapid oscillations develops before and after the pulse. This is illustrated very well in a recent paper by Rothenberg and Grischkowsky [7], where the authors demonstrate this phenomena both experimentally (using short laser pulses in optical fibers) and numerically (using direct numerical solution of defocusing NLS). Both the experimental and the numerical results look very much like the sort of pictures we have found in our studies of the semiclassical limit of defocusing NLS. Since the equation governing the evolution of

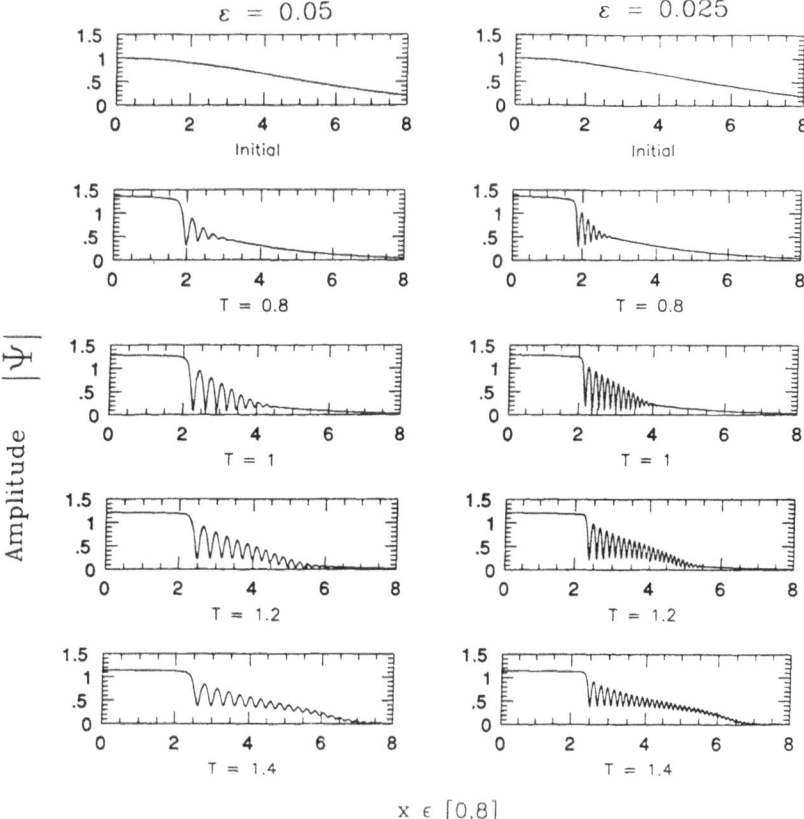

Figure 2. Defocusing nonlinearity; the initial "momentum" $S_x = -\tanh x$. The amplitude $|\psi|$ is displayed as a function of x, at several times, for two values of $\epsilon = 0.05, 0.025$.

x–space   κ–space

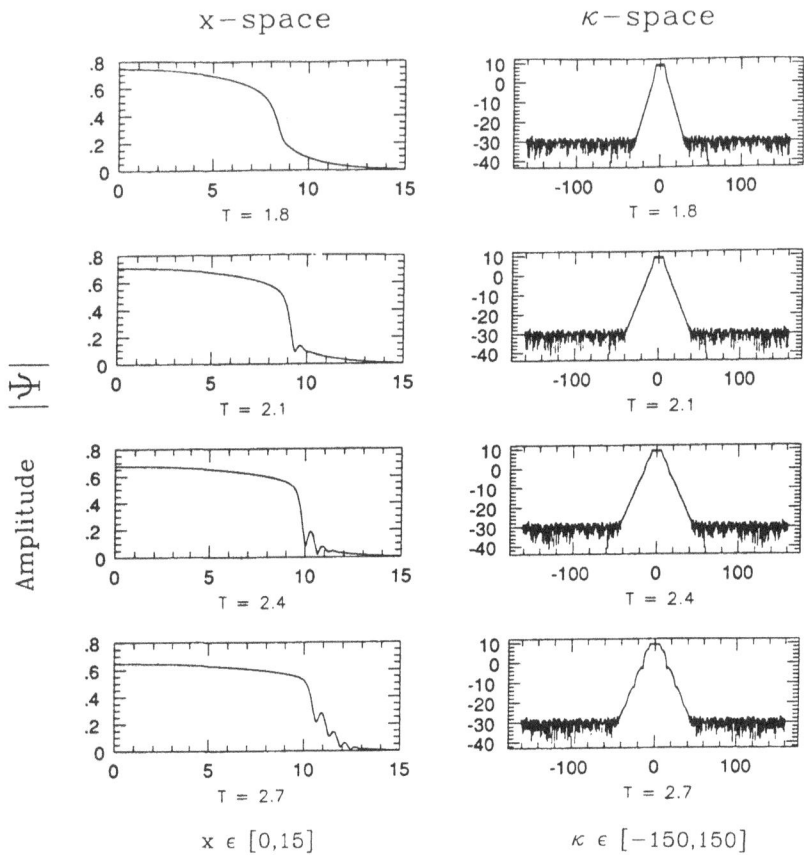

Figure 3.   Defocusing nonlinearity; the initial amplitude is gaussian with no initial "momentum"($S_x = 0$). Displayed are the amplitude $|\psi|$ *vs* x, and the power spectrum $\ln(|\hat{\psi}|^2)$ *vs* k.

short pulses in optical fibers is the semiclassical scaling of NLS we have *experimental* evidence supporting the semiclassical theory. The main discrepency between experiment and theory is the presence of some oscillations *after* the edge of the pulse. In [7] it is suggested that these oscillations are due to the Raman correction which begins to become significant for *psec* pulses. It would be interesting to explore this possibility, and determine if this is indeed the case.

3 Focusing NLS

3.1 The Modulational Instability

The theory of the semiclassical limit of defocusing NLS is clearly well in hand. What goes wrong when we try to apply this same analysis to the focusing case? In Section 2.2 we applied the naive geometric optics ansatz $\psi = A(x,t) \exp(iS(x,t)/\epsilon)$ to NLS,

$$i\epsilon\psi_t + \epsilon^2\psi_{xx} + \beta|\psi|^2\psi = 0, \tag{3.1}$$

to derive the modulation equations

$$\begin{pmatrix} U \\ \rho \end{pmatrix}_t = \begin{pmatrix} -U & +2\beta \\ -\rho & -U \end{pmatrix} \begin{pmatrix} U \\ \rho \end{pmatrix}_x. \tag{3.2}$$

We pointed out that this system is exactly that of $1 - D$ gas dynamics with a certain equation of state. For the focusing case ($\beta > 0$), $\frac{\partial P}{\partial \rho} < 0$ and the strange equation of state causes the gas to clump up. We argued that this should be interpreted as a manifestation of the well known modulational instability. If we carry out the modulational instability calculation of the plane wave solution of NLS we find that the wave numbers of the unstable modes satisfy

$$\kappa_{unstable} \leq \frac{\sqrt{\beta}A}{\epsilon}. \tag{3.3}$$

This means that A wants to develop structure on the $\sqrt{\beta}/\epsilon$ scale. We need to include this structure in our ansatz if we want to correctly model the semiclassical limit of the focusing case. In the case where β is not small the scale of the instability is the same as the scale of the oscillatory structure of the plane wave, and we have no natural separation of scales. Therefore we choose to study the case of small nonlinearity $\beta = \epsilon$. In this case the new structure is on the $O(1/\sqrt{\epsilon})$ scale and we maintain a separation of scales.

3.2 The Focusing Modulation Equations

We begin with the focusing NLS

$$i\epsilon\psi_t \quad + \quad \epsilon^2\psi_{xx} + \epsilon|\psi|^2\psi = 0 \tag{3.4}$$

$$\psi(x,0) \quad = \quad A(\frac{x}{\sqrt{\epsilon}}, x) \exp(i\frac{S(x)}{\epsilon}). \tag{3.5}$$

Since the modulational instability will force structure to appear on the $x/\sqrt{\epsilon}$ scale we should allow this variation into our initial data. Physically we now have three scales: *(i)* x - laboratory scale; *(ii)* x/ϵ, the scale of the rapid oscillations; and *(iii)* an intermediate $x/\sqrt{\epsilon}$ scale of the instability. If we make the anstaz

$$\psi(x,t) \quad = \quad A(X, x, T, t) \exp(i\frac{S(x,t)}{\epsilon}) \tag{3.6}$$

$$X \quad = \quad \frac{x}{\sqrt{\epsilon}} \tag{3.7}$$

$$T \quad = \quad \frac{t}{\sqrt{\epsilon}}. \tag{3.8}$$

we get the following system of equations

$$S_t \quad + \quad S_x^2 = 0 \tag{3.9}$$

$$A_T \quad + \quad 2S_x A_X = 0 \tag{3.10}$$

$$i(A_t + 2S_x A_x) \quad + \quad A_{XX} + iS_{xx}A + |A|^2 A = 0. \tag{3.11}$$

We claim that these three equations describe the small ϵ behavior for times before the break time. There are several comments worth making. First of all it is important to realize that these equations capture both the pre-breaking evolution of the structure on the $\sqrt{\epsilon}$ scale and the onset of breaking. Secondly it should be noted that despite the fact that we weakened the effect of the non-linearity this is still a *nonlinear* theory - the nonlinearity influences the evolution over time scales of order 1 (in t, the unscaled variable). Finally notice that in the case $S_x(x,0) = constant$ the system of equations(3.9-3.11) is basically a restatement of the Galilean invariance of NLS. We can look at the above system as an extension of the Galilean invariance to the case where S varies slowly on the X scale. This system describes, in this weakly nonlinear situation, *(i)* the saturation of the modulational instability; *(ii)* the interaction between the x and $x/\sqrt{\epsilon}$ scales and *(iii)* the onset of wave-breaking.

The focusing equations have some noticable differences from the defocusing equations. In terms of $U = 2S_x$ the eikonal equation for this weakly nonlinear case becomes

$$U_t + UU_x = 0. \tag{3.12}$$

U satisfies the inviscid Burgers' equation. Under the initial conditions of nontrivial phase that we used for the defocusing case ($U(x,0) = -2\tanh(x)$), this Burger's equation predicts that the caustic region for the weakly focusing case has a single central region where characteristics intersect. When we run numerical experiments on the focusing NLS we see the onset of rapid ($O(\epsilon)$) oscillations in just such a region of space-time.

3.3 Numerics

We have run a series of numerical experiments on the focusing NLS to attempt to confirm our ideas. We integrated NLS using the split-step algorithm detailed in section 2.4 of this paper. All of these experiments are performed on the focusing NLS, and all but the first are in the case of weak nonlinearity discussed earlier. In all but the first experiment the dispersion relation for the weakly nonlinear equation is given by $\omega = \epsilon \kappa^2 - |A|^2$. For the focusing case the typical band of significant κ is much wider than in the analogous defocusing case - the typical 'largest significant wavenumber' is close to 100, and the amplitudes are large, typically $|A|_{max} \approx 4$. While the resolution is not as good as in the defocusing case, we have $2\pi/\omega\delta t \approx 20$, or 20 time-steps per temporal period, which is still more than adequate.

3.3.1 Experiment 1 – Focusing Instability

Experiment 1 shows the failing of the naive two-scale ansatz when applied to the focusing NLS equation. From the modulational instability argument we know that the width of the band of unstable modes is $\propto \epsilon^{-1}$, with corresponding growth rates $\propto \epsilon^{-.5}$. This experiment shows the evolution of the power spectrum under focusing NLS with WKB type initial data and order 1 nonlinearity. Notice that the high wavenumber modes grow on extremely short time scales, much faster than that on which the (elliptic) modulation equations evolve. Obviously the two scale ansatz misses some important behavior.

3.3.2 Experiment 2 – Weak Nonlinearity Before Breaking

Experiment 2 shows the evolution of weakly focusing NLS with initial data

$$A(X,x,0) \quad = \quad (1 + 0.2\cos^2(X))\exp(-x^2) \tag{3.13}$$

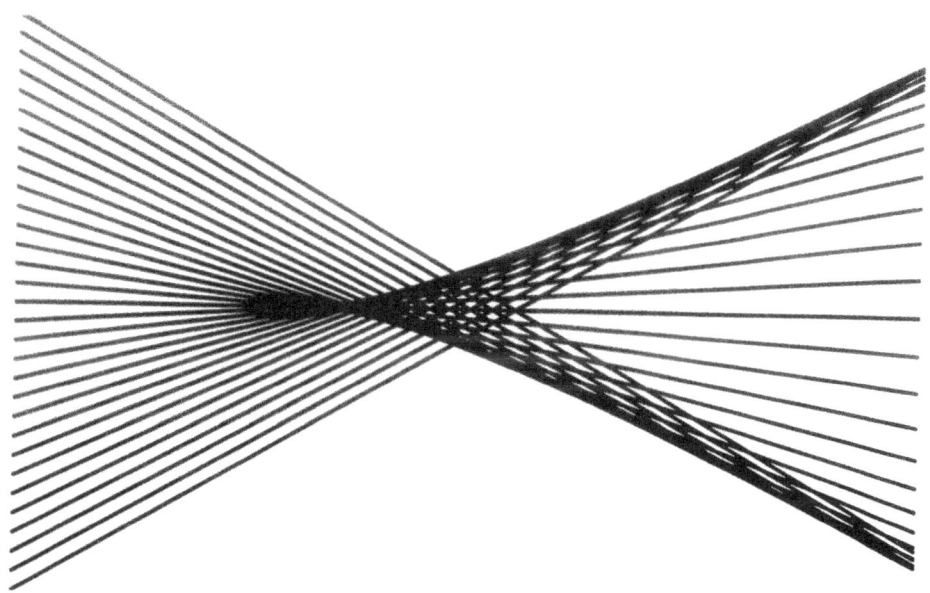

Figure 4. Characteristics for Burgers' Equation with $-\tanh(x)$ initial data

$$S_x(x,0) = -\tanh(x) \qquad (3.14)$$

for $\epsilon = .05$ and $\epsilon = 0.025$, along with corresponding power spectra. Note that the band of unstable modes is smaller than in the strongly nonlinear case, and grows more slowly, but the behavior is still very different from that of the linear Schrödinger equation, which leaves $|\psi(k)|$ invariant. According to our Burgers' theory the time of first breaking is $t = .5$. Notice that the last picture, at time $t = 0.51$, seems to show evidence of the onset of breaking.

3.3.3 Experiment 3 – Scaled Variables

According to our theory the amplitude A of the solution satifies

$$A_T + 2S_x A_X = 0 \qquad (3.15)$$
$$i(A_t + 2S_x A_x) + A_{XX} + iS_{xx}A + |A|^2 A = 0, \qquad (3.16)$$

so that $A^\epsilon = A(x/\sqrt{\epsilon}, t/\sqrt{\epsilon}, x, t)$. This means that $A^{2\epsilon} = A(x/\sqrt{2\epsilon}, t/\sqrt{2\epsilon}, x, t)$. So if we compare $A^{2\epsilon}(\sqrt{2}x, \sqrt{2}t)$ to $A^\epsilon(x, t)$ we find that the dependence on the fastest scales $x/\sqrt{\epsilon}, t/\sqrt{\epsilon}$ is the same. (The dependence on the x, t scales is different, however.) In Experiment 3 we plot A^ϵ against scaled $A^{2\epsilon}$. It seems apparent that the controlling behavior is correct – the maxima and minima move together. It would be interesting (though computationally more difficult) to compare slices of A^ϵ and $A^{2\epsilon}$ taken at fixed distances along the Burgers' characteristics.

3.3.4 Experiment 4 – Post-Breaking

This experiment is meant to give a qualitative idea of the behavior of the solutions in the post breaking regime. The post breaking regime seems to be characterized by a centrally located region with a great deal of small structure, a relatively quiet region in the wings, and a region of large $|\psi|$ marking the transition between the other two regions. We interpret this central region as being the region of multi-valuedness of the eikonal equation, and the region of large $|\psi|$ as (approximately) marking the boundary of this region of multi-valuedness. This

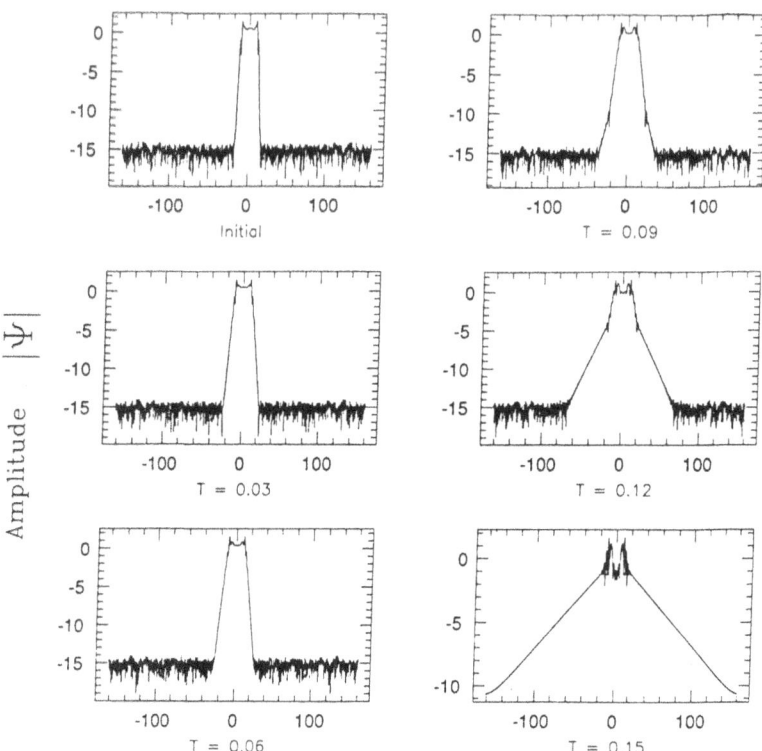

Figure 5. Focusing,strong 0(1) nonlinearity. Displayed are the power spectrum $\ln(|\hat{\psi}|^2)$ *vs* k.

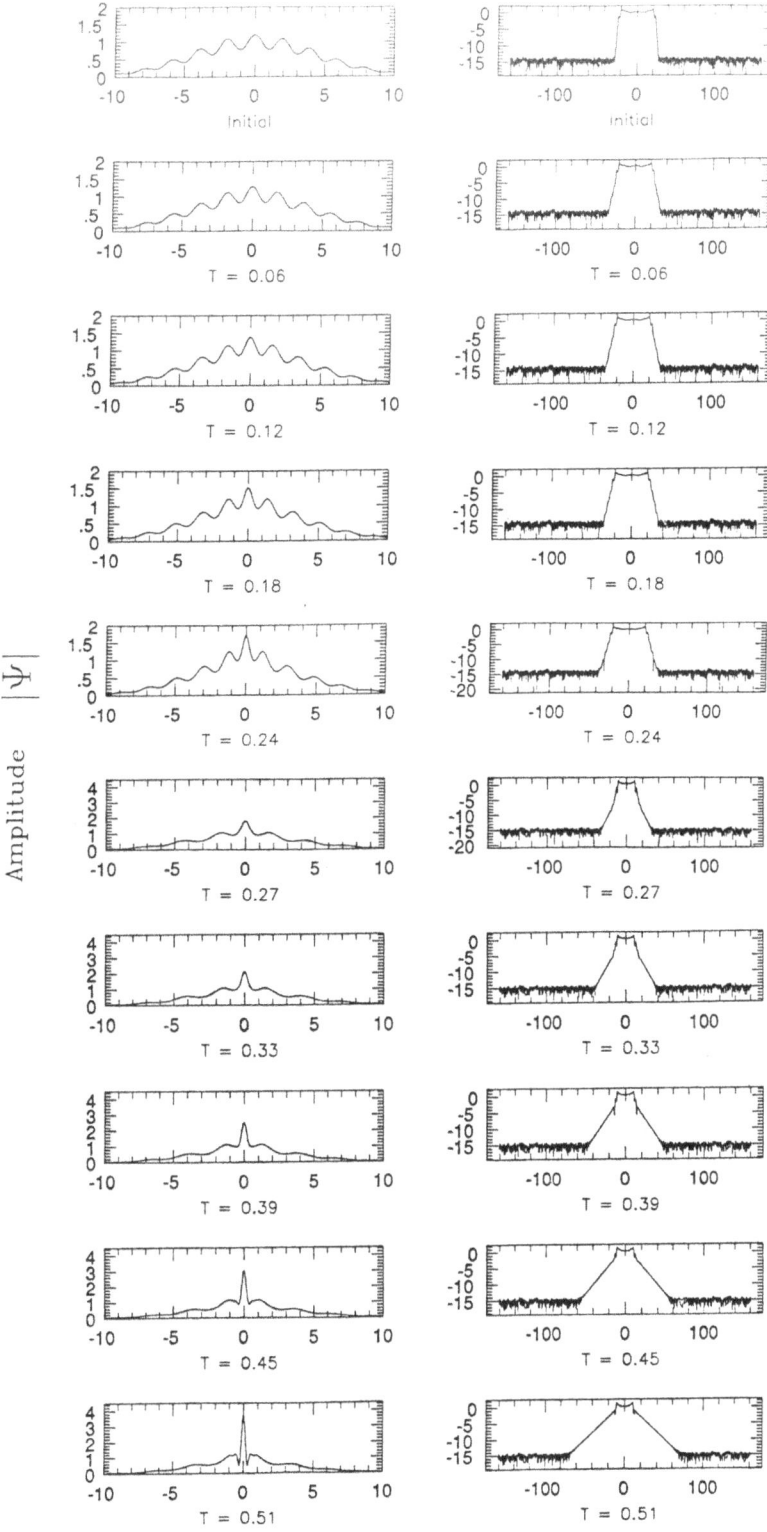

Figure 6. Focusing, weak $O(\epsilon)$ nonlinearity. Displayed are the amplitude $|\psi|$ *vs* x, and the power spectrum $\ln(|\hat{\psi}|^2)$ *vs* k.

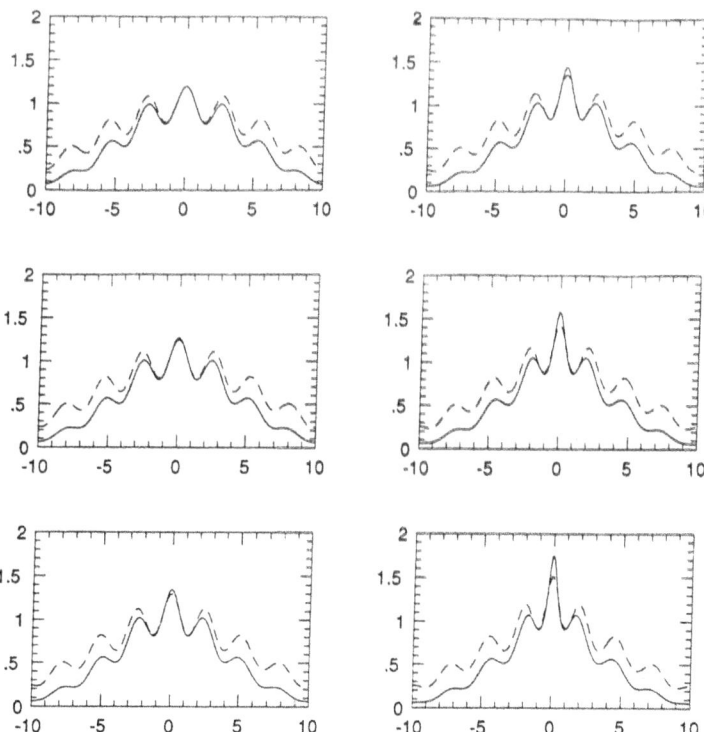

Figure 7. The amplitude $|\psi|$ vs scaled x, for the data of Experiment 2 (figure 6). The details of the scaling are described in the text.

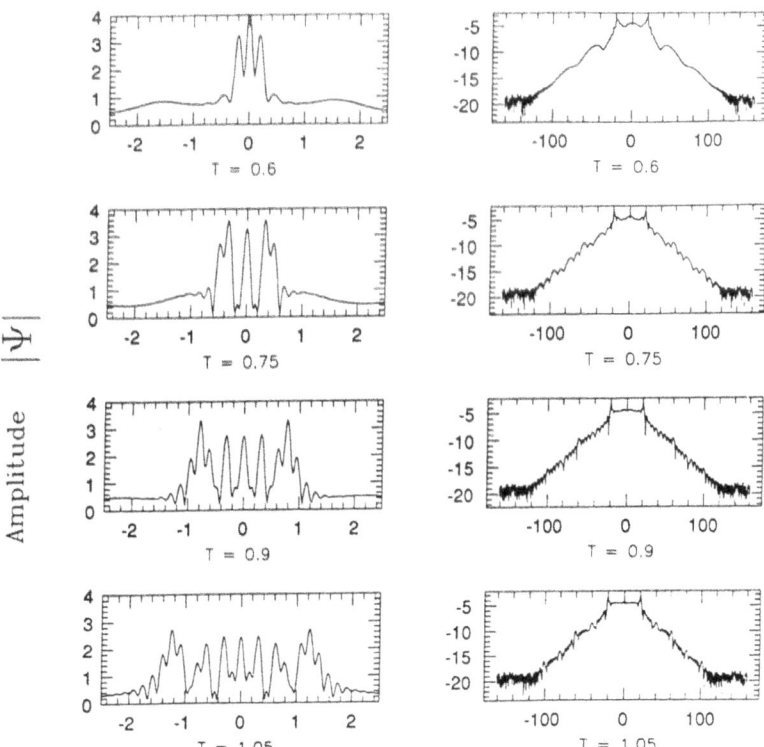

Figure 8. Focusing, weak $0(\epsilon)$ nonlinearity. The experiment is that of Experiment 2 (figure 6), but for times beyond break time.

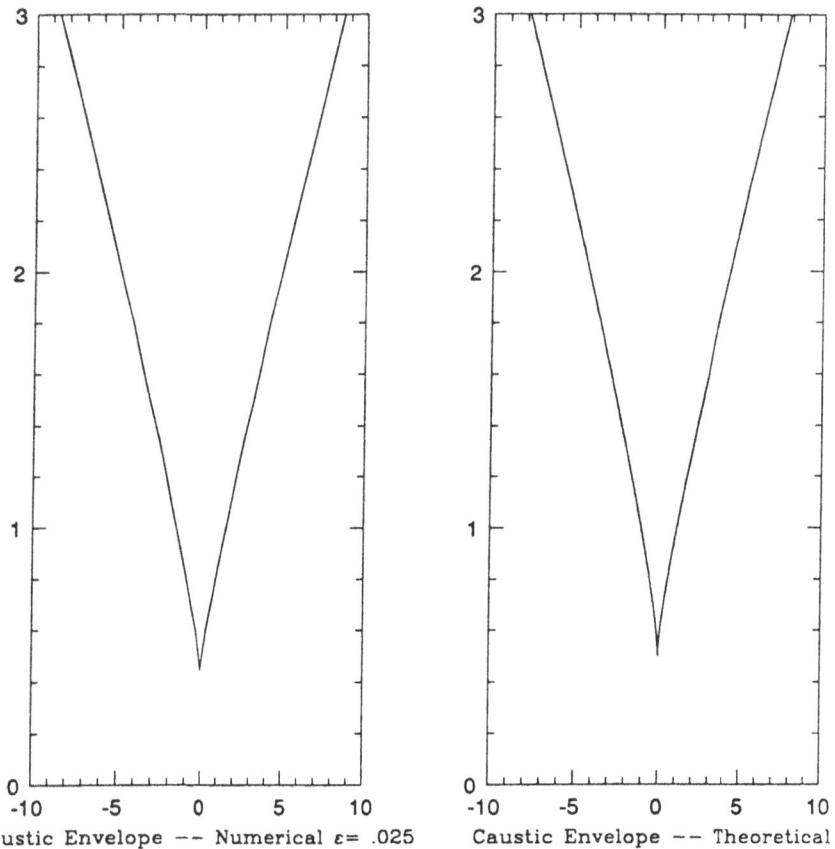

Figure 9. The caustic envelop in space-time: (a) Estimated from numerical data as described in text. (b) Theoretical prediction from the eikonal equation.

interpretation, as we will see in the next experiment, gives good agreement between theory and experiment. The data is shown in Figure 8. Notice the stability of the power spectrum from break time to the final time of the experiment.

3.3.5 Experiment 5 – The Caustic Region

This last experiment is an attempt to determine the caustic envelope experimentally, and to compare the experimental results with the theoretical predictions. As we have seen in the previous experiment the post-breaking solutions are characterized by a centrally located region of small scale stucture bounded by a small region of large $|\psi|$. We have tried to crudely estimate the extent of the caustic region by tracking the maxima – the time of first break is taken to be the time when the central maxima splits into two, and the boundaries of the caustic are taken to be the locations of these maxima. We should stress that while we should expect the amplitude to be large near the caustic there is no real reason to expect the maxima to occur exactly on the caustic envelope. A good analogy would be with WKB theory for the (time independent) Schrodinger equation. In this (ODE) case the maxima of the wave function do not occur at the turning points, but are displaced by a small amount (related to the zeroes of the Airy function.) When we compare the experimental caustic with the theoretical curve obtained from the characteristics of the eikonal equation we find very good agreement, though the experimental curve seems to consistently over-estimate the size of the region.

4 Concluding Remarks and an Intriguing Conjecture

In the focusing, but weakly nonlinear, case, the intensity of the NLS field develops a region of short wavelength, finite amplitude oscillations which might be interpreted as a "sea of solitons". The transition between this sea of rather violent oscillatory behavior and a region of quiescent behavior is very sharp, with a transition boundary that is well described by a caustic curve in $x - t$ space that can be calculated from the eikonal equation for the linear NLS equation.

In *nonlinear* geometrical optics, the eikonal equation for the phase of the wave depends nonlinearly upon the amplitude of the wave; the eikonal and transport equations of nonlinear geometrical optics form a coupled system. In contrast, in *weakly nonlinear* geometric optics, the amplitude dependence is removed from the eikonal equation and appears as an extra nonlinear term in the transport equation. The eikonal equation can then be solved independently of the amplitude; it's solution, as we have seen, locates the caustic curves which delineate the oscillatory from quiescent regions of space-time.

In the focusing case the modulational instability forces structure in the wave at an intermediate spatial scale. Mathematically, modulation on this intermediate scale necessitates the retention of second derivatives in the transport equation. The nonlinear evolution equation which results is itself a variable coefficient nonlinear Schroedinger equation. That a variable coefficient NLS equation describes the evolution of long waves in the exact NLS system is a consequence of the Galilean invariance of the exact system. It would be interesting to derive this feature of NLS from a "renormalization group approach", where NLS should arise as a "fixed point" of the appropriate dynamical system.

We summarize the weakly nonlinear behavior of the focusing NLS equation in the form of a 'theorem' whose validity is indicated by both our formal calculations and our numerical experiments:

Theorem 1 *Let $\psi^\epsilon(x,t)$ denote the unique solution of the initial value problem*

$$
\begin{aligned}
i\epsilon\psi_t \quad + \quad & \epsilon^2\psi_{xx} + \epsilon\psi\psi^*\psi = 0 \\
\psi(x, t = 0) \quad = \quad & A_{in}(\frac{x}{\sqrt{\epsilon}}; x)\exp[\frac{i}{\epsilon}S_{in}(x)].
\end{aligned}
$$

Then, for $0 \leq t < t_b$ and for sufficiently small ϵ,

$$\psi^\epsilon(x,t) \simeq b(\frac{\zeta(x,t)}{\sqrt{\epsilon}}, t; x) \exp[\frac{i}{\epsilon} S(x,t)] \qquad (4.1)$$

Here $S(x,t)$ is the unique solution of the eikonal equation

$$S_t + S_x^2 = 0 \qquad (4.2)$$
$$S(x, t = 0) = S_{in}(x), \qquad (4.3)$$

which, for $0 \leq t < t_b$ (the time of first breaking), can be constructed from the characteristics $\zeta(x,t)$ defined implicitly by

$$x = \zeta + 2P_{in}(\zeta)t \qquad P_{in} \equiv S_{in,x}.$$

The function $b(z,t;x)$ is an approximation to the solution $B(z,t)$ of the equation

$$i(B_t + P_x B) + \zeta_x^2 B_{zz} + \sqrt{\epsilon}\zeta_{xx}B + BB^*B = 0,$$
$$B(z, t = 0) = A_{in}(z, \sqrt{\epsilon}z),$$

where the coefficients are functions of z, defined by

$$P_x \equiv \partial_x P_{in}(\zeta(x,t))$$
$$\zeta_x \equiv \partial_x \zeta(x,t)$$
$$\zeta_{xx} \equiv \partial_{xx}\zeta(x,t),$$

all evaluated at $\zeta(x,t) = \sqrt{\epsilon}z$. The approximation $b(z,t;x)$ of $B(z,t)$ is obtained by 'freezing' these coefficients. Thus the function $b(z,t;x)$ satisfies

$$i(b_t + P_x b) + \zeta_x^2 b_{zz} + bb^*b = 0$$
$$b(z, t = 0; x) = A_{in}(z; x),$$

with the coefficients given by

$$P_x = \frac{P_{in}'}{1 + 2P_{in}'t},$$
$$\zeta_x^2 = \frac{1}{(1 + 2P_{in}'t)^2},$$
$$P_{in} \equiv P_{in}(\zeta(x,t)).$$

This "theorem", whose validity has only been established with formal asymptotics, provides information about the small ϵ behavior of ψ^ϵ. From representation (4.1), one can deduce the structure of certain weak limits. (Here, even "prebreaking", the convergence as $\epsilon \to 0$ is only weak because of the rapid oscillations on the intermediate scale of the modulational instability.) For example, if representation (4.1) is valid,

$$|\psi^\epsilon|^2 \longrightarrow \lim_{L \to \infty} \frac{1}{2L} \int_{-L}^{L} |b(y,t;x)|^2 dy. \qquad (4.4)$$

A similar expression exists for the weak limit of $\epsilon(\psi^*\psi_x - \psi\psi_x^*)$, while the quantity $(b^*b_z - bb_z^*)$ is higher order in ϵ. Thus, the asymptotic representation (4.1) gives prebreaking information about both the weak limit and its evolution.

Certainly one of the most important features of the reduced equations is that second derivatives have been retained because of the modulational instability. First, we extend the definition of the Riemann invariants, including terms of higher order in ϵ:

$$\Gamma_+^\epsilon = -P^\epsilon + \sqrt{-2\epsilon}\, a$$
$$\Gamma_-^\epsilon = -P^\epsilon - \sqrt{-2\epsilon}\, a,$$

where $A = a\exp(i\alpha)$ and $P^\epsilon \equiv \partial_x(S + i\epsilon\alpha)$. Then the exact evolution of $\Gamma_\pm$ is given by

$$\Gamma^\epsilon_{+,t} - (\frac{3}{2}\Gamma^\epsilon_+ + \frac{1}{2}\Gamma^\epsilon_-)\Gamma^\epsilon_{+,x} = -\epsilon^2\partial_x(\frac{a_{xx}}{a})$$
$$\Gamma^\epsilon_{-,t} - (\frac{3}{2}\Gamma^\epsilon_- + \frac{1}{2}\Gamma^\epsilon_+)\Gamma^\epsilon_{-,x} = -\epsilon^2\partial_x(\frac{a_{xx}}{a}),$$

which is closed since $a \equiv (\Gamma^\epsilon_+ - \Gamma^\epsilon_-)/\sqrt{-2\epsilon}$. Note that these exact evolution equations contain the regularizing derivative terms.

Much of this conference focused upon first order equations for certain "invariant differentials" [11] of the form

$$\Omega^{(1)}_t + \Omega^{(2)}_x = 0.$$

It is intriguing to conjecture the existence of a modified form of these modulation equations for such 'invariant differentials" involving second derivatives, a form which would be equivalent to the prebreaking Riemann invariant equations; perhaps something like

$$\Omega^{(1)}_t + \Omega^{(2)}_x = \Lambda_{xx}.$$

The form of the generator of the conservation laws before averaging certainly makes the existence of such an invariant form plausible. If such an invariant form of modulation equations exists pre-breaking, it is likely to extend to a post-breaking form, and would constitute a new form of evolution for the modulation differentials.

Acknowledgement: We wish to acknowledge many extremely useful conversations with David Levermore throughout the course of this work; and one useful conversation with Warren MacEvoy. We also acknowledge support from the DOD in the form of a Graduate Fellowship; as well as grants from AFOSR and NSF.

References

[1] S. Jin, C. D. Levermore, and D. W. McLaughlin. Semi-Classical Limits of Defocusing NLS. *Preprint*, 1991.

[2] S. Jin. *Semi-Classical Limits of Defocusing NLS*. PhD thesis, University of Arizona, 1991.

[3] P. D. Lax and C. D. Levermore. The Small Dispersion Limit of the Korteweg-de Vries Equation I. *Comm. Pure and Applied Math.*, 36:253–290, 1983.

[4] P. D. Lax and C. D. Levermore. The Small Dispersion Limit of the Korteweg-de Vries Equation II. *Comm. Pure and Applied Math.*, 36:571–593, 1983.

[5] P. D. Lax and C. D. Levermore. The Small Dispersion Limit of the Korteweg-de Vries Equation III. *Comm. Pure and Applied Math.*, 36:809–830, 1983.

[6] G. B. Whitham. Linear and Nonlinear Waves. *Wiley-Interscience, New York*, 1974.

[7] J. E. Rothenberg and D. Grischkowsky. Observation of the Formation of an Optical Intensity Shock and Wave Breaking in the Nonlinear Propagtion of Pulses in Optical Fibers. *Physical Review Letters*, 62(5):531–534, 1989.

[8] N.Tzoar and M. Jain. Self-Phase Modulation in Long-Geometry Optical Waveguides. *Physical Review A*, 23(3):1266–1270, 1981.

[9] J. E. Rothenberg. Observation of the Buildup of Modulational Instability from Wave Breaking. *Optics Letters*, 16(1):18–20, 1990.

[10] D. W. McLaughlin, D. Muraki, and O. Wright. Optical Shocking and the Propagation of Polarized Light. in preparation 1991.

[11] H. Flaschka, M. G. Forest, and D. W. McLaughlin. Multiphase Averaging and the Inverse Spectral Solution of the Korteweg-de Vries Equation. *Comm. Pure Appl. Math*, 33:739–784, 1980.

A NUMERICAL STUDY OF NEARLY INTEGRABLE
MODULATION EQUATIONS

M. Gregory Forest[1,2] and Amarendra Sinha [2]

Department of Mathematics
The Ohio State University
Columbus, OH 43210

Abstract. Previous numerical studies of the damped, driven sine-Gordon equation with spatially periodic boundary conditions have identified various low-dimensional attractors and bifurcation phenomena. These attractors are fully nonlinear (i.e., order one amplitude) space-time structures which have been independently measured (using the spectral transform) in terms of integrable, sine-Gordon modes. Based on these direct measurements, we posit a leading order approximation to the perturbed flow in terms of modulated sine-Gordon wavetrains. Our goal here is to present *dynamical simulations of two-phase perturbed sine-Gordon modulation equations* and to compare these predictions with the results of direct pde simulations using Ed Overman's codes.

I.   FORMULATION, BACKGROUND AND MOTIVATION

We consider the damped, driven sine-Gordon equation with even, periodic boundary conditions:

$$q_{tt} - q_{xx} + \sin q = \epsilon[-\alpha q_t + \Gamma \cos(\omega_d t)] \tag{I.1a}$$

$$0 \le x \le L \tag{I.1b}$$

$$e^{iq(x,t)} = e^{iq(x+L,t)} \tag{I.1c}$$

$$q(x,t) = q(-x,t). \tag{I.1d}$$

Several detailed numerical studies have been reported on this system [1], indicating various interesting attractors and bifurcation phenomena. We shall fix all parameters,

---

[1]   Research supported by NSF DMS 88-03465, 91-04806.

[2]   We acknowledge computer support on the CRAY YMP from the Ohio Supercomputer Center.

except the driving amplitude $\epsilon\Gamma$, as follows:

$$\epsilon\alpha = .04, \quad \omega_d = .87, \quad L = 12. \tag{I.1e}$$

Our goal here is to use modulated integrable modes to resolve the attractor dynamics of (I.1a–e) *in the pre-chaotic range of* Γ, $0 \leq \epsilon\Gamma < .07$. Namely, for $.04 < \epsilon\Gamma < .07$, we have observed [1] solutions of (I.1) with the following characteristics:

 (i) the spatial structure consists of a localized coherent wave;

 (ii) the time flow consists of a modulation about periodicity of frequency $\omega_d = .87$.

Using the direct spectral transform [2], we measure the integrable sine-Gordon mode content in the field $(q^\epsilon(x, t_n), q_t^\epsilon(x, t_n))$ at each time step t_n. We find there exists initial data such that:

 (i) the spatial structure $(q^\epsilon(x, t_n), q_t^\epsilon(x, t_n))$ is well-approximated by a two-phase, even, periodic, sine-Gordon breather train;

 (ii) the two-phase approximation is robust for $t_n \gg 1$, consisting of a modulation in the breather wavetrain parameters without significant amplitude in additional nonlinear modes.

Remark: We focus in this paper on the pre-chaotic regime of (I.1a–e), with large enough driver amplitude to sustain non-homogeneous spatial structure, $.04 < \epsilon\Gamma < .07$. For smaller values of $\epsilon\Gamma$, the waveform goes flat, while for larger values of $\epsilon\Gamma$, there is a bifurcation to chaos along with an additional spatial mode of significant amplitude.

This study represents a first nontrivial step in using integrable modes as a basis for resolving attractor dynamics in nearly integrable pdes. The standard application of linear, orthogonal bases (e.g., Fourier modes) is appealing because the derivation is straightforward and the numerical integration is routine for each finite mode truncation. It is therefore possible to systematically explore the order of truncation (i.e., number of modes) until the dynamics is robust to retaining more modes. One can thereby effectively cover the attractors and explore various dynamical issues, such as bifurcations to chaos. We refer to [1, 3] for such studies.

The present study, however, represents an alternative mode resolution, in terms of integrable modes, of nearly integrable attractors. The drawback is that the derivation of the corresponding amplitude or modulation equations is nontrivial [4]; the numerical simulation of these modulation equations is likewise nontrivial, requiring significant background code development of the inverse spectral transform [5]. Fortunately, these drawbacks can be overcome for specific studies such as this one.

The motivation here, we emphasize, is not for engineering purposes. Rather, we focus on integrable modes in order to retain important geometrical and analytical features of the integrable phase space. As the small parameter (ϵ) goes to zero in these modulation equations, we recover integrability at each order of mode truncation, i.e., for every number of phases. Likewise, for pde initial data nearby an unstable integrable solution, these integrable mode truncations preserve the underlying homoclinic structure which has been identified as a primary culprit in perturbed bifurcations to chaos [1].

This study is *nontrivial* because we truncate on two-phase, modulated breather trains, which have non-trivial spatial and temporal dependence. It is only the first step, however, because this two-phase ansatz does not capture homoclinic structure. Therefore, *we only seek to model the pre-chaotic regime in this paper*. The accurate resolution of pre-chaotic attractors is necessary, nonetheless, to establish that these near-integrable methods are successful up to the chaotic regime. From these positive results, we can then move forward to the more difficult regimes.

II. THE MODULATION EQUATIONS FOR TWO-PHASE PERTURBED SINE-GORDON BREATHER TRAINS

Based on direct numerical studies described above, we posit an approximate solution of (I.1a–e) of the form:

$$q^\epsilon(x,t) = q_2(\theta_1, \theta_2; \{E_k(t)\}_{k=1}^4) + \mathcal{O}(\epsilon), \tag{II.1}$$

where the two *phases* θ_1, θ_2 satisfy even, periodic spatial constraints,

$$(\theta_1)_x = (\theta_2)_x = \kappa_1 = \kappa_2 = \frac{2\pi}{L} = \text{ constant}. \tag{II.2}$$

The *dynamics* of the two-phase ansatz (II.1) consists of the *fast phase equations*

$$(\theta_1)_t = \omega_1(\{E_k(t)\}_{k=1}^4), \tag{II.3}$$
$$(\theta_2)_t = \omega_2(\{E_k(t)\}_{k=1}^4) = -\omega_1,$$

coupled with the *energy modulation equations* (given below) which govern the spectral parameters $\{E_k(t)\}_{k=1}^4$. The derivation of these modulation equations [4] is nonstandard, relying heavily on integrable structure. We shall need some additional background in order to state these modulation equations, but defer to [4] for their derivation.

First, the phase equations (II.2,3) can be integrated to yield

$$\theta_1 = \kappa_1 x + \int_0^t \omega_1(\tilde{t})d\tilde{t} + \theta_1^0, \tag{II.4}$$

$$\theta_2 = \kappa_1 x - \int_0^t \omega_1(\tilde{t})d\tilde{t} - \theta_1^0,$$

where the phase constants θ_1^0, θ_2^0 are constrained by $\theta_2^0 = -\theta_1^0$ to maintain evenness in x. Next, the leading order waveform, q_2 in (II.1), is given by an integrable theta function solution of sine-Gordon [6,5]:

$$q_2(\theta_1, \theta_2; \{E_k\}_{k=1}^4) = 2i \ln\left[\frac{\Theta(\vec{\ell}(\theta_1, \theta_2))}{\Theta^*(\vec{\ell}(\theta_1, \theta_2))}\right], \tag{II.5a}$$

where $\Theta(\vec{\ell})$ is the Riemann theta function of two complex variables $\vec{\ell} = (\ell_1, \ell_2)$:

$$\Theta(\ell_1, \ell_2) = \sum_{\vec{m} \in \mathbb{Z} \times \mathbb{Z}} \exp\{i\pi\langle B\vec{k}, \vec{m}\rangle + 2\pi i\langle\vec{\ell}, \vec{m}\rangle\}, \tag{II.5b}$$

with $\langle\vec{u}, \vec{v}\rangle$ denoting the usual inner product of two-vectors, B is the 2×2 normalized period matrix of a genus two, hyperelliptic Riemann surface, $(E, R(E))$, with branch points at $0, \infty$ and the spectral parameters $\{E_k\}_{k=1}^4$,

$$R^2(E) = E \prod_1^4 (E - E_k). \tag{II.5c}$$

The explicit dependence of $\vec{\ell}$ on θ_1, θ_2 and $\{E_k\}_1^4$ for these two-phase breather trains is

$$\vec{\ell} = (\vec{e}_1 - 2\vec{B}_1)\theta_1 + (\vec{e}_2 - 2\vec{B}_2)\theta_2,$$
$$\vec{B}_j = \text{column } j \text{ of } B, \tag{II.6}$$
$$(\vec{e}_j)_k = \delta_{jk}.$$

41

The final ingredients are the explicit definition and form of B as well as an explicit parametrization of the two-phase, even and periodic set of spectral parameters $\{E_k\}_{k=1}^4$. The following simple periodic spectrum satisfies these constraints:

$$\left\{ E_j = \frac{1}{16} e^{i\phi_j}; \quad \phi_3 = -\phi_1, \phi_4 = -\phi_2, \quad 0 < \phi_1 < \phi_2 < \pi \right\}. \tag{II.7}$$

This two-parameter family, (II.7), is parametrized by (ϕ_1, ϕ_2), or equivalently, (κ_1, ω_1), with $\kappa_2 = \kappa_1$ and $\omega_2 = -\omega_1$. The fixed period constraint, $L = 12 = 2\pi/\kappa_1$, reduces the spectral parameter count to one, so that ϕ_1 *is the free modulation parameter for this family*. The remaining parameter is the initial phase constant, θ_1^0, equation (II.4).

Summary: There is *one modulational degree of freedom* in the spectral parameters $\{E_k\}_{k=1}^4$ for this 2-phase, even, fixed period family of breather trains, together with one arbitrary initial phase constant, θ_1^0.

This implies *we need one modulation ode*, for ϕ_1 defined in (II.7), to close the dynamical system with the phase equation (II.3) for this ansatz. If we define

$$Z_j = \frac{1}{2} \left(E_j + \frac{1}{16^2 E_j} \right) = \frac{1}{16} \cos(\phi_j), \quad j = 1, 2, \tag{II.8a}$$

then the modulation odes for ϕ_j are [4, 7]

$$(Z_j)_t = \frac{(Z_j - \frac{1}{16})\langle T_1 \rangle_t}{(Z_j - \chi)}, \quad j = 1 \text{ or } 2, \tag{II.8b}$$

$$\langle T_1 \rangle = \frac{1}{L} \int_0^L \frac{1}{32} [(q_t + q_x)^2 - 2\cos q] dx, \tag{II.8c}$$

where $\chi(Z_j)$ is uniquely specified below in terms of ϕ_1 (or Z_1), and $q \equiv q_2(\theta_1, \theta_2)$ as given in (II.5) is inserted in the *averaged density* (II.8c), which is an integrable conserved quantity. Finally, the holomorphic period matrix B and meromorphic period data χ are defined as follows.

Let ψ_1, ψ_2 denote holomorphic differentials,

$$\psi_j = (C_{j_1} E + C_{j_2}) \frac{dE}{R(E)}, \quad j = 1, 2, \tag{II.9}$$

and $\Omega^{(x)}$ denote a meromorphic differential

$$\Omega^{(x)} = \left(-\frac{1}{2} E^3 - 16\alpha_1 E^2 + \alpha_1 E + \frac{1}{2 \cdot 16^2} \right) \frac{dE}{ER(E)}, \tag{II.10a}$$

with $R(E)$ defined in (II.5c),

$$\alpha_1 = -\frac{1}{32} \left[\frac{1}{16} + \chi \right]. \tag{II.10b}$$

Figure 1 depicts a basis of cycles on the Riemann surface $(E, R(E))$, denoted a_1, a_2, b_1, b_2.

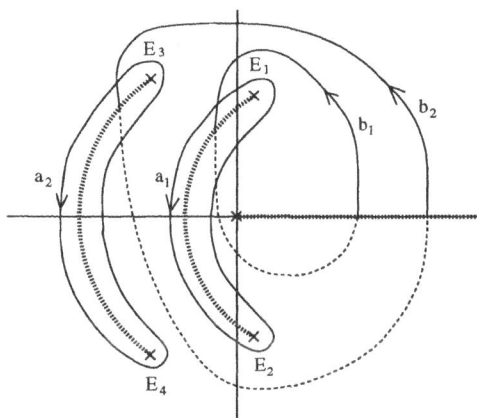

Figure 1. The canonical cycles for the two-phase Riemann surface.

Then χ is fixed by the condition

$$\oint_{a_j - 2b_j} \Omega^{(x)} = 0, \quad j = 1 \text{ or } 2. \tag{II.10c}$$

The normalization constants $C_{jk}(\{E_k\}_{k=1}^4)$ are uniquely determined by

$$\oint_{a_i} \psi_j = \delta_{ij}, \quad i, j = 1, 2, \tag{II.11a}$$

and then the columns $\vec{B}_k$ of B are defined by

$$\oint_{b_k} \vec{\psi} = \vec{B}_k, \quad \vec{\psi} = (\psi_1, \psi_2)^t. \tag{II.11b}$$

For the branch point symmetry depicted in (II.7), the 2×2 period matrix B inherits the following symmetry [5]

$$B = \begin{pmatrix} 1/2 & 0 \\ 0 & 1/2 \end{pmatrix} + i \begin{pmatrix} \alpha & \alpha \\ \alpha & \beta \end{pmatrix}, \qquad \alpha, \beta > 0. \tag{II.11c}$$

III. A RECALL OF AVERAGING PREDICTIONS FROM THE ENERGY-PHASE MODULATION EQUATIONS

In a previous paper [7], we used the method of averaging to predict frequency and phase locked solutions from the odes (II.3), (II.8). This result is achieved by supposing the breather frequency ω_1 locks to the driver frequency, $\omega_1 = \omega_d = .87$, and then the free phase shift constant θ_1^0 is picked out by the condition that (II.8b) averages to zero over one period, $T = 2\pi/\omega_1$, of the perturbation. This phase locking condition for θ_1^0 simplifies to (with $\omega_d = .87$)

$$\sin(\omega_d \theta_1^0) = -\frac{\alpha}{\Gamma} \frac{C_1}{C_2} \in [-1, 1], \tag{III.1a}$$

$$C_1 = \int_0^{2\pi/\omega_d} \int_0^L q_t(q_t + q_x) dx\, dt, \tag{III.1b}$$

$$C_2 = \int_0^{2\pi/\omega_d} \int_0^L q_t \sin(\omega_d t) dx\, dt, \tag{III.1c}$$

43

and the constants C_1, C_2 are uniquely specified with q replaced in (III.1) by q_2 above with $\omega_1 = \omega_d$. The apparent dependence on θ_1^0 in (III.1b,c) is removed by the average in x and t.

For $\omega_d = .87$, the phase locking condition (III.1) immediately implies that for fixed damping coefficient $\epsilon\alpha = .04$, there is a minimum driver amplitude, $\epsilon\Gamma^*$, necessary to sustain the locked breather train,

$$\Gamma^* = \frac{\alpha C_1}{C_2}, \qquad \Gamma > \Gamma^*, \tag{III.2}$$

for which there are two phase locked constants,

$$(\theta_1^0)^+ = \frac{1}{\omega_d} \sin^{-1}\left(\frac{-\alpha C_1}{\Gamma C_2}\right) \tag{III.3}$$

$$(\theta_1^0)^- = \pi/\omega_d - (\theta_1^0)^+.$$

For $\omega_d = .87$, $C_1/C_2 \approx 1.223$.

These existence predictions will be clarified below by determining the evolution of the modulation equations (II.3,8) with initial data given by the frequency-phase locked prediction, (III.1–3):

$$\omega_1(0) = .87, \tag{III.4}$$

$$\theta_1(x,0) = \kappa_1 x + (\theta_1^0)^{\pm}.$$

For $\Gamma < \Gamma^*$, we anticipate and confirm non-existence of these modulated breather trains. For $\Gamma > \Gamma^*$, we anticipate and confirm existence, and then explore robustness by varying the initial data (III.4).

IV. NUMERICAL INTEGRATION OF THE PERTURBED MODULATION EQUATIONS

Section II defines the "energy" (II.8) and "phase" (II.3) modulation odes, while the remaining formulas (II.1, 2, 4–7, 9–11) define the ingredients necessary to implement the numerical integration of this class of two-phase, even, periodic breather trains.

Here we report the results for fixed damping coefficient, $\epsilon\alpha = .04$, fixed driver frequency, $\omega_d = .87$, and fixed spatial length, $L = 12$. These parameters are chosen to agree with earlier studies [1]. The results were similar for other reasonable values (e.g., $L > 2\pi$, $\omega < 1$, and $\epsilon\alpha \ll 1$ etc.) of these parameters. However, this has not been studied extensively.

The remaining three parameters which are varied in the following experiments consist of : (1) the *driver amplitude*, $\epsilon\Gamma$; (2) the "energy" *initial data*, $\phi_1(0)$ (or equivalently, the initial frequency $\omega_1(0)$ of the pde Cauchy data, $(q_2(x,0), (q_2)_t(x,0))$); and (3) the "phase" *initial data*, $\theta_1^0 = \theta_1(0)$.

For the numerical integration of the modulation equations, we modify (I.1a) by a translation in time to

$$q_{tt} - q_{xx} + \sin q = \epsilon[-\alpha q_t - \Gamma \cos(\omega_d(t + t_0))], \tag{IV.1}$$

where $t_0 \in [0, 2\pi/\omega_d)$. The quantity $\langle \mathcal{T}_1 \rangle_t$ is computed along the perturbed flow (IV.1), and is given by:

$$\langle \mathcal{T}_1 \rangle_t = \frac{1}{16L} \int_0^L (q_x + q_t)(\epsilon\alpha q_t + \epsilon\Gamma \cos(\omega_d(t + t_0)))dx, \tag{IV.2}$$

where $q = q_2$ is the modulated integrable two-phase solution (II.5) with $E_j = E_j(t), j = 1, \cdots, 4$, and we take $\theta_1^0 = 0$ in (II.4). Thus for fixed values of the parameters $\epsilon\Gamma$ and t_0, the only dependent variable $\phi_1(t)$ (or $\phi_2(t)$) is uniquely specified by the initial data $\phi_1(0)$ (or $\phi_2(0)$). However, the function $\phi_2(\phi_1)$ (or $\phi_1(\phi_2)$) is a complicated nonlinear function defined implicitly by the spatial length and the evenness symmetry. Due to the difficulty in computing this function, we solve equation (II.8) for $j = 1$ *and* 2, with an initial data $(\phi_1(0), \phi_2(0))$ satisfying the periodicity requirements, which can be obtained by a search in the parameter space. The modulation equations then preserve the fixed period constraint. The constancy of the spatial period is then monitored and serves as a check on the codes. In fact, the function $\phi_2(\phi_1)$ can be determined numerically by solving the modulation equations with $\Gamma = 0$ and some positive and negative values of $\epsilon\alpha$. The results are shown in figure 2.

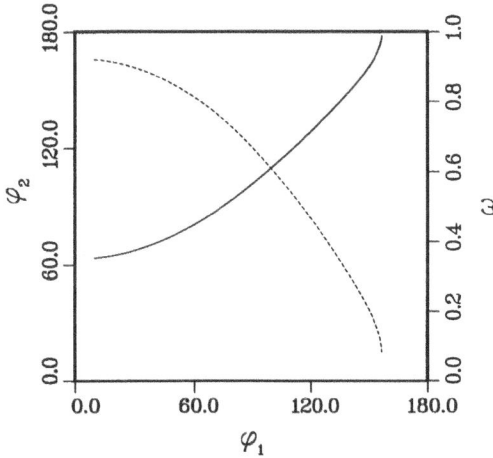

Figure 2. We plot ϕ_2 (solid curve) and ω_1 (dashed curve) as functions of ϕ_1 for the two-phase solutions with $L = 12$.

We now proceed to describe the numerical experiments. For the first set of experiments we fix [5] the initial data $(\phi_1(0), \phi_2(0)) \approx (42.2550°, 71.9234°)$ so that $L = 12$ and $\omega_1(0) = \omega_d = .87$. For $\epsilon\Gamma \in (0, .07)$ we vary the relative phase $t_0 \in [0, 2\pi/\omega_d) \approx [0, 7.2225)$, with the following conclusions:
(a) there is a critical driver amplitude, $\epsilon\Gamma^*$, necessary to sustain the two-phase modulations for $t \gg 1$.
(b) for $\epsilon\Gamma > \epsilon\Gamma^*$, there is a range of $t_0 \in [0, 2\pi/\omega_d)$ such that the two-phase modulations persist;
(c) there is a reasonable agreement (to $\mathcal{O}(\epsilon)$) with the averaging predictions and the direct simulation of the pde.

To present these results, it is convenient to define a quantity, $T_f(t_0)$, associated with each integration of (II.3, 8). $T_f(t_0)$ is *the lifetime of the integration*, i.e., the maximum time until the integration of (II.3, 8) violates the two-phase ansatz. $T_f(t_0)$ *finite* implies that the spectral parameter, $E_1(t) = \frac{1}{16}e^{i\phi_1(t)}$, approaches the real axis with $\phi_1 \to 0$ (the limit $\phi_2 \to \pi$ is rare in the presence of damping, i.e., $\epsilon\alpha > 0$). This limit corresponds to zero amplitude in one of the two nonlinear modes, and the wavetrain $q_2(x, t)$ approaches an x-independent limit of a uniform pendulum chain. *Physically*, finite $T_f(t_0)$ corresponds to the pendulum chain being unable to sustain the localized spatial structure and thereby relaxes to the uniform state. We refer to

[5] for integrable descriptions of the two-phase to one-phase limit within this class of solutions. Likewise, $T_f(t_0) = \infty$ implies the breather modulation equations evolve within this two-phase class for all time. These results are therefore best represented in the graph of $1/T_f$ as a function of t_0.

Figure 3 depicts the graph of $1/T_f(t_0)$ for $\epsilon\Gamma = .04$. Since the graph is bounded away from zero, we conclude *there are no phase constants for which the breather modulation can be sustained.* Therefore, $\epsilon\Gamma = .04$ is below the threshold value. Figure 4 shows the dynamical evolution of the initial data for $t_0 = 2.0$, indicating the finite time departure from this two-phase class.

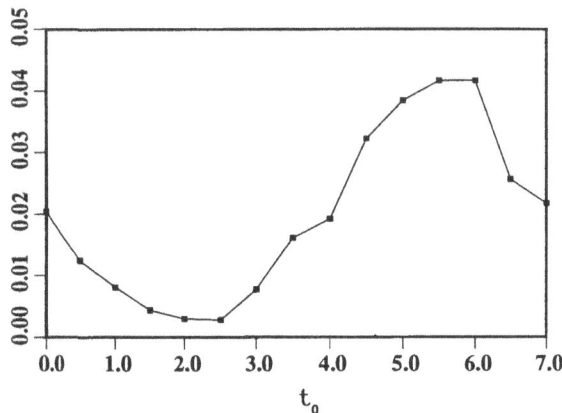

Figure 3.   The graph of $1/T_f(t_0)$ for $\epsilon\Gamma = .04$.

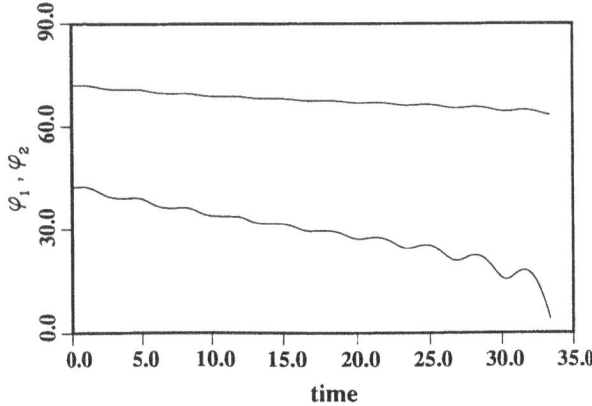

Figure 4.   The solution $\phi_1(t), \phi_2(t)$ of the modulation equations for $\epsilon\Gamma = .04$, and $t_0 = 2.0$.

To identify the critical driving amplitude, we graph $1/T_f(t_0)$ for $\epsilon\Gamma = .0488$ and .0490 in figures 5 and 6 respectively. Figure 6 then shows that *a window of phases opens for which the two-phase modulating breather train persists for all time.* We remark the existence results of [7] predict a critical amplitude $\epsilon\Gamma^* \approx .0489$.

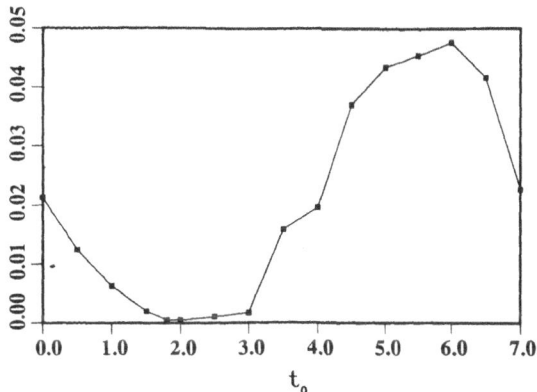

Figure 5. The graph of $1/T_f(t_0)$ for $\epsilon\Gamma = .0488$.

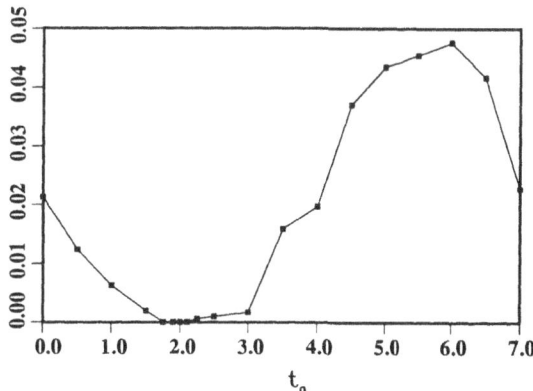

Figure 6. The graph of $1/T_f(t_0)$ for $\epsilon\Gamma = .0490$.

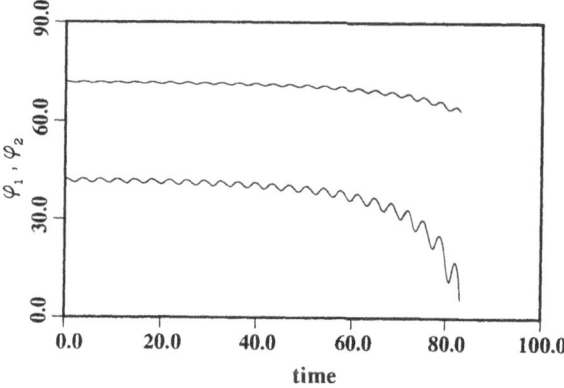

Figure 7. The solution $\phi_1(t), \phi_2(t)$ of the modulation equations for $\epsilon\Gamma = .05$, and $t_0 = 1.536$.

For a more detailed study of the modulation equations, we take $\epsilon\Gamma = .05$. In this case, the stable range of phases (i.e. $T_f(t_0) = \infty$) is $1.54 \lesssim t_0 \lesssim 2.62$. We recall, the phase-locking prediction from [7], $t_0^+ \approx 2.041$ and $t_0^- \approx 1.57$. Figure 7 shows the dynamical evolution of $\phi_1(t), \phi_2(t)$ for $t_0 = 1.536$ just *outside* the stable range; figure 8 depicts the evolution for $t_0 = 1.58$ just *inside* the stable range; and figure 9 depicts the evolution for the averaging prediction $t_0 = 2.041$. Figure 10 depicts the pde approximation $q^\epsilon(x,t) \approx q_2(x,t; E_j(t))$ given by this modulated breather train.

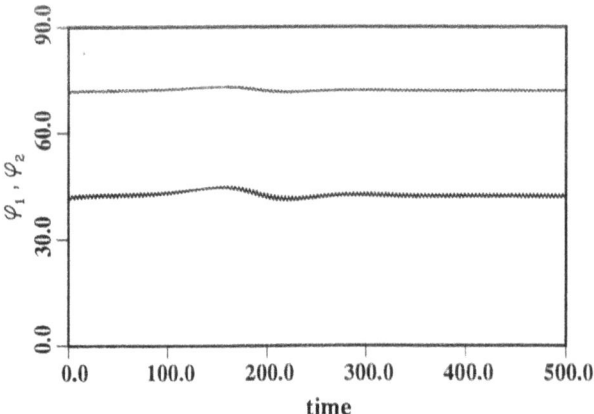

Figure 8. The solution $\phi_1(t), \phi_2(t)$ of the modulation equations for $\epsilon\Gamma = .05$, and $t_0 = 1.58$.

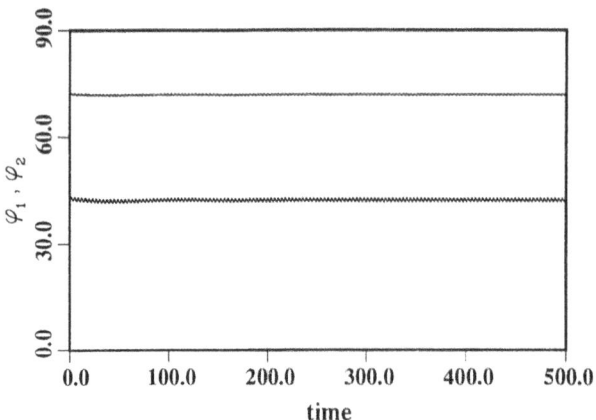

Figure 9. The solution $\phi_1(t), \phi_2(t)$ of the modulation equations for $\epsilon\Gamma = .05$, and $t_0 = 2.041$.

In order to test the validity of the two-phase approximation, we compare these results with the direct numerical integration, using Ed Overman's code, of the perturbed pde (IV.1) for the equivalent initial data $q_2(x, 0), (q_2)_t(x, 0)$. We refer to [1] for a detailed numerical study of the perturbed pde. First we note the critical driving amplitude $\epsilon\Gamma^* \approx .0489$ compares well with the perturbed pde bifurcation value [1, 8] of $\epsilon\Gamma \approx .05$. Secondly, we numerically integrate the pde (IV.1) with $\epsilon\Gamma = .05$ and initial data given by the two-phase solution. As t_0 is varied, we find that the

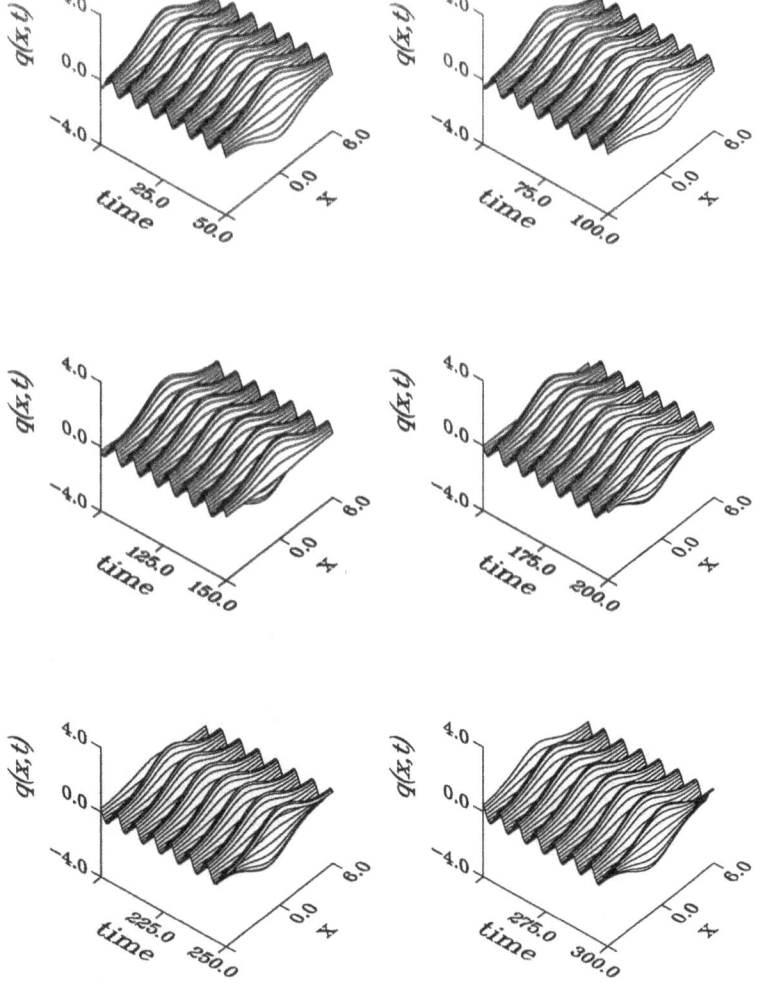

Figure 10. The pde approximation $q^{\epsilon}(x,t)$ given by the modulated breather train for $\epsilon\Gamma = .04$ and $t_0 = 2.0$.

pde solution approaches either an x-independent state or a breather train and these ranges are approximately the same as that of the two-phase modulation equations.

For t_0 in the stable range, the solution of the perturbed pde approaches a frequency and phase-locked breather train. In [7], the averaging prediction was shown to agree (to $\mathcal{O}(\epsilon)$) with the perturbed pde. To study the asymptotic behavior of the modulation equations, we consider

$$h(s) = \int_{T}^{T+2\pi/\omega_d} \int_{0}^{L} |q^{\epsilon}(x,t;t_0) - q_2(x,t+s-t_0)|^2 dt dx$$

where $T \gg 1$, and q_2 is the integrable two-phase solution with $L = 12$ and $\omega_1 = \omega_d = .87$. We find that $h(s)$ has a minimum at $s \approx 2.027$, for all t_0 in the stable range. Figures 11 and 12 depict the comparison of the two wavetrains for $t_0 = 1.58$ and $t_0 = 2.5$ respectively. This evidence supports the notion of an attracting breather state, with these various wavetrains in the domain of attraction.

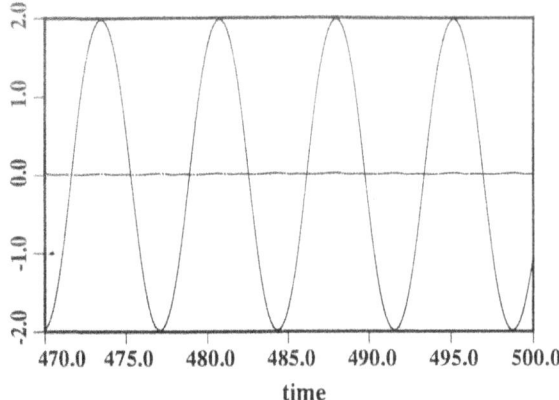

Figure 11. Here we compare the pde $q^{\epsilon}(x, t; t_0)$ and the two-phase breather $q_2(x, t + s - t_0)$ with $L = 12, \omega_1 = \omega_d = .87$ and $\epsilon\Gamma = .05, t_0 = 1.58, s = 2.027$. The solid and dashed curves represent the pde approximation and the breather respectively for $x = 0$. The dotted and chain-dotted curves are the L_2 and L_∞ norms respectively of the difference.

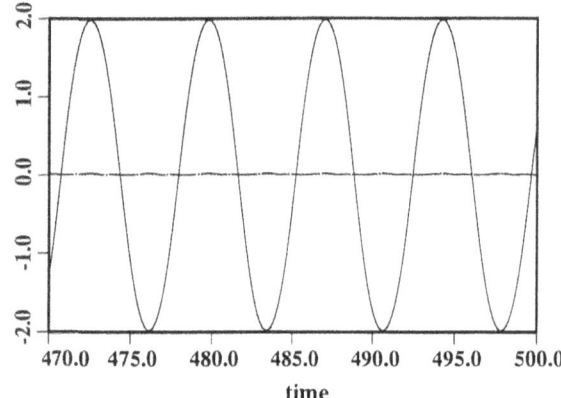

Figure 12. Here we show a comparison similar to that in figure 11, but for $t_0 = 2.5$. The results are similar for all t_0 in the stable range.

For a second experiment, we fix $\epsilon\Gamma = .05$ and $t_0 = 2.0$ in the stable range, and we vary $\phi_1(0)$, or equivalently, the initial frequency (see figure 2). The goal here is to explore the robustness of the breather train attractor to variations in the frequency of the initial data. Figure 13 depicts the graph of $1/T_f$ as a function of the initial frequency. Figure 14 shows the dynamical evolution with initial data $(\phi_1(0), \phi_2(0)) = (37.7610°, 70.1538°)$ which corresponds to frequency initial data $\omega_1(0) = .88113$. As shown in figure 15, the evolution for $\omega_1(0)$ in the stable range asymptotically approaches the *same* modulated breather train as in the first experiment.

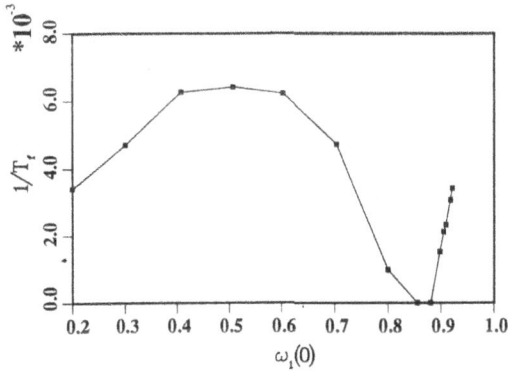

Figure 13. The graph of $1/T_f$ as a function of the initial frequency $\omega_1(0)$ for $\epsilon\Gamma = .05$.

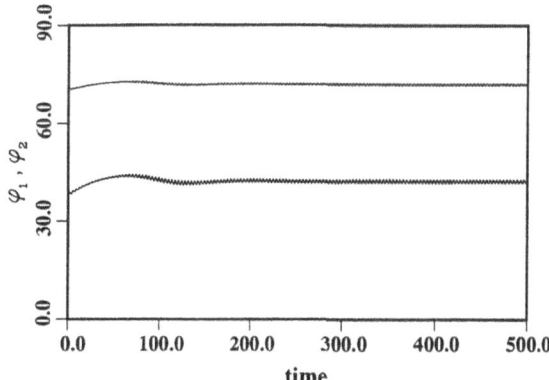

Figure 14. The solution $\phi_1(t), \phi_2(t)$ of the modulation equations for $\epsilon\Gamma = .05$, and $t_0 = 2.0$ with initial initial condition given by $(\phi_1(0), \phi_2(0)) = (37.7610°, 70.1538°)$ with $\omega_1(0) = .88113$.

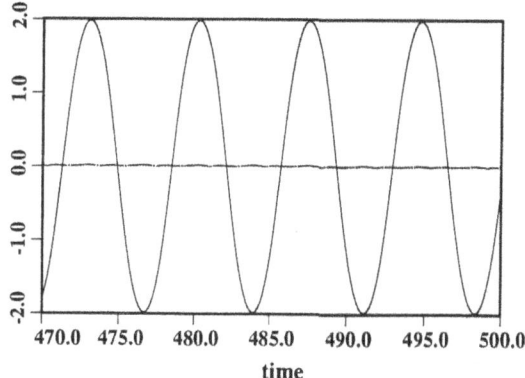

Figure 15. Here we provide a comparison of the pde approximation determined by the solution shown in figure 14 with the *same* breather considered in figure 11.

51

CONCLUSION

These numerical results satisfy two goals. First, we have shown that perturbed modulation equations of integrable pdes can be made practically useful. Second, we have shown these integrable mode approximations yield dynamical agreement with the direct pde results in the pre-chaotic regime. It remains to be seen if these modulation equations will be practical for dynamical resolution of chaotic attractors.

ACKNOWLEDGMENTS

We thank Ed Overman for allowing us to use his numerical codes. We also acknowledge consistent and valuable discussions with Dave McLaughlin.

REFERENCES

[1] A. R. Bishop, M. G. Forest, D. W. McLaughlin and E. A. Overman II, A quasiperiodic route to chaos in a near-integrable pde, *Physica D* 23:293(1986); A quasiperiodic route to chaos in a near-integrable pde: homoclinic crossings , *Physics Letters A* 127:335 (1988).

[2] A. R. Bishop, D. W. McLaughlin and E. A. Overman II, Coherence and chaos in the driven damped sine-Gordon equation: measurement of soliton spectrum, *Physica D* 19:1 (1986).

[3] A. R. Bishop, R. Flesch, M. G. Forest, D. W. McLaughlin and E. A. Overman II, Correlations between chaos in a perturbed sine-Gordon equation and a truncated modal system, *SIAM J. Math. Anal.* 23:1511 (1990); C. Xiong, Ph. D. Thesis, The Ohio State University (1991).

[4] N. M. Ercolani, M. G. Forest, D. W. McLaughlin and A. Sinha, Fully nonlinear modal equations for nearly integrable PDEs, *Journal of Nonlinear Science*, to appear.

[5] R. Flesch, M. G. Forest and A. Sinha, Numerical inverse spectral transform for the periodic sine-Gordon equation: theta function solutions and their linearized stability , *Physica D* 48:169 (1991).

[6] N. M. Ercolani and M. G. Forest, The geometry of real sine-Gordon wavetrains, *Comm. Math. Phys.* 99:1 (1985).

[7] M. G. Forest, S.-P. Sheu and A. Sinha, Frequency and phase locking of spatially periodic perturbed sine-Gordon breather trains, *SIAM J. on Appl. Math.*, 52:746 (1992).

[8] E. A. Overman, Private communication.

THE QUASICLASSICAL LIMIT

OF THE INVERSE SCATTERING PROBLEM METHOD

V.V. Geogjaev

Institute of Oceanology
Moscow

INTRODUCTION

Some hydrodynamical type equations can be obtained as the limits of the equations which are integrable by the ISP method. In this paper the variant of the ISP method for solving such equations is presented. First we shall show how it works for KdV equation and after that perform the same procedure in more detail for Benney equations.

THE DISPERSIONLESS KdV EQUATION

As the simplest case let us take KdV equation

$$u_t - 6uu_x + u_{xxx} = 0$$

One can obtain it's so-called quasiclassical limit by substituting $\varepsilon\partial_x$ for ∂_x, $\varepsilon\partial_t$ for ∂_t and taking $\varepsilon \to 0$. This corresponds to very slowly changing potential u. Thus we obtain:

$$u_t - 6uu_x = 0 \tag{1}$$

For the simplicity we shall consider only the case $u > 0$. We also suppose that u has a compact support, i.e. there exists an x_0 such that for $x > x_0$ $u \equiv 0$. The latter condition can be simply avoided by the regularization of integrals.

Let us take the L-equation for KdV

$$\psi_{xx} + (\lambda^2 - u)\psi = 0$$

and look what will be with it in the quasiclassical limit. First, by making a substitution $\psi = exp \left[t \int_{x_0}^{x} \chi \ dx \right]$ we obtain the

Singular Limits of Dispersive Waves, Edited by
N.M. Ercolani et al., Plenum Press, New York, 1994

Riccatty equation:

$$t\chi_x - \chi^2 + \lambda^2 - u = 0$$

To take quasiclassical limit of it we neglect χ_x

$$-\chi^2 + \lambda^2 - u = 0 \qquad\qquad (2)$$

and get two values of χ

$$\chi = \pm\sqrt{\lambda^2 - u}$$

which corresponds to the quasiclassical limit of the Jost functions

$$\psi_{1,2} = exp\left[\pm t \int_{x_0}^{x} \sqrt{\lambda^2 - u}\ dx\right]$$

This approximation for Jost functions is correct with logarithmic accuracy (i.e. it gives the correct value not of ψ but of $ln\ \psi$) everywhere except such λ for which between x and x_0 there is a reflecting point at which $\lambda^2 - u = 0$.

Now we introduce the function $\Omega(\lambda)$ which plays the role of the scattering data. Let $x(\lambda)$ be such that $u(x(\lambda)) = \lambda^2$.

$$\Omega(\lambda, t) = \int_{x_0}^{x(\lambda)} \sqrt{\lambda^2 - u}\ dx$$

This function is defined only on the real axis where $x(\lambda)$ exists.

It is easy to see that the quasiclassical limit of the reflecting coefficient for KdV is $r = -exp(2t\omega)$ (except such λ for which $r = 0$). Hence $\Omega(\lambda)$ has the simple dependence on time t:

$$\Omega(\lambda, t) = \Omega(\lambda, 0) + 4\lambda^3 t \qquad\qquad (3)$$

This can be easily proved directly.

Now we have the following scheme for solving the Cauchy problem for (1). We can find $\Omega(\lambda, 0)$ from the initial data, then find $\Omega(\lambda, t)$ and then using $\Omega(\lambda, t)$ find $u(x, t)$.

To perform the latter part of this scheme we introduce the following function S ('action')

$$S = \int_{x_0}^{x} \sqrt{\lambda^2 - u}\ dx \qquad\qquad (4)$$

and consider its analytical properties. It is analytic everywhere except a pole with residue $(x - x_0)$ at $\lambda = \infty$ and a cut at the real axis between $-\sqrt{u}$ and $\sqrt{u}$ at which $Re\ S(\lambda) = \Omega(\lambda, t)$.

So we have

$$S=(x-x_0)\sqrt{\lambda^2-u}+\frac{1}{\pi t}\int_{-\sqrt{u}}^{\sqrt{u}}\frac{\Omega(\lambda^*,t)}{\lambda^2-\lambda^{*2}}\sqrt{\frac{\lambda^2-u}{\lambda^{*2}-u}}\;d\lambda^* \tag{5}$$

This is not the definition of S because we don't know the value of u yet. To define it we use the fact that (5) has as a rule infinite derivative at $\lambda=\sqrt{u}$ but the derivative of (4) is finite because the radical has been integrated. The condition for the derivative of (5) to be finite at $\lambda=\sqrt{u}$ is

$$\frac{2}{\pi}\int_0^{\sqrt{u}}\frac{\partial\Omega(\lambda,t)}{\partial\lambda}\frac{d\lambda}{\sqrt{\lambda^2-u}}+x-x_0=0 \tag{6}$$

This is the connection between $u(x,t)$ and $\Omega(\lambda,t)$ from which we can find $u(x,t)$ as an implicit function.

It is convenient to substitute (2) into (6):

$$\frac{2}{\pi}\int_0^{\sqrt{u}}\frac{\partial\Omega(\lambda,0)}{\partial\lambda}\frac{d\lambda}{\sqrt{\lambda^2-u}}+x-x_0+6ut=0 \tag{7}$$

This and the following formulas (14) corresponds to Tsarev's formulas[1].

THE BENNEY EQUATIONS

Now let us look how this scheme works in the case of the Benney equations:

$$\partial_t\eta_k+\partial_x(\eta_k v_k)=0$$
$$\partial_t v_k+v_k\partial_x v_k+\partial_x\sum_{l=1}^N\eta_l=0 \tag{8}$$

Zakharov[2] has shown that these equations are the quasiclassical limit of the vector nonlinear Schrödinger equation. By taking quasiclassical limit of L-A pair the overdetermined system of two equations for function $\mu(\lambda,x,t)$ can be obtained[2,3]:

$$\mu+\sum_{l=1}^N\frac{\eta_l}{\mu+v_l}=\lambda \tag{9}$$

$$\partial_t\mu=\partial_x(\frac{1}{2}\mu^2+\sum_{l=1}^N\eta_l) \tag{10}$$

(8) is the compatibility condition for (9) and (10). This means that if $\vec{\eta}(x,t)$ and $\vec{v}(x,t)$ are the solution of (8) then $\mu(\lambda,x,t)$ obtained from (9) is the solution of (10).

Equation (9) is the quasiclassical limit of L-equation and corresponds to equation (2) for KdV. Let us study it in detail. It defines in implicit form the function $\mu(\lambda)$ which is analytic on the $(N+1)$-sheets Riemann surface. This surface we denote by G and the value of μ at sheet number k we denote by $\mu^{(k)}$. Numbers of sheets we choose so that

$$\mu^{(k)} \to -\upsilon_k + O(\tfrac{1}{\lambda}), \qquad k=1,2\ldots N$$

$$\mu^{(0)} \to \lambda + O(\tfrac{1}{\lambda})$$

We denote branch points of $\mu(\lambda)$ by λ_k, $k=1,2\ldots 2N$, and values of μ at these points as μ_k.
μ_k can be found from equation

$$1 + \sum_{l=1}^{N} \frac{\eta_l}{(\mu+\upsilon_l)^2} = 0 \qquad (11)$$

λ_k can be found by substituting μ_k into (9).

If we consider μ, λ_k and μ_k as functions of $\vec{\eta}$ and $\vec{\upsilon}$ rather then of x and t we can define them from (9) and (11). For example, for $N=1$

$$\mu = \frac{\lambda-\upsilon}{2} \pm \sqrt{\frac{(\lambda+\upsilon)^2}{4} - \eta}$$

$$\lambda_{1,2} = -\upsilon \pm 2\sqrt{\eta}$$

$$\mu_{1,2} = -\upsilon \pm \sqrt{\eta}$$

We cannot write explicit formulas for $\mu(\lambda,\vec{\eta},\vec{\upsilon})$, $\lambda_k(\vec{\eta},\vec{\upsilon})$ and $\mu_k(\vec{\eta},\vec{\upsilon})$ if $N>2$ because of the high degree of the algebraic equations (9) and (11). But in spite of this we can consider them as given functions defined by (9) and (11).
Let us examine the Riemann surface G.
We denote numbers of sheets coinciding at λ_k as m_k and n_k, $m_k<n_k$. So $\mu^{(m_k)}(\lambda_k) = \mu^{(n_k)}(\lambda_k) = \mu_k$.
The shape of G is the following. If all λ_k are real all the sheets from 1 to N have only two branch points and shares them with the sheet 0 where they are glued. In this case $m_k=0$, $n_k=\frac{k}{2}$ for even k and $n_k=\frac{k+1}{2}$ for odd k.
If there are some couples of complex λ_k they are shared by corresponding sheets of numbers other than 0.
Fig. 1 and fig. 2 represents surface G for $N=2$ for these two cases. Cuts are shown by double lines.

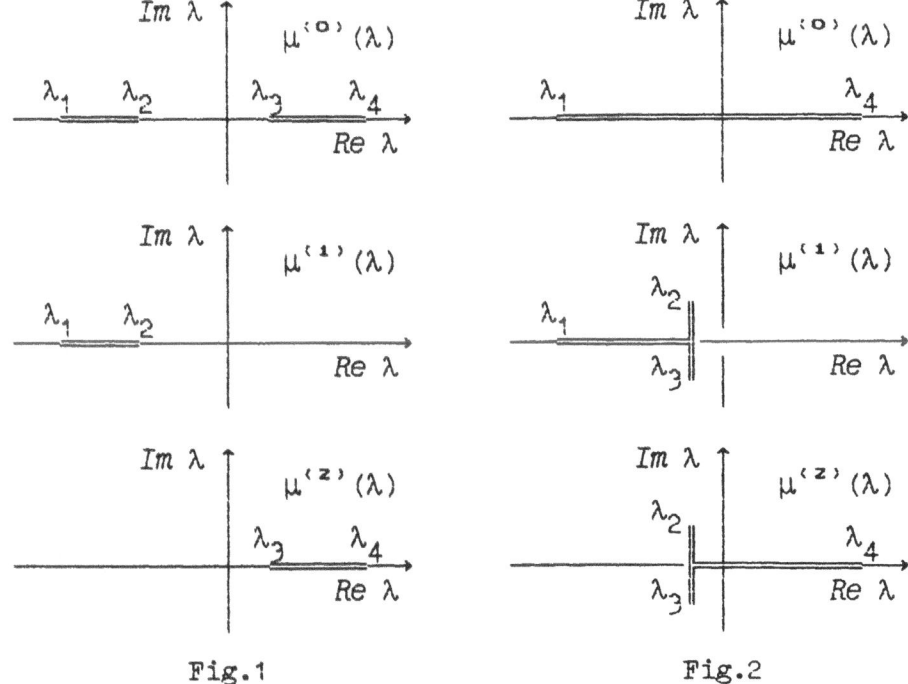

Fig.1   Fig.2

For the sake of simplicity we should consider $\tilde{\eta}$ and $\tilde{v}$ such that there exists an x_0 such as $\tilde{\eta}(x,t)=const$ and $\tilde{v}(x,t)=const$ for $x>x_0$. The simplest case is $\eta_k(x_0)=0$. Then

$$\mu^{(0)}(\lambda,x_0)\equiv\lambda,$$
$$\mu^{(k)}(\lambda,x_0)\equiv-u_k, \quad k=1,2\ldots N,$$
$$\lambda_1(x_0)=\mu_1(x_0)=\lambda_2(x_0)=\mu_2(x_0)=-u_1,$$
$$\lambda_3(x_0)=\mu_3(x_0)=\lambda_4(x_0)=\mu_4(x_0)=-u_2, \text{ etc.}$$

If this is not the case, the regularization of integrals is needed.

Now let us define $2N$ functions which play the role of the reflecting coefficients

$$\Omega_k(\lambda,t)=\int_{x_k(\lambda)}^{x_0}(\mu^{(m_k)}-\mu^{(n_k)})dx \qquad (12)$$

Here $x_k(\lambda)$ is the inverse function to $\lambda_k(x)$. If $x_k(\lambda)$ is multivalued then $\Omega_k(\lambda,t)$ is multivalued too. This doesn't change the following procedure.

If all λ_k are real, each Ω_k is defined on the part of the real axis. If some λ_k are complex then the following procedure needs corresponding Ω_k to be defined at some region of the plane λ. To get this we have to restrict ourselves by analytic $\tilde{\eta}$ and $\tilde{v}$. Then we can continue Ω_k analytically.

Ω_k has simple dependence on time t:

$$\Omega_k(\lambda,t)=\Omega_k(\lambda,0)+$$

$$+\frac{1}{2}\left[\left[\mu^{(m_k)}(\lambda,x_0)\right]^2-\left[\mu^{(n_k)}(\lambda,x_0)\right]^2\right]t \tag{13}$$

This can be proved by differentiating of (12) by t and using (10).

While x changes $\lambda_k(x)$ moves. We define $2N$ contours Γ_k on the surface G which are the tracks of λ_k. Each contour Γ_k originates at the point $\lambda_k(x_0)$ then continues to the point $\lambda_k(x)$ and then returns to the point $\lambda_k(x_0)$ at the another sheet.

Below we prove that

$$-\frac{1}{2\pi i}\sum_{l=1}^{2N}\int_{\Gamma_l}\frac{\partial\Omega_l(\lambda,0)}{\partial\lambda}\frac{d\lambda}{\mu(\lambda,\vec{\eta}(x,t),\vec{v}(x,t))-\mu_k(\vec{\eta}(x,t),\vec{v}(x,t))}=$$

$$=x-x_0-\mu_k(\lambda,x_0)t, \qquad k=1,2\ldots2N \tag{14}$$

This is the main result of the article. These $2N$ relations (enumerated by k) give us the possibility to solve the Cauchy problem for (8). After obtaining $\Omega_k(\lambda,0)$ from the initial conditions we can take integrals in (14) and get $2N$ algebraic relations which gives $\vec{\eta}(x,t)$ and $\vec{v}(x,t)$ as implicit functions.

Let us also write regularized variants of (12) and (14) for the case $x_0=\infty$.

$$\Omega_k(\lambda,t)=\int_{x_k(\lambda)}^{x_0}\left[(\mu^{(m_k)}(\lambda,x,t)-\mu^{(n_k)}(\lambda,x,t))-(\mu^{(m_k)}(\lambda,x_0)-\right.$$

$$\left.-\mu^{(n_k)}(\lambda,x_0))\right]dx-(\mu^{(m_k)}(\lambda,x_0)-\mu^{(n_k)}(\lambda,x_0))x_k(\lambda) \tag{12a}$$

$$-\frac{1}{2\pi i}\sum_{l=1}^{2N}\int_{\Gamma_l}\frac{\partial\Omega_l(\lambda,0)}{\partial\lambda}\frac{d\lambda}{\mu(\lambda,\vec{\eta}(x,t),\vec{v}(x,t))-\mu_k(\vec{\eta}(x,t),\vec{v}(x,t))}=$$

$$=x-\mu_k(\lambda,x_0)t, \qquad k=1,2\ldots2N \tag{14a}$$

To prove (14) we should consider function

$$S(\lambda,x,t)=\int_{x}^{x_0}\mu(\lambda,x^*,t)dx^*$$

It is analytic on Riemann surface G with $\lambda_k(x)$ branch points everywhere except contours Γ_k and simple pole at $\lambda=\infty$ at zero sheet where $S=(x_0-x)\lambda+O(\frac{1}{\lambda})$. At Γ_k S has jump Ω_k. So we have

$$S=(x-x_0)\mu(\lambda,\vec{\eta},\vec{v})+$$

$$+\frac{1}{2\pi i}\sum_{l=1}^{2N}\int_{\Gamma_l}\frac{\Omega_l(\lambda^*,0)}{\mu(\lambda^*,\vec{\eta},\vec{v})-\mu(\lambda,\vec{\eta},\vec{v})}\frac{\partial\mu(\lambda^*,\vec{\eta},\vec{v})}{\partial\lambda^*}d\lambda^* \qquad (15)$$

As $\mu(\lambda)=\mu_k+O(\sqrt{\lambda-\lambda_k})$ at the branch points and S is the integral of μ, S has finite derivative at $\lambda=\lambda_k$. By differentiating of (15) we get

$$\frac{\partial\omega}{\partial\lambda}\Big|_{\lambda=\lambda_k}=\frac{\partial\mu}{\partial\lambda}\Big|_{\lambda=\lambda_k}\Bigg[x-x_0+$$

$$+\frac{1}{2\pi i}\sum_{l=1}^{2N}\int_{\Gamma_l}\frac{\partial\Omega_l(\lambda,t)}{\partial\lambda}\frac{d\lambda}{\mu(\lambda,\vec{\eta}(x,t),\vec{v}(x,t))-\mu_k(\vec{\eta}(x,t),\vec{v}(x,t))}\Bigg]$$

As $\frac{\partial\mu}{\partial\lambda}\Big|_{\lambda=\lambda_k}=\infty$ at $\lambda=\lambda_k$ the brackets should be equal to zero

$$\frac{1}{2\pi i}\sum_{l=1}^{2N}\int_{\Gamma_l}\frac{\partial\Omega_l(\lambda,t)}{\partial\lambda}\frac{d\lambda}{\mu(\lambda,\vec{\eta}(x,t),\vec{v}(x,t))-\mu_k(\vec{\eta}(x,t),\vec{v}(x,t))}+$$

$$+x-x_0=0$$

Substituting (13) here we prove (14).

REFERENCES

1. Tsarev S.P. - DAN, 1985, 282, pp.534-537.
2. Zakharov V.E. - Functional Anal. Appl., 1980,14, pp.15-24.
3. Couperschmidt B.A., Manin Yu.I. - Functional Anal. Appl., 1977,11, pp.31-42 and 1978, 12 pp.25-37.
4. Geogjaev V.V. - DAN, 1985, 284, pp.1093-1097.
5. Geogjaev V.V. - Theor. Math. Phys., 1987, 73, pp.255-263.

SOLVING DISPERSIONLESS LAX EQUATIONS

John Gibbons - Dept. of Mathematics, Imperial College, London, SW7 2BZ, U.K.

Yuji Kodama - Dept. of Mathematics, Ohio State University, Columbus, OH 43210

Abstract

Geogdzhaev's method is used to derive the solution to the initial value problem of any dispersionless Lax equation. The particular case of the dispersionless Boussinesq equation is worked out in detail and possible generalisations are considered.

1 Introduction

There has been much progress recently on Hamiltonian systems of the form:

$$\frac{\partial \lambda_i}{\partial t} = u_i^{\,j}(\underline{\lambda}) \frac{\partial \lambda_j}{\partial x} \tag{1}$$

Whenever such a system admits a change of variables which diagonalises the matrix $u_i^{\,j}$, so that the equation can be reduced to Riemann invariant form,

$$\frac{\partial \lambda_i}{\partial t} = v_i(\underline{\lambda}) \frac{\partial \lambda_i}{\partial x} \tag{2}$$

and it also possesses non-trivial symmetries in the same diagonal form, then it can be solved exactly. Two important examples of such systems are Whitham's equations [1,2] and Benney's equations [3,4,5,6]. In [7] Tsarev showed how such diagonalisable systems can be solved; he gave the generalised hodograph solution

$$x + v_i(\underline{\lambda})t = w_i(\underline{\lambda}) \tag{3}$$

Here the functions w_i satisfy a set of *linear* equations; these are the compatibility conditions between (2) and a symmetry of it:

$$\frac{\partial \lambda_i}{\partial s} = w_i(\underline{\lambda}) \frac{\partial \lambda_i}{\partial x} \tag{4}$$

In [8] Geogdzhaev showed how the classical solution to the simplest such system:

$$\frac{\partial u}{\partial t} = u \frac{\partial u}{\partial x} \tag{5}$$

the dispersionless KdV equation, could be considered as the appropriate limit of the solution of the KdV via the inverse scattering transform (but see also the work of Lax and Levermore [9]). There and in [10] he also considered the Zakharov reduction of the Benney hierarchy

$$\frac{\partial u_i}{\partial t} = u_i \frac{\partial u_i}{\partial x} + \sum_{j=1}^{N} \frac{\partial h_j}{\partial x}$$

$$\frac{\partial h_i}{\partial t} = \frac{\partial (u_i h_i)}{\partial x} \tag{6}$$

This system may be considered as the quasi-classical, or dispersionless, limit of the N-component vector NLS equation. In [11] we showed how these results could be interpreted in terms of canonical transformations.

Here we consider another reduction of the Benney hierarchy, the dispersionless Lax equations [12].

Singular Limits of Dispersive Waves, Edited by
N.M. Ercolani et al., Plenum Press, New York, 1994

2 Dispersionless Lax Equations

The Benney hierarchy admits many reductions. Among the simplest are the following, derived by the Gel'fand–Dikii 'fractional power' ansatz. We denote:

$$\Lambda = p^N + A_0 p^{N-2} + \cdots + A_{N-2} \tag{7}$$

and then construct the Hamiltonians, polynomial in p and the $(N-1)$ variables $A_i(x)$:

$$H_m = ((\Lambda)^{m/N})_+ \tag{8}$$

Here the fractional power is to be regarded as a formal series in p, with leading term p^m, and the subscript $_+$ denotes its polynomial part. The equations of motion of the hierarchy are then

$$\frac{\partial \Lambda}{\partial t_m} = \{H_m, \Lambda\} \tag{9}$$

Here the braces denote the canonical Poisson bracket with respect to x and p. We note that the right-hand side is identically zero if N divides m. It is useful to denote $\lambda = \Lambda^{1/N}$. Below we will perform a canonical transformation to variables where λ is the new momentum; then the new Hamiltonian is just λ^m.

It is necessary to suppose that the polynomial $\Lambda(p)$ has N distinct real zeroes, $p_1 > p_2 > \ldots > p_N$; it will have $(N-1)$ distinct real turning points between these, and we may see, from (9), that the values of Λ at these turning points are Riemann invariants. We denote the stationary value of Λ found between p_{k+1} and p_k as Λ_k, and the corresponding value of p as $\tilde{p}_k$. It is convenient to suppose that as $x \to -\infty$ each of these Riemann invariants tends monotonically to zero, so that in this limit $\Lambda \to p^N$, and hence $(p - \lambda) \to 0$.

The generating function of the canonical transformation we need is

$$S(x, \lambda) = \int_{-\infty}^{x} (p(x', \lambda) - \lambda) dx' + \lambda x \tag{10}$$

(always supposing the potentials A_i tend to zero sufficiently rapidly that this integral converges). We then have

$$p(x, \lambda) = \frac{\partial S}{\partial x} \tag{11}$$

while ξ, the canonical conjugate of λ, is given by

$$\xi = \frac{\partial S}{\partial \lambda} \tag{12}$$

We will be able to rewrite the equations of motion in the new variables below, but first it is important to make the definition of S as precise as possible. We choose λ to be analytic except on the real p-axis, between p_N and p_1. Just above the real p-axis, between p_{k+1} and p_k, λ has the argument $k\pi/N$; just below the cut, it has the argument $-k\pi/N$. In the λ-plane, this segment of the real p-axis thus appears as a *pair* of cuts, stretching along the rays $\arg(\lambda) = \pm k\pi/N$, as far as the branch points λ_k and λ_k^*, where

$$\lambda_k^N = \Lambda_k \tag{13}$$

which was defined above. This choice of λ satisfies

$$\lambda = p + O(1/p) \tag{14}$$

as $|p| \to \infty$. The inverse function $p(x, \lambda)$ is then seen to be analytic for all λ not on the rays

$$\arg(\lambda) = \pm k\pi/N \tag{15}$$

where $k = 1, \ldots, N-1$. On either side of these cuts, $p(x, \lambda)$ is real, between p_{k+1} and p_k; at the branch point λ_k, p takes the value $\tilde{p}_k$. Since the Riemann invariants Λ_k approach zero monotonically as $x \to -\infty$, so do the $2(N-1)$ branch points λ_k and λ_k^*. Therefore, for any λ on the ray $\arg(\lambda) = k\pi/N$, either $p(x, \lambda)$ is analytic for all x, or there is some unique value of x, $x_k^*(\lambda)$ say, such that $\lambda_k(x_k^*) = \lambda$. Since, for $x > x_k^*(\lambda)$, $p(x, \lambda)$ is real, we find that the imaginary part of S is independent of x:

$$Im(S(x, \lambda)) = \int_{-\infty}^{x_k^*(\lambda)} Im(p(x', \lambda) - \lambda) dx' + x_k^*(\lambda) Im(\lambda) \tag{16}$$

3 The Time Evolution

The time dependence of $p(x, \lambda)$ is as follows:

$$\frac{\partial p}{\partial t_m} = \frac{\partial}{\partial x} H_m(p(x, \lambda), x) \tag{17}$$

We thus obtain the Hamilton-Jacobi equation:

$$\frac{\partial S}{\partial t_m} = H_m(p(x, \lambda), x) - \lambda^m = H_m(\frac{\partial S}{\partial x}, x) - \lambda^m \tag{18}$$

Therefore if $\arg(\lambda) = k\pi/N$ and $x > x_k^*(\lambda)$, $H_m(p(x, \lambda))$ is real, and we hence have:

$$\frac{\partial}{\partial t_m} Im(S(x, \lambda)) = -Im(\lambda^m) \tag{19}$$

It is thus rather more convenient to consider, instead of S, the function

$$\Phi = S + \lambda'' \, t_m \tag{20}$$

since $Im(\Phi)$ is time-independent on the cuts. Now, since p and Φ are both analytic away from the cuts, so is the expression

$$\Psi = \Phi - H_m(p, x)t_m - px \tag{21}$$

Since the term $H_m t_m + px$ is real on the cuts, evidently $Im(\Psi) = Im(\Phi)$ there. Finally we note that, since $S = \lambda x + O(1/\lambda)$,

$$\Psi = O(1/\lambda) \tag{22}$$

as $|\lambda| \to \infty$.

4 The Inverse Problem

Since Ψ is analytic away from the cuts, and tends to zero at infinity, we have, by Cauchy's theorem,

$$\Psi(x, \lambda') = \cdot \frac{1}{2\pi i} \oint_\Gamma \frac{\Psi(x, \lambda)}{p(x, \lambda) - p(x, \lambda')} \frac{\partial p}{\partial \lambda} \, d\lambda \tag{23}$$

Here Γ is a contour which encircles the cuts anti–clockwise, but does not enclose λ'. We may write $\Gamma = \Gamma_- - \Gamma_+$, where Γ_+ is in the upper half λ-plane, Γ_- in the lower. Clearly we may replace Ψ by Φ in the integral without changing its value, as the difference between them is an entire function of p. Thus we obtain:

$$\Phi(x, \lambda') - H_m(p(x, \lambda'), x)t_m - p(x, \lambda')x =$$
$$-\frac{1}{2\pi i} \oint_\Gamma \frac{\Phi(x, \lambda)}{p(x, \lambda) - p(x, \lambda')} \frac{\partial p}{\partial \lambda} \, d\lambda \tag{24}$$

Differentiating with respect to λ', and denoting $\partial H_m/\partial p$ as v_m, we get:

$$\frac{\partial \Phi}{\partial \lambda'} - \frac{\partial p}{\partial \lambda'}(x + v_m(p(x, \lambda'), x)t_m) =$$

$$-\frac{\partial p}{\partial \lambda'} \frac{1}{2\pi i} \oint_\Gamma \frac{\Phi}{(p(x, \lambda) - p(x, \lambda'))^2} \frac{\partial p}{\partial \lambda} \, d\lambda =$$

$$-\frac{\partial p}{\partial \lambda'} \frac{1}{2\pi i} \oint_\Gamma \frac{\partial \Phi}{\partial \lambda} \frac{d\lambda}{p(x, \lambda) - p(x, \lambda')} =$$

$$\frac{\partial p}{\partial \lambda'} \frac{1}{\pi} \int_{\Gamma_+} \frac{\partial Im\Phi}{\partial \lambda} \frac{d\lambda}{p(x, \lambda) - p(x, \lambda')} \tag{25}$$

Here we have first integrated by parts, and then collapsed Γ onto the cuts, noting that $\Phi|_{\Gamma_+} = \Phi^*|_{\Gamma_-}$. At the $(N-1)$ points $\lambda_k(x)$, the derivative $\partial \lambda/\partial p$ vanishes, while $\partial \Phi/\partial \lambda$ is bounded. Thus the residue $\partial \Phi/\partial p$ vanishes also. We therefore obtain:

$$x + v_m(\tilde{p}_k(x), x)t_m = -\frac{1}{\pi} \int_{\Gamma_+} \frac{\partial Im\Phi}{\partial \lambda} \frac{d\lambda}{p(x, \lambda) - \tilde{p}_k(x)} \tag{26}$$

Here the left-hand side is precisely Tsarev's generalised hodograph formula, while the right-hand side has no x-dependence, except through the potentials $A_i(x)$. The equations may thus be regarded as a set of $N-1$ equations, depending on the parameters x and t_m, for the $N-1$ potentials A_i.

5 The Dispersionless Boussinesq Equation

This, after the dispersionless KdV, is the simplest system in this class; however it is sufficiently complicated to illustrate the method. The equation of motion is, from (10):

$$\Lambda = p^3 + A_0 p + A_1$$

$$\frac{\partial \Lambda}{\partial t_2} = \{p^2 + \frac{2}{3} A_0, \Lambda\} \tag{27}$$

or, more explicitly:

$$\frac{\partial A_0}{\partial t_2} = 2 \frac{\partial A_1}{\partial x}$$

$$\frac{\partial A_1}{\partial t_2} = -\frac{2}{3} A_0 \frac{\partial A_0}{\partial x} \tag{28}$$

The function $p(x, \lambda)$ is given by:

$$p = \sqrt[3]{(\lambda^3 - A_1 + \sqrt{(\lambda^3 - A_1)^2 + 4(A_0/3)^3})/2}$$

$$+ \sqrt[3]{(\lambda^3 - A_1 - \sqrt{(\lambda^3 - A_1)^2 + 4(A_0/3)^3})/2} \tag{29}$$

The branch cuts for the two functions $\lambda(p)$ and $p(\lambda)$, as well as the contour Γ, are shown below.

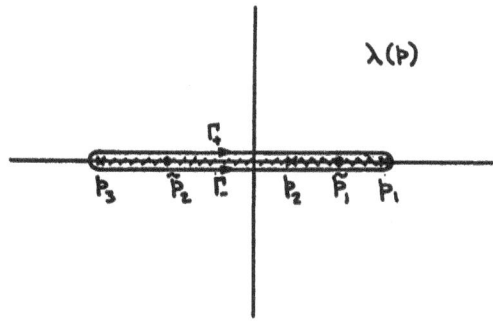

Figure 1. The branch cuts of the function $\lambda(p)$ and the contours $\Gamma_\pm$ in the p-plane.

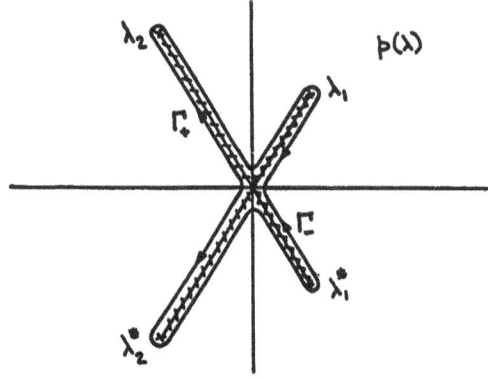

Figure 2. The branch cuts of the function $p(\lambda)$ and the contours $\Gamma_\pm$ in the λ-plane.

The two characteristic speeds are:

$$\tilde{p}_1 = \sqrt{-A_0/3}$$

$$\tilde{p}_2 = -\sqrt{-A_0/3} \tag{30}$$

while the corresponding branch points in the upper half λ-plane are:

$$\lambda_1 = \alpha \sqrt[3]{-2(-A_0/3)^{3/2} - A_1}$$

$$\lambda_2 = \alpha^2 \sqrt[3]{-2(-A_0/3)^{3/2} + A_1} \tag{31}$$

where α denotes $\exp(i\pi/3)$. We assume that $A_0 < 0$, so that the system is hyperbolic, and further that $4A_0^3 + 27A_1^2 < 0$, so that the cubic has three distinct real roots.

We then obtain:

$$\Phi = \int_{-\infty}^{x} (p(x',\lambda) - \lambda)dx' + \lambda x$$

$$x \pm \sqrt{-A_0/3}\, t_2 = \frac{1}{\pi} \int_{-\infty+ie}^{\infty+ie} \frac{1}{p(x,\lambda) \mp \sqrt{-A_0/3}}\, d(Im\Phi(\lambda(p,x))) \tag{32}$$

In the inversion formula, we note that Φ is real except between the roots of $\Lambda(p)$.

6 Conclusions

Here we have seen how the Cauchy problem for dispersionless Lax equations may be solved in an effective, explicit way by considering the generating function of a canonical transformation. Although the technical details of the method depend on the specific problem studied, the principle is the same in each case. It is therefore reasonable to hope that similarly effective solutions may be obtained to other open problems in this class. Some important examples are:

1. The Benney equations. These are equivalent to the Vlasov equation:

$$\frac{\partial f}{\partial t_2} + p\frac{\partial f}{\partial x} - \frac{\partial A_0}{\partial x}\frac{\partial f}{\partial p} = 0$$

$$A_n = \int_{-\infty}^{\infty} p^n f\, dp \tag{33}$$

Here, in some cases, (with $f < 0$) we can obtain an equation like (25), but it is not clear how we can use this to obtain the solution of the problem.

2. The Whitham equations.

$$\frac{\partial \lambda_i}{\partial t_n} = \frac{\omega_n}{\omega_1}(\lambda_i,\lambda)\frac{\partial \lambda_i}{\partial x} \tag{34}$$

Here the ω_i are meromorphic differentials on the Riemann surface

$$\mu^2 = \prod_0^{2g}(\lambda - \lambda_i) \tag{35}$$

The differential ω_1 plays a role analogous to the momentum p, and we may construct a generating function S as above. The inversion problem is much more difficult, however, as the Cauchy kernel no longer takes a simple form.

3. The Zabolotskaya–Khokhlov equation.

$$\frac{\partial^2 A_0}{\partial t_3 \partial x} + \frac{1}{2}\frac{\partial^2 A_0^2}{\partial x^2} = \frac{\partial^2 A_0}{\partial t_2^2} \tag{36}$$

Although this is intimately connected with the Benney hierarchy, it is not itself a member of the class of systems (1); instead it arises as a consistency condition between different equations in the Benney hierarchy. Also its initial value problem is essentially different; only the moment A_0 is given, throughout the (x, t_2)-plane. It has been shown [13] that the initial value problem for this system is equivalent to the inverse scattering problem for the Hamilton-Jacobi equation

$$\frac{\partial S}{\partial t_2} = (\frac{\partial S}{\partial x}) + 2A_0(x, t_2) \tag{37}$$

This difficult problem is the classical limit of the inverse problem for the time-dependent Schrödinger equation, solved by Manakov [14].

7 References

[1] G.B. Whitham, Linear and Nonlinear Waves, Wiley-Interscience, New York, (1974).

[2] H. Flaschka, M.G. Forest and D.M. McLaughlin, Comm. Pure and Appl. Math. $\underline{33}$(1980)739.

[3] D.J. Benney, Stud. Appl. Math. $\underline{52}$ (1973) 45.

[4] B.A. Kupershmidt and Yu.I. Manin, Funct. Anal. App. $\underline{11}$ (1978) 188.
Funct. Anal. Appl. $\underline{12}$ (1978) 20.

[5] V.E. Zakharov, Funct. Anal. Appl., $\underline{14}$ (1980) 89.

[6] J. Gibbons, Physica $\underline{3}$D (1981) 50.

[7] S.P. Tsarev, Dokl. Akad. Nauk. SSSR $\underline{282}$ (1985) 534.

[8] V.V. Geogdzhaev, Dokl. Akad. Nauk. SSSR $\underline{284}$ (1985) 1093.

[9] C.D. Levermore and P.D. Lax, Proc. Nat. Acad. Sci. U.S.A. $\underline{76}$ (1979) 3602.

[10] V.V. Geogdzhaev, Teor. Mat. Fiz. $\underline{73}$ (1987) 255.

[11] J. Gibbons and Y. Kodama, Nonlinear World Vol.1, Proceedings of IV Workshop on Nonlinear
and Turbulent Processes in Physics, World Scientific, Singapore, (1990), p.166.

[12] J. Gibbons and Y. Kodama, Phys. Lett. $\underline{135A}$ (1989) 167.

[13] J. Gibbons, Dynamical Problems in Soliton Systems, ed. S. Takeno,
Springer-Verlag (1985), p. 36.

[14] S.V. Manakov, Physica $\underline{3}$D (1980) 420.

NONISOSPECTRAL SYMMETRIES OF THE KDV EQUATION
AND THE CORRESPONDING SYMMETRIES
OF THE WHITHAM EQUATIONS

P.G. Grinevich

L.D.Landau Institute for Theoretical Physics
Kosygina 2, Moscow, 117940, GSP-1, USSR

0. INTRODUCTION

In our paper we construct a new infinite family of symmetries of the Whitham equations (averaged Korteveg-de-Vries equation). In contrast with the ordinary hydrodynamic-type flows these symmetries are nonhomogeneous (i.e. they act nontrivially at the constant solutions), are nonlocal, explicitly depend upon space and time coordinates and form a noncommutative algebra, isomorphic to the algebra of the polynomial vector fields in the plain.

We will study the averaged Korteveg-de-Vries (KdV) equation. But the main results of our paper can be easily extended to other averaged systems, associated with the integrable by the inverse scattering transform $1 + 1$ equations.

We will assume original KdV to be normalized as:

$$u_t = \frac{1}{4}u_{xxx} - \frac{3}{2}uu_x. \qquad (0.1)$$

We will also need a $2 + 1$ - dimensional generalization of KdV - the Kadomtsev - Petviashvili (KP) equation

$$(u_t - \frac{1}{4}u_{xxx} + \frac{3}{2}uu_x)_x = 3\alpha^2 u_{yy}, \ \alpha^2 = \pm 1. \qquad (0.2)$$

Let us remind the definition of the averaged KdV equations in the classical one-phase case. Equation (0.1) possesses a three-parametric family of moving waves

$$u(x,t) = 2\mathcal{P}(x - vt|g_2, g_3) - v/6. \qquad (0.3)$$

($\mathcal{P}(x,|g_2,g_3)$ denotes the Weierstrass $\mathcal{P}$ - function). We may try to construct asymptotical KdV solutions of the modulated wave type

$$u(x,t) = 2\mathcal{P}(x - v(X,T)t + \phi(X,T)|g_2(X,T),g_3(X,T)) - v(X,T)/6 + \epsilon u_1, \qquad (0.4)$$

where $X = \epsilon x$, $T = \epsilon t$, ϵ is a small parameter (X and T are called "slow" variables). Assuming the correcting term u_1 to be a bounded function as $x \sim 1/\epsilon, t \sim 1/\epsilon$ we can prove that the functions $g_2(X,T)$, $g_3(X,T)$, $v(X,T)$ satisfy some hydrodynamical-type equations:

$$\frac{\partial}{\partial T}\begin{pmatrix} g_2(X,T) \\ g_3(X,T) \\ v(X,T) \end{pmatrix} = \hat{V}(g_2, g_3, v) \cdot \frac{\partial}{\partial X}\begin{pmatrix} g_2(X,T) \\ g_3(X,T) \\ v(X,T) \end{pmatrix} \qquad (0.5)$$

where the matrix $\hat{V}(g_2, g_3, v)$ can be expressed via elliptic functions.

The averaged KdV (0.5) may be transformed to the Riemann diagonal form

$$\frac{\partial}{\partial T}\begin{pmatrix} r_1(X,T) \\ r_2(X,T) \\ r_3(X,T) \end{pmatrix} = \begin{pmatrix} v_1(\vec{r}) & 0 & 0 \\ 0 & v_2(\vec{r}) & 0 \\ 0 & 0 & v_3(\vec{r}) \end{pmatrix} \frac{\partial}{\partial X}\begin{pmatrix} r_1(X,T) \\ r_2(X,T) \\ r_3(X,T) \end{pmatrix}. \qquad (0.6)$$

The diagonalizing change of variables (Whitham [1]) is the following : the Riemann invariants r_i, $i = 1, 2, 3$ are the roots of the polynomial $R(\lambda) = 4(\lambda + v/6)^3 - g_2(\lambda + v/6) - g_3$, $r_1 \leq r_2 \leq r_3$. The matrix $\hat{V}$ in the diagonal form reads as:

$$v_1(\vec{r}) = \frac{r_1 + r_2 + r_3}{3} - \frac{2}{3}(r_2 - r_1)\frac{K(s)}{K(s) - E(s)}$$

$$v_2(\vec{r}) = \frac{r_1 + r_2 + r_3}{3} - \frac{2}{3}(r_2 - r_1)\frac{(1 - s^2)K(s)}{E(s) - (1 - s^2)K(s)} \qquad (0.7)$$

$$v_3(\vec{r}) = \frac{r_1 + r_2 + r_3}{3} + \frac{2}{3}(r_3 - r_1)\frac{(1 - s^2)K(s)}{E(s)}$$

where $s^2 = (r_2 - r_1)/(r_3 - r_1)$, $E(s)$, $K(s)$ are elliptic integrals.

Averaging procedure can be treated as a nonlinear analog of the WKB method. It can also be applied to the multiphase solutions of soliton equations (nonlinear superpositions of moving waves) (Flaschka - Forest - McLaughlin [2], Lax - Levermore [3]). Averaged KdV equations are known as Whitham equations. Averaged equations associated with soliton systems are also known as equations of slow modulations or as equations of the soliton lattice dynamics.

It is well-known that the original KdV equation is an integrable in the Liouville sense hamiltonian system. For averaging (at least in the one-phase case) the integrability is not necessary but additional structures of the original equations are usually inherited in the associated averaged equations. (The connection between the original structures and the averaged ones may be very nontrivial). The Whitham equations, associated with the KdV solutions with arbitrary number of phases have the following properties.

1) They can be presented in the hamiltonian form (Dubrovin - Novikov [4]).

2) They can be written in the Riemann diagonal form (Flaschka - Forest - McLaughlin [2]). For the hydrodynamical-type systems with more then 2 components the existence of the Riemann invariants is a very nontrivial property.

3) They have an infinite number of conservation laws in involution (i.e. an infinite number of mutually commuting symmetries). The full set of integrals consist of $2g + 1$ infinite families where g is the number of phases. One of these families is formed by the averaged higher KdV flows, the other $2g$ have no direct analogs in the KdV theory. These $2g$ additional families were constructed by S.P.Tsarev [5] for $g = 1$ and by B.A.Dubrovin [6] for all g.

4) They can be integrated by the generalized hodograph transform (Tsarev [7], [5]). In fact Tzarev proved that any hamiltonian hydrodynamical-type $1 + 1$ system

possessing Riemann invariants is integrable. Algebro-geometrical interpretation of the hodograph transform was suggested by I.M.Krichever [8].

The fact that the KdV equation has an infinite set of commuting symmetries (higher KdV) is well-known. They can be expressed via recursion operator Λ

$$\frac{\partial u}{\partial t_{2n+1}} = \frac{d}{dx}\left(-\frac{\Lambda}{4}\right)^n u = -2\frac{d}{dx}\left(-\frac{\Lambda}{4}\right)^{n+1} \cdot 1. \tag{0.8}$$

where $\Lambda = -\partial_x^2 + 2\partial_x^{-1}u\partial_x + 2u$, $x = t_1$, $t = t_3$. But it is less known that the group of symmetries of KdV is much wider (the same is valid for all integrable by the inverse scattering transform systems). In our parer the averaged KdV symmetries, associated with the following KdV symmetries will be studied

$$\frac{\partial u}{\partial \beta_{2n}} = -2\frac{d}{dx}\left(-\frac{\Lambda}{4}\right)^{n+1} (\sum_{k=0}^{\infty}((2k+1)t_{2k+1}\left(-\frac{\Lambda}{4}\right)^k) \cdot 1. \tag{0.9}$$

Let all the the higher times t_k, $k > 3$ be equal to zero. Then for $n = -2, 0, 2$ we have:

$$\partial u/\partial \beta_{-2} = 3tu_x - 2 \qquad \text{(Galilean).} \tag{0.10}$$

$$\partial u/\partial \beta_0 = 3t(\frac{1}{4}u_{xxx} - \frac{3}{2}uu_x) + xu_x + 2u \qquad \text{(Scaling).} \tag{0.11}$$

$$\partial u/\partial \beta_2 = 3t(\frac{1}{16}u^{\mathrm{v}} - \frac{5}{8}uu_{xxx} - \frac{5}{4}u_xu_{xx} - \frac{15}{8}u^2u_x) + x(\frac{1}{4}u_{xxx} - \frac{3}{2}uu_x) + u_{xx} - 2u^2 - \frac{1}{2}u_x\partial_x^{-1}u. \tag{0.12}$$

We call the symmetries (0.9) nonisospectral because they change the spectrum of the auxiliary linear problem (see section 2) or Virasoro because they form a noncommutative algebra, isomorphic to the algebra of the polynomial vector fields in the complex plane (the full Virasoro algebra does emerge in the KP theory). The ordinary higher KdV are isospectral in this sense.

First examples of the nonisospectral integrable equations (without connection with the symmetry problem) were constructed by F.Calogero - A.Degasperis [9],V.A.Belinskii - V.E.Zakharov [10], Maison [11], G.Baruchy - T.Redge [12]. The fact that we can generate new KdV symmetries by applying the recursion operator to the scaling transformation was pointed out by N.H.Ibragimov - A.B.Shabat in [13], but they did not pay much attention to this fact because they studied in [13] only local symmetries. Some nonisospectral KP symmetries were constructed by H.H.Chen, Y.C.Lee, J.E.Lin [14], F.J.Schwarz [15]. A general approach based on the so-called mastersymmetries was developed by B.Fuckssteiner and W.Oevel [16]. A very convenient method of constructing nonisospectral symmetries for the integrable by the inverse scattering transform equations was developed by A.Yu.Orlov and E.I.Schulman [17]. An analog of formula (0.9) for the KP equation was found by A.S.Fokas and P.M.Santini [18].

All these papers were dedicated to the algebraic theory of the nonisospectral symmetries or to the vanishing in the infinity case. The periodic (finite-gap) theory needs a special consideration. I.M.Krichever and S.P.Novikov calculated the action of a part of such symmetries on the finite-gap KP solutions in [19] (their subalgebra was associated with a given KP solution and did not vary the spectral curve). They posed also the problem for the full KP analog of the algebra (0.9).

The action of the symmetries (0.9) on the finite-gap KdV solutions was calculated by A.Yu.Orlov and the author in [20], [21]. Similar results were also obtained in [20], [21] for the KP equation. It was shown that this action has a natural geometrical interpretation (see paragraph 1.3).

Late 80'ies a number of links between the nonlinear integerable equations and the conformal field theory was discovered. For example the finite-gap τ-function in the KP theory coincides with the determinant of the $\bar{\partial}$ operator on the appropriate bundle (see, for example [20]). This determinant plays the central role in the string theory. The matrix models of the two-dimensional gravity give us another important example [22]. The partition function for the one-matrix model in the double scale limit coincides with the τ-function, corresponding to a special KdV solution, determined by the constraint ([22]):

$$\frac{\partial u}{\partial \beta_{-2}} = 0. \tag{0.13}$$

In the conformal field theory the algebra of the vector fields in the circle and its central extension - the Virasoro algebra play the key role. A subalgebra of the algebra of nonisospectral symmetries corresponds to them in the KP theory (the central extension does emerge if we calculate the action of these symmetries at the τ-function) (see [34], [20]) .

Similar results for the two-dimensional topological quantum theory were obtained by I.M.Krichever [23]. He proved that the averaged KP equations are connected with the topological models and calculated the action of the Virasoro symmetries on these equations in terms of the averaged τ-function. Applications of the Whitham equations to the topological models were also considered by B.A.Dubrovin [6].

In our paper we use the results of [23], [20], [21] to calculate the action of non-isospectral symmetries on the averaged KdV in terms of the Riemann invariants. We will assume the averaged KdV to be written in the Flaschka - Forest - MkLaughlin form. The plan of our parer in the following. In the first section we remind the necessary definitions and results from the Riemann surfaces theory, including the deformations of the complex structures via the vector fields action. In the section 2 we remind the necessary results from the periodic KdV and KP theory including the action of the nonisospectral symmetries on the finite-gap solutions. The section 3 is dedicated to the averaged KdV equations. We remind the construction of the Abel Whitham hierarchy and present a new noncommutative set of symmetries.

1. THE RIEMANN SURFACES THEORY. SOME DEFINITIONS AND RESULTS

In this section we remind the constructions from the Riemann surfaces theory we will use later ([24] - [26], [6], [20], [33]).

In our paper the words "Riemann surface" will always denote a compact nondegenerate Riemann surface. Such surfaces may be characterized as close Riemann surfaces of finite genus or as algebraic Riemann surfaces. We will consider two main classes - the general finite-gap surfaces (they correspond to the KP quasiperiodic solutions) and the hyperelliptic Riemann surfaces with a branch point in the infinity (they correspond to the quasiperiodic KdV solutions).

A Riemann surface Γ is called hyperelliptic if there exists a meromorphic function E on Γ with two simple poles or with one second-order pole. This function maps Γ to the complex plane, the covering $E : \Gamma \to \bar{C}$ is two-sheeted so Γ is isomorphic to

$$a) \quad Y^2 = (E - E_1) \ldots (E - E_{2g+2}) \quad \text{or} \tag{1.1.a}$$

$$b) \quad Y^2 = (E - E_1) \ldots (E - E_{2g+1}) \tag{1.1.b}$$

respectively (here g is the genus of Γ). The branch points of Γ over E are $E_1, \ldots, E_{2g+2}$ in the case a or $E_1, \ldots, E_{2g+1}, \infty$ in the case b. In our paper only the hyperelliptic surfaces of the type b will be considered because only they are related to the KdV theory.

1.1. Cycles On The Riemann Surfaces

Any compact nondegenerate Riemann surface is topologically equivalent to the sphere with a finite number of handles. This number is called genus, we will denote it g.

The canonical basis of 1-cycles in Γ consists of $2g$ elements $a_1, \ldots, a_g, b_1, \ldots, b_g$ such, that

$$a_i \circ a_j = b_i \circ b_j = 0, \quad a_i \circ b_j = \delta_{ij}. \qquad (1.2)$$

where $\circ$ denotes the intersection number. (Of course there are infinitely many nonhomotopical choices of canonical basises.) The surface Γ can be described as the result of gluing the sides of a $4g$-sided polygon (the picture corresponds to $g = 2$)

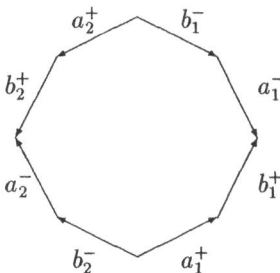

in the following way: $a_i^+ \leftrightarrow a_i^-$, $b_i^+ \leftrightarrow b_i^-$. (All the vertices are glued together). The sides $a_i^\pm$, $b_i^\pm$ correspond to the cycles a_i, b_i respectively.

Let Γ be a hyperelliptic surface - a two-sheeted covering over the E-plane with the branch points ∞, E_1, $\ldots$, E_{2g+1} and the cuts $(-\infty, E_1)$, (E_2, E_3), $\ldots$, (E_{2g}, E_{2g+1}). Then we can choose the following canonical basis

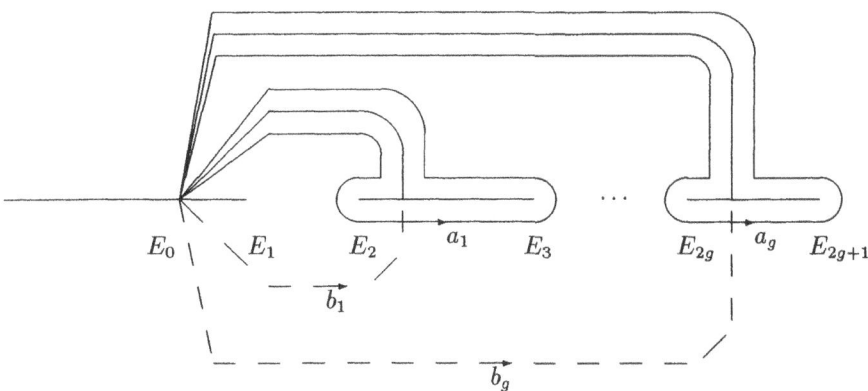

where the point E_0 correspond to the vertices of the polygon.

1.2. Differentials On The Riemann Surfaces

Let us have a marked point ∞ in our Riemann surface Γ with a local parameter z. We use the notation ∞ because this point corresponds to the infinite value of energy in the spectral theory. The local parameter z is a function defined in the neighbourhood of ∞ such, that $z(\infty) = 0$, $dz(\gamma) \mid_{\gamma=\infty} \neq 0$. For convenience we introduce an additional function $\lambda(\gamma) = 1/z(\gamma)$. For the hyperelliptic surfaces we will always assume $E(\infty) = \infty$ and $E = -\lambda^2$.

For us the following objects in Γ will be necessary.

1) Canonical basis of holomorphic differentials $\omega_1, \ldots, \omega_g$. Differentials ω_j are determined by the normalization conditions

$$\oint_{a_j} \omega_i = \delta_{ij}. \tag{1.3}$$

The matrix (B_{ij})

$$B_{ij} = \oint_{b_j} \omega_i. \tag{1.4}$$

is called Riemann matrix. It is symmetrical $(B_{ij} = B_{ji})$, the imaginary part of the Riemann matrix Im B_{ij} is positive definite. The coefficients of the Tailor series for ω_j in ∞ we will denote q_{jk}^H.

$$\omega_j = \sum_{m \geq 1} q_{jm}^H \lambda^{-m-1} d\lambda \tag{1.5}$$

2) Meromorphic differentials Ω_j with the only pole in the point ∞ such, that

$$\Omega_k = d(\lambda^k) + O(1), \quad \oint_{a_i} \Omega_k = 0. \tag{1.6}$$

We denote

$$\oint_{b_j} \Omega_k = (U_k)_j, \quad \vec{U}_k = ((U_k)_1, \ldots, (U_k)_g)$$

$$\Omega_k = d(\lambda^k) + \sum_{m \geq 1} Q_{km} \lambda^{-m-1} d\lambda. \tag{1.7}$$

The Riemann bilinear relations (see (1.16) below) result in

$$(U_j)_k = -2\pi i q_{kj}^H, \quad Q_{ij} = Q_{ji}. \tag{1.8}$$

If Γ is a hyperelliptic curve then $\Omega_{2k} = d((-E)^k)$, $(U_{2k})_l = 0$, $Q_{kl} = 0$ if at least one of the indexes k, l is even.

3) Multivalued holomorphic differentials ω_k^i, σ_k^i (they are defined *only* in the *hyperelliptic* Riemann surfaces). These differentials are determined by the following properties

$$\Delta_{a_j} \omega_k^i = \delta_{ij} d(E^k), \quad \Delta_{b_j} \omega_k^i = 0, \tag{1.9}$$

$$\Delta_{b_j} \sigma_k^i = -\delta_{ij} d(E^k), \quad \Delta_{a_j} \sigma_k^i = 0, \tag{1.10}$$

$$\oint_{a_j} \omega_k^i = \oint_{a_j} \sigma_k^i = 0. \tag{1.11}$$

Here Δ_{a_i} and Δ_{b_i} mean the increment of the differentials when going along the cycles a_i and b_i respectively. We will denote

$$\oint_{b_j} \omega_k^i = (U_k^i)_j, \quad \oint_{b_j} \sigma_k^i = (V_k^i)_j \tag{1.12}$$

$$\omega_k^i = \sum_{m \geq 1} Q_{km}^i \lambda^{-m-1} d\lambda, \quad \sigma_k^i = \sum_{m \geq 1} R_{km}^i \lambda^{-m-1} d\lambda, \tag{1.13}$$

4) Quasimomentum 1-differential dp, meromorphic in Γ with the only pole in the point ∞ such, that

$$dp = -id(\lambda) + O(1), \quad \text{Im} \oint_{a_j} dp = \text{Im} \oint_{b_j} dp = 0.$$

5) Algebra of the holomorphic vector fields in the punctured neighbourhood of ∞ with the basis

$$l_i = \lambda^{i+1} \partial_\lambda. \tag{1.14}$$

These fields have the following commutators

$$[l_i, l_j] = (j - i)l_{i+j}. \tag{1.15}$$

In the hyperelliptic case we will consider a subalgebra, generated by the elements with even indexes. All elements of this subalgebra are single-valued in the E-plane (in the neighbourhood of infinity).

$$L_i = l_{2i} = 2(-1)^i E^{i+1} \partial_E.$$

The deformations of Γ, generated by these elements preserve the hyperelliptic structure. (The action of the vector fields on the Riemann surfaces will be described below in the section 1.3).

6) Holomorphic 2-differentials in Γ. These differentials can be written in the local coordinates as $\omega^{(2)} = \omega^{(2)}(z)(dz)^2$, the space of such differentials is $3g - 3$ - dimensional as $g > 1$, 1 - dimensional as $g = 1$ and empty as $g = 0$. We have a standard overdetermined full set in this space $\omega_{ij}^{(2)} = \omega_i \omega_j$ where ω_i are holomorphic 1-differentials.

7) Riemann bilinear relations.

Let Ω_1, Ω_2 be multivalued meromorphic differentials in Γ such, that

a) Ω_1, Ω_2 have no branch points and are locally single-valued.

b) All the residues of Ω_1, Ω_2 are equal to 0.

c) $d^{-1}\Delta_{a_i}\Omega_j$, $d^{-1}\Delta_{b_i}\Omega_j$ are single-valued meromorphic functions in Γ without singularities on the cycles. We assume all these functions to be equal to 0 in the point of intersection of all cycles (this point correspond to the vertices of the polygon).

Then we have the following relation:

$$2\pi i \sum \mathrm{res}(d^{-1}\Omega_1)\Omega_2 = \sum_i \left\{ \oint_{a_i} \Omega_1 \oint_{b_i} \Omega_2 - \oint_{a_i} \Omega_2 \oint_{b_i} \Omega_1 + \oint_{a_i} (d^{-1}\Delta_{b_i}\Omega_2)\Omega_1 \right\} +$$

$$+ \sum_i \left\{ -\oint_{a_i} (d^{-1}\Delta_{b_i}\Omega_1)\Omega_2 + \oint_{b_i} (d^{-1}\Delta_{a_i}\Omega_1)\Omega_2 - \oint_{b_i} (d^{-1}\Delta_{a_i}\Omega_2)\Omega_1 \right\} + \tag{1.16}$$

$$+ \sum_i \left\{ -\oint_{a_i} (d^{-1}\Delta_{b_i}\Omega_1)(\Delta_{b_i}\Omega_2) - \oint_{b_i} (d^{-1}\Delta_{a_i}\Omega_2)(\Delta_{a_i}\Omega_1) \right\}.$$

All the terms in the right-hand side of (1.16) are correctly defined because of c). The property b) guaranties the correctness of the left-hand side.

Remark. All the residues of Ω_1, Ω_2 are equal to 0 so we have

$$\sum \mathrm{res}(d^{-1}\Omega_1)\Omega_2 = -\sum \mathrm{res}(d^{-1}\Omega_2)\Omega_1 \tag{1.17}$$

8) We will use the following scalar product, introduced in [27] and generalized in [23], [6]. Let the 1 - forms Ω_1, Ω_2 be linear combinations of differentials, defined in the points 1-3. Then

$$V_{\Omega_1 \Omega_2} = \mathrm{res}\,(d^{-1}\Omega_1)\Omega_2 - \frac{1}{2\pi i} \sum_i \oint_{a_i} \Omega_1 \oint_{b_i} \Omega_2 +$$

$$+ \frac{1}{2\pi i} \sum_i \oint_{a_i} (d^{-1}\Delta_{b_i}\Omega_1)\Omega_2 - \frac{1}{2\pi i} \sum_i \oint_{b_i} (d^{-1}\Delta_{a_i}\Omega_1)\Omega_2 \tag{1.18}$$

The brackets $(\ )_+$ in (1.18) denote the singular part of a function

$$\left(\sum_{m=-\infty}^{m=+\infty} a_k \lambda^k\right)_+ = \sum_{m=1}^{m=+\infty} a_k \lambda^k \tag{1.19}$$

Comparing (1.17) with the Riemann relations (1.16) we see that the product (1.18) is symmetrical

$$V_{\Omega_1 \Omega_2} = V_{\Omega_2 \Omega_1}. \tag{1.20}$$

(The last two terms in (1.16) are equal to 0.)

Simple direct calculations result in:

$$V_{\Omega_k \Omega_l} = Q_{kl} = Q_{lk}$$

$$V_{\omega_k \Omega_l} = q_{kl}^H$$

$$V_{\omega_k^\alpha \Omega_l} = Q_{kl}^\alpha = -\frac{1}{2\pi i} \oint_{b_\alpha} (E^k - E_0^k) \Omega_l$$

$$V_{\sigma_k^\alpha \Omega_l} = R_{kl}^\alpha = -\frac{1}{2\pi i} \oint_{a_\alpha} (E^k - E_0^k) \Omega_l$$

$$V_{\omega_k \omega_l} = -\frac{1}{2\pi i} B_{ik}$$

$$V_{\omega_k \omega_l^\alpha} = -\frac{1}{2\pi i}(U_l^\alpha)_k = -\frac{1}{2\pi i} \oint_{b_\alpha} (E^l - E_0^l) \omega_k \tag{1.21}$$

$$V_{\omega_k \sigma_l^\alpha} = -\frac{1}{2\pi i}(V_l^\alpha)_k = -\frac{1}{2\pi i} \oint_{a_\alpha} (E^l - E_0^l) \omega_k$$

$$V_{\omega_k^\alpha \omega_l^\beta} = -\frac{1}{2\pi i} \oint_{b_\alpha} (E^k - E_0^k) \omega_l^\beta$$

$$V_{\omega_k^\alpha \sigma_l^\beta} = -\frac{1}{2\pi i} \oint_{b_\alpha} (E^k - E_0^k) \sigma_l^\beta$$

$$V_{\sigma_k^\alpha \sigma_l^\beta} = -\frac{1}{2\pi i} \oint_{a_\alpha} (E^k - E_0^k) \sigma_l^\beta$$

The following property of the scalar product $V_{\Omega_1 \Omega_2}$ is important for the averaged equations theory.

Lemma 1.1 Let Γ be a hyperelliptic Riemann surface with the branch points ∞, $E_1, \ldots, E_{2g+1}$. Then the scalar products $V_{\Omega_k \Omega_l}$, $V_{\omega_k \Omega_l}$, $V_{\omega_k^\alpha \Omega_l}$, $V_{\sigma_k^\alpha \Omega_l}$, $V_{\omega_k \omega_l}$, $V_{\omega_k \omega_l^\alpha}$, $V_{\omega_k \sigma_l^\alpha}$ can be expressed via branch points $E_1, \ldots, E_{2g+1}$ and the normalization point E_0 in terms of the hyperelliptic integrals (elliptic integrals as $g = 1$).

Proof of the Lemma. The forms ω_k, Ω_k in the hyperelliptic case have the following representations

$$\omega_k = \frac{\left(\sum_{l=1}^g A_{kl} E^{l-1}\right) dE}{\sqrt{-4(E - E_1)\ldots(E - E_{2g+1})}}$$

$$\Omega_k = \frac{\left(-k(-E)^{g+\frac{k-1}{2}} + \sum_{l=1}^g C_{kl} E^{l-1}\right) dE}{\sqrt{-4(E - E_1)\ldots(E - E_{2g+1})}}. \tag{1.22}$$

The coefficients A_{kl}, C_{kl} are uniquely determined by the normalization conditions

$$\oint_{a_j} \omega_i = \delta_{ij}, \quad \oint_{a_j} \Omega_i = 0. \tag{1.23}$$

Equations (1.23) are equivalent to a linear system on A_{kl}, C_{kl}. All the coefficients of this system are hyperelliptic integrals, the system is non-degenerate (see [25]) and we can solve it. The coefficients of the Tailor series in ∞ for ω_k, Ω_k can be algebraically expressed via A_{kl}, C_{kl}, E_j. The remark that all the integrals

$$\oint (E^k - E_0^k)\omega_l, \quad \oint (E^k - E_0^k)\Omega_l \tag{1.24}$$

are hyperelliptic completes the proof.

1.3. Deformations Of The Riemann Surfaces And The Riemann Problem: The Moduli Space

All the surfaces of genus g are topologically equivalent. But if $g \geq 1$ then they may be different as complex manifolds. Consider the simplest case $g = 1$ of complex tori. There exists an unique holomorphic differential ω_1 such that

$$\oint_a \omega_1 = 1. \tag{1.25}$$

The parameter

$$\tau = \oint_b \omega_1, \quad \text{Im } \tau > 0 \tag{1.26}$$

is correctly defined if we have a fixed basis of 1 - cycles a, b in Γ. The surface Γ can be described as a factor of the complex plane by the group of shifts, generated by:

$$z \to z + 1, z \to z + \tau. \tag{1.27}$$

Let Γ, Γ' be complex tori (g=1) with the basises of 1 - cycles a, b and a', b' respectively. Then a complex map $f : \Gamma \to \Gamma'$ such that $f(a) = a'$, $f(b) = b'$ exists if and only if $\tau = \tau'$.

Parameter τ depends on the choice of basic cycles. Let us have two basises a, b and $\tilde{a}$, $\tilde{b}$ respectively. Then there exists a matrix $\begin{pmatrix} \alpha & \beta \\ \gamma & \delta \end{pmatrix} \in sl(2, \mathbf{Z})$ such, that

$$\begin{pmatrix} \tilde{a} \\ \tilde{b} \end{pmatrix} = \begin{pmatrix} \alpha & \beta \\ \gamma & \delta \end{pmatrix} \begin{pmatrix} a \\ b \end{pmatrix}, \quad \tilde{\tau} = \frac{\gamma + \delta\tau}{\alpha + \beta\tau}. \tag{1.28}$$

The group $sl(2, \mathbf{Z})$ consists of all 2×2 integer matrices such that $\alpha\delta - \beta\gamma = 1$.

Summing all of this we have:

Two complex (one-dimensional in the complex sense) tori with the parameters τ and τ' are isomorphic if and only if there exists a matrix

$$\begin{pmatrix} \alpha & \beta \\ \gamma & \delta \end{pmatrix} \in sl(2, \mathbf{Z}) \quad \text{such, that} \quad \tau' = \frac{\gamma + \delta\tau}{\alpha + \beta\tau}. \tag{1.29}$$

The space of all 1-dimensional complex tori is called the moduli space for $g = 1$ or the space of all complex structures on the surfaces of genus $g = 1$. We have proved that the moduli space for $g = 1$ is the factor of the complex upper-plane Im $\tau > 0$ by the group $sl(2, \mathbf{Z})$.

It is possible to consider the moduli space of all Riemann surfaces with a set of marked points and the moduli space of the Riemann surfaces with a set of marked points and a local parameter in one of them. The last space is infinite-dimensional.

The moduli space of the Riemann surfaces with $g > 1$ is a $3g - 3$-dimensional complex manifold. We do not want to discuss its properties in details. But the following construction is very important for us.

Let Γ be a Riemann surface with a marked point ∞, a local parameter $z = 1/\lambda$ in the neighbourhood of ∞, S be a small contour surrounding ∞, $v = v(\lambda)d\lambda$ be a holomorphic vector field in the vicinity of the contour $U(S)$. Then we can construct a family of new Riemann surfaces Γ_α depending on a parameter α (see [25]). We will assume α to be sufficiently small.

The contour S splits Γ to a small disk D, containing the point ∞ and an open Riemann surface $\Gamma \backslash D$ ($\Gamma \backslash D$ denotes the set of point γ such, that $\gamma \in \Gamma$, $\gamma \notin D$). We can cover Γ by two regions Γ_+, Γ_- such, that:

    1) $D \subset \Gamma_-$.
    2) $\Gamma \backslash D \subset \Gamma_+$.
    3) $\Gamma_+ \cap \Gamma_- = U(S)$.

If γ is a point of $U(S)$ then we will denote the correspondent points in Γ_+ and Γ_- γ_+ and γ_- respectively. Γ may be treated as the result of gluing Γ_+ to Γ_- $f : \Gamma_+ \to \Gamma_-$ $f(\gamma_+) = \gamma_-$. Let us introduce a new gluing function f_α by formula

$$f_\alpha(\gamma_+) = \exp(\alpha v)\gamma_-, \tag{1.30}$$

where $\exp(\alpha v)\gamma_-$ denotes the shift of the point γ_- via the vector field v after the lapse of the time α. Then we can define a new Riemann surface Γ_α as the result of gluing Γ_+ to Γ_- via the function f_α.

The Riemann surfaces Γ and Γ_α are constructed from the same parts Γ_+ and Γ_-. Thus we have a map $\mathbf{E} : \Gamma \to \Gamma_\alpha$, coinciding with identical maps $\Gamma_+ \to \Gamma_+$ on $\Gamma \backslash D$ and $\Gamma_- \to \Gamma_-$ on D. Map $\mathbf{E}$ has a jump on the contour S. If we have some marked points in Γ or local parameters in some of them then we will map these objects by $\mathbf{E}$.

Let us have a infinitesimal transformation (α is infinitely small) and Δ be a holomorphic tensor field in Γ_α. The map $\mathbf{E}$ carries Δ from Γ_α to Γ so we can treat Δ as a tensor field in Γ with a jump on the contour S. The boundary values of Δ on S Δ_+ and Δ_- satisfy the following relation

$$\Delta_+ - \Delta_- = \alpha L_v \Delta, \tag{1.31}$$

where L_v denotes the Lie derivative

$$L_{v(\lambda)\partial_\lambda}\, g(\lambda)(d\lambda)^\alpha = (v(\lambda)g'(\lambda) + \alpha v'(\lambda)g(\lambda))\, (d\lambda)^\alpha \tag{1.32}$$

(α is infinitely small so we can write Δ_+ in the right-hand side of (1.31) as well as Δ_-).

Formula (1.31) gives us a very convenient method for calculation the variations of tensor object via complex structure variations: we calculate the right-hand side of (1.31) and solve the Riemann problem in the appropriate functional class. Riemann problem is one of the basic objects in the soliton theory so this approach allows us to connect these deformations of the Riemann surfaces with the non-isospectral symmetries (see section 2.2).

In the hyperelliptic case our construction has the following interpretation. Let Γ be a hyperelliptic surface over the E-plane with the branch points ∞, $E_1, \ldots, E_{2g+1}$, $v = v(E)\partial_E$ be a vector field in the E-Plane, nonsingular for all $E \neq \infty$, the local parameter in the point ∞ be $z = 1/\lambda$, $\lambda = \sqrt{-E}$.

Then Γ_α is a hyperelliptic Riemann surface with the branch points ∞, $\exp(\alpha v)E_1$, $\ldots$, $\exp(\alpha v)E_{2g+1}$ and the local parameter $z = 1/\lambda$, $\lambda = \sqrt{-E}$. The map $\mathbf{E}$ carries the point $E_\pm$ in Γ to the point $\exp(\alpha v)E_\pm$ in Γ_α if the point $E_\pm$ is located outside the neighbourhood of ∞ surrounded by a small contour S and carries the point $E_\pm$ in Γ to the point $E_\pm$ in Γ_α if the point $E_\pm$ is located inside the neighbourhood of ∞ (the sign $\pm$ means the upper or lower sheet).

If we have a family of deformed hyperelliptic Riemann surfaces Γ_α and a function on this family $f(E_\pm, \alpha)$ then we have two differential operators

$$\partial_\alpha f = \frac{\partial f}{\partial \alpha} \tag{1.33}$$

and

$$D_\alpha f = \frac{\partial f}{\partial \alpha} + L_v f, \tag{1.34}$$

where L_v is the Lie derivative (1.32) in the first argument. In contrast with ∂_α the operator D_α can be generalized to arbitrary Riemann surfaces and can be treated as a connection generated by the map E.

The following property of the connection D_α is very important for us:

Lemma 1.2. Let $\Delta(E_\pm, \alpha)$ be a holomorphic tensor field in $\Gamma_\alpha \backslash \infty$ for all α. Then $D_\alpha \Delta(E_\pm, \alpha)$ is nonsingular outside the neighbourhood of ∞ (for ∂_α in is not valid in the branch points).

While calculating the nonisospectral symmetries we will need the derivatives of the holomorphic differentials by the complex structures. In the hyperelliptic case we have:

Lemma 1.3. Let Γ be a hyperelliptic Riemann surface, $L_n = l_{2n} = 2(-1)^n E^{n+1} \partial_E$ be a holomorphic vector field in the E-plane, $i \geq -1$, Γ_α be the deformation of Γ via the vector field v. Then

$$D_\alpha \Omega_k = k\Omega_{2n+k} + \sum_{m=1}^{2n-1} Q_{km} \Omega_{2n-m}, \tag{1.35}$$

$$D_\alpha \omega_k = \sum_{m=1}^{2n-1} q_{km}^H \Omega_{2n-m}, \tag{1.36}$$

$$D_\alpha \omega_k^\alpha = 2(-1)^n k\omega_{n+k}^\alpha + \sum_{m=1}^{2n-1} Q_{km}^\alpha \Omega_{2n-m}, \tag{1.37}$$

$$D_\alpha \sigma_k^\alpha = 2(-1)^n k\sigma_{n+k}^\alpha + \sum_{m=1}^{2n-1} R_{km}^\alpha \Omega_{2n-m}, \tag{1.38}$$

Proof of the Lemma 1.3. In accord with the rule described above we have to solve the following Riemann problem:

Let Δ be one of the differentials Ω_k, ω_k, ω_k^α, σ_k^α. Then we have to construct a pair of differentials Δ_+, Δ_- such that

1) $\Delta_+ - \Delta_- = L_v \Delta$.
2) Δ_- is defined and nonsingular in the neighbourhood of ∞.
3) Δ_+ is nonsingular for all $E_\pm \neq \infty$.
4) $\oint_{a_i} \Delta_+ = 0$.
5) Δ_+ is single-valued as $\Delta = \Omega_k$ or $\Delta = \omega_k$,

$$\Delta_{a_i} \Delta_+ = 2(-1)^n k\Delta_{a_i} \omega_{n+k}^\alpha, \quad \Delta_{b_i} \Delta_+ = 0 \text{ as } \Delta_+ = \omega_k^\alpha$$

$$\Delta_{b_i} \Delta_+ = 2(-1)^n k\Delta_{a_i} \sigma_{n+k}^\alpha, \quad \Delta_{a_i} \Delta_+ = 0 \text{ as } \Delta_+ = \sigma_k^\alpha$$

Then $D_\alpha \Delta = \Delta_+$.

In fact we are looking for a differential with the properties 3-5 in the finite part of Γ and the prescribed singularity $L_v \Delta$ in ∞. But such differential can be easily constructed as a linear combination of Ω_k, ω_k, ω_k^α, σ_k^α. This completes the proof.

Corollary 1. ([25]).

$$\frac{\partial B_{ij}}{\partial \alpha} = \oint_S v\omega_i\omega_j. \tag{1.39}$$

where the product $v(\lambda)\partial_\lambda \, \omega_i(\lambda)d\lambda \, \omega_j(\lambda)d\lambda$ is the 1-form $v(\lambda)\omega_i(\lambda)\omega_j(\lambda)d\lambda$. This formula is valid for general Riemann surfaces as well as for hyperelliptic ones. We assume that the contour S goes around the point ∞ counterclockwise.

Corollary 2. The vector field v do not vary the complex structure of Γ if and only if all the integrals

$$\oint_S v\omega_i^{(2)}$$

is are equal to 0. Here $\omega_i^{(2)}$ is the basis of holomorphic 2-forms. It proves that the moduli space is $3g - 3$ dimensional as $g > 1$ and 1 - dimensional as $g = 1$.

1.4. Riemann Theta-Functions.

Let B_{ij} be a complex symmetrical $g \times g$ matrix such, that Re b_{ij} is negative definite, $\vec{z}$ be a complex g-component vector. Then the Riemann theta-function can be defined as an infinite sum ([26])

$$\theta(\vec{z}|b_{ij}) = \sum_{m_1,\ldots,m_g} \exp\left\{\frac{1}{2}\sum_{kj} b_{kj}m_km_j + \sum_k z_km_k\right\}, \tag{1.40}$$

where m_k, $k = 1,\ldots,g$ are arbitrary integers. This sum converges for all $\vec{z}$. The theta function has the following periodicity properties

$$\theta(z_1, z_2,\ldots,z_k + 2\pi i,\ldots,z_g|b_{ij}) = \theta(z_1, z_2,\ldots,z_k,\ldots,z_g|b_{ij}). \tag{1.41}$$

$$\theta(z_1 + b_{1k}, z_2 + b_{2k},\ldots,z_g + b_{gk}|b_{ij}) = \theta(z_1, z_2,\ldots,z_g|b_{ij})\exp\{-b_{kk}/2 - z_k\}. \tag{1.42}$$

The zeros ot the theta function are described by the following Lemma (see [26]).

Lemma 1.4. Let Γ be a Riemann surface with a marked point γ_0, B_{ij} be the matrix of periods (1.4), $b_{ji} = 2\pi i B_{ji}$, $\vec{A}(\gamma)$ be the Abel transform, i.e. $\vec{A}(\gamma)$ is a multivalued map $\Gamma \to \mathbf{C}^g$ determined by the formula

$$(\vec{A}(\gamma))_k = 2\pi i \int_{\gamma_0}^\gamma \omega_k, \quad k = 1,\ldots,g. \tag{1.43}$$

Then there exists a g-dimensional complex vector $\vec{K} = \vec{K}(\Gamma,\gamma_0)$ such that the function

$$\varphi(\gamma) = \theta(\vec{A}(\gamma) - \vec{A}(\gamma_1) - \ldots - \vec{A}(\gamma_g) + \vec{K}|b_{ij}), \quad \gamma \in \Gamma. \tag{1.44}$$

has one of the following properties:
1) $\varphi(\gamma) \equiv 0$ or
2) $\varphi(\gamma)$ has exactly g zeros in the points $\gamma_1, \ldots, \gamma_g$.
$\vec{K}$ is called Riemann constants vector.

2. PERIODIC THEORY OF THE KORTEVEG - DE - VRIES AND KADOMTSEV - PETVIASHVILI EQUATIONS

In this section we will remind some facts from the KdV and KP theory. Two topics are the most interesting for us: periodic (quasiperiodic) finite-gap theory and the action of the nonisospectral symmetries on the finite-gap solutions.

Finite-gap KdV solutions can be treated as nonlinear superpositions of the moving waves. Such solutions are the basic objects for the averaging procedure. The have been constructed in the papers of S.P.Novikov, B.A.Dubrovin, V.B.Matveev, A.R.Its, P.Lax, H.McKean and P. van Moerbeke (see book [28] for more detailed description and references). Finite-gap KP solutions were constructed by I.M.Krichever [29]. The direct periodic problem for KP is more complicated (the results, obtained by I.M.Krichever can be found in [29]).

The first step in constructing averaged nonisospectral symmetries is the following: the action of these symmetries on the finite-gap solution is calculated. This problem was solved by A.Yu.Orlov and the author in [20], [21]. We remind some results of these parers. We use the representation for nonisospectral symmetries, suggested by A.Yu.Orlov and E.I.Schulman [17] as the most convenient for us.

2.1. KdV And KP Theory, Integration And Isospectral Symmetries, Periodic Theory, Baker-Akhiezer Function And Cauchy Kernel

The theory of the KdV equation

$$u_t = \frac{1}{4}u_{xxx} - \frac{3}{2}uu_x. \tag{2.1}$$

is based upon the existence of the following representation. Let

$$L = -\partial_x^2 + u(x,t), \; A = \partial_x^3 - \frac{3}{4}(u\partial_x + \partial_x u) \tag{2.2}$$

be ordinary differential operators depending on an extra parameter t. Then the function $u(x,T)$ satisfy (2.1) if and only if the following relation takes place

$$\partial L/\partial t = [L, A]. \tag{2.3}$$

Representation (2.3) is called Lax pair or $L - A$ pair for KdV.

One of the first results of the soliton theory was the existence of infinitely many mutually commuting KdV symmetries. They can be written via the recursion operator

$$\frac{\partial u}{\partial t_{2n+1}} = \frac{d}{dx}\left(-\frac{\Lambda}{4}\right)^n u = -2\frac{d}{dx}\left(-\frac{\Lambda}{4}\right)^{n+1} \cdot 1, \tag{2.4}$$

where

$$\Lambda = -\partial_x^2 + 2\partial_x^{-1}u\partial_x + 2u \tag{2.5}$$

or in the Lax form

$$\partial L/\partial t_{2n+1} = [L, A_{2n+1}], \tag{2.6}$$

where

$$A_{2n+1} = \{(-L)^{\frac{2n+1}{2}}\}_+, \tag{2.7}$$

$(-L)^{\frac{2n+1}{2}}$ denotes a formal pseudodifferential operator, i.e a series in ∂_x with finite number of positive terms and infinite number of negative, $\{\ \}_+$ denotes the differential part (see [31]). We mark the times by odd indexes to have unified notations for KdV and KP. Here $t_1 = x$, $t_3 = t$.

Equations (2.3), (2.6) result in the following property: the spectrum of L does not depend upon the times t_3, t_5, ..., t_{2k+1}, ... This is the reason why we call these symmetries isospectral.

KdV equation is integrated by the inverse scattering transform, i.e. we consider the "scattering data" for L as a new variable instead of $u(x)$. (We write "scattering data" in quotation marks to stress that these data coincides with the physical scattering data only for some functional classes of potentials). Because of the isospectral property the evolution law for the "scattering data" is very simple (see [28]). The map from u to the scattering data for the vanishing in the infinity potentials in the small amplitude limit coincides with the Fourier transform and can be treated as its nonlinear analog.

The realization of this scheme depends on the functional class of the potential. Let us remind the scheme for the periodic case

$$u(x + \Pi, t) = u(x, t), \qquad (2.8)$$

where $u(x, t)$ is a real nonsingular potential.

We consider the following spectral problems for L:

a) Main problem $L\psi = E\psi$, ψ is bounded in x.

b) Auxiliary problem $L\psi_a = E\psi_a$, $\psi_a(0) = \psi_a(\Pi) = 0$.

The spectrum of the main problem consists of a set of intervals $[E_1, E_2]$, $[E_3, E_4]$, $\ldots$, $[E_{2n-1}, E_{2n}]$, $\ldots$ where $E_1 < E_2 \leq E_3 < E_4 \leq E_5 \ldots$, $E_{2n+1} - E_{2n} \to 0$ as $n \to \infty$.

The spectrum of the auxiliary problem consists of an infinite number of points $d_1 < d_2 < d_3 < \ldots$ located in the gaps $d_1 \in [E_2, E_3]$, $d_2 \in [E_4, E_5]$, $d_3 \in [E_6, E_7], \ldots$.

The Bloch eigenfunction $\psi(x, E)$ normalized by the conditions $\psi(x + \Pi, E) = \exp(\Pi i p(E)) \, \psi(x, E)$ and $\psi(0, E) = 1$ is meromorphic on a two-sheeted Riemann surface Γ over the E-plane with branch points $E_1, E_2, \ldots, \infty$, and has simple poles in the points $\gamma_1, \gamma_2, \ldots$, such that the projection of γ_n to the E-plane coincides with d_n.

The function $p(E_{\pm})$ is defined in Γ and is called quasimomentum.

The spectrum corresponding to a general potential has infinite number of gaps. But for us the so-called finite-gap case when $E_{2n} = E_{2n+1}$ for all $n > g$ is the most important. (The infinite-genus case can also be studied [32] but the answers are much more complicated). Finite-gap potentials are dence in the space of all periodic potentials.

The inverse problem data in the finite-gap case is the following:

1) $2g + 1$ real numbers $E_1, E_2, \ldots, E_{2g+1}$, $E_1 < E_2 < \ldots < E_{2g+1}$. The points E_k are the boundary point of the spectrum.

2) g points $\gamma_1, \ldots, \gamma_g$ in a hyperelliptic Riemann surface Γ with the branch points ∞, $E_1, E_2, \ldots, E_{2g+1}$, such, that $E(\gamma_k) \in [E_{2k}, E_{2k+1}]$ where E is the projection to the E-plane.

This data uniquely determines the potential $u(x)$.

The flows (2.3), (2.6) do not move the branch points E_k but they shift the divisor γ_k. The evolution of the points γ_k can be described in terms of ordinary differential equations, derived by B.A.Dubrovin (see [28]). This system is nonlinear, but it has a very nice explicit solution.

Lemma 2.1. (see [28]). Let $\vec{A}(\gamma)$ be the Abel transform defined in the paragraph 1.4. Then

$$\vec{A}(\gamma_1) + \vec{A}(\gamma_2) + \ldots + \vec{A}(\gamma_g) = \vec{A}_0 + x\vec{U}_1 + t\vec{U}_3 + t_5\vec{U}_5 + \ldots, \qquad (2.9)$$

where $\vec{A}_0$ is some constant vector, the vectors U_k are defined by (1.7).

Inverse transform to (2.9) can be written in terms of the theta-functions. The answer is given by the A.R.Its - V.B.Matveev formula (see [28], [33]).

$$u(x, t_3, t_5, \ldots) = -2\partial_x^2 \log \theta(\vec{V}_0(\gamma_1, \ldots, \gamma_g) + x\vec{U}_1 + t\vec{U}_3 + t_5\vec{U}_5 + \ldots |b_{ij}) + C(\Gamma), \quad (2.10)$$

where $V_0(\gamma_1, \ldots, \gamma_g)$, $C(\Gamma)$ are some constants, $b_{ij} = 2\pi i B_{ij}$, B_{ij} is the Riemann matrix (1.4).

General finite-gap solutions are quasiperiodic. The characterization of periodic solutions in terms of the inverse data is a complicated problem.

A slightly different approach to the inverse problem is more convenient in some situations (see [28]). Instead of normalizing $\psi(E,0) = 1$ for all t we consider a function $\Psi(\gamma, \vec{t})$, $\gamma \in \Gamma$, $\vec{t} = (t_1, t_3, t_5, \ldots)$, $t_1 = x$, $t_3 = t$ such that

1) $L\Psi(\gamma, \vec{t}) = E(\gamma)\Psi(\gamma, \vec{t})$.

2) $\Psi(\gamma, x + \Pi, t_3, t_5, \ldots) = \exp(i\Pi p(\gamma))\Psi(\gamma, x, t_3, t_5, \ldots)$

3) $\Psi(\gamma, 0, 0, 0, \ldots) = 1$

4) $(\partial_{t_{2n+1}} - A_{2n+1})\Psi(\gamma, \vec{t}) = 0$. (Symmetries (2.6) is mutually commuting so the condition 4 is self-consistent).

$\Psi(\gamma, \vec{t})$ is called Baker-Akhiezer function. It has the following analytical properties:

Pr.1) $\Psi(\gamma, \vec{t})$ is meromorphic in $\Gamma \backslash \infty$.

Pr.2) $\Psi(\gamma, \vec{t})$ has simple poles in the points $\gamma_1, \ldots, \gamma_g$ and no other singularities in $\Gamma \backslash \infty$.

Pr.3) $\Psi(\gamma, \vec{t})$ has an essential singularity as $\gamma \to \infty$

$$\Psi(\lambda, \vec{t}) = \exp(\Theta(\gamma, \vec{t}))[1 + \chi_1(\vec{t})/\lambda + \chi_2(\vec{t})/\lambda^2 + \ldots] \quad (2.11)$$

where in the correspondent to KdV hyperelliptic case $\lambda = \sqrt{-E}$, $\Theta(\gamma, \vec{t}) = \lambda x + \lambda^3 t + \lambda^5 t_5 + \ldots$.

Lemma 2.2. The properties Pr.1 -Pr.3 uniquely determined the function $\Psi(\gamma, \vec{t})$. It can be expressed in terms of the theta-functions (see, for example review [33])

$$\Psi(\gamma, \vec{t}) = \frac{\theta(\sum t_k \vec{U}_k + \vec{A}(\gamma) - \vec{A}(\gamma_1) - \ldots - \vec{A}(\gamma_g) + \vec{K})\theta(-\vec{A}(\gamma_1) - \ldots - \vec{A}(\gamma_g) + \vec{K})}{\theta(\sum t_k \vec{U}_k - \vec{A}(\gamma_1) - \ldots - \vec{A}(\gamma_g) + \vec{K})\theta(\vec{A}(\gamma) - \vec{A}(\gamma_1) - \ldots - \vec{A}(\gamma_g) + \vec{K})} \cdot$$

$$\cdot \exp(\sum t_k \int^\gamma \Omega_k). \quad (2.12)$$

Here $\vec{K}$ is the vector of Riemann constants (see paragraph 1.4), the integrals are normalized by $\int^\gamma \Omega_k = \lambda^k + o(1)$.

Comparing (2.12) with

$$u(\vec{t}) = 2\partial_x \chi_1(\vec{t}) \quad (2.13)$$

we can easily derive (2.12).

Now we will remind the finite-gap inverse scattering transform for KP [29].

$$(u_t - \frac{1}{4}u_{xxx} + \frac{3}{2}uu_x)_x = 3u_{yy}. \quad (2.14)$$

Let Γ be *arbitrary* Riemann surface of genus g with a marked point ∞, a local parameter $1/\lambda$ in ∞ and g marked points $\gamma_1, \ldots, \gamma_g$. Then for the data of general position there exists a unique function $\Psi(\gamma, \vec{t})$, $\gamma \in \Gamma$, $\vec{t} = (t_1, t_2, t_3, t_4, t_5, \ldots)$, $t_1 = x$, $t_2 = y$, $t_3 = t$ with the analytic properties Pr.1 - Pr.3 (for KP $\Theta(\gamma, \vec{t}) = \lambda x + \lambda^2 y + \lambda^3 t + \lambda^4 t_4 + \lambda^5 t_5 + \ldots$). Representation (2.12) is valid for general Riemann surfaces but we have to sum in (2.12) by even indexes as well as by odd ones.

For all k there exists a unique ordinary differential operator $\tilde{A}_n$ such that

$$(\partial_{t_n} - \tilde{A}_n)\Psi(\gamma, \vec{t}) = 0, \quad \tilde{A}_n = \partial_x^n + \ldots, \quad \tilde{A}_2 = -L. \quad (2.15)$$

The compatibility conditions

$$[(\partial_{t_2} - \tilde{A}_2), (\partial_{t_k} - \tilde{A}_k)] = 0 \quad (2.16)$$

give the Lax pair for KP as $k = 2$ and isospectral symmetries as $k > 3$. The potential

$u(\vec{t})$ is given by slightly changed (2.10) - the sum is over all indexes - odd and even. y - independent KP solutions satisfy KdV.

We need also the Baker-Akhiezer conjugate differential $\Psi^+(\gamma, \vec{t})$. Its analytical properties are the following

Pr.1') $\Psi^+(\gamma, \vec{t})$ is a holomorphic in $\Gamma \backslash \infty$ 1-differential.

Pr.2') $\Psi^+(\gamma, \vec{t})$ has simple zeros in the points $\gamma_1, \ldots, \gamma_g$.

Pr.3') $\Psi^+(\gamma, \vec{t})$ has an essential singularity as $\gamma \to \infty$

$$\Psi(\lambda, \vec{t}) = d\lambda \, \exp(-\Theta(\gamma, \vec{t}))[1 + \chi_1^+(\vec{t})/\lambda + \chi_2^+(\vec{t})/\lambda^2 + \ldots] \qquad (2.17)$$

where $\Theta(\gamma, \vec{t}) = \lambda x + \lambda^2 y + \lambda^3 t + \lambda^4 t_4 + \lambda^5 t_5 + \ldots$.

$$(\partial_{t_n} + \tilde{A}_n^+)\Psi^+(\gamma, \vec{t}) = 0, \qquad (2.18)$$

where $\tilde{A}_n^+$ is the formal conjugate to $\tilde{A}_n$, i.e. $(a(x)(\partial_x)^n)^+ = (-\partial_x)^n a(x)$. The following ortogonality properties are important:

Lemma 2.3. 1) Let S be a small contour surrounding the point ∞. Then

$$\oint_S \Psi(\gamma, \vec{t})\Psi^+(\gamma, \vec{t}') = 0 \qquad (2.19)$$

for all $\vec{t}, \vec{t}'$.

2) Let $p(\gamma) = \int^\gamma dp$ where dp is the quasimomentum differential defined in the section 1, $G(\gamma)$ be a contour in Γ consisting of all points γ' such that $\operatorname{Im} p(\gamma') = \operatorname{Im} p(\gamma)$. Then in the contour $G(\gamma)$ the following relation is valid

$$\int_{-\infty}^{+\infty} \Psi(\gamma, \vec{t})\Psi^+(\gamma', \vec{t}) = 2\pi i \delta(\gamma - \gamma'). \qquad (2.20)$$

We have to calculate deformations of the Baker - Akhiezer function. For this purpose we need an appropriate Cauchy kernel.

Lemma 2.4. ([20]). The Cauchy - Baker - Akhiezer kernel $\omega(\gamma, \gamma', \vec{t})$ with the following properties:

1) $\omega(\gamma, \gamma', \vec{t})$ is a function in γ and a 1-form in γ'.

2) $\omega(\gamma, \gamma', \vec{t})$ is meromorphic function of γ in $\Gamma \backslash \infty$ with simple poles $\gamma_1, \ldots, \gamma_g, \gamma'$.

3) As a function of γ' $\omega(\gamma, \gamma', \vec{t})$ is meromorphic in $\Gamma \backslash \infty$ with one pole γ and zeros in the points $\gamma_1, \ldots, \gamma_g$.

4) $\omega(\gamma, \gamma', \vec{t}) = o(\exp(\Theta(\gamma, \vec{t})))$ as $\gamma \to \infty$.

5) $\omega(\gamma, \gamma', \vec{t}) = o(\exp(-\Theta(\gamma', \vec{t})))d\lambda$ as $\gamma' \to \infty$.

6) $\omega(\gamma, \gamma', \vec{t}) \sim \frac{d\lambda'}{2\pi i(\lambda' - \lambda)}$ as $\lambda \to \lambda'$.

can be written in the following form

$$\omega(\gamma, \gamma', x, y, t, \ldots) = \frac{1}{2\pi i} \int_{\pm\infty}^x \Psi(\gamma, x', y, t, \ldots)\Psi^+(\gamma', x', y, t, \ldots)dx'. \qquad (2.21)$$

If $\gamma' \notin G(\gamma)$ the sing in limit of integration in (2.21) is uniquely determined by the convergence condition. For $\gamma' \in G(\gamma)$ the property (2.20) guaranties correctness.

First formula similar to (2.21) was obtained by I.M.Krichever and S.P.Novikov in [19] for systems with discrete x.

2.2. Nonisospectral Symmetries: Action On The Finite-Gap Solutions

We have pointed out in the introduction that KdV equation possesses the following set of symmetries

$$\frac{\partial u}{\partial \tau_{2n}} = -2\frac{d}{dx}\left(-\frac{\Lambda}{4}\right)^{n+1}\left(\sum_{k=0}^{\infty}((2k+1)t_{2k+1}\left(-\frac{\Lambda}{4}\right)^{k}\right) \cdot 1. \tag{2.22}$$

They can be written in much more convenient form suggested by A.Yu.Orlov and E.I.Schulman [17].

$$\frac{\partial u}{\partial \tau_{2n}} = -2\frac{d}{dx}\ res\mid_{\lambda=\infty} (\lambda^{2n+1}\partial_{\lambda}\Psi(\lambda,\vec{t}))\Psi^{+}(\lambda,\vec{t}), \tag{2.23}$$

Ordinary higher KdV equations (2.4) have similar representation

$$\frac{\partial u}{\partial t_{2n+1}} = -2\frac{d}{dx}\ res\mid_{\lambda=\infty} \lambda^{2n+1}\Psi(\lambda,\vec{t})\Psi^{+}(\lambda,\vec{t}). \tag{2.24}$$

The formulas (2.22), (2.24) can be treated in the following way. Let us substitute the asymptotical expansions (2.11), (2.17) to the linear problem

$$L\Psi(\lambda,\vec{t}) = -\lambda^{2}\Psi(\lambda,\vec{t}), \quad L\Psi^{+}(\lambda,\vec{t}) = -\lambda^{2}\Psi^{+}(\lambda,\vec{t}). \tag{2.25}$$

¿From (2.25) we can calculate all the coefficients $\chi_k(\vec{t})$, $\chi_k^+(\vec{t})$ via $u(\vec{t})$ (in nonlocal form). Then we substitute them to (2.23), (2.25). All the exponents in $\Psi\Psi^+$ are reduced and the residue can be explicitly calculated. We obtain a close system on $u(\vec{t})$ (may be nonlocal).

Lemma 2.5. Equations (2.23), (2.25) coincides with (2.22), (2.4) respectively.

Proof of the Lemma. For small n we can check it by direct calculations. Then we apply the following identity

$$\Lambda(\Psi(\lambda,\vec{t})\Psi^{+}(\lambda,\vec{t})) = -4\lambda^{2}(\Psi(\lambda,\vec{t})\Psi^{+}(\lambda,\vec{t})). \tag{2.26}$$

(it is a direct consequence of (2.25)).

For the KP equation we can write two-parametric set of symmetries ([17])

$$\frac{\partial u}{\partial \tau_{nm}} = -2\frac{d}{dx}\ res\mid_{\lambda=\infty} (\lambda^{n}\partial_{\lambda}^{m}\Psi(\lambda,\vec{t}))\Psi^{+}(\lambda,\vec{t}) \tag{2.27}$$

but only the symmetries with $m = 0, 1$ are compatible with the finite-gap structure.

Theorem 2.1. Let Γ be a Riemann surface with a marked point ∞, a local parameter $1/\lambda$ in ∞, a set of points $\gamma_1, \ldots, \gamma_g$ in Γ where g is the genus of Γ, $l_n = \lambda^{n+1}\partial_{\lambda}$ be a vector field in the punctured neighbourhood of ∞. Consider the deformation of Γ generated by the field l_n (it was described in the paragraph 1.3). Then the correspondent variation of the KP solution, constructed by this data reads as

$$\frac{\partial u}{\partial \tau_n} = -2\frac{d}{dx}\ res\mid_{\lambda=\infty} (\lambda^{n+1}\partial_{\lambda}\Psi(\lambda,\vec{t}))\Psi^{+}(\lambda,\vec{t}) \tag{2.28}$$

i.e. it coincides with a symmetry (2.27) such that $m = 1$.

Proof of the theorem. In the paragraph 1.3 we have explained that the calculation of the Baker-Akhiezer function variation is equivalent to the following Riemann problem on the contour S, surrounding the point ∞

$$(\delta\Psi(\lambda,\vec{t}))_{+} - (\delta\Psi(\lambda,\vec{t}))_{-} = \lambda^{n+1}\partial_{\lambda}\Psi(\lambda,\vec{t}). \tag{2.29}$$

Solution of (2.29) reads as

$$\delta\Psi(\lambda,\vec{t}) = \oint_S \omega(\lambda,\mu,\vec{t})\mu^{n+1}\partial_\mu\Psi(\mu,\vec{t}). \tag{2.30}$$

where $\omega(\lambda,\mu,\vec{t})$ is the Cauchy - Baker - Akhiezer kernel defined in the Lemma 2.4. Expanding (2.30) as $\lambda \to \infty$ and using (2.20) we obtain (2.28).

Corollary 1. Symmetries (2.23) act on the finite-gap KdV solutions as

$$\partial E_s/\partial\tau_{2n} = 2(-1)^n E_s^{n+1}, \ \partial E(\gamma_k)/\partial\tau_{2n} = 2(-1)^n E^{n+1}(\gamma_k). \tag{2.31}$$

i.e. all the spectral data is shifted via the vector field $2(-1)^n E^{n+1}\partial/\partial E$ (see paragraph 1.3).

3. WHITHAM EQUATIONS: THE FULL ABEL HIERARCHY AND NONISOSPECTRAL SYMMETRIES

The KdV equation (2.3) and the symmetries (2.6) (they are called higher KdV) form a commutative set of flows. This set is called KdV hierarchy.

Averaged KdV hierarchy was constructed in [2], [3]. We will not discuss how the averaged KdV equations can be derived from the original ones and so we will only remind the answer.

The starting point for the averaging procedure is the space of all g-gap KdV solutions. Such solutions are parameterized by the branch points $E_1, \ldots, E_{2g+1}$ and the points $\gamma_1, \ldots, \gamma_g$ in Γ. The branch points E_k are integrals of motion and the points γ_k play the role of phases (see for example [28] and references therein).

If we consider a slow modulated wave-type solution then the points E_k slowly depend on coordinate and times $E_k = E_k(X, T, T_5, \ldots)$ where $X = \epsilon x$, $T = \epsilon t$, $T_5 = \epsilon t_5$, $\ldots$, $\epsilon \ll 1$. Functions $X, T, T_5, \ldots$ are called slow variables.

For averaging any full set of integrals can be used. But direct calculations for $g = 1$ (see [1]) shows that the variables E_k result in the simplest form of the averaged equations so it is very natural to use them for higher genera.

The averaged KdV hierarchy can be written in the following form, suggested by Flaschka - Forest - McLaughlin [2]

$$\frac{\partial E_k}{\partial T_{2n+1}} = w_k^{2n+1}(E_1, \ldots, E_{2g+1})\frac{\partial E_k}{\partial X}, \tag{3.1}$$

where

$$w_k^{2n+1}(E_1, \ldots, E_{2g+1}) = \frac{\Omega_{2n+1}(E_k)}{\Omega_1(E_k)}, \tag{3.2}$$

Ω_k are the differentials defined in the paragraph 1.2. We see that the flows (3.1) have the Riemann diagonal form.

The averaged KdV equations have wider symmetry group then the original KdV. For example the scaling transform $X \to \alpha X$, $T \to \alpha T$ has no analogs in original equations.

¿From the results of S.P.Tzarev [5], [7] it was known that the Whitham equations have $2g + 1$ infinite series of symmetries but the averaged KdV hierarchy gives us only one of them. The full set of symmetries was constructed by S.P.Tzarev in [5] for $g = 1$ and B.A.Dubrovin for all g. It can be written as

$$\frac{\partial E_k}{\partial T_n^{a_i}} = \frac{\omega_n^i(E_k)}{\Omega_1(E_k)}\frac{\partial E_k}{\partial X},$$

$$\frac{\partial E_k}{\partial T_n^{b_i}} = \frac{\sigma_n^i(E_k)}{\Omega_1(E_k)} \frac{\partial E_k}{\partial X}, \qquad (3.3)$$

$$\frac{\partial E_k}{\partial T_i^H} = \frac{\omega_i(E_k)}{\Omega_1(E_k)} \frac{\partial E_k}{\partial X},$$

where $i = 1, \ldots, g$, $n = 1, 2, \ldots$. Let us denote

$$w_k^{a_i n} = \frac{\omega_n^{a_i}(E_k)}{\Omega_1(E_k)}, w_k^{b_i n} = \frac{\omega_n^{b_i}(E_k)}{\Omega_1(E_k)}, w_k^{H n} = \frac{\omega_i(E_k)}{\Omega_1(E_k)}. \qquad (3.4)$$

(see the definitions in the paragraph 1.2).

We have four families of differentials, times and velocities. To avoid too long notations we will use the following agreement:

Ω_α^F may denote *any* of the differentials Ω_k, ω_i, ω_k^i, σ_k^i; $w_s^{F\alpha}$ and T_α^F are the correspondent velocities and times, $Q_{\alpha\beta}^F = V_{\Omega_\alpha^F \Omega_\beta^F}$ where V is the scalar product defined in the paragraph 1.2.

The following statements are important for us.

Lemma 3.1 ([2], [6]). The flows (3.1), (3.3) can be written as

$$\frac{\partial \Omega_1}{\partial T_\alpha^F} = \frac{\partial \Omega_\alpha^F}{\partial X}. \qquad (3.5)$$

Here we always assume

$$\frac{\partial}{\partial T_\alpha^F} = \frac{\partial}{\partial T_\alpha^F}\Big|_{E=\text{const}}.$$

Proof of the Lemma. The differentials in the both sides of (3.5) have the following properties.

1) They are single-valued in Γ.

2) Their integrals by a_j-cycles are equal to 0.

3) They have second-order poles in the branch points $E_1, \ldots, E_{2g+1}$ and no other singularities. All their residues are equal to zero.

4) Let $z = \sqrt{E - E_k}$ be local parameter in the neighbourhood of the branch point E_k. Then the singular part of both differentials is equal to $\Omega_\alpha^F(z)/2z^2$.

Properties 1-4 uniquely determined a differential so the left-hand side is equal to the right-hand one.

Lemma 3.2 ([2], [6]). All the symmetries (3.1), (3.3) are mutually commuting.

Proof of the Lemma. Consider the differentials $\partial \Omega_\alpha^F / \partial T_\beta^F$, $\partial \Omega_\beta^F / \partial T_\alpha^F$. They satisfy the properties 1-3 of the Lemma 3.1 and have the same singularities in the branch point so they are equal

$$\frac{\partial \Omega_\alpha^F}{\partial T_\beta^F} = \frac{\partial \Omega_\beta^F}{\partial T_\alpha^F}.$$

and we have

$$\frac{\partial}{\partial T_\alpha^F} \frac{\partial}{\partial T_\beta^F} \Omega_1^F = \frac{\partial \Omega_\alpha^F}{\partial T_\beta^F} = \frac{\partial \Omega_\beta^F}{\partial T_\alpha^F} = \frac{\partial}{\partial T_\beta^F} \frac{\partial}{\partial T_\alpha^F} \Omega_1^F.$$

Lemma 3.3 ([8], [23], [6]). The scalar products $Q_{\alpha\beta}^F$ satisfy the following relation

$$\frac{\partial Q_{\alpha\beta}^F}{\partial T_\gamma^F} = \frac{\partial Q_{\alpha\gamma}^F}{\partial T_\beta^F}. \qquad (3.6)$$

Thus we can define the function

$$Q_\alpha^F(\vec{T}^F) - \int_{\vec{0}}^{\vec{T}^F} \sum_\beta Q_{\alpha\beta}^F dT_\beta^F. \qquad (3.7)$$

Here the vector $\vec{T}^F$ contains all the times.

Proof of the Lemma. From the formulas (1.21) we see that the coefficients $\partial Q^F_{\alpha\beta}/\partial T^F_\gamma$ can be expressed as integrals of expansion coefficients for the differential $\partial \Omega^F_\beta/\partial T^F_\gamma$ Applying the Lemma 3.2 we finish the proof.

Lemma 3.4 Let us shift the branch points E_s of the surface Γ and the normalization point E_0 via a vector field $l_{2k} = 2(-1)^k E_{k+1} \partial_E$

$$\frac{\partial E_s}{\partial \tau_{2n}} = 2(-1)^n E_s^{n+1}, \quad s = 0, 1, \ldots, 2g+1. \tag{3.8}$$

Then we have

$$\frac{\partial w_s^{2k+1}}{\partial \tau_{2n}} = (2k+1)w_s^{2k+2n+1} - w_s^{2n+1}w_s^{2k+1} + \sum_{m=0}^{n-1}\left[\left(\frac{\partial}{\partial T_{2k+1}} - w_s^{2k+1}\frac{\partial}{\partial X}\right)Q_{2m+1}\right]w_s^{2n-2m-1}. \tag{3.9}$$

$$\frac{\partial w_s^{Hk}}{\partial \tau_{2n}} = -w_s^{2n+1}w_s^{Hk} + \sum_{m=0}^{n-1}\left[\left(\frac{\partial}{\partial T_k^H} - w_s^{Hk}\frac{\partial}{\partial X}\right)Q_{2m+1}\right]w_s^{2n-2m-1}. \tag{3.10}$$

$$\frac{\partial w_s^{a_ik}}{\partial \tau_{2n}} = 2(-1)^n k w_s^{a_ik+n} - w_s^{2n+1}w_s^{a_ik} + \sum_{m=0}^{n-1}\left[\left(\frac{\partial}{\partial T_k^{a_i}} - w_s^{a_ik}\frac{\partial}{\partial X}\right)Q_{2m+1}\right]w_s^{2n-2m-1}. \tag{3.11}$$

$$\frac{\partial w_s^{b_ik}}{\partial \tau_{2n}} = 2(-1)^n k w_s^{b_ik+n} - w_s^{2n+1}w_s^{b_ik} + \sum_{m=0}^{n-1}\left[\left(\frac{\partial}{\partial T_k^{b_i}} - w_s^{b_ik}\frac{\partial}{\partial X}\right)Q_{2m+1}\right]w_s^{2n-2m-1}. \tag{3.12}$$

(The function Q_j was defined in Lemma 3.3).

The Lemma is proved by direct calculations using (1.35) - (1.38) and (3.2), (3.4).

Lemma 3.5. Let us have a pair of flows

$$\frac{\partial E_s}{\partial \tau} = R_s(E_s) + \left(\sum_\alpha f_\alpha(\vec{T})w_s^{F\alpha}\right)\frac{\partial E_s}{\partial X}, \tag{3.13}$$

$$\frac{\partial E_s}{\partial T^F_\gamma} = w_s^{F\gamma}\frac{\partial E_s}{\partial X}. \tag{3.14}$$

Then the flows (3.13) and (3.14) commute if and only if the following compatibility condition holds:

$$\frac{\partial w_s^{F\gamma}}{\partial \tau} = \sum_\alpha\left[\left(\frac{\partial}{\partial T^F_\gamma} - w_s^{F\gamma}\frac{\partial}{\partial X}\right)f_\alpha(\vec{T})\right]w_s^{F\alpha} \tag{3.15}$$

where $\partial w_s^{F\gamma}/\partial \tau$ denotes the variations of velocities via the shift

$$\frac{\partial E_s}{\partial \tau} = R_s(E_s). \tag{3.16}$$

This Lemma is proved by simple direct calculation.

Now we have prepared everything to formulate and prove our main result.

Theorem 3.1. The flows

$$\frac{\partial E_s}{\partial \tau_{2n}} = 2(-1)^n E_s^{n+1} + \left\{\sum_{k\geq 0}(2k+1)T_{2k+1}w_s^{2k+2n+1}\right\}\frac{\partial E_s}{\partial X} +$$

$$+ \left\{2(-1)^n\sum_{k\geq 0}k(T_k^{a_i}w_s^{a_ik+n} + T_k^{b_i}w_s^{b_ik+n}) + \sum_{0\leq k<n}Q_{2k+1}w_s^{2n-2k-1}\right\}\cdot\frac{\partial E_s}{\partial X}. \tag{3.17}$$

commute with the whole Whitham hierarchy (3.1), (3.3). The nonlocal functions Q_j were defined in the Lemma 3.3. All the partial derivatives of Q_j can be expressed via branch points E_s and normalization point E_0 in terms of hyperelliptic integrals (see Lemma 1.1).

Comparing the Lemmas 3.4 and 3.5 we prove this theorem.

ACKNOWLEDGMENTS

The author is very grateful to B.A.Dubrovin, I.M.Krichever and S.P.Tzarev for useful discussions.

REFERENCES

1. G.B.Whitham. Linear and nonlinear waves. N.Y., Wiley, (1974).

2. H.Flaschka, G.Forest, D.W.MkLaughlin. Commun. Pure Appl. Math. $\underline{33}$, no.6 (1980).

3. Peter D.Lax, C.David Levermore. Commun. Pure Appl. Math., $\underline{36}$, pp.253-290, 571-593, 809-830 (1983).

4. B.A.Dubrovin, S.P.Novikov. Sov. Math. Doklady $\underline{27}$, p.665 (1983); Russian Math. Surveys 44:6 (1989), p.35.

5. S.P.Tzarev. Math. USSR Izvestiya, $\underline{54}$, no. 5 (1990).

6. B.A.Dubrovin. Hamiltonian formalism of Whitham-type hierarchies and topological Landau-Ginzburg models. Preprint 1991.

7. S.P.Tzarev. Soviet. Math. Dokl. $\underline{31}$, pp.488-491 (1985).

8. I.M.Krichever. Funct. Anal. Appl. $\underline{11}$ (1988), p.15.

9. F.Calogero, A.Degasperis. Lett. Nuov. Cim. $\underline{22}$, p.420 (1978).

10. V.A.Belinskii, V.E.Zakharov. JETP $\underline{12}$, p.1953 (1978).

11. Maison. Phys. Rev. Lett. $\underline{41}$, p.521 (1978).

12. G.Baruccy, T.Redge J. Math. Phys. $\underline{18}$ No 6, p.1149 (1976).

13. N.H.Ibragimov, A.B.Shabat. Dokl. Akad. Nauk SSSR $\underline{244}$ (1979), p.1.

14. H.H.Chen, Y.C.Lee, J.E.Lin. Physica 9D, $\underline{9}$, No 3, pp.439-445 (1983).

15. F.J.Schwarz. J. Phys. Soc. Japan, $\underline{51}$ No 8, p. 2387 (1982).

16. B.Fuckssteiner. Progr. Thoer. Phys., $\underline{70}$, pp. 1508-1522 (1983); W.Oevel, B.Fuckssteiner Phys. Lett. $\underline{88}$A, p.323 (1982).

17. A.Yu.Orlov, E.I.Schulman. Additional Symmetries of the Integrable Systems and Conformal Algebra Representations. Preprint IA and E, No 217, Novosibirsk (1984); A.Yu.Orlov, E.I.Schulman. Theor. Math. Phys., $\underline{64}$, pp. 323-327 (1985) (Russian); A.Yu.Orlov, E.I.Schulman. Additional Symmetries for 2-D Integrable Systems. Preprint IA and E, No 277, Novosibirsk (1985); A.Yu.Orlov, E.I.Schulman. Lett. Math. Phys. $\underline{12}$, pp. 171-179 (1986). A.Yu.Orlov in Proceedings Int. Workshop "Plasma theory and ..." in Kiev 1987, World Scientific (1988) p.116-134.

18. A.S.Fokas, P.M.Santini. Commun. Math. Phys. $\underline{116}$, p.449 (1088).

19. I.M.Krichever, S.P.Novikov. Funct. Anal. i Prilog. $\underline{21}$, No 2, pp. 46-63 (1987); $\underline{21}$, No 4, pp. 47-61 (1987); $\underline{23}$, No 1, pp. 41-56 (1989).

20. P.G.Grinevich, A.Yu.Orlov. Virasoro action on Riemann surfaces, Grassmannians, det $\bar{\partial}_j$ and Segal-Vilson τ function. in "Problmes of modern quantum field theory", ed. A.A.Belavin, A.U.Klimyk, A.B.Zamolodchikov, Springer 1989.

21. P.G.Grinevich, A.Yu.Orlov. Funct. Anal. Appl. $\underline{24}$ no. 1, (1990); Higher symmetries of KP equation and the Virasoro action on Riemann surfaces. in "Nonlinear evolution equations and dynamical systems", ed. S.Carillo, O.Ragnisco, Springer 1990.

22. A.Gerasimov, A.Marshakov, A.Mironov, A.Morozov, A.Orlov. Nuclear Phys. B 357, pp. 565-618 (1991).

23. I.M.Krichever. "The dispersionless Lax equations and topological minimal models". Preprint 1991 Torino; Lecture at the Sakharov memory congress (1991).

24. G.Springer. Introduction to Riemann surfaces. Addison - Wesley Publishing Company, Inc. Reading, Massachusetts, USA (1957).

25. M.Shiffer, D.K.Spencer. Functionals of finite Riemann surfaces. Princeton, New Jersey (1953).

26. D.Mumford. Tata lectures on Theta. Birkhäuser, Boston - Basel - Stuttgart (1983).

27. R.Dijkgraaf, H.Verlinder, E.Verlinder. "Topological strings in $d < 1$". Princeton preprint PUPT - 1024, IASSNS-HEP - 90/71.

28. V.E.Zakharov, S.V. Manakov, S.P. Novikov, L.P. Pitaevsky. Soliton theory. Plenum, New York, 1984.

29. I.M.Krichever. Docl. Akad. Nauk. SSSR, 227, No 2, pp. 291-294 (1977).

30. I.M.Krichever. Uspekhy. Mat. Nauk. 44, No 2, pp. 121-184 (1989).

31. I.M.Gel'fand, L.A.Dikii. Funct. Anal. Pril. 10, No 4, pp. 13-29 (1976).

32. H.McKean, E.Trubowitz. Commun. Pure. Appl. Math. 29, pp. 143-226 (1976).

33. B.A.Dubrovin. Uspekhi. Math. Nauk. 36, No 2, pp. 11-40 (1981).

34. N.Kawamoto, Yu.Namikava, A.Tsuchiya, Ya.Yamada. Commun. Math. Phys. 116, No. 2, pp.247-308 (1988).

BREAKING PROBLEM IN DISPERSIVE HYDRODYNAMICS

Gennady A. El', Alexander V. Gurevich, and
Alexander L. Krylov

P.N. Lebedev Physical Institute
Academy of Sciences of the USSR
Moscow

Whitham modulation equations[1] are investigated very actively last time.
This is caused by theoretical aspects of general theory of hydrodynamic type
systems[2], as well as by the importance of physical applications, most interesting of which are nondissipative shock waves (NSW). It is well known that
wave breaking of the simple Riemann wave[3] in dispersive hydrodynamics leads
to the appearance of continuously expanding region filled with undamped
small-scale nonlinear oscillations. This is NSW. First the shock wave problem
in dispersive hydrodynamics - Gurevich-Pitaevsky (GP) problem - was considered in Ref.4, where analytic solution of the problem of an initial discontinuity decay in KdV hydrodynamics was obtained and numerical analysis of the
simple wave breaking was done. Then this topic was developed in Refs.5-10.
The important property of modulation sets for KdV, NLS and SG-equations which
was effectively used in these works is the possibility to represent them in
characteristic (Riemann) form.[11-13] Partially this fact allowed to introduce
important concept of quasi-simple wave and receive some exact NSW-solu-
tions.[8,9]

In the present work, with the help of generalized hodograph method (GHM)
proposed by Tsarev in Ref.14, the general solution of wave breaking problem
for KdV equation is constructed in an explicit form, the same is done for defocusing NLS equation. We consider KdV equation in form:

$$\partial_t u + u \partial_x u + \partial^3_{xxx} u = 0, \qquad (1)$$

and defocusing NLS equation

$$2i \partial_t \psi + \partial^2_{xx} \psi - 2|\psi|^2 \psi = 0. \qquad (2)$$

There are some aspects of modulations for SG-equation

$$\partial_{tt}^2 \varphi - \partial_{xx}^2 \varphi + \sin\varphi = 0 \qquad\qquad (3)$$

considered in Appendix .

A simple method which allowed to transform Tsarev equations for KdV into scalar form in Ref.15 is used to obtain general system of linear second order equations containing no elliptic integrals. For KdV and NLS this system reduces to Euler-Poisson type equation set and was recently obtained in Refs.16, 17 by a direct calculation. The general solution of this system was obtained in the beginning of the century by Eisenhart.[18] GP-problem, however, demands special investigations to use this solution.

The most important property of Tsarev equations and scalar system is their "two-dimensional" nature: any two equations (i,j) can be resolved independently of the others for r_k = const (k ≠ i,j). Due to this fact it's possible to restrict oneself in solving two-dimensional Goursat problem, i.e., essentially, to use the common hodograph transform $(x,t) \rightarrow (r_i, r_j)$ to construct three-dimensional solution. This solution depends on two arbitrary functions describing monotonic initial data of the wave breaking problem and has a singularity (discontinuity of the second derivative), if these functions are different. For KdV, corresponding equations and some important solutions were obtained in Ref.15. Breaking of analytic profile, and quasisimple waves characterized by one initial function are also considered. For "analytic breaking", the solution becomes a polynomial.

GP-PROBLEM AND GENERALIZED HODOGRAPH METHOD

A simple Riemann wave is described by the equation

$$\partial_t r + V(r)\partial_x r = 0, \qquad\qquad (4)$$

which has a solution

$$x - V(r)t = W(r) \qquad\qquad (5)$$

where $W(r)$ is the inverse of the initial profile $r = r_0(x)$.

KdV-breaking

Let the start of breaking takes place for t=0, x=0, r=0 (Fig.1a) and

$$r(x,0) = \begin{cases} r_0^+(x) \le 0 & \text{for } x \ge 0, \\ r_0^-(x) > 0 & \text{for } x < 0. \end{cases} \qquad\qquad (6)$$

The inverse function is (Fig.1b)

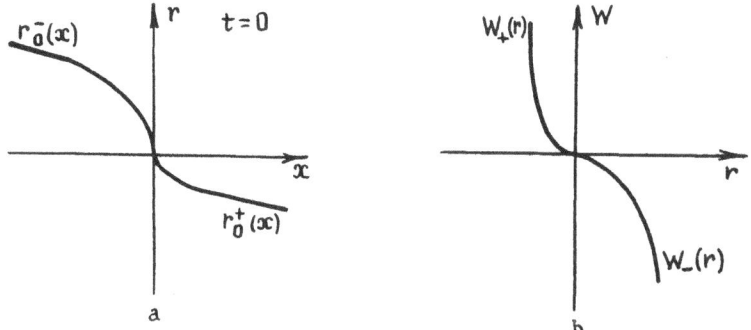

Fig.1. Initial GP-problem data (a) and inverse function (b)

$$W(r) = \begin{cases} W_+(r) & \text{for } r \le 0 \\ W_-(r) & \text{for } r > 0 \end{cases} \qquad (7)$$

We consider here monotonic initial data with the only breaking point. After the wave-breaking of (6), (7) in KdV dispersive hydrodynamics ($V(r)=r$), the solution in the NSW region is described by three functions $r_3(x,t) \ge r_2(x,t) \ge r_1(x,t)$ (Fig.2) that satisfy Whitham modulation system (one-phase averaging)

$$\partial_t r_i + V_i(r)\partial_x r_i = 0, \quad i=1,2,3 \qquad (8)$$

where the group velocities $V_i(r)$ can be represented in a "potential" form[15]

$$V_i = \frac{\partial_i \omega}{\partial_i k} = U - \lambda \frac{\partial_i U}{\partial_i \lambda} \qquad (9)$$

Here

$$U = \frac{\omega}{k} = \frac{1}{3}\sum_{j=1}^{3} r_j , \qquad (10a)$$

$$\lambda = \frac{2\pi}{k} = 6^{1/2} \int_{r_1}^{r_2} d\tau \; [\prod_{j=1}^{3} (\tau - r_j)]^{-1/2} \qquad (10b)$$

- phase velocity and the wavelength.

The representation (9) takes place in a general case and follows from the wave number k conservation law

$$\partial_t k + \partial_x \omega = 0, \qquad (11)$$

which is the consequence of (8).[19]

The solution $r_j(x,t)$ ($j=1,2,3$) describing NSW evolution satisfies the conditions of junction with "external" solution $r(x,t)$ of (1) on unknown boundaries of the NSW: $x=x^-(t)$ - trailing front, and $x=x^+(t)$ - leading front (Fig.2):

$$r_3(x^-,t) = r_-(x^-,t) \quad \text{for } r_2 = r_1 ,$$
$$r_1(x^+,t) = r_+(x^+,t) \quad \text{for } r_2 = r_3 \tag{12}$$

- GP-conditions.

For investigations of such problems the hodograph transform proved its effectiveness. First it was used by Gurevich, Krylov, and Masur in Ref.9 for studying the quasisimple wave with changing r_2, r_3 and constant r_1 (wave propagation along undisturbed medium). Three (and more) dimensional GP-problems (8), (12) can be investigated by the generalized hodograph method (GHM)[14,2] in following brief version.

Vector generalization of the Riemann wave (5) for the system (8) is

$$x - V_i(r)t = W_i(r), \quad i = 1,2...N \tag{13}$$

(N is an order of the system (8)), where $W_i(r)$ cannot be taken arbitrary, but must satisfy the compatibility conditions (Tsarev equations):

$$\frac{\partial_i W_j}{W_i - W_j} = \frac{\partial_i V_j}{V_i - V_j}, \quad i \neq j . \tag{14}$$

The general solution of (14) determines all hydrodynamic symmetries

$$\partial_\tau r_i + W_i(r)\partial_x r_i = 0, \tag{15}$$

commuting with (8), that is $\partial^2_{\tau t} r_i = \partial^2_{t\tau} r_i$.

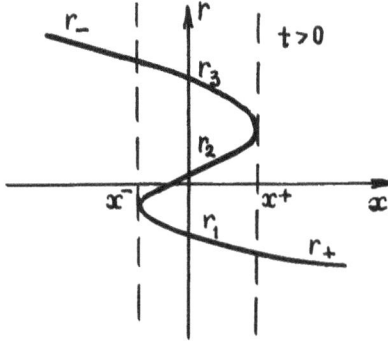

Fig.2. Riemann invariants as functions of x in NSW.

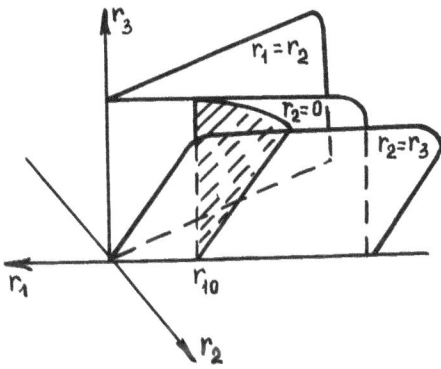

Fig.3. The region of determination in r-space.

For the KdV-system, (14) is determined in the region $r_1 \leq r_2 \leq r_3$, $r_1 \leq 0$, $r_3 \geq 0$, between $r_2 = r_1$ – plane (trailing front), and $r_2 = r_3$ – plane (leading front) (Fig.3). GP-conditions in r-space take the extremely simple form

$$W_1(r_1, 0, 0) = W_+(r_1) ,$$
$$W_3(0, 0, r_3) = W_-(r_3) .$$
(16)

They provide joining the solutions of (14) with the corresponding branches of W(r) from (7) along the fronts: it's easy to obtain from (14) and the explicit expressions for $V_i(r)$ that

$$(\partial_2 + \partial_3)W_1\Big|_{r_2=r_3} = 0, \quad (\partial_1 + \partial_2)W_3\Big|_{r_2=r_1} = 0$$
(17)

for W_1 and W_3 limited; that is values W_1 carry out invariantly along the trailing front $r_2 = r_3$ from r_1-axis, and corresponding conclusions take place for W_3 and the leading front.

So instead of nonlinear problem (8), (12) with boundary conditions on unknown boundaries, we have in r-space the linear system (14) with linear boundary conditions (16).

The most important property of (14) is its two-dimensional structure: for any i,j (i≠j) for r_k=const (k≠i,j) it can be considered (and really explicitly resolved) – this will be effectively used later.

NLS-breaking

The NLS equation (2) by the change of variables $\psi = \rho^{1/2}\exp(i\varphi)$, $\partial_x\varphi = \upsilon$ can be written in the hydrodynamic form[8]

93

$$\partial_t \rho + \partial_x(\rho v) = 0,$$

$$\partial_t v + v\partial_x v + \partial_x \rho - \frac{1}{4}\partial_x\left(\frac{\partial^2_{xx}\rho}{\rho} - \frac{(\partial_x\rho)^2}{2\rho^2}\right) = 0. \qquad (18)$$

Whitham equations for (18) have the form (8) for $r_j; j=0,1,2,3$, $r_3 \geq r_2 \geq r_1 > r_0$ (Fig.4). The characteristic velocities as well as before are given by (9), where

$$U = \frac{1}{4}\sum_{j=0}^{3} r_j, \qquad (19a)$$

$$\lambda(r) = 2^{1/2}\int_{r_2}^{r_3} \left[-\prod_{j=0}^{3}(\tau - r_j)\right]^{-1/2} d\tau \qquad (19b)$$

The role of external equations is played now by Euler hydrodynamics equations with $\gamma = 2$. The simple wave is given by $r_0 =$const. Changing dependent variables $r'_i = \frac{3}{4}(r_i - r_0)$ (i=1,2,3) and passing to moving reference system $x' = x - r_0 t$, we receive in x',t-coordinates three-dimensional (r'_1, r'_2, r'_3) GP-problem, which is analogous to one considered in previous section.

SCALAR POTENTIAL AND LINEAR EQUATIONS WITH NO ELLIPTIC INTEGRALS

There is considerable difficulty in direct researching of the equations (14) for KdV or NLS, as the right parts of them represent a rather complicated combinations of the elliptic integrals (see (9), (10), (19)). There exists, however, the procedure that allows to transform the system (14) into a linear system of the second order equations containing no elliptic integrals.

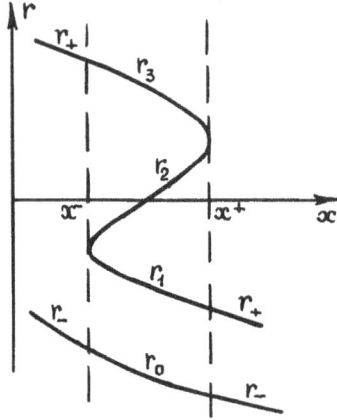

Fig.4. Riemann invariants of NLS averaged system as functions of x for t>0.

First note one important consequence of (14): using the potential re-presentation (9) it's easy to obtain

$$\partial_j(W_i \partial_i k) = \partial_i(W_j \partial_j k),$$ (20)

i.e., unknown functions W_i also permits the potential representation

$$W_i = \frac{\partial_i(kf)}{\partial_i k} = f - \lambda \frac{\partial_i f}{\partial_i \lambda}$$ (21)

where f - some function, which has the meaning of generalized phase velocity. It follows from the wave number conservation law for the flow commuting with (8)

$$\partial_\tau k + \partial_x(kf) = 0.$$ (22)

Some of (22) have the natural meaning of the wave number conservation laws for higher equations of the corresponding hierarchy. Introducing theRiemann varia-bles into (22) we receive (21).

Formulas (11), (12) allows us to scalarize the compatibility equations (14): we have one unknown function f instead of N functions W_j ($j=1,2,\ldots N$). This function satisfies the second order equation system:

$$\frac{\partial_i(\partial_j f/\partial_j U)}{\dfrac{\partial_i f}{\partial_i U} - \dfrac{\partial_j f}{\partial_j U}} = \frac{\partial_i(\partial_j \lambda/\partial_j U)}{\dfrac{\partial_i \lambda}{\partial_i U} - \dfrac{\partial_j \lambda}{\partial_j U}}, \qquad i \neq j.$$ (23)

For the KdV and NLS equations due to (10a), (19a), the system (23) is reduced to more simple one[15]

$$\frac{\partial^2_{ij} f}{\partial_i f - \partial_j f} = \frac{\partial^2_{ij} \lambda}{\partial_i \lambda - \partial_j \lambda}.$$ (24)

Remark. This reduction generally takes place if $U(r)$ satisfies the same sys-tem:

$$\frac{\partial^2_{ij} U}{\partial_i U - \partial_j U} = \frac{\partial^2_{ij} \lambda}{\partial_i \lambda - \partial_j \lambda}$$

From the integral representation (10b), (19b) for $\lambda(r)$, it easily follows for the KdV and NLS that

$$\frac{\partial^2_{ij}\lambda}{\partial_i\lambda - \partial_j\lambda} = \frac{1}{2(r_i-r_j)} \ , \tag{25}$$

and so unknown function $f(r)$ in both cases satisfies the same system:

$$E_{ij}f = 0, \quad i,j = 1, \ \ldots \ N, \quad i \neq j,$$

where $\hspace{10cm}$ (26)

$$E_{ij} = \partial^2_{ij} - (\partial_i - \partial_j)/2(r_i - r_j)$$

- Euler-Poisson operator.[20]

It's important to emphasize that the system (26) is not universal: for example, SG-modulation equations reduce to the different form (also containing no elliptic integrals) - see Appendix.

Further on we'll consider the wave breaking problem for the KdV hydrodynamics boundary conditions for (26), taking into account (16), have the form

$$f(0,0,r_3) = f_-(r_3) = \frac{1}{2} r_3^{-1/2} \int_0^{r_3} x^{-1/2} W_-(x)dx < 0,$$

$$\hspace{8cm} (27)$$

$$f(r_1,0,0) = f_+(r_1) = \frac{1}{2} (-r_1)^{-1/2} \int_0^{-r_1} x^{-1/2} W_+(-x)dx > 0 \ .$$

GENERAL SOLUTION OF GP-PROBLEM

The general solution of three-dimensional system (26) was found in 1918 by Eisenhart[18] and has the form

$$f = \int_0^{r_1} \frac{\varphi_1(\mu)}{R^{1/2}(\mu)} d\mu + \int_0^{r_2} \frac{\varphi_2(\mu)}{R^{1/2}(\mu)} d\mu + \int_0^{r_3} \frac{\varphi_3(\mu)}{R^{1/2}(\mu)} d\mu \tag{28}$$

where $\varphi_i(x)$ - arbitrary functions, $R(\mu) = \prod_{j=1}^{3} (\mu-r_j)$. From the integral representation, we easily obtain the solution of GP-problem for the antisymmetric initial data described by the only function $W(x)$; from (27) we have in this case

$$f_-(z) = -f_-(-z) = \frac{1}{2} z^{-1/2} \int_0^z x^{-1/2} W(x)dx \tag{29}$$

Then the solution is symmetric

$$f(r_1,r_2,r_3) = \int_0^{r_3} \frac{\phi(\tau)d\tau}{[-\prod_{i=1}^{3}(\tau-r_i)]^{1/2}} + \int_{r_1}^{0} \frac{\phi(-\tau)d\tau}{[\prod_{i=1}^{3}(\tau-r_i)]^{1/2}} \qquad (30)$$

where

$$\phi(z) = \frac{1}{2\pi} \int_0^z \frac{W(x)}{(z-x)^{1/2}}dx \ . \qquad (31)$$

In general case, however, we have two different hydrodynamic regimes $r_+(x,t)$ and $r_-(x,t)$ joined by the Whitham region (NSW) and now we'll construct such solution. First we'll find the solution $f(r_1,0,r_3)$ of the Goursat problem

$$E_{13}f = 0; \quad f(0,0,r_3) = f_-(r_3), \quad f(r_1,0,0) = f_+(r_1), \qquad (32)$$

where $f_\pm(z)$ are given by (27).
The solution of (32) is

$$f(r_1,0,r_3)=\int_0^{r_3} \frac{\phi_-(\tau)d\tau}{[\tau(r_3-\tau)(\tau-r_1)]^{1/2}} + \int_{r_1}^{0} \frac{\phi_+(-\tau)d\tau}{[-\tau(r_3-\tau)(\tau-r_1)]^{1/2}} \qquad (33)$$

$\equiv G(r_1,r_3)$, where

$$\phi_\pm(z) = \frac{1}{2\pi} \int_0^z \frac{W_\pm(x)}{(z-x)^{1/2}} dx, \ z>0; \quad W_+(x) = -W_+(-x) \qquad (34)$$

To find dependence of the solution on r_2 we consider the function $G(r_1, r_3)$ as the data for the Goursat problem in planes r_1=const ($r_2 \geq 0$), and r_3=const ($r_2 \leq 0$). For example in every plane r_1=r_{10}=const (Fig.5a and

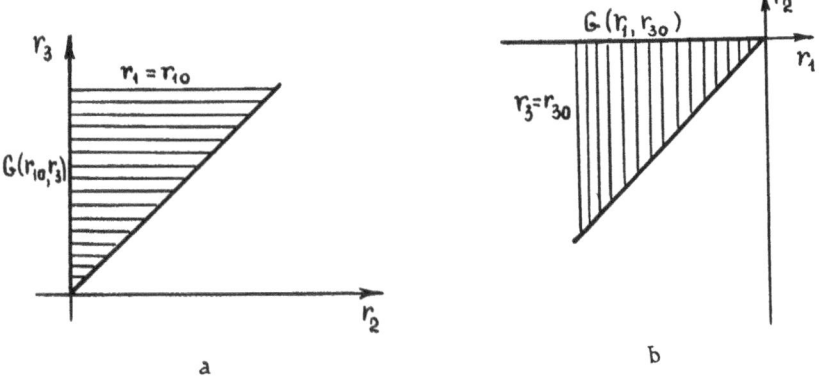

Fig.5. Integration domains in planes a – r_1=const, b – r_3=const.

the shaded domain on Fig.3) we have the problem ($r \geq 0$)

$$E_{23}f(r_{10},r_2,r_3)=0, \quad f(r_{10},0,r_3)=G(r_{10},r_3),$$

$$f(r_{10},r,r) \text{ is finite} \tag{35}$$

The only solution of the problem is

$$f(r_{10},r_2,r_3) = \int_{r_2}^{r_3} \frac{\phi_-(\tau)d\tau}{[(\tau-r_{10})(\tau-r_2)(r_3-\tau)]^{1/2}} +$$

$$+ \int_{r_{10}}^{0} \frac{\phi_+(-\tau)d\tau}{[(\tau-r_{10})(r_2-\tau)(r_3-\tau)]^{1/2}} . \tag{36}$$

The solution for $r_2 \leq 0$ (Fig.5b) can be found by the similar way. Complete solution of the problem (26), (27) is ultimately ($r_1 < 0$, $r_3 > 0$)

$$f(r_1,r_2,r_3) = \begin{cases} \displaystyle\int_{r_2}^{r_3} \frac{\phi_-(\tau)d\tau}{[-\prod\limits_{i=1}^{3}(\tau-r_i)]^{1/2}} + \int_{r_1}^{0} \frac{\phi_+(-\tau)d\tau}{[\prod\limits_{i=1}^{3}(\tau-r_i)]^{1/2}} & \text{for } r_2 \geq 0 \quad (37a) \\[4ex] \displaystyle\int_{r_1}^{r_2} \frac{\phi_+(-\tau)d\tau}{[\prod\limits_{i=1}^{3}(\tau-r_i)]^{1/2}} + \int_{0}^{r_3} \frac{\phi_-(\tau)d\tau}{[-\prod\limits_{i=1}^{3}(\tau-r_i)]^{1/2}} & \text{for } r_2 \leq 0. \quad (37b) \end{cases}$$

It is easy to show that $\partial_2 f|_{r_2=0}$ is continuous (this requirement follows from the smoothness of $r_2(x,t)$ and (13), (21)). Nevertheless the plane $r_2=0$ is singular as the solution is described by the different formulas for $r_2 > 0$ and for $r_2 < 0$. Such weak discontinuity takes place always if the initial data is not analytic in the point of wave breaking . If $W(r)$ is an analytic function, (37a,b) turn into (30).

The curves $x^{\pm}(t)$ (the boundaries of the NSW in a physical plane) are the multiple characteristics of (8) and can be found from the system consisting of (13) on the boundaries and equations $dx^{\pm}/dt = v^{\pm}$ ($v^{\pm}$ are multiple characteristic velocities on $\pm$ fronts).

We consider now important special case – the wave breaking of the anti-symmetric initial profile $r_0(x) = -|x|^{1/q}\text{sgn } x$, $q > 1$, arbitrary. The sought solutions then are self-similar: $r_i = t^{\gamma}l_i(x/t^{\gamma+1})$, $\gamma = 1/(1-q)$. The solution in r-space can be found with the help of (37), (34), but the more simple way to obtain it is to use directly the uniform solutions of (26) in the form

$$f = r_i^q \, \phi(-q, \; 1/2; \; 1/2-q; \; r_j/r_i),$$

where $\phi(a,b; \; c; \; z)$ is a solution of the according hypergeometric equation (see for details Ref.15).

For q = M-integer, odd (analytic data) the solution is polynomial:

$$f = P_M(r) = - \frac{2^M M!}{(2M-1)!!(2M+1)} \sum_{m+n+l=M} \frac{(\frac{1}{2})_m (\frac{1}{2})_n (\frac{1}{2})_l}{m!n!l!} \; r_1^m \, r_2^n \, r_3^l \; , \quad (38)$$

$$(a)_n = \Gamma(a+n)/\Gamma(a)$$

If q = N – integer, even (nonanalytic data), (38) evidently does not describe wave breaking problem. The solution then takes the form

$$f(r) = \begin{cases} P_N(r) - A_N \displaystyle\int_{r_2}^{r_3} \frac{\tau^{N+1/2} d\tau}{[(\tau-r_1)(\tau-r_2)(r_3-\tau)]^{1/2}} & \text{for } r_2>0 \\[4ex] -P_N(r) + A_N \displaystyle\int_{r_1}^{r_2} \frac{(-\tau)^{N+1/2} d\tau}{[(\tau-r_1)(\tau-r_2)(\tau-r_3)]^{1/2}} & \text{for } r_2<0 \end{cases} \qquad (39)$$

where

$$A_N = \frac{2}{\pi} \sum_{k=1}^{N} \frac{(-1)^k}{2N-2k+1} \; \frac{N!}{k!(N-k)!} \; ,$$

and it has the weak discontinuity for $r_2=0$, while the first derivative is continuous.

QUASISIMPLE WAVE, NONMONOTONIC INITIAL DATA AND SOLITON WAVE

A concept of quasisimple wave was introduced by Gurevich and Krylov in Ref.8, and it was used in Ref.9 for description of the breaking of Riemann wave propagating along the undisturbed medium, that accords to $r_o^+(x)=0$. Then NSW is described by two Whitham equations for r_2, r_3 (and $r_1 \equiv 0$). This reduced Whitham system can be studied by the common hodograph transform $(x,t) \Rightarrow (r_2, r_3)$ which is the special case of GHM. In Ref.9 the quasisimple waves were studied for monotonic as well as for localized initial perturbations (Fig.6). In this last case the hodograph transform becomes two-sheeted. All these solutions can be obtained now in an explicit form.

Monotonic initial data

Putting $r_1 \equiv 0$ in the general solution (37) we get the integral representation for quasisimple wave

$$f(0,r_2,r_3) = \int_{r_2}^{r_3} \frac{\phi(\tau)\,d\tau}{[\tau(r_3-\tau)(r_2-\tau)]^{1/2}} \tag{40}$$

Here $\phi(\tau)$ is given by (31) as before.

Uniform solution describing wave breaking of the profile $r_0^-(x) = (-x)^{1/q}$, $q>1$ is

$$f(0,r_2,r_3) = - \frac{r_3^q\, \pi^{1/2}\Gamma(1+q)}{2\Gamma(q+3/2)} F(-q,\ 1/2;\ 1;\ 1-r_2/r_3), \tag{41}$$

where $F(a,b;\ c;\ z)$ – Gauss hypergeometric function. For $q = M$–integer, the hypergeometric function becomes a polynomial

$$f(0,r_2,r_3) = - \frac{2^M M!}{(2M-1)!!(2M+1)} \sum_{n+l=M} \frac{(\tfrac{1}{2})_n (\tfrac{1}{2})_l}{n!\,l!} r_2^n\, r_3^l, \tag{42}$$

Nonmonotonic (localized) initial data

The aspect that is new in comparison with the monotonic case is that the function $W(r)$ is two-valued, and as a consequence, the hodograph transform $(x,t) \rightarrow (r_2,r_3)$ is two-sheeted. The independent variables r_2, r_3 vary on two sheets, $0 \le r_2 \le r_3$, $0 \le r_3 \le 1$ (we take max $r_0(x) = 1$). Sheet I corresponds to the "nose" part of the initial data (the part to the right of the maximum), while sheet II corresponds to the "tail" region. The characteristics on both sheets have the same equations $r_2 = $ const, $r_3 = $ const,

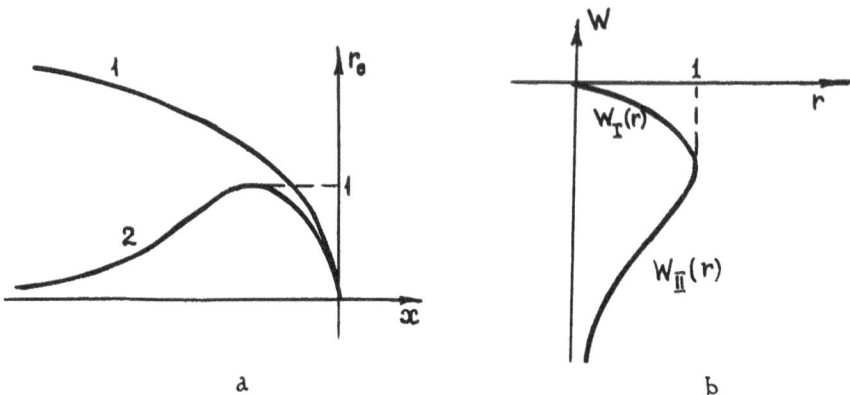

a   b

Fig.6.a Initial perturbation: 1 – monotonically increasing; 2 – nonmonotonic (localized); b Inverse function for localized perturbation.

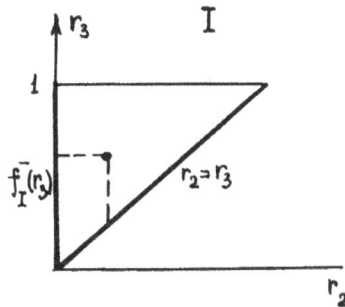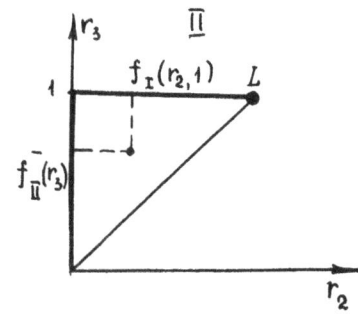

Fig.7. Regions of dependence in the problems on the sheets I and II for the system (26). The thick lines show those part of the boundary from which the data "carried over".

but the semi-characteristics, specifying the basins of influence, are different (Fig.7).

The plan for obtaining the desired solution is natural. First we find the solution on sheet I and then we find the solution on sheet II, using continuous splicing with the solution on sheet I at $r_3=1$. Because the system (26) is hyperbolic and the splicing line is a characteristic, this procedure is correct.[20]

The solution on sheet I is described by the integral (40), which is regular on the diagonal $r_2 = r_3$. On the sheet II the solution is a sum of the solution (40) and the integral

$$f^*(r_2, r_3) = \int_0^{r_2} \frac{\phi^*(\tau)d\tau}{[\tau(r_2-\tau)(r_3-\tau)]^{1/2}} \ , \tag{43}$$

where $\phi^*(\tau)$ is determined from a condition

$$f_I(r_2,1) = f_{II}(r_2,1) \tag{44}$$

Integral (44) vanishes on the trailing front $r_2=0$; it is not regular on the diagonal $r_2 = r_3$ (the problem on sheet II is the common Goursat problem – see Fig.7).

The corresponding picture in (x,t)-plane is depicted in Fig.9. Regions I and II correspond to sheets I and II in (r_2,r_3) plane, the curve MN corresponds to the line of splicing of the sheets and on it $r_3 = 1$. At the time t_N the tail of the NSW reaches the maximum $r = 1$ of the "outer" solution (the point M). At time t_N the head solution achieves its greatest amplitude. After this it does not grow, and it moves with a constant velocity. The distance between the solitons near the leading front increases here linearly with

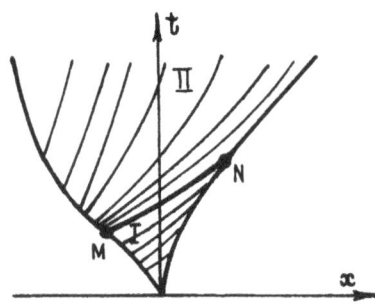

Fig.8 Region of the NSW in the x,t plane. Thin lines are $r_3(x,t)$ = const.

time.The correspondent stright-line segment of the boundary of the NSW region in the (x,t) plane is carried over by the hodographtransformation to the point r_2=1, r_3=1 on sheet II (point L onFig.8). The other points of the right boundary of the sheet are not reached at finite x and t - these are solitons of smaller amplitude that are realized only asymptotically. This asymptotic is called in Ref.9 the soliton wave.

The soliton wave is a solution of the modulation equations for r_2=r_3=r. In the considered case r_1=0, the soliton wave is described by one equation

$$\partial_t r + \frac{2}{3} r \partial_x r = 0,\qquad (45)$$

having the solution

$$x = \frac{2}{3} rt + x_0(r) \ .\qquad (46)$$

Comparison of the solution (46) with the corresponding limit of the hodograph solution (13) gives us

$$x_0(r) = W_2(0,r,r) \ .\qquad (47)$$

A concept of the soliton wave may be introduced in three-dimensional (in r-space) case too. Solitons then move along the slowly varying hydrodynamic flow. The system for KdV soliton wave now consists of two equations

$$\partial_t r + (\frac{1}{3} r_1 + \frac{2}{3} r)\partial_x r = 0,$$

$$\partial_t r_1 + r_1 \partial_x r_1 = 0.\qquad (48)$$

It means geometrically the possibility to consider hodograph equations in the leading front plane r_2=r_3=r. It should be noted that this two-dimensional reduction takes place in integrable as well as in nonintegrable systems of dispersive hydrodynamics.[21]

APPENDIX

SG-hodograph equations with no elliptic integrals

SG modulation equations have the Riemann form (8) (N = 2).[12] We'll use the potential representation (9) for the characteristic velocities, in which

$$U = \frac{16(r_1 r_2)^{1/2} + 1}{16(r_1 r_2)^{1/2} - 1} \quad , \tag{A.1}$$

$$\lambda = \frac{16(r_1 r_2)^{1/2}}{16(r_1 r_2)^{1/2} - 1} \int_{r_2}^{0} \frac{d\tau}{(-\tau(\tau-r_1)(\tau-r_2))^{1/2}} \quad , \tag{A.2}$$

$$r_1 < r_2 < 0.$$

Substituting (A.1), (A.2) into the general system (23) we obtain after some manipulations the scalar second order equation containing no elliptic integrals

$$\partial_{12}^2 f = \frac{1}{2[(1-16(r_1 r_2)^{1/2}]} \left(\frac{\partial_1 f}{r_2} + \frac{\partial_2 f}{r_1} \right) + \frac{1}{2(r_1 - r_2)} (\partial_1 f - \partial_2 f). \tag{A.3}$$

This equation may be resolved separating the variables if introduce

$$z = \frac{1}{1-16(r_1 r_2)^{1/2}} \quad , \quad y = \frac{r_1 + r_2}{2(r_1 r_2)^{1/2}}$$

instead of r_1, r_2 (cf. Ref.10).

ACKNOWLEDGEMENTS

The authors thank S.P. Novikov, N.G. Mazur, and V.V. Khodorovsky for useful discussions.

REFERENCES

1. G.B.Whitham, Non-linear dispersive waves, Proc. Roy. Soc. A283, 238(1965).
2. B.A.Dubrovin, and S.P.Novikov, Hydrodynamics of weakly deformed solution lattices, Russian Math. Surveys 44: 6, 29 (1989) [in Russian].

3. L.D.Landau and L.M.Lifshitz, "Hydrodynamics" [in Russian], Nauka (1988).

4. A.V.Gurevich and L.P.Pitaevsky, Nonstationary structure of nondissipative shock wave, Sov. Phys. JETP 65, 291 (1974).

5. P.D.Lax and C.D.Levermore, The small dispersion limit for the Korteweg – de Vries Equation I, II, and III, C.P.A.M., 36, 253, 571, 809 (1983).

6. V.V.Avilov, S.P.Novikov, Evolution of the Whitham's zone in KdV theory, Soviet Math. Dokl. 32, 366 (1987).

7. G.V. Potemin, Algebro-geometric construction of self-similar solutions of the Whitham equations, Russian Math. Surveys 43: 5, 252 (1988).

8. A.V.Gurevich and A.L.Krylov, Dissipationless shock waves in media with positive dispersion, Sov. Phys. JETP 65: 5, 944 (1987).

9. A.V.Gurevich, A.L.Krylov and N.G.Mazur, Quasisimple waves in Korteweg – de Vries hydrodynamics, Sov. Phys. JETP 68: 5, 966 (1989).

10. A.V.Gurevich, N.I.Gershenzon, A.L.Krylov, and N.G.Mazur, About the solutions of the SG-equation by the modulated waves method, Sov. Phys. Dokl. 305: 3, 593 (1989) [in Russian].

11. H.Flashka, M.G.Forest, D.W. Mc Laughlin, Multiphase averaging and the inverse spectral solution of the Korteweg – de Vries equation, Comm. Pure Appl. Math. 33: 6, 739 (1980).

12. M.G.Forest, D.W. Mc Laughlin, Modulation of Sinh – and Sine-Gordon wavetrains, Siam J. of Appl. Math. 68: 1, 11 (1983).

13. M.V.Pavlov, Nonlinear Schrödinger equation and Bogolubov-Whitham method of averaging, Theor. Math. Phys 71: 3, 351 (1987) [in Russian].

14. S.P.Tsarev, Poisson brackets and one-dimensional Hamiltonian systems of hydrodynamic type, Soviet Math. Dokl. 31, 488 (1985).

15. A.V.Gurevich, A.L.Krylov, and G.A.El', Riemann wave breaking in dispersive hydrodynamics, JETP Lett. 54: 2, 102 (1991).

16. V.R.Kudashev, and S.E.Sharapov, Inheritage of KdV symmetries under the Whitham averaging and hydrodynamic type symmetries of the Whitham equations, Theor. Math. Phys. 87: 1, 40 (1991) [in Russian].

17. V.R.Kudashev, S.E.Sharapov, Hydrodynamic symmetries for Whitham equations for nonlinear Schrödinger equation, Preprint of I.V.Kurchatov Institute of Atomic Energy IAE-5263/6 (1990).

18. L.P.Eisenhart, Triply conjugate systems with equal point invariants, Annals of Math. 120: 4, 262 (1918).

19. G.B.Whitham, "Linear and Nonlinear Waves", Wiley (1974).

20. R.Courant and D.Hilbert, "Methods of mathematical physics", Vol.II, New York (1962).

21. A.V.Gurevich, A.L.Krylov, and G.A.El', Nonlinear modulated waves in dispersive hydrodynamics, Sov. Phys. JETP 71: 5 (1990).

WHITHAM DEFORMATIONS

OF TWO - DIMENSIONAL LIOVILLE TORI

V.Yu. Novokshenov

Institute of Mathematics
Urals Branch of the Acad. Sci. of the USSR
USSR, 450000 Ufa, Chernyshevski str. 112

0. INTRODUCTION

The finite-band integration method for Hamiltonian systems arising in classical mechanics deals with the inverse spectral problem for the Sturm-Lioville operator

$$L = \frac{d^2}{dt^2} + u(t) \tag{1}$$

the spectrum of which consists of finite number of allowed bands (see [1,2]). We restrict ourselves to the case of two bands since most of interesting examples (Kowalewski top, Neumann system, etc.) are covered by it. The Hamiltonian then takes the form

$$H = p_1 p_2 + V(q_1, q_2), \tag{2}$$

where $q_1 = u$, $q_2 = u'' - \frac{5}{2}u^2$, $p_1 = q_1'$, $p_2 = u'$ and the potential V is bi-quadratic in q_1, q_2 (for details see [3,4]). There is the first integral J_2 commuting with $J_1 = H$ so the Hamiltonian system (2) is completely integrable and its phase flow covers the invariant torus $J_1 = c_1$, $J_2 = c_2$ almost periodically. Our aim is to construct a class of perturbed Hamiltonians $H_\varepsilon = H + \varepsilon H_1 + \ldots$, $\varepsilon \ll 1$, which are close to the completely integrable one in the following sense. The phase flow of the perturbed motion keeps to be almost periodic in t and undergoes slow (at a speed $O(\varepsilon)$) evolution together with new "invariant" torus $J_1 = c_1(\varepsilon t)$, $J_2 = c_2(\varepsilon t)$. That means an adiabatic perturbation producing no resonances or any chaotic motion of the perturbed system. The deformation of the Lioville torus and the flow on it is constructed explicitly in terms of finite-band integration technique. The construction provides transformation of initial torus $J_1 = c_1(\tau_-)$, $J_2 = c_2(\tau_-)$ to any (with minor restrictions) given torus $J_1 = c_1(\tau_+)$, $J_2 = c_2(\tau_+)$. The form of approximate solution resembles the well-known Kuzmak – Whitham averaging ansatz [5,6] for multi-periodic dynamical systems. The construction is illusrated in Section 4 by concrete example of the Neumann system.

A number of examples of such deformations were produced first by an asymptotic integration of nonlinear P.D.E. with step-like initial data [7-9] or nonlinear O.D.E. like Painleve equations [10,11]. A process of averaging of perturbed Hamiltonian system with generic perturbations $H_1, \ldots$ gives rise to another example of Whitham deformations which is studied below in Section 5.

1. INTEGRATION METHOD

Let us first sketch briefly the finite-band integration procedure (see, for example [1-3]). Consider a hyperelliptic curve Γ of genus 2

$$\mu^2 = P_5(\lambda) = (\lambda - \lambda_1) \ldots (\lambda - \lambda_5), \tag{3}$$

where the branch points $\lambda_1 > \lambda_2 > \ldots > \lambda_5$ are real-valued and the bands $(-\infty, \lambda_5]$, $[\lambda_4, \lambda_3], [\lambda_2, \lambda_1]$ form the spectrum of the Sturm-Lioville operator L.

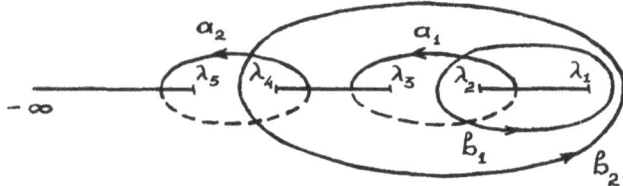

Fig. 1. The spectrum of the operator L and the basis of cycles a, b.

Choose the canonical basis of cycles on Γ as shown at Fig.1 and construct a basis of normalized holomorphic differentials $d\omega_1(p), \quad d\omega_2(p)$

$$\int_{a_j} d\omega_k(p) = 2\pi i \delta_{jk}, \qquad B_{jk} = \int_{b_j} d\omega_k(p). \tag{4}$$

Here p is a point of Γ "hanging" over λ on the sheets $C \setminus (\lambda_5, \lambda_4) \cup (\lambda_3, \lambda_2)$. The Riemann theta-function with characteristics on Γ has standard form [13]

$$\theta \begin{bmatrix} \alpha \\ \beta \end{bmatrix} (z|B) = \sum_{m \in Z^2} exp \left[\frac{1}{2} \langle B(m + \alpha), m + \alpha \rangle + \langle z + 2\pi i \beta, m + \alpha \rangle \right], \tag{5}$$

$$B = \{B_{jk}\}, \qquad \theta \begin{bmatrix} 0 \\ 0 \end{bmatrix} (z|B) = \theta(z).$$

Let $d\Omega$ be an abelian differential of the second kind on Γ with a principal part $d(\sqrt{\lambda})$ at infinity

$$d\Omega = \frac{\lambda^2 + a\lambda + b}{2\sqrt{P_5(\lambda)}} d\lambda, \tag{6a}$$

$$\int_{a_j} d\Omega = 0, \qquad U_j = \int_{b_j} d\Omega, \qquad j = 1, 2. \tag{6b}$$

The Lioville torus of the Hamiltonian system (2) coinsides with the Jacobian $\mathbf{Jac}(\Gamma)$ and the dynamical variables are expressed through the potential of L operator

$$u(t) = 2\partial_t^2 ln \ \theta(Ut + z_0), \tag{7}$$

where $z_0 = (z_{01}, z_{02})$ is arbitrary constant vector.

The eigenfunction Ψ is meromorphic on $\Gamma \setminus \infty$ and exponential as $p \to \infty$

$$\Psi(p,t) = \frac{\theta(A(p) + Ut + z_0)\theta(z_0)}{\theta(A(p) + z_0)\theta(Ut + z_0)} e^{t\Omega(p)}, \tag{8}$$

where $A(p) = \left(\int\limits_{\infty}^{p} d\omega_1, \int\limits_{\infty}^{p} d\omega_2 \right)$ – the Abelian transform,

$$\Omega(p) = \int\limits_{\infty}^{p} d\Omega = \sqrt{\lambda} + + O(\lambda^{-\frac{1}{2}}), \quad \lambda \to \infty. \tag{9}$$

The zeros of Ψ lie at the points $p_1 = (\gamma_1, \sqrt{P_5(\gamma_1)}), p_2 = (\gamma_2, \sqrt{P_5(\gamma_2)})$, while the real variables γ_1, γ_2 satisfy the Dubrovin equations

$$\frac{d\gamma_1}{dt} = \frac{2i\sqrt{P_5(\gamma_1)}}{\gamma_1 - \gamma_2}, \qquad \frac{d\gamma_2}{dt} = \frac{2i\sqrt{P_5(\gamma_2)}}{\gamma_2 - \gamma_1}, \tag{10}$$

which are equivalent to the initial Hamiltonian system.

The dynamical variables γ_1, γ_2 are linked with the potential $u(t)$ through the constraints

$$\gamma_1 + \gamma_2 = u, \qquad \gamma_1 \gamma_2 = \frac{1}{8}(3u^2 + u'') + \frac{1}{2} \sum_{i<j} \lambda_i \lambda_j. \tag{11}$$

2. DEFORMATIONS OF INTEGRABLE HAMILTONIAN

Introduce small deformation parameter $\varepsilon \ll 1$ and define "slow time" $\tau = \varepsilon t$. Let the branch points of Γ depend on slow time

$$\lambda_j = \lambda_j(\tau), \quad j = 1, 2 \ldots, 5,$$

while the other components of Ψ - function are defined exactly as above. The velocity vector $U(\tau)$ needs to be altered since the t-derivative of $z = Ut$ in dynamical variables (7) is now

$$\frac{dz}{dt} = U + \tau U_\tau = U + O(1), \quad \tau > 0.$$

It is clear that the Hamiltonian (2) deformed in such a way is no longer close to the initial one. In order to overcome this obstacle let us define another Abelian differential $d\Omega_n, \quad n = 1, 2, \ldots,$ such that

$$a) \quad \Omega_n(p, \tau) = \int\limits_{p_0}^{p} d\Omega_n = -\lambda^{2n-\frac{1}{2}} + c_1 \lambda^{2n-\frac{3}{2}} + \ldots + c_{2n-1} \lambda^{\frac{1}{2}} + O(\lambda^{-\frac{1}{2}}), \quad p \to \infty \tag{12a}$$

$$\partial_\tau c_j = 0, \quad j = 1, \ldots, 2n, \qquad d\Omega_n(p, \tau) = \frac{Q_{2n+1}(\lambda)}{2\sqrt{P_5(\lambda)}} d\lambda,$$

where $Q_{2n+1}(\lambda)$ is polynomial of order $2n + 1$.

$$b) \qquad \int\limits_{a_1} d\Omega_n = \int\limits_{a_2} d\Omega_n = 0, \tag{12b}$$

c) *The polynomial $Q_{2n+2}(\lambda)$ has three real-valued zeroes and the other $2n - 2$ zeroes do not lie on the real axis.*

The existence of such a differential is proved in Appendix.

Define the deformed Ψ–function as follows

$$\Psi(p,t) = \frac{\theta(A(p) + U(\tau)t + V(\tau)\epsilon^{-1} + z_0)\ \theta(z_0)}{\theta(A(p) + z_0)\ \theta(U(\tau)t + V(\tau)\epsilon^{-1} + z_0)}e^{t\Omega(p,\tau) + \epsilon^{-1}\Omega_n(p,\tau)}, \qquad (13)$$

where $V = \left(\int_{b_1} d\Omega_n, \int_{b_2} d\Omega_n\right)$, and all other components of Ψ –function coinside with those of equation (8). Remind that now $A = A(p,\tau)$, $B_{ij} = B_{ij}(\tau)$ since the Riemann surface Γ depends on τ.

Assume that deformation of $\Gamma = \Gamma(\tau)$ is governed by the following Whitham equation

$$\left(\ \tau d\Omega(p,\tau) + d\Omega_n(p,\tau)\right)\Big|_{\lambda\,=\,\lambda_j(\tau)} \quad \partial_\tau\lambda_j(\tau) = 0, \quad j = 1, 2, \ldots, 5. \qquad (14)$$

The equations (14) turn to be a self-similar reduction of the well-known quasilinear Whitham system appearing in the modulation theory for evolution equations of KdV type (see [14]). The principal reason for the choice of deformation in the form (13) follows from two basic properties of Abelian integrals in (12).

Theorem 1. *If the branch points satisfy the Whitham equations (14) then the following equations hold*

$$\tau\ \partial_\tau d\Omega(p,\tau) + \partial_\tau d\Omega_n(p,\tau) = 0, \qquad (15a)$$

$$\tau\ \partial_\tau U + \partial_\tau V = 0 \qquad (15b)$$

Proof. Consider the differential $d\Phi = \tau\ \partial_\tau d\Omega + \partial_\tau d\Omega_n$.

Due to hyperllipticity of $\Gamma(\tau)$ the differentials $d\Omega, d\Omega_n$ near branch point have the form

$$d\Omega = \frac{q(\lambda)}{\sqrt{\lambda - \lambda_j(\tau)}}, \quad d\Omega_n = \frac{q_n(\lambda)}{\sqrt{\lambda - \lambda_j(\tau)}}, \quad \lambda \to \lambda_j, \qquad (16)$$

where $q,\ q_n$ are analytic as $\lambda \to \lambda_j$, so for $d\Phi$ one has the asymptotics

$$d\Phi = \frac{\tau d\Omega(p,\tau) + d\Omega_n(p,\tau)}{2(\lambda - \lambda_j(\tau))}\partial_\tau\lambda_j(\tau) +$$

$$+ O\left(\frac{d\lambda}{\sqrt{\lambda - \lambda_j(\tau)}}\right), \quad \lambda \to \lambda_j.$$

The equation (14) yields now that the differential $d\Phi$ is Abelian, i.e. has asymptotics (17) at any branch point. The normalisation conditions (6b),(12b) provide all a-periods of $d\Phi$ to be zero and the asymptotic behavior at infinity (6a),(12a) means that $d\Phi$ is regular

$$d\Phi = O(\lambda^{-\frac{3}{2}}d\lambda), \quad \lambda \to \infty.$$

Putting these facts together we have (see [13])

$$d\Phi = 0, \qquad (17)$$

Integrating $d\Phi$ over b-cycles one immediately gets the equation (15). The theorem is proved.

Consider now the action of the operator L on new "eigenfunction" $\Psi(p, t, \varepsilon)$
(12)

$$L\Psi = (\partial_t^2 + u(t, \varepsilon) - \lambda)\Psi + 2\varepsilon\partial_t\partial_\tau\Psi + O(\varepsilon^2). \tag{18}$$

Theorem 2. *The potential $u(t, \varepsilon)$ in the equation (18) is given by the formula similar to (7)*

$$u(t, \varepsilon) = 2\partial_t^2 \ln\theta(U(\tau)t + V(\tau)\varepsilon^{-1} + z_0). \tag{19}$$

Proof. The proof of (18),(19) goes straightforward following the well-known Kriche-ver's scheme (see [1]). First, note that Ψ-function (13) is well-defined on Γ, mero-morphic on $\Gamma \setminus \infty$, and has an asymptotics at infinity

$$\Psi(p, t, \varepsilon) = \left(1 + \frac{\xi_1}{k} + \frac{\xi_2}{k^2} + \dots\right) \exp\left(kt + q(k)\varepsilon^{-1}\right), \tag{20}$$

where $k = \sqrt{\lambda}$ is local parameter, $q(k) = k^{4n-1} + c_1 k^{4n-2} + \dots + c_{2n-1}$ – non-vanishing terms of asymptotics (12a). The function (20) satisfies the equation

$$(\partial_t^2 + u(t, \varepsilon) - \lambda)\Psi = O\left(\frac{1}{k}\right) \exp\left(kt + q(k)\varepsilon^{-1}\right), \tag{21}$$

$$u(t, \varepsilon) = -2\partial_t\ \xi_1(t, \varepsilon). \tag{22}$$

Since the divisor of Ψ is non-special and $(L - \lambda)\Psi\ \exp\left(kt + q(k)\varepsilon^{-1}\right)$ vanishes at infinity the right-hand side of (20) is zero due to uniqueness of Baker-Akhiezer function. The formula (18) for the potential follows directly from (22). Calculating the complete t - derivative one gets the equation (18). The theorem is proved.

3. STRUCTURE OF THE WHITHAM DEFORMATION

Consider now the properties of solutions of the Whitham system (14).

$$\left(\tau d\Omega(p, \tau) + d\Omega_n(p, \tau)\right)\Big|_{\lambda\,=\,\lambda_j(\tau)} \quad \partial_\tau\lambda_j(\tau) = 0, \quad j = 1, 2, \dots, 5. \tag{23}$$

The equations (23) are exactly a self-similar reduction of the well-known quasilin-ear Whitham system proposed by H.Flaschka, G.Forest and D.McLaughlin [14] for asymptotic integration of KdV equation

$$\left(\partial_T - S_j(\lambda_1, \dots, \lambda_{2N+1})\partial_X\right)\lambda_j(X, T) = 0, \quad j = 1, 2, \dots, 2N + 1. \tag{24}$$

where $S_j = -d\Omega_2(\lambda)/d\Omega_1(\lambda)\Big|_{\lambda\,=\,\lambda_j}$, Ω_1, Ω_2 are Abelian integrals multiplied by t and x in the Ψ - function formula for KdV equation, $T = \varepsilon t, X = \varepsilon x$.

A generalisation of the hodograph method was proposed by S.P.Tsarev [12] for an exact integration of the system (24) . However the self-similar system (23) is much simpler one, so it is possible to give a complete description of its solutions in a way discussed in [7-9]. Following a notation of these papers we call a branch point λ_j a *moving point* if $\partial_\tau\lambda_j \neq 0$. It is clear that every equation (23) has two solutions $\tau = -\Omega_n/\Omega$ and $\partial_\tau\lambda_j = 0$. Remind that all branch points are assumed to be real-valued.

A basic fact of monotonicity of the deformation is established by the following

Theorem 3. *There exists $\tau_0 > 0$ such that if the point $\lambda_j(\tau)$ is moving for $0 < \tau < \tau_0$, then*

1) *all other branch points are immovable*

$$\partial_\tau \lambda_i(\tau) = 0, \quad i \neq j.$$

2) *the point λ_j moves from right to left along the real axis*

$$\partial_\tau \lambda_i(\tau) < 0,$$

Proof. The proof is very similar to those given in [9]. Consider the function

$$\gamma(p) = \frac{\tau \; d\Omega + d\Omega_n}{d\lambda}$$

For any τ it admits a representation $\gamma(p) = (\lambda^{2n+1} + \ldots)/\sqrt{P_5(\lambda)}$, where at the numerator stands a real-valued polynomial. The normalisation conditions (6b), (12b) yield that $\gamma(p)$ has zero at each of the lacunas $(\lambda_2, \lambda_3), (\lambda_4, \lambda_5)$. For $\tau = 0$ all its other $2n$ zeros are not lying on the real axis due to the condition $c)$ on differential $d\Omega_n$. Therefore it exists $\tau_0 > 0$ such that for all $0 < \tau < \tau_0$ this property would still hold.

Suppose that there are *two* moving points λ_i and λ_j at a certain moment $0 < \tau < \tau_0$. It means that $\gamma|_{\lambda = \lambda_i} = \gamma|_{\lambda = \lambda_j} = 0$, i.e. the function $\gamma(p)$ has 4 real-valued zeroes which is impossible by definition of $\gamma(p)$.

In order to prove the property 2) denote $G(\lambda \mid \{\lambda_j\}) = d\Omega_n/d\Omega$ and suppose that $\lambda = \lambda_i(\tau)$ is moving. Then the Whitham system (23) yields

$$\tau = -\; G(\lambda_i \mid \{\lambda_j\}), \tag{25}$$

$$\partial_\tau \lambda_i(\tau) = -\frac{1}{\partial_{\lambda_i} G(\lambda \mid \{\lambda_j\})}, \tag{26}$$

where all other branch points are immovable due to the property 1).

Estimate the complete derivative

$$\frac{d}{d\lambda_i} G(\lambda_i \mid \{\lambda_j\}) = \partial_\lambda G(\lambda \mid \{\lambda_j\})\big|_{\lambda = \lambda_i} + \partial_{\lambda_i} G(\lambda \mid \{\lambda_j\})\big|_{\lambda = \lambda_i}. \tag{27}$$

The first term here is positive since the equation (25) has only one solution and $G(\lambda \mid \{\lambda_j\}) \to +\infty$ as $\lambda \to +\infty$. In order to estimate the second term (27) note that the identity (16) and assumption $\partial_\tau \lambda_j \neq 0$ yield

$$\tau \; \partial_{\lambda_i} d\Omega + \partial_{\lambda_i} d\Omega_n + \frac{\tau \; d\Omega + d\Omega_n}{2(\lambda - \lambda_i)} = 0$$

Taking here a limit as $\lambda \to \lambda_i$ one has

$$G(\lambda \mid \{\lambda_j\})\big|_{\lambda = \lambda_i} = -\frac{1}{2}\partial_{\lambda_i} G(\lambda \mid \{\lambda_j\})\big|_{\lambda = \lambda_i}.$$

Hence the right-hand side of (27) is positive which proves the property 2). The theorem is proved.

With the help of this theorem it is possible to establish the existence and uniqueness of Whitham deformation of a curve $\Gamma = \Gamma(\tau)$, satisfying the boundary conditions

$$\Gamma(\tau_1) = \Gamma_1, \quad \Gamma(\tau_2) = \Gamma_2. \tag{28}$$

110

In contrast to Whitham deformations of curves generated by finite- band solutions of evolution equations (see [7,9]) , the finite-dimensional dynamical system (10) puts on a constraint on Γ

$$\text{genus of } \Gamma(\tau) = 2 \quad \text{for any} \quad \tau \geq 0. \tag{29}$$

Hence there are impossible here the scenarios of "creation" and "annihilation" of spectrum bands considered in [7] which are related to a possibility to exitate any arbitrary number of degrees of freedom in evolution equation. The condition (29) together with the dynamics of branch points formulated above in Theorem 3 essentially narrow a class of admissible boundary conditions (28). One simple example of boundary conditions assuring the existence of Whitham deformation is given by the following

Theorem 4. *Let the sets of branch points $\{\lambda_j^1\}$ and $\{\lambda_j^2\}$ corresponding to the boundary curves Γ_1 and Γ_2 satisfy the condition*

$$\lambda_1^1 > \lambda_1^2 > \lambda_2^1 > \lambda_2^2 > \ldots > \lambda_5^1 > \lambda_5^2. \tag{30}$$

Then for any given differential $d\Omega_n$ there exists the only one deformation $\Gamma = \Gamma(\tau)$ with boundary conditions (30).

Proof. According to the Theorem 2 all $\lambda_j(\tau)$ can move only one by one and strictly from right to left so that

$$G(\lambda_i) > G(\lambda_j) \quad \text{as} \quad \lambda_i > \lambda_j.$$

Therefore the equation $\tau = - G(\lambda_i)$ means that first the point λ_1 would be shifted at the moment $\tau_1 = - G(\lambda_1^1 \mid \{\lambda_j^1\})$. The motion ends as $\lambda_1(\tau)$ reaches the boundary point λ_1^2. Next the point λ_2^1 begins to move within interval $[\lambda_2^1, \lambda_2^2]$ and so on. All moving points $\lambda_i(\tau)$ are determined uniquely from the equations (25) where all $\lambda_i \neq \lambda_j$ are immovable. The deformation ends at the moment $\tau = - G(\lambda_5^2 \mid \{\lambda_j^2\})$. The theorem is proved.

The process of deformation described above can be illusrated as follows.

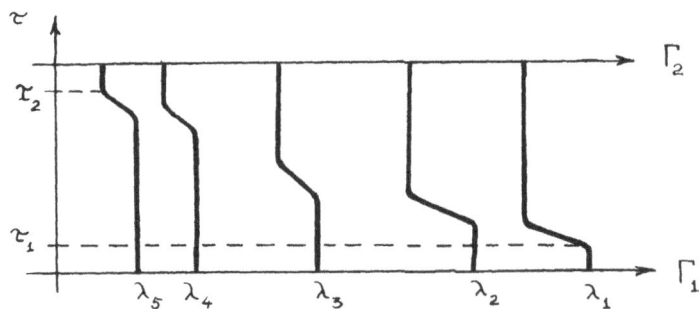

Fig.2. Monotonic Whitham deformation.

4. EXAMPLE

We consider here an important example of system integrable by finite-band technique discussed above in section 1. It is the Neumann system describing a motion of a particle on two-dimensional sphere [1,2]

$$x^2 = \sum_{i=1}^{3} x_i^2 = 1 \tag{31}$$

111

under the action of quadratic potential

$$U(x) = \frac{1}{2} \sum_{i=1}^{3} a_i x_i^2, \quad a_i = \text{const.}$$

The Hamiltonian has the form

$$H = \frac{1}{2} \sum_{i=1}^{3} (a_i x_i^2 + y_i^2)$$

providing together with constraint $x^2 = 1$ the equations of motion

$$\frac{d^2 x_i}{dt^2} = -a_i x_i + u(t) x_i, \qquad i = 1, 2, 3,$$

where $u(t)$ is Lagrangian multiplier arising due to the constraint (31). It can be calculated explicitly so the system takes the form

$$\frac{d^2 x_i}{dt^2} = -a_i x_i + x_i \sum_{k=1}^{3} \left(a_k x_k^2 - \left(\frac{dx_k}{dt} \right)^2 \right). \tag{32}$$

The functions

$$J_k = x_k^2 + \sum_{i \neq j} \frac{(x_k y_i - x_i y_k)^2}{a_i - a_k}, \qquad k = 1, 2, 3. \tag{33}$$

are integrals of motion in involution for the system (32).

The Neumann system admits an exact solution in terms of Riemann theta- functions of genus 2. Following [2,15] consider three eigenfunctions x_1, x_2, x_3 of the operator (1) $L = d^2/dt^2 - u(t)$ with eigenvalues $-a_1, -a_2, -a_3$ respectively

$$L x_i = -a_i x_i, \qquad i = 1, 2, 3. \tag{34}$$

The potential $u(t)$ here is supposed to provide purely two-band spectrum for the operator L in such a way that

$$\lambda_5 = -a_3 < \lambda_4 < \lambda_3 = -a_2 < \lambda_2 < \lambda_1 = -a_1.$$

and the Riemann surface Γ is defined by the equation (3) $\mu^2 = P_5(\lambda) = (\lambda - \lambda_1) \ldots (\lambda - \lambda_5)$. The solutions x_1, x_2, x_3 of the equation (34) are reconstructed from the Baker-Akhiezer function $\Psi(\lambda, t)$ (8) by normalisation [15]

$$x_i(t) = \alpha_i \Psi(-a_i, t), \quad \alpha_i = \left(\prod_{i \neq j} (a_i - a_j) \right)^{-\frac{1}{2}}.$$

This implies the constraint (31) to be true [15] together with an explicit formulae for dynamical variables

$$x_1(t) = \alpha_1 \frac{\theta[(0, 1/2), (0, 0)](tU + z_0)\theta(z_0)}{\theta[(0, 1/2), (0, 0)](z_0)\theta(tU + z_0)},$$

$$x_2(t) = \alpha_2 \frac{\theta[(1/2, 0), (0, 1/2)](tU + z_0)\theta(z_0)}{\theta[(1/2, 0), (0, 1/2)](z_0)\theta(tU + z_0)}, \tag{35}$$

$$x_3(t) = \alpha_3 \frac{\theta[(0,0),(1/2,1/2)](tU + z_0)\theta(z_0)}{\theta[(0,0),(1/2,1/2)](z_0)\theta(tU + z_0)}.$$

Here z_0 is arbitrary two-dimensional vector, U is the same as in (6b) and theta
-functions with characteristics are defined in (5).

Consider now an adiabatic transformation of the system (31) mapping its initial
Lioville torus defined by the integrals of motion $J_k = J_k(\tau_-)$ into some given torus
with new values of integrals $J_k = J_k(\tau_+) > J_k(\tau_-)$, $k = 1, 2, 3$, where τ_-, τ_+ are
any moments of slow time. This poses boundary conditions (30) for the branch points
$\lambda_j = \lambda_j(\tau)$, $j = 1, 2, \ldots, 5$, since they can be expressed directly through J_k (see
[15])

$$J_k = \frac{\prod\limits_{j=0}^{2}(\lambda_{2k} - \lambda_{2j+1})}{\prod\limits_{j \neq k}(\lambda_{2k} - \lambda_{2j})}.$$

The transformation of the Lioville tori now can be produced by Whitham deformation
(23). Its existence and uniqueness is guaranteed by Theorem 4 of Section 3.

The approximate solutions (35) are constructed in a way described in Section 2.
Making changes of arguments in (35)

$$tU + z_0 \quad \longmapsto \quad tU(\tau) + \varepsilon^{-1}V(\tau) + z_0.$$

and adding a second order term in ε we get an asymptotic ansatz

$$x_i(t) = \Theta_i(z + z_0(\tau) \mid \tau) + \varepsilon \Delta_i(z + z_0(\tau) \mid \tau) + O(\varepsilon^2), \quad i = 1, 2, 3. \tag{36}$$

where $z = tU(\tau) + \varepsilon^{-1}V(\tau)$, $z_0 = (z_{01}, z_{02})$, Θ denotes the right-hand side of
(35) and $\mid \tau)$ means that theta's and delta's depend on τ through the branch points
$\lambda_j(\tau)$. . The first order terms Δ_j satisfy linearized system

$$L\Delta_i + a_i \Delta_i - 2\Theta_i \sum_{k=1}^{3}(a_k \Theta_k \Delta_k - \partial_t \Theta_k \partial_t \Delta_k) =$$

$$= -2\partial_{t\tau}^2 \Theta_i - 2U_1 \partial_{z_1} \Theta_i z'_{01}(\tau) - 2U_2 \partial_{z_2} \Theta_i z'_{02}(\tau), \quad i = 1, 2, 3. \tag{37}$$

We used here estimate $\sum \Theta_k \partial_\tau \Theta_k = O(\varepsilon)$ following from (36) and the constraint
(31). Let Δ_i^0 be the solutions of homogeneous system (37). It can be shown [16] that
they are two-periodic in t with periods U_1, U_2 and can be expressed through the
Ψ-functions of L-operator.

The solvability condition of the system (37) in class of slowly growing at infinity
functions, $\Delta_i(t) = o(t)$, $t \to \infty$, yields an averaged over two periods functions in
right-hand side of (37) to be zero:

$$\frac{1}{2\pi^2} \int\limits_{0}^{2\pi} \int\limits_{0}^{2\pi} \left(-2\frac{\partial^2 \Theta_i}{\partial t \partial \tau} - 2U_1 \frac{\partial \Theta_i}{\partial z_1} z'_{01}(\tau) - \right.$$

$$\left. -2U_2 \frac{\partial \Theta_i}{\partial z_2} z'_{02}(\tau)\right)\Delta_i^0 dt_1 dt_2 = 0, \quad i = 1, 2. \tag{38}$$

where $t_1 = tU_1 + z_{01}$, $t_2 = tU_2 + z_{02}$. Note also that it is sufficient to take only two of
three equations (37) because the third function Δ_3 depends upon the others due to the
constraint (31). Since the solutions (35) x_1, x_2 are linearly independent the system
(38) for $z_{01}(\tau)$, $z_{02}(\tau)$ is non-degenerate so it has at least local solution.If there exists

global solution at the interval of slow time (τ_-, τ_+) then the construction of the principal term of asymptotics (36) is complete. It defines a smooth transformation of two-dimensional Lioville torus of the Neumann system (32) from initial to final ground states together with a phase flow on it which locally (for fixed τ) satisfies the system (32).

5. PERTURBATIONS OF DUBROVIN EQUATIONS

The deformation theory of an integrable system described above can be applied to asymptotic integration of a perturbed system

$$\left(\frac{d\gamma_1}{dt}\right)^2 = -\frac{4P_5(\gamma_1)}{(\gamma_1 - \gamma_2)^2} + \varepsilon f_1(\gamma_1, \gamma_2), \qquad (39a)$$

$$\left(\frac{d\gamma_2}{dt}\right)^2 = -\frac{4P_5(\gamma_2)}{(\gamma_2 - \gamma_1)^2} + \varepsilon f_2(\gamma_1, \gamma_2), \qquad (39b)$$

where f_1, f_2 are any smooth functions. Since the solutions of unpertubed system (10) γ_1, γ_2 are periodic in t so are the perturbations f_1, f_2 of the system (39). The well-known Kolmogorov- Arnold- Moser theory predicts that almost all initial conditions for the system (39) provide smooth quasiperiodic trajectories on its deformed Lioville torus. A construction of the solutions can be obtained through Kuzmak - Whitham averaging method [5,6]. An asymptotic ansatz has to be chosen through the deformed potential $u(t, \varepsilon)$ (19) substituted to formulae (11) for γ_1, γ_2.

$$\gamma_k = \gamma_k(z + z_0(\tau) \mid \tau) + \varepsilon \delta_k(z + z_0(\tau) \mid \tau) + \ldots, \quad k = 1, 2, \qquad (40)$$

where $z = tU(\tau) + \varepsilon^{-1}V(\tau)$, $z_0 = (z_{01}, z_{02})$. Here the notation $\mid \tau)$ underlines a fact that gamma's and delta's depend upon the slow variable $\tau = \varepsilon t$ through the branch points λ_j.

The first order terms in ε satisfy linearized system (39)

$$L_k(\gamma_1, \gamma_2)\delta_k - \frac{2i\sqrt{P_5(\gamma_k)}}{(\gamma_1 - \gamma_2)^2}\delta_k = \frac{f_k}{\gamma_k'} - \partial_\tau \gamma_k - z_{0k}'(\tau)\gamma_k', \quad k = 1, 2, \qquad (41)$$

where L_1, L_2 are linearized on γ_1, γ_2 operators (10).

$$L_1 = \frac{d}{dt} - \frac{iP_5'(\gamma_1)}{(\gamma_1 - \gamma_2)\sqrt{P_5(\gamma_1)}} - \frac{2i\sqrt{P_5(\gamma_1)}}{(\gamma_1 - \gamma_2)^2},$$

$$L_2 = \frac{d}{dt} + \frac{iP_5'(\gamma_2)}{(\gamma_1 - \gamma_2)\sqrt{P_5(\gamma_2)}} - \frac{2i\sqrt{P_5(\gamma_2)}}{(\gamma_1 - \gamma_2)^2},$$

In order to get the second term of approximation (40) to be small i.e. $\varepsilon\delta_k = o(1)$ it is necessary to keep zero an average over two periods of γ_1, γ_2 of right-hand side of equations (41). Let δ_1^0, δ_2^0 be solutions of the homogeneous system (32), $t_1 = tU_1 + z_{01}$, $t_2 = tU_2 + z_{02}$ and γ_k are periodic with respect to t_k. Then according to an averaging procedure [5] the following scalar products have to be zero

$$\frac{1}{2\pi^2}\int_0^{2\pi}\int_0^{2\pi}\left(f_k(\gamma_1, \gamma_2)(\gamma_k')^{-1} - \partial_\tau \gamma_k - z_{0k}'(\tau)\gamma_k'\right)\delta_k^0 dt_1 dt_2 = 0, \quad k = 1, 2.$$

114

Changing variables $(t_1, t_2) \mapsto (\gamma_1, \gamma_2)$ here with the help of equations (31) one gets nonlinear equations in slow variable τ for the phase shifts z_{01}, z_{02}

$$z'_{0k}(\tau) = \int\limits_{\lambda_2}^{\lambda_3} \int\limits_{\lambda_4}^{\lambda_5} \left(\frac{f_k(\gamma_1, \gamma_2)}{\gamma'_k} - \partial_\tau \gamma_k \right) \delta_k^0 \frac{(\gamma_1 - \gamma_2)^2}{\sqrt{P_5(\gamma_1)} \sqrt{P_5(\gamma_2)}} d\gamma_1 d\gamma_2$$

$$\bigg/ \int\limits_{\lambda_2}^{\lambda_3} \int\limits_{\lambda_4}^{\lambda_5} \delta_k^0 \frac{(\gamma_1 - \gamma_2)}{\sqrt{P_5(\gamma_{3-k})}} d\gamma_1 d\gamma_2, \tag{42}$$

since γ_1, γ_2 are running within allowed bands of spectrum $\lambda_2 < \gamma_1 < \lambda_3, \quad \lambda_4 < \gamma_2 < \lambda_5$.

The following theorem shows a necessity of the Whitham equations (14) for the branch points λ_j governing the principal term of asymptotics (40).

Theorem 5. *The averaging formula (42) for phase shifts is true if the Whitham equations for the branch points hold*

$$\left(\tau d\Omega(p, \tau) + d\Omega_n(p, \tau) \right)\Big|_{\lambda = \lambda_j(\tau)} = 0 \qquad j = 1, 2, \ldots, 5. \tag{43}$$

Proof. Consider a behavior of the first integrand of the numerator (42) near the limit of integration λ_j. It has an asymptotics

$$\frac{f_k(\gamma_1, \gamma_2)}{P_5(\gamma_k)} = \frac{\text{const}}{\lambda_j - \gamma_k} + O(1), \qquad \gamma_k \to \lambda_j.$$

Apply now the expression of gamma's through the Ψ - function (13) (see [1])

$$\frac{\partial_t \Psi(p, t)}{\Psi(p, t)} = \frac{\gamma'_k}{\lambda - \gamma_k} + O(1), \qquad \gamma_k \to \lambda. \tag{44}$$

Since γ_k is a simple zero of the Ψ -function the integral (42) is non-singular if $d\Psi(p, t) \big|_{\lambda \to \lambda_j} = 0$. Differentiating the exponent in formula (13) in a way used in the proof of Theorem 1 one immediately gets the Whitham equations (43). The theorem is proved.

The equation (43) means that every branch point λ_j has to be moving at any moment of time. Hence the construction of of auxiliary Abelian differential $d\Omega_n$ has to be changed since the conditions a) – c) at Section 2 provide the monotonic properties of the branch points (see Theorem 3). We would not go into further details here leaving them for separate investigation.

APPENDIX

Construction of the differential $d\Omega_n$ satisfying the conditions a) – c).

Let $\lambda_1(\tau), \lambda_2(\tau), \ldots, \lambda_5(\tau)$ be given real-valued branch points of the curve $\Gamma(\tau)$. We look for a differential $d\Omega_n$ in the form

$$\frac{d\Omega_n}{d\lambda} = -\frac{\lambda^{2n+1} + \kappa_1 \lambda^{2n} + \ldots + \kappa_{2n-1} \lambda^2 + \gamma_1 \lambda + \gamma_2}{\sqrt{P_5(\lambda)}},$$

where all coefficients κ_j, γ_j are real-valued. At infinity one has an asymptotics

$$\frac{d\Omega_n}{d\lambda} = -\lambda^{2n - \frac{3}{2}} + (\kappa_1 + p_1)\lambda^{2n - \frac{5}{2}} + \ldots + (\kappa_{2n-1} + p_{2n-1})(\lambda^{-\frac{1}{2}}) + O(\lambda^{-\frac{3}{2}}),$$

115

where $p_1, \ldots, p_{2n}$ are real polynomials of variables $\lambda_1, \lambda_2, \ldots, \lambda_5$. In order to satisfy the condition a) we put

$$c_j = \kappa_j + p_j, \quad \partial_r c_j = 0, \quad j = 1, 2, \ldots, 2n - 1.$$

The condition b) can be reached by a choice of coefficients γ_1, γ_2 remaining free . Clearly the system of algebraic equations $\int_{a_1} d\Omega_n = \int_{a_2} d\Omega_n = 0$ is linear with respect to γ_1, γ_2 having the matrix of coefficients

$$\left\{ \int\limits_{a_j} \frac{\lambda^{k-1}}{\sqrt{P_5(\lambda)}} d\lambda \right\}, \quad j, k = 1, 2,$$

the determinant of which is non-zero.

Since the real polynomial $Q_{2n+1}(\lambda)$ at a numerator of $d\Omega_n/d\lambda$ is of order $2n + 1$ and has two real-valued zeros there has to be at least one extra zero at the real axis. In order to make its other $2n - 2$ zeros to be complex-valued one can use a simple property of real polynomials: there is a zero of the derivative between any two real-valued zeros of the polynomial. Let us demand now that the derivative $\partial_\lambda^2(\lambda^{2n+1} + \kappa_1 \lambda^{2n} + \ldots + \kappa_{2n-1} \lambda^2)$ has no real-valued zeros. It puts on some condition of inequality-type on the arbitrary constants $c_1, c_2, \ldots, c_{2n-1}$ entering the coefficients κ_j in a linear way.

The differential $d\Omega_n$ constructed in such a way satisfy all the conditions a) – c).

REFERENCES

1. Dubrovin B.A., Krichever I.M., Novikov S.P., in: Modern problems in mathematics. Fundamental developments. Vol. 4 (Dynamical systems 4) Itogi Nauki i Techniki. Moscow: VINITI pp. 179 - 285 (1985).
2. Moser J. , Progr. in Math., 8, pp.233 - 289 (1980).
3. Bobenko A.I. , LOMI Preprint P - 10 - 87 , Leningrad (1987).
4. Dubrovin B.A. , Usp. Math. Nauk 31:2, pp. 12 - 80 (1981).
5. Kuzmak G.E. , Prikl. Mat. i Mech. 23:1 , pp. 515 - 525 (1959).
6. Whitham G.B. Linear and nonlinear waves, Wiley - Interscience, New York (1974).
7. Bikbaev R.F., Novokshenov V.Yu. , in: Proceedings of III Intern. Kiev Work shop, Vol. 1, pp.32 - 35 (1988).
8. Levermore C.D. , Comm.Partial Diff. Equat., 13:4. pp.495 - 511 (1988).
9. Bikbaev R.F. , Algebra and Anal., 2:3, pp.131 - 143 (1990).
10. Novikov S.P. , Funct.Anal. i Priloz., 24:4, pp.43-53 (1990).
11. Novokshenov V.Yu. , Doklady AN SSSR, 311:2 , pp.288 - 291 (1990).
12. Tsarev S.P. , Doklady AN SSSR, 282:3, pp.534 - 537 (1985).
13. Fay J.D. Theta functions on Riemann surfaces, Lecture Notes in Math.. 352, Springer-Verlag, New York (1973).
14. Flaschka H., Forest G., McLaughlin D. , Comm. Pure Appl. Math., 33:6, pp. 739 - 784 (1980).
15. Veselov A.P. , Funct. Anal. i Priloz., 14:1, pp. 48-50 (1980).
16. Marchenko V.A. Sturm-Lioville operators and their applications, Naukova Dumka, Kiev (1977).

HIGHER RANK DARBOUX TRANSFORMATIONS

Geoff Latham[1] and Emma Previato[2]

Center for Mathematics and its Applications, ANU, Canberra
ACT 2601, Australia, *and*
MSRI and Boston University, Boston, MA 02215, USA

Introduction

The Darboux transformation has been discovered several times in history (cf. references in [EK] and in [G3]): the reason for this is its deep geometric significance. In this note we explore some geometric aspects of the transformation and their relevance to the difficult problems of describing explicitly higher–rank commutative algebras of ordinary differential operators and KP flows.

Let us begin by recalling the definition. For definiteness, we work with coefficients that are complex analytic functions in a neighborhood of $x = 0$ (but we shall not change the notation when encountering an occasional pole at $x = 0$), and we say "eigenfunction" or "self–adjoint" in a formal, local sense, not in the stronger sense of topological vector spaces. We consider a differential operator $L = \partial^n + u_{n-2}(x)\partial^{n-2} + \cdots + u_0(x)$, where $\partial = d/dx$ (such a normalization of the first two coefficients is customary in the theory: it presupposes the fact that the leading coefficient of the operator in question is nonzero at $x = 0$), and a nonzero eigenfunction $\psi(x, \lambda)$ for which $L\psi = \lambda\psi$. Then $\tilde{L} = \left(\partial - \psi'/\psi\right) L \left(\partial - \psi'/\psi\right)^{-1}$ is by definition a Darboux transformation of L. The calculation is performed in the ring $\mathcal{P}$ of formal pseudodifferential operators, but the result belongs to the smaller ring $\mathcal{D}$ of differential operators (which is Euclidean) because $L - \lambda$ is divisible on the right by $\partial - \psi'/\psi$; indeed, they both annihilate ψ.

Burchnall and Chaundy discovered the geometric significance of a special case of the Darboux transformation, which they called transference. They asked the following question: classify and give explicit formulas for all commutative subalgebras $\mathcal{A}$ of the ring $\mathcal{D}$ which contain a normalized element L as given above, and that are not simply polynomial rings $\mathbf{C}[B]$ for some $B \in \mathcal{A}$. It turns out that the answer is significantly different depending on whether the rank of $\mathcal{A}$ is 1, or higher, where rank $\mathcal{A} = r = \gcd\{\text{order } B, B \in \mathcal{A}\}$. This rank is also the dimension of the vector space V_P of common eigenfunctions for $\mathcal{A}$, in the following sense: let P be a point of $X_0 = \operatorname{Spec}\mathcal{A}$, which is an affine curve, namely $P: \mathcal{A} \to \mathbf{C}$ is a ring homomorphism; $\psi \in V_P$ if and only if $B\psi = P(B)\psi$, $\forall B \in \mathcal{A}$, cf. [W]. Actually we can only assert that $\dim V_P = r$ for smooth points $P \in X_0$, however, the sheaf obtained by taking the dual V_P^* to be the fibre over each point, and completing the standard trivialization $s_0, \ldots, s_{r-1}$ $(s_j: \psi(x) \mapsto \psi^{(j)}(0))$

[1] gal851@vulcan.anu.edu.au
[2] ep@cs.bu.edu

across P_∞, is a coherent torsion free sheaf over the compact curve $X = X_0 \cup P_\infty$, and hence a rank r vector bundle $\mathcal{E}$ if X_0 is smooth. In the rank 1 case, Burchnall and Chaundy gave an essentially complete (for generic curves, that is) answer to their problem: the common eigenfunction of $\mathcal{A}$ is essentially unique, and for such $\psi(x, P) \in V_P$, provided $\psi(0, P) \neq 0$ (generically true in P), they defined the transference on $\mathcal{A}$ as the Darboux transformation by ψ. Notice that each operator $B \in \mathcal{A}$ is divisible on the right by $A = \partial - \psi'/\psi$, thus one obtains an isomorphic ring $\tilde{\mathcal{A}} \subset \mathcal{D}$ after conjugation by A. Burchnall and Chaundy proved that conversely, every ring $\tilde{\mathcal{A}}$ (normalized as we said) isomorphic to $\mathcal{A}$ can be obtained by a sequence of at most g transferences, where $g = \text{genus}\, X$, and they interpreted the transference as the abelian sum $P + P_1 + \cdots + P_g$, where $P_1 + \cdots + P_g$ is the divisor corresponding to the line bundle $\mathcal{E}$. In particular, the "isospectral" $\mathcal{A}$'s for which the $\text{Spec}\,\mathcal{A}$ are isomorphic curves are parametrized by $\text{Jac}\, X$ (there is a theta divisor where the coefficients of the corresponding $\mathcal{A}$'s acquire poles), and the coefficients of the elements of $\mathcal{A}$ can be written using Riemann's theta function.

There is a PDE aspect to Darboux transformations too, although in this context they are traditionally called Bäcklund transformations. Indeed, x-translation preserves the spectral curve $\text{Spec}\,\mathcal{A}$, but there are other deformations that also have this property, they are called the KP flows. These flows are defined by the equations

$$\partial_{t_j} \mathcal{L} = [(\mathcal{L}^j)_+, \mathcal{L}], \qquad j = 2, 3, \cdots,$$

where $\mathcal{L} = \partial + u_{-1}(\mathbf{t})\partial^{-1} + u_{-2}(\mathbf{t})\partial^{-2} + \cdots \in \mathcal{P}$, $\mathbf{t}$ denotes the sequence $t_1 = x$, t_2, t_3, $\cdots$ of time parameters, and $(\ )_+$ is the operator–part obtained by deleting the negative powers of ∂ in the expansion. In our deformation problem for the algebra $\mathcal{A}$ we take the initial value of $\mathcal{L}$, at $t_j = 0$ for $j > 1$, to be for example $L^{1/n}$. Just as elements of $\mathcal{A}$ represent functions regular on X_0, $\mathcal{L}$ represents a local parameter z for which $z^{-1}(P_\infty) = 0$, by the equation $\mathcal{L}\psi = z\psi$, $\psi \in V_P$. An $\mathcal{L}$ that satisfies the time–deformations written above is called a "solution to the KP hierarchy". A Darboux transformation with respect to a fixed value of the parameter, z_0 say, so that $\mathcal{L}\psi = z_0\psi$, takes $\mathcal{L}$ into another solution $\tilde{\mathcal{L}}$. This can be seen by using the equivalent formulation for the KP flows:

$$\partial_{t_j} \mathcal{L} = [-S^{-1}\partial_{t_j} S, \mathcal{L}],$$

where $S = 1 + s_1(\mathbf{t})\partial^{-1} + s_2(\mathbf{t})\partial^{-2} + \cdots \in \mathcal{P}$ is such that $S\mathcal{L}S^{-1} = \partial$. Since the Darboux transformation is $\tilde{\mathcal{L}} = A\mathcal{L}A^{-1}$ (where A depends on $\mathbf{t}$ the way ψ does), we can take $\tilde{S} = \partial S A^{-1}$ and check directly that

$$\partial_{t_j} \tilde{\mathcal{L}} = [-\tilde{S}^{-1}\partial_{t_j} \tilde{S}, \tilde{\mathcal{L}}]$$

holds for all j. Let us note that S is the inverse of the "wave operator", in the sense that the Baker–Akhiezer function for $\mathcal{L}$ is $\psi(\mathbf{t}, z) = S^{-1}e^{xz}$. The Baker–Akhiezer function corresponds to an element W of the Grassmannian Gr, cf. [SW]; this correspondence associates to any W in the big cell a KP solution, and can be thought of as a dressing by W of the trivial solution ∂. Thus, the Darboux transformations are special kinds of dressings. Lastly, let us bring together the geometry of the transference, as described in the previous paragraph, and the PDE theory, following [EK]. If we begin with a commutative algebra $\mathcal{A}$, we may use for the Darboux transformation an eigenfunction of L which is not common to the entire algebra; in this case to describe the corresponding solution $\tilde{\mathcal{L}}$, we want to study the maximal–commutative subalgebra which contains $\tilde{\mathcal{A}} = A\mathcal{A}A^{-1} \cap \mathcal{D}$. It is by such generalized transference that Ehlers and Knörrer manage to add or subtract a singular point to the original curve X and produce the soliton/rational solutions of the KdV equation.

In higher rank, both the corresponding ODO and PDE problems have not, so far, been answered explicitly: the correspondence between the higher–rank ring $\mathcal{A}$ and the rank r bundle $\mathcal{E}$ over X is obscured by the deformation data, which includes $r-1$ arbitrary functions. Only when $g = 1$, the parameter space of semistable rank r bundles with fixed determinant also has dimension $r - 1$, and this is probably the reason why the answer has been found: for genus 1 and rank 2, all commutative rings of ODO's have been explicitly described, cf. [KN], [D], [G1]. The KP deformations become much easier to describe if viewed as a "trivial solution" (which encodes the $r - 1$ functional parameters) dressed by an element of the Grassmannian, cf. [PW], but then the relationship with the geometry becomes utterly unexplicit. Again for $g = 1$ and $r = 2$, [KN] and [G2] find a criterion (KN, for Krichever–Novikov equation) on the arbitrary function that enters their formulas, which is equivalent to the first three flows of the KP hierarchy for their $\mathcal{L}$ operators, however, no explicit solutions of this condition are known except in degenerate cases (i.e. X becomes singular).

In view of this, we consider in this note the $g = 1$, $r = 2$ case, although the general framework and some of our considerations will hold for any genus and rank. We study transference and generalized transference, namely Darboux. Our study is directed toward two main open questions: first, (i) given a rank 2 commutative algebra $\mathcal{A}$, when is the rank true as opposed to fake, by which we mean that there exists an odd–order operator that commutes with $\mathcal{A}$? (equivalently, because X is smooth, when is $\mathcal{A}$ maximal–commutative?); and second, (ii) what is the significance of the KN equation and how to solve it? We have partial answers to these questions, which we now describe. First we calculate the effect of transference on the geometric data, in the spirit of Burchnall–Chaundy and Ehlers–Knörrer: not surprisingly, transference associated to a point P is again the abelian sum. The advantage here is, in rank r the j–th flows are abelian flows if j is a multiple of r, so in our situation we can identify the t_2 flow as the transference. These general facts are the subject of §1. In §2, we give some answers to questions (i) and (ii) in the singular case. We show that if X is the cuspidal curve, a suitable transference takes an L of order 4 into the square of a second order operator. Thus, our solution was obtained by generalized transference of a KdV ring. The rank is fake if and only if there is an odd order operator that commutes with the original KdV operator: this can be detected by the form of the KdV solution, and as a bonus we find the answer to a question of Grünbaum's, namely one of his examples is true rank 2. From this we get too an explanation of the relationship between the singular KN equation and the KdV equation derived in [SSY] (see also [DS]): the KN equation has been derived for the rank 2 deformation and the [SSY] correspondence is nothing but the Darboux transformation that takes us back to the KdV operator. In a concluding section we illustrate a question, in the Burchnall-Chaundy spirit, which remains open: exactly which rank 2 algebras $\tilde{\mathcal{A}}$ isomorphic to $\mathcal{A}$ can be reached by a sequence of generalized transferences? We give an idea of what happens by looking at a fake case.

There are of course many other aspects of the Darboux transformation worth generalizing to higher rank. For example, the Ercolani–McKean interpretation of Fay's trisecant identity in terms of Darboux transformations [EMcK]; and current work by Adler and van Moerbeke [AvM] which uses Darboux transformations to describe certain strata of the Grassmannian.

1. General results

As we said in the introduction, a Darboux transformation is the conjugation of an operator L by the first order operator A that annihilates an eigenfunction ψ of L. A is unique up to multiplication by a function (the invertible elements of $\mathcal{D}$) and we use the form $A = \partial - \psi'/\psi$ to preserve the normalization of L. This form is the reason for assuming $\psi(0) \neq 0$ (cf. also [EK]): if ψ is the common eigenfunction of a rank 1 algebra $\mathcal{A}$, this means that the divisor $P_1 + \cdots + P_g$ associated to the line bundle defined by $\mathcal{A}$ does not contain the point P corresponding to ψ.

If again ψ is the rank 1 common eigenfunction of a rank 1 algebra $\mathcal{A}$ corresponding to the point P, then $A = \partial - \psi'/\psi$ is the gcd of all $B - P(B)$, $B \in \mathcal{A}$, and this is how Burchnall and Chaundy defined the transference: $B - P(B) = B_1 A$, $\tilde{B} - P(B) = AB_1$. The natural generalization is the following:

1.1 Definition. Let $\mathcal{A}$ be a commutative subalgebra of $\mathcal{D}$ of rank r, P a (smooth) point of $X_0 = \mathrm{Spec}\,\mathcal{A}$ which is not part of the divisor determined by the section $s_0 \wedge \ldots \wedge s_{r-1}$ (cf. Introduction), and A the monic r-th order gcd of $B - P(B)$, $B \in \mathcal{A}$. The transference of the ring $\mathcal{A}$ is by definition the set of conjugates $\tilde{\mathcal{A}} = A\mathcal{A}A^{-1}$.

From now on we look at the simplest possible case: X an elliptic curve. In rank 1 this means that $\mathcal{A}$ contains an operator L of order two and an operator B of order three (in rank 2 these become 4 and 6 resp.) and they can be normalized so as to satisfy the equation: $\mu^2 = \lambda^3 + \frac{1}{4}g_2\lambda + \frac{1}{4}g_3$, for $L = \lambda, B = \mu$.

Let us recall what happens in rank 1:

1.2 Proposition. (cf. [D], [EK]) The generators of $\mathcal{A}$ have the form:
$L = \partial^2 - 2\wp(x - e)$, $B = \partial^3 - 3\wp(x - e)\partial - \frac{3}{2}\wp'(x - e)$. The divisor of the section s_0 is $P_1 = (\wp(-e), \frac{1}{2}\wp'(-e))$. If $P = (\wp(f), \frac{1}{2}\wp'(f))$, the result of transference by P is:
(i) at the $\wp$–function level, $e \mapsto e - f$;
(ii) at the sheaf level, $\mathcal{O}(P_1) \mapsto \mathcal{O}(P_1) \otimes \mathcal{O}(-P) \otimes \mathcal{O}(P_\infty)$;
(iii) at the Grassmannian level, where a local trivialization of the sheaf at P_∞ is also needed, or equivalently at the Baker–Akhiezer function level,
$\psi(x, Q) \mapsto \psi(x, P)^{-1} \{\psi(x, Q), \psi(x, P)\} / \{\psi(x, Q), \psi(x, P)\}|_{x=0}$, interpreted in the limit as $Q \to P$.

In rank 2 things are complicated by the introduction of an arbitrary function and it is necessary to recall the explicit formulas. We actually record the calculation in both Grünbaum's and Krichever–Novikov's formalisms, because the former has the advantage of involving only differential polynomials in the arbitrary function (as opposed to $\wp$) and the latter has the advantage of making the geometry more transparent. As we recall the conversion between the two formalisms, we note a limit process that takes the nonself–adjoint case into the self–adjoint case (these are solved separately in [G1], but treated equally in [KN]).

1.3 Review and comparison. The generators in rank 2 have the form: $L = \left(\partial^2 + \frac{1}{2}c_2\right)^2 + c_1\partial + \partial c_1 + c_0$, $B = \left(L^{3/2}\right)_+$ where:

$$c_0 = -g^2 + \sigma_1 g + \sigma_0,$$
$$c_1 = g',$$
$$c_2 = \frac{-2g'g''' + g''^2 - g^4 + 2\sigma_1 g^3 + 6\sigma_0 g^2 + r_1 g + r_0}{2g'^2}.$$

The function $g(x)$ is arbitrary provided $g' \not\equiv 0$ (of course one has to rule out singularities at $x = 0$ for the condition $L \in \mathcal{D}$ to hold), and the constants σ_0, σ_1, r_0, r_1 determine the coefficients g_2, g_3 of the curve. In the self–adjoint case $c_1 \equiv 0$, the solution becomes:

$$c_2 = -\frac{c_0'''}{c_0'} + \frac{1}{2}\frac{c_0''^2}{c_0'^2} + \frac{c_0^3 + p_1 c_0 + p_0}{c_0'^2}$$

with $c_0(x)$ the arbitrary (nonconstant) function and the curve is given by: $\mu^2 = \lambda^3 + \frac{1}{4}p_1\lambda - \frac{1}{8}p_0$. The comparison with [KN] is given in [G1] for the nonself–adjoint case:

$$\sigma_0 = -2\wp(\gamma_0),$$
$$\sigma_1 = \wp''(\gamma_0)/\wp'(\gamma_0),$$
$$r_0 = 0,$$
$$r_1 = 4\wp'(\gamma_0),$$
$$g(x) = \wp'(\gamma_0)/\big(\wp(\gamma_0) - \wp((c(x)))\big),$$

where γ_0 is some constant, and the points of the divisor of the section $s_0 \wedge s_1$ are given by $P_1 = (\wp(-\gamma_1(0)), \frac{1}{2}\wp'(-\gamma_1(0)))$ and $P_2 = (\wp(-\gamma_2(0)), \frac{1}{2}\wp'(-\gamma_2(0)))$, where $\gamma_1(0) = \gamma_0 + c(0)$ and $\gamma_2(0) = \gamma_0 - c(0)$ (cf. 1.6). We will simply write γ_1 for $\gamma_0 + c(x)$ and γ_2 for $\gamma_0 - c(x)$. In the self–adjoint case, the conversion is given by $c_0 = -2\wp(c(x))$; notice this is the case $\gamma_0 = 0$. The comparison can be made more general:

$$c_0 = -\wp(\gamma_1) - \wp(\gamma_2),$$
$$c_1 = c'\big(\wp(\gamma_2) - \wp(\gamma_1)\big),$$
$$c_2 = -\Big(\frac{c''}{c'} - 2c'\Phi\Big)' - \frac{1}{2}\Big(\frac{c''}{c'} - 2c'\Phi\Big)^2 - \frac{1}{2c'^2},$$

where $\Phi = \zeta(\gamma_1) - \zeta(\gamma_2) - \zeta(2c)$ and in the limit,

$$\lim_{\gamma_0 \to 0} \Phi = 2\zeta(c) - \zeta(2c) = -\frac{1}{2}\frac{\wp''(c)}{\wp'(c)},$$
$$c_2 = -\Big(\frac{c''}{c'} + c'\frac{\wp''(c)}{\wp'(c)}\Big)' - \frac{1}{2}\Big(\frac{c''}{c'} + c'\frac{\wp''(c)}{\wp'(c)}\Big)^2 - \frac{1}{2c'^2}.$$

1.4 Proposition. Let $\mathcal{A}$ be as in 1.3 and $P = \big(\wp(\rho), \frac{1}{2}\wp'(\rho)\big)$ be a point of X_0 different from P_1 and P_2. Let $G = \partial^2 + g_1\partial + g_0$ be the gcd of $B - P(B)$, $B \in \mathcal{A}$ (note: g_0 and g_1 depend on x and ρ). Then under the transference $\mathcal{A} \mapsto \tilde{\mathcal{A}} = G\mathcal{A}G^{-1}$:

(i) the geometric data of $\mathcal{A}$, namely the bundle $\mathcal{E}$ and the parabolic structure at P_∞, are tensored by $\mathcal{O}(-P) \otimes \mathcal{O}(P_\infty)$, and the deformation of the bundle, which is given by $\Psi_x\Psi^{-1}$ with

$$\Psi = \begin{pmatrix} \psi_0(x,Q) & \psi_1(x,Q) \\ \partial\psi_0(x,Q) & \partial\psi_1(x,Q) \end{pmatrix}$$

for any basis ψ_0, ψ_1 of simultaneous solutions of $B\psi_i = Q(B)\psi_i$ (for example the Baker vector), changes into the one given by

$$\tilde{\Psi} = \begin{pmatrix} G\psi_0 & G\psi_1 \\ \partial(G\psi_0) & \partial(G\psi_1) \end{pmatrix}.$$

(ii) If the rank of $\mathcal{A}$ is fake, let $\mathcal{B}$ be a maximal commutative subalgebra of $\mathcal{D}$ which contains $\mathcal{A}$. $\mathcal{B}$ corresponds to a curve Γ and a line bundle $\mathcal{E}_\Gamma$ over it, and for a suitable choice of a local parameter w near infinity on Γ, the inclusion $\mathcal{A} \hookrightarrow \mathcal{B}$ corresponds to a double covering $\pi: \Gamma \to X$, $w \mapsto w^2 = z$, $\pi_* \mathcal{E}_\Gamma = \mathcal{E}$. Then transference by G can be achieved by two transferences on Γ, by the points T_1 and T_2 whose image under π is P.

(iii) It is known that the rj-th KP deformations of a rank r algebra have the effect of twisting by $\exp(z^j t_{rj})$ the transition matrix of the bundle. Over an elliptic curve for $j = 1$, this has the effect of translating by t_r the abelian coordinate of the corresponding divisor.

Proof. (i) can be proved as in [EK]; indeed, by definition the bundle and parabolic structure can be found by determining the module structure of $C[\partial]$ over the rings $\mathcal{A}$ and $\tilde{\mathcal{A}}$; as in Lemma 4 and Lemma 5 of [EK], one identifies the finite sections of the $\tilde{\mathcal{A}}$-module $C[\partial]$ with the sections of the direct image of the $\mathcal{A}$-module $C[\partial]$ which, paired with two independent solutions of $G\psi = 0$, give zero. The statement on the deformation also follows from the definition. (ii) means simply that there is an odd order operator in $\mathcal{B}$, C say, and a factorization $G = A_2 A_1$, for which $L - w^2$ is divisible on the right by G and $C - w$ is divisible by A_1 but not G: $L - w^2 = L_1 A_2 A_1$, $C - w = C_1 A_1$; then, $A_1 L_1 A_2 - w^2$ and $A_1 C_1 + w$ are divisible on the right by A_2, so this second step achieves the G transference for $\mathcal{A}$. All these operators can be expressed using the Baker function of Γ. (iii) is just a remark which will reappear in the next proposition.

Before making our next statement let us introduce some notation and formulas: under the assumption made in Proposition 1.4, let

$$L - \lambda = Q\,G, \qquad Q = \partial^2 - g_1 \partial + q_0, \qquad \tilde{L} - \lambda = G\,Q.$$

First we remark that since the coefficient g_1 is $-\partial \log W$, where W is the wronskian of a basis φ_1, φ_2 of solutions of $G\varphi = 0$, then

$$\tilde{c}_2 = c_2 - 4g_1' = c_2 + 4\partial^2 \log\{\varphi_1, \varphi_2\}.$$

Next, since we look for an explicit expression after transference of the solution data (namely $\tilde{g}$, $\tilde{\gamma}_1$, $\tilde{\gamma}_2$ and in more detail also $\tilde{c}_0$, $\tilde{c}_1$, $\tilde{c}_2$), we have to calculate the coefficients g_0, g_1, q_0 in terms of the original g, c_0, c_1, c_2, γ_1, γ_2 and $\lambda = \wp(\rho)$, $\mu = \frac{1}{2}\wp'(\rho)$.

The calculation of Q and G is straightforward and rests on the Euclidean algorithm (cf. [L]). To find the new $\tilde{g}$, which is only determined up to an additive constant, we will have to express the new $\tilde{c}_1$ as a derivative. Notice that this is a priori clear from: $\tilde{c}_1 - c_1 = q_0' + g_1'' - g_0'$, although we took a more direct approach to find $\tilde{c}_1$, for which it was not necessary to know q_0 explicitly. We report this intermediate direct result: introduce for convenience

$$U(x) = \tfrac{1}{2}c_1^2 c_2 + \tfrac{1}{2}c_1 c_1'' - \tfrac{1}{4}c_1'^2,$$
$$V(x; \lambda) = (\lambda + \tfrac{1}{2}c_0)^2 + U(x),$$

then using the form 1.3 of the generators

$$\tilde{g} = g + \big\{2\mu(\lambda + \tfrac{1}{2}c_0) - U'(\lambda + \tfrac{1}{2}c_0)/c_1 + c_0'U/c_1\big\}V^{-1} + K_0$$

for some constant K_0.

1.5 Proposition. With notation as above, the result of transference for a rank 2 algebra $\mathcal{A}$ is the following:

(i) in the Grünbaum formalism,

$$\tilde{g} = K_0 - K_1 + \frac{K_2}{g + K_1}$$

(cf. the proof for the significance of the constants K_0, K_1, K_2 which depend on λ, μ and σ_0, σ_1, r_0, r_1, in that they give the curve). If L was self–adjoint,

$$\tilde{g} = \frac{2\mu}{\lambda + \frac{1}{2}c_0}$$

or, if the new $\tilde{L}$ is also self–adjoint, namely $(\lambda, \mu) = (\lambda, 0)$ is a branch point,

$$\tilde{c}_2 = c_2 + 4\partial^2 \log(\lambda + \tfrac{1}{2}c_0),$$

$$\tilde{c}_0 = -2\lambda + \frac{2h'(\lambda)}{\lambda + \frac{1}{2}c_0}, \qquad \text{where} \qquad h(\lambda) = \lambda^3 + \tfrac{1}{4}g_2\lambda + \tfrac{1}{4}g_3.$$

(ii) In the Krichever–Novikov formalism,

$$\tilde{\gamma}_1 = \gamma_0 + \rho + c(x), \qquad \tilde{\gamma}_2 = \gamma_0 + \rho - c(x).$$

(iii) After n transferences by points with coordinates $\rho_1, \cdots, \rho_n$, the result on the coefficient c_2 (which under deformation is the solution of the KP equation) is the following:

$$\tilde{c}_2 = c_2 - 4\partial^2 \log\left(\frac{\sigma(\gamma_1)\,\sigma(\gamma_2)}{\sigma(\gamma_1 + \rho_1 + \cdots + \rho_n)\,\sigma(\gamma_2 + \rho_1 + \cdots + \rho_n)}\right).$$

Proof. The assertions in (i) for the self–adjoint case are a direct adaptation of the calculation reported before the proposition. Before explaining the significance of the K's and proving the linear fractional transformation formula for $\tilde{g}$, we turn to (ii) and make the conversion into curve parameters. Using the formulas 1.3, we obtain for the calculation of $\tilde{g}$:

$$\tilde{g} = g + \frac{1}{2}\frac{\wp'(\rho) - \wp'(\gamma_2)}{\wp(\rho) - \wp(\gamma_2)} + \frac{1}{2}\frac{\wp'(\rho) - \wp'(\gamma_1)}{\wp(\rho) - \wp(\gamma_1)} + K_0,$$

$$= \frac{\wp'(\gamma_0)}{\wp(\gamma_0) - \wp(c)} + \zeta(\rho + \gamma_2) + \zeta(\rho + \gamma_1) - \zeta(\gamma_1) - \zeta(\gamma_2) - 2\zeta(\rho) + K_0,$$

where the second equality is a consequence of the well–known addition formulas for the Weierstrass functions;

$$\frac{1}{2}\frac{\wp'(u) - \wp'(v)}{\wp(u) - \wp(v)} = \zeta(u + v) - \zeta(u) - \zeta(v),$$

$$\frac{\wp'(u)}{\wp(u) - \wp(v)} = \zeta(u + v) + \zeta(u - v) - 2\zeta(u).$$

To find the conversion as in 1.3, we recall that $\gamma_1 = \gamma_0 + c$, $\gamma_2 = \gamma_0 - c$, differentiate the expression for $\tilde{g}$ and use a further identity:

$$\wp(\gamma_2) - \wp(\gamma_1) = \frac{\wp'(\gamma_0)\wp'(c)}{\left(\wp(\gamma_0) - \wp(c)\right)^2}.$$

to get that $\tilde{c}_1 = \tilde{g}' = c'(\wp(\gamma_2 + \rho) - \wp(\gamma_1 + \rho))$, which shows that we can put

$$\tilde{g} = \frac{\wp'(\gamma_0 + \rho)}{\wp(\gamma_0 + \rho) - \wp(c)}$$

by taking the additive constant to be

$$K_0 = 2\zeta(\gamma_0) + 2\zeta(\rho) - 2\zeta(\gamma_0 + \rho) = -\frac{\wp'(\gamma_0) - \wp'(\rho)}{\wp(\gamma_0) - \wp(\rho)}.$$

This proves (ii). Now we turn to (i) whose derivation may be of some independent interest for it brings out some differential equations satisfied by the coefficients of G. Using the direct formula for $\tilde{g}$ again, and using a direct calculation (Euclidean division) for expressing g_0, g_1, q_0, we arrive at the identities:

$$g_0 = \tfrac{1}{2}c_2 + \tfrac{1}{2}(g - \tilde{g} + K_0) - \tfrac{1}{2}\frac{c_1'}{c_1}g_1,$$

$$q_0 = \tfrac{1}{2}c_2 + \tfrac{1}{2}(\tilde{g} - g - K_0) - \tfrac{1}{2}\frac{c_1'}{c_1}g_1 - g_1'.$$

Substituting these in the following expression for c_2 which comes from comparing the coefficients of ∂^2 in $L - \lambda = Q\,G$,

$$c_2 = g_0 + 2g_1' - g_1^2 + q_0,$$

we obtain a Riccati equation for g_1:

$$g_1' - \frac{c_1'}{c_1}g_1 - g_1^2 = 0$$

which we solve:

$$g_1 = \frac{-c_1}{g + K_1} = -\partial \log(g + K_1).$$

Substituting the above expressions for g_0, q_0 into the ∂–term of the relation $L - \lambda = Q\,G$:

$$c_2' + 2c_1 = g_1 q_0 + g_1'' - g_1 g_1' - g_0 g_1 + 2g_0'$$

and using the Riccati equation for g_1, we obtain a first order equation for $\tilde{g}$:

$$c_1 = g_1(\tilde{g} - g - K_0) - \tilde{c}_1,$$

which we solve:

$$\tilde{g} = K_0 - K_1 + \frac{K_2}{g + K_1}.$$

To determine the constants, first we compute by direct substitution

$$g_1 = c'(\zeta(\gamma_2 + \rho) - \zeta(\gamma_2) - \zeta(\gamma_1 + \rho) + \zeta(\gamma_1)),$$

$$= \partial \log\left(\frac{\sigma(\gamma_1)\sigma(\gamma_2)}{\sigma(\gamma_1 + \rho)\sigma(\gamma_2 + \rho)}\right),$$

which is also of independent interest because it is the negative of the logarithmic derivative of the wronskian. From the two expressions derived so far for g_1, we need only determine K_1 so that $g + K_1$ is a constant multiple of the reciprocal of the ratio of σ-functions given in the last logarithmic derivative. Thus if we take

$$K_1 = 2\zeta(\gamma_0) + \zeta(\rho) - \zeta(2\gamma_0 + \rho) = -\frac{\wp'(\gamma_0) - \wp'(\rho)}{\wp(\gamma_0) - \wp(\rho)} - \frac{\wp'(\gamma_0 + \rho)}{\wp(\gamma_0 + \rho) - \wp(\gamma_0)}$$

the two expressions for g_1 agree. To see this, we use the alternative expression for $g = \zeta(\gamma_1) + \zeta(\gamma_2) - 2\zeta(\gamma_0)$ and the identity

$$\frac{\sigma(2y)\sigma(u + v)\sigma(u - v)}{\sigma(u + y)\sigma(u - y)\sigma(v + y)\sigma(v - y)} = \zeta(u + y) - \zeta(u - y) - \zeta(v + y) + \zeta(v - y)$$

with $y = \gamma_0$, $u = \gamma_0 + \rho$ and $v = c$, to deduce

$$g + K_1 = \left(\frac{\sigma(2\gamma_0)}{\sigma(2\gamma_0 + \rho)\sigma(\rho)}\right)\left(\frac{\sigma(\gamma_1 + \rho)\sigma(\gamma_2 + \rho)}{\sigma(\gamma_1)\sigma(\gamma_2)}\right).$$

Next if we let

$$K_2 = -\frac{\sigma(2\rho + 2\gamma_0)\sigma(2\gamma_0)}{\sigma^2(\rho)\sigma^2(\rho + 2\gamma_0)} = \wp(\rho + 2\gamma_0) - \wp(\rho)$$

and use the above identity with $y = \gamma_0 + \rho$, $u = \gamma_0$ and $v = c$, together with the last expression for $g + K_1$, we see that

$$\frac{K_2}{g + K_1} = \zeta(\gamma_1 + \rho) + \zeta(\gamma_2 + \rho) - \zeta(\rho) - \zeta(2\gamma_0 + \rho).$$

This combined with $K_0 - K_1 = \zeta(\rho) - 2\zeta(\gamma_0 + \rho) + \zeta(2\gamma_0 + \rho)$ shows that, by the earlier addition formula,

$$K_0 - K_1 + \frac{K_2}{g + K_1} = \zeta(\gamma_1 + \rho) + \zeta(\gamma_2 + \rho) - 2\zeta(\gamma_0 + \rho) = \tilde{g}$$

for these values of the K's. Lastly, to prove the formula (iii) we compute by induction the effect of j transferences by the points $\left(\wp(\rho_k), \tfrac{1}{2}\wp'(\rho_k)\right)_{1 \le k \le j}$:

$$\tilde{g}_1^{(j-1)} = \partial \log\left(\frac{\sigma(\gamma_1 + \rho_1 + \cdots + \rho_{j-1})\,\sigma(\gamma_2 + \rho_1 + \cdots + \rho_{j-1})}{\sigma(\gamma_1 + \rho_1 + \cdots + \rho_j)\,\sigma(\gamma_2 + \rho_1 + \cdots + \rho_j)}\right),$$
$$\tilde{c}_2^{(j)} = \tilde{c}_2^{(j-1)} - 4\partial\,\tilde{g}_1^{(j-1)} = c_2 - 4\partial\left(\tilde{g}_1^{(j-1)} + \cdots + \tilde{g}_1 + g_1\right),$$

and putting these together gives (iii).

1.6 Remarks. The statements of 1.4 and 1.5 translate into each other: indeed, the original divisor of $\mathcal{A}$, namely the zeroes of $s_0 \wedge s_1$, is given by the poles of $\partial \log\{\varphi_1, \varphi_2\} = -g_1$ (for $x = 0$ and $(\lambda_0, \mu_0) = (\wp(\rho_0), \tfrac{1}{2}\wp'(\rho_0))$ a point of X_0): from the formula for g_1 given in the proof of 1.5, we see that $\rho_0 = -\gamma_1$ and $-\gamma_2$ resp., give the points P_1 and P_2. As predicted by 1.5 (i), the points of the new divisor are given by $P_1 - P + P_\infty$, $P_2 - P + P_\infty$, with parameters $-\gamma_1 - \rho$, $-\gamma_2 - \rho$. The statement 1.4 (iii) about the KP flows is made transparent by Grünbaum's formula for the corresponding $g(x, t_2, t_3) = \wp'(t_2)/(\wp(t_2) - \wp(c(x, t_3)))$, cf. [G2]. Notice also that in the fake case 1.4 (ii) and (iii) say that this particular double transference corresponds to the t_2 flow for the line bundle $\mathcal{E}_\Gamma$ upstairs; it is already known that simple transference by the w–eigenfunction corresponds to its $x = t_1$ flow.

We make a few comments on generalized transference which will be useful in §2. The useful class of rational/soliton solutions of KdV can be obtained by successive applications of a rather simple Darboux transformation, as explained in [AM], [BC], [EK]. These transformations need not be transferences in the sense that the chosen eigenfunction ψ of L is common to the algebra $\mathcal{A}$. Nevertheless let us note:

1.7 Remark. If the rank of $\mathcal{A}$ is true (or fake), so is the rank of $\tilde{\mathcal{A}} = A\mathcal{A}A^{-1} \cap \mathcal{D}$. This is proved by observing that if B is an operator that commutes with L, then $A(BL)A^{-1} \in \tilde{\mathcal{A}}$ and order $BL = $ order $B + $ order L which does not affect the gcd(order B, order L).

We emphasize some differences between the smooth and singular cases for X_0. Let us first recall the situation for rank 1: $\mathcal{A}$ is maximal–commutative in $\mathcal{D}$ if and only if $\mathcal{E}$ is what is called a "maximal" sheaf in [SW]. The curve X_0 could still be singular (though the "least–singular" that gives the same solution) and $\mathcal{E}$ might not be a line bundle. The simplest example is given by a nice construction in [BC] which we recall for later use:

1.8 Construction of non "reducible" algebras from reducible ones.
(Burchnall and Chaundy call reducible an algebra of type $\mathbf{C}[Z]$, where Z is some element of $\mathcal{D}$.) The idea is to start with $\mathbf{C}[\partial]$ and do the Darboux transformation by an eigenfunction of ∂^3 but not of ∂, ∂^2, namely x^2 (the Darboux point of $X_0 = \mathbf{A}^1$ is thus 0). The resulting $\tilde{\mathcal{A}}$ is isomorphic to $\mathbf{C}[z^3, z^4, z^5]$ and the common eigenspace of $\tilde{\mathcal{A}}$ at all points of Spec $\mathbf{C}[z^3, z^4, z^5]$ has dimension 1 except at 0, where it is spanned by the transform of 1 and x.

This is also the example given in [SW], where however it is remarked that if X_0 has planar singularities, e.g. it is hyperelliptic, then all maximal sheaves are line bundles. This is also remarked by [EK] in their situation:

1.9 Remark. Let $\mathcal{A} = \mathbf{C}[L, B]$ be maximal–commutative, order $L = 2$ and order $B = 2g + 1$. Then the common eigenspace of L, B is 1–dimensional for all points $P \in X_0$, even the singular ones.

In rank 2 (and, for our considerations, an $\mathcal{A}$ generated by an L of order 4 and a B of order 6), we can produce a jump in dimension by using the Burchnall–Chaundy construction.

1.10 Remark. There is a rank 2 algebra $\mathcal{A}$ (fake rank, however) whose common eigenspace has dimension 2 for each point of Spec $\mathcal{A}$ except 0, where the dimension is 3. $\mathcal{A}$ is produced from $\mathbf{C}[\partial^4, \partial^6]$ by a Darboux transformation by the eigenfunction x^3.

Finally, we summarize the effect of Darboux transformations on the Ehlers–Knörrer algebras: if $\mathcal{A}$ is as in 1.9, and $P \in$ Spec $\mathcal{A}$ (not contained in the divisor of $\mathcal{E}$), then;
  (i) if ψ is an eigenfunction of L but not of B, $\tilde{\mathcal{A}}$ is again maximal and Spec $\tilde{\mathcal{A}}$ is singular,
  (ii) if ψ is an eigenfunction of L and B, there are two cases: if P is smooth, $\tilde{\mathcal{A}} \cong \mathcal{A}$ is again maximal, but if P is singular $\tilde{\mathcal{A}} \cong \mathcal{A}$ is not maximal.
We will treat an analog of (ii) in the next section.

2. Singular case: some answers

In this section we let the curve: $\mu^2 = \lambda^3 + \frac{1}{4}g_2\lambda + \frac{1}{4}g_3$ be singular and we give an interpretation of Grünbaum's example (c) of a KP solution in [G2], we find its true rank and generalize this observation so as to turn all cuspidal operators L into squares of KdV operators: this explains the transformation between KN equations (singular case) and KdV given in [SSY].

Let's recall that the following fact was proved in [KN] and in [G2] with a different method: if $g(x, t_2, t_3) = \wp'(t_2)/(\wp(t_2) - \wp(c(x, t_3)))$ then $c_2(x, t_2, t_3)$ as defined in 1.3 in terms of g satisfies the KP equation:

$$\frac{3}{4} \frac{\partial^2 c_2}{\partial t_2^2} = \frac{\partial}{\partial x} \left(\frac{\partial c_2}{\partial t_3} - \frac{1}{4} \frac{\partial^3 c_2}{\partial x^3} - \frac{3}{4} c_2 \frac{\partial c_2}{\partial x} \right) \qquad \text{(KP)}$$

if and only if c satisfies the KN equation:

$$\frac{\partial c}{\partial t_3} = \frac{1}{4} \frac{\partial^3 c}{\partial x^3} + \frac{3}{8} \frac{1 - (\partial^2 c/\partial x^2)^2}{\partial c/\partial x} - \frac{3}{2} \wp(2c) \left(\frac{\partial c}{\partial x} \right)^3. \qquad \text{(KN1)}$$

We note that the $\wp$-function can be removed from this last equation by the transformation $v = \wp(c)$, as in [SSY], and the result is:

$$\frac{v_{t_3}}{v_x} = \frac{1}{4} \{v, x\} + \frac{3}{2} \frac{v^3 + \frac{1}{4} g_2 v + \frac{1}{4} g_3}{v_x^2} \qquad \text{(KN2)}$$

where $\{v, x\}$ denotes the Schwarzian derivative: $v_{xxx} v_x^{-1} - \frac{3}{2} v_{xx}^2 v_x^{-2}$. Grünbaum in [G2] gives three examples of solutions of the KN equation for the cuspidal case $g_2 = g_3 = 0$. He shows by Darboux transformation in [G3] that the first two examples (a) and (b) can be transformed into "reducible" rings, namely to rings that contain the square of a KdV operator; these KdV operators are actually constant-coefficient, so in particular he shows that the original solutions have fake rank 2 (for all values of the parameters t_2, t_3). We recall his example (c):

$$c(x, t_3) = \left((1 + \tfrac{3}{2} t_3) x \right)^{1/2},$$

$\gamma_0 = t_2$, corresponding to $P_0 = (\wp(\gamma_0), \frac{1}{2} \wp'(\gamma_0)) = \left(t_2^{-2}, -t_2^{-3} \right)$

which gives the KP solution,

$$c_2(x, t_2, t_3) = -\frac{4x}{3t_3 + 2} - \frac{4(3t_3 + 2)^2}{\left((3t_3 + 2)x - 2t_2^2 \right)^2}.$$

If we use the transference formula for the point $-P_0 = (t_2^{-2}, t_2^{-3})$ we obtain,

$$\tilde{c}_2 = -\frac{4x}{3t_3 + 2} - \frac{4}{x^2}.$$

The only reason for "eliminating" t_2 was to produce a self–adjoint operator L; for fixed values of t_2, t_3, we regard it as an ODO and investigate the rank of the algebra $\mathcal{A} = \mathbb{C}[L, B]$ where as usual $B = (L^{3/2})_+$.

Our first step is to produce a KdV solution from the KN solution, following [SSY], our second step is to explain this is a transference that takes L to the square of a KdV operator; then we generalize this construction to all singular curves in analogy to 1.8.

2.1 The transformation. If the polynomial

$$h(v) = v^3 + \tfrac{1}{4} g_2 v + \tfrac{1}{4} g_3$$

has repeated roots, as [SSY] say, a linear-fractional transformation of v will lower the degree of h in the (KN2) equation. In our case, $g_2 = g_3 = 0$, $u = 1/v$ satisfies the equation:

$$\frac{u_{t_3}}{u_x} = \tfrac{1}{4}\{u, x\} + \tfrac{3}{2}\frac{u}{u_x^2}.$$

Note that $u = (k + \tfrac{3}{2}t_3)x$ solves this equation for any $k \in \mathbf{C}$; $k = 1$ corresponds to Grünbaum's example, indeed

$$v = \frac{1}{u} = \wp(c) = \frac{1}{c^2} = \frac{1}{(1 + \tfrac{3}{2}t_3)x}.$$

Next, the position

$$\varphi \overset{\text{def}}{=} -\frac{u_{xxx}}{u_x} + \frac{1}{2}\frac{u_{xx}^2}{u_x^2} - \frac{2u}{u_x^2},$$

gives a solution of the KdV equation, again as predicted in [SSY]:

$$\varphi_{t_3} = \tfrac{1}{4}\varphi_{xxx} + \tfrac{3}{4}\varphi\varphi_x.$$

For our example,

$$\varphi = -\frac{2x}{k + \tfrac{3}{2}t_3}.$$

2.2 The transference. To interpret the (KN2)$\leftrightarrow$(KdV) transformation of 2.1, let us first assume only $g_3=0$ and perform the transference of a self–adjoint operator L using the point $(\lambda, \mu) = (0, 0)$. The result is $\tilde{L} = (\partial^2 + \tfrac{1}{2}\tilde{c}_2)^2 + \tilde{c}_0$ where, according to 1.5(i),

$$\tilde{c}_2 = c_2 + 4\partial^2 \log c_0,$$
$$\tilde{c}_0 = \frac{g_2}{c_0}$$

so that if $g_2 = 0$ also, $\tilde{L}$ is a perfect square. Moreover, if we set $c_0 = -2v$, the solution formulas of 1.3 give

$$\tilde{c}_2 = -\frac{v_{xxx}}{v_x} + \frac{1}{2}\frac{v_{xx}^2}{v_x^2} - \frac{2v^3}{v_x^2} + 4\partial^2 \log v = \varphi$$

(the second equal sign is obtained by letting $u = 1/v$ in the definition of φ.) Hence,

$$\tilde{L} = \left(\partial^2 + \tfrac{1}{2}\varphi\right)^2$$

and, as expected, φ evolves according to the KdV flows for $\mathcal{L} = \left(\partial^2 + \tfrac{1}{2}\varphi\right)^{1/2}$. Notice, finally, that φ solves the KdV equation which is exactly the t_2–independent version of (KP) above.

2.3 True rank. We apply the construction/observation 2.2 to Grünbaum's example $c = ((k + \tfrac{3}{2}t_3)x)^{1/2}$, $g_2 = g_3 = 0$. We let $k = 0$ to economize (the statements are the same for all k). Then

$$\tilde{L} = \left(\partial^2 - \frac{2x}{3t_3}\right)^2$$

and $\varphi = -4x/3t_3$ is a KdV solution of the Airy type discussed in [DG]. The centralizer of $L_1 = \partial^2 - 2x/3t_3$ is just the polynomial ring in L_1, because if it contained an element of odd order, φ would be of hyperelliptic (possibly singular) type. Conclusion: since by

1.7 transference does not change the trueness of the rank, the original algebra $\mathcal{A}$ has true rank 2 (for all nonzero values of the parameter t_3).

2.4 General singular case: Discussion. Let the curve X_0 be singular, i.e.

$$\mu^2 = (\lambda - e_1)^2(\lambda - e_2), \quad 2e_1 + e_2 = 0.$$

We can use a transference at the singular point to take away the singularity as in case (ii) of the Ehlers-Knörrer construction (end of §1); the nice fact is that the desingularized $\mathcal{B}$ that contains $\tilde{A}$ will be "reducible" in the sense of 1.8 (because there is only one even order, namely 2, that does not appear among the orders of the elements of $\mathcal{A}$). Or we could, more transparently perhaps, have gone the other way: take *any* KdV operator L_1, consider the rank 2 ring $\mathbf{C}[L_1^2, L_1^3]$, and apply the same idea of 1.8, namely take a (double) Darboux transformation by an eigenfunction of L_1^2 which is not an eigenfunction of L_1: this yields the general singular rank 2 algebra. We give the details next:

2.5 Proposition. Any L such that the associated rank 2 algebra $\mathcal{A}$ has singular spectrum can be transferenced to the (translate of the) square of a KdV operator.

Proof. Let the curve be given by $\mu^2 = h(\lambda)$, where,

$$\begin{aligned} h(\lambda) &= (\lambda - e_1)(\lambda - e_2)(\lambda - e_3), \\ &= \lambda^3 + \tfrac{1}{4}g_2\lambda + \tfrac{1}{4}g_3, \\ 0 &= e_1 + e_2 + e_3. \end{aligned}$$

Consider a self-adjoint L (for simplicity only). The transference by the point $(e_i, 0)$ gives, according to 1.5,

$$\tilde{c}_2 = c_2 + 4\partial^2 \log(e_i + \tfrac{1}{2}c_0),$$

$$\tilde{c}_0 = -2e_i + \frac{2h'(e_i)}{e_i + \tfrac{1}{2}c_0}.$$

Setting $c_0 = -2v$ in the first of these equations and using the solution formula 1.3, we obtain,

$$\tilde{c}_2 = -\frac{v_{xxx}}{v_x} + \frac{1}{2}\frac{v_{xx}^2}{v_x^2} - 2\frac{h(v)}{v_x^2} + 4\partial_x^2 \ln(v - e_i).$$

One now easily checks that by letting $\varphi = \tilde{c}_2$,

$$-\tfrac{1}{4}\varphi_{xxx} - \tfrac{3}{4}\varphi\varphi_x + \varphi_{t_3} = 6\frac{h'(e_i)v_x}{(v - e_i)^2},$$

whenever v solves (KN2),

$$v_{t_3} = \tfrac{1}{4}\left(v_{xxx} - \tfrac{3}{2}\frac{v_{xx}^2}{v_x}\right) + \tfrac{3}{2}\frac{h(v)}{v_x}.$$

Thus if $h(v)$ has a repeated root at e_i, i.e. $h'(e_i) = 0$, then

$$\tilde{L} = \left(\partial^2 + \tfrac{1}{2}\varphi\right)^2 - 2e_i.$$

2.6 Remark. From Proposition 2.5 and the equation for φ, it follows that any rational solution of the singular (KN2) equation corresponds to a rational KdV solution.

3. Questions

In the spirit of Burchnall and Chaundy, we would like to know how large a class of "isospectral" solutions we obtain by Darboux transformation in rank 2. The answer we got in 1.5 shows that for the special case of "transference" (which is a double Darboux) we change the point parameter $\gamma_0 \mapsto \gamma_0 + \rho$ and not the arbitrary function c; however, by performing a single Darboux transformation that preserves the curve, namely by one common eigenfunction, we do expect $c(x)$ to change (cf. 1.4(ii)); we have not been able to find an explicit solution to this problem, although it should be possible, by pushing further the investigation of the connection matrix $\Psi_x \Psi^{-1}$ (cf. [K] and 1.4(i)). We feel that it would be an important step toward answering the difficult question of true rank, (i) in the introduction, more precisely (in the nonsingular case): for which functions $c(x)$ is the rank 2 algebra $C[L, B]$ maximal commutative? What we could do was to take a look at a fake and singular case, use the well known τ–function solution, and compute the effect of a single Darboux transformation. Roughly speaking, the function $c(x)$ in the two Darboux steps of the transference changes by translations $x \mapsto x - \frac{1}{k}$, $x \mapsto x + \frac{1}{k}$, which are inverses of each other. More precisely,

3.1 Remark. The Schur KP solutions transform under Darboux according to the KP vertex operator (a fact known to the Kyoto school, cf. [JM]).

The following example is a Schur solution of KP in the spirit of Grünbaum's (a) in [G2], for the differential operator $L = (\partial^2 + \frac{1}{2}c_2)^2 + c_1\partial + \partial c_1 + c_0$. This operator is one of a commuting pair (of orders 4 and 6) when,

$$c_2 = 4\partial^2 \log W(p_3, p_2) = 4\partial^2 \log(\frac{1}{12}x^4 + t_2^2 - xt_3),$$
$$c_1 = \partial\partial_{t_2} \log W(p_3, p_2),$$
$$g = \partial_{t_2} \log W(p_3, p_2) - 2/t_2$$
$$= \left(-2t_2^{-3}\right) \left(t_2^{-2} - \left(xt_3 - \frac{1}{12}x^4\right)^{-1}\right)^{-1},$$

where $p_3 = t_3 + xt_2 + \frac{1}{6}x^3$, $p_2 = t_2 + \frac{1}{2}x^2$, $p_1 = x$, $p_0 = 1$ are the first four Schur polynomials (cf. also 3.2 below) and W denotes the wronskian determinant. This L can be obtained by gcd transference from the self–adjoint operator which has $v = 1/(xt_3 - \frac{1}{12}x^4)$ using the point $P_0 = (t_2^{-2}, -t_2^{-3})$. Now we perform a Darboux transformation using the point $P = (k^4, k^6)$, $k \neq 0$, on the singular curve $\mu^2 = \lambda^3$.

L is the fourth power of the pseudo $\mathcal{L}$ obtained from ∂ by two applications of the Darboux (transference) process, cf. [G3]. Hence the function,

$$\psi \stackrel{\text{def}}{=} \frac{W(p_3, p_2, \varphi)}{W(p_3, p_2)} = \frac{\begin{vmatrix} p_3 & p_2 & 1 \\ p_2 & p_1 & k \\ p_1 & p_0 & k^2 \end{vmatrix}}{\begin{vmatrix} p_2 & p_3 \\ p_1 & p_2 \end{vmatrix}} e^{kx + k^2 t_2 + k^3 t_3},$$

where $\varphi = \exp(kx + k^2 t_2 + k^3 t_3)$ is the eigenfunction of ∂, is an eigenfunction of L and satisfies,

$$L\psi = k^4\psi.$$

Let f denote ψ'/ψ, then we have,

$$L - k^4 = \left(\partial^3 + f\partial^2 + (c_2 + 3f' + f^2)\partial + 3f'' + 5ff' + f^3 + c_2f + c_2' + 2c_1\right)(\partial - f),$$

and then defining $\tilde{L}$ by,

$$\tilde{L} - k^4 = (\partial - f)(\partial^3 + f\partial^2 + (c_2 + 3f' + f^2)\partial + 3f'' + 5ff' + f^3 + c_2 f + c_2' + 2c_1),$$

we find that,

$$\tilde{c}_2 = c_2 + 4f',$$
$$\tilde{c}_1 = c_1 + \tfrac{1}{2}c_2' + 2ff' + f''.$$

The second of these equations implies that,

$$\tilde{g} = g + \tfrac{1}{2}c_2 + f^2 + f' + K_0,$$

for some constant K_0. Denote by $\epsilon(k^{-1})$ the *vector* $(1/k, 1/(2k^2), 1/(3k^3), \cdots)$, and let $\mathbf{t}$ denote the assemblage of KP variables $(x, t_2, t_3, \cdots)$, then by $\mathbf{t} - \epsilon(k^{-1})$ we mean the collection $(x - 1/k, t_2 - 1/(2k^2), t_3 - 1/(3k^3), \cdots)$. Using the explicit forms for c_2 and ψ we get,

$$\tilde{c}_i(\mathbf{t}) = c_i(\mathbf{t} - \epsilon(k^{-1})), \qquad i = 1, 2, 3,$$
$$\tilde{g}(\mathbf{t}) = g(\mathbf{t} - \epsilon(k^{-1})).$$

To obtain this formula for $\tilde{g}$, we must take $K_0 = 2/t_2 - k^2 - 2/(t_2 - 1/(2k^2))$. Notice that $\tilde{\psi}(\mathbf{t}) = \psi(\mathbf{t} - \epsilon(k^{-1}))$ satisfies $\tilde{L}\tilde{\psi} = k^4\tilde{\psi}$ and this is exactly the action of the KP vertex operator. Notice finally that $L\psi_i = k^4\psi_i$ for $\psi_0(x; k) := \psi(x; k)$, $\psi_1(x; k) := \psi(x; ik)$, $\psi_2(x; k) := \psi(x; -k)$, $\psi_3(x; k) := \psi(x; -ik)$. The common eigenfunctions satisfying $G\psi_i = 0$ are given by $i = 0, 2$. A second Darboux transformation by the other (transformed) common eigenfunction $\tilde{\psi}_2(\mathbf{t}) = \psi_2(\mathbf{t} - \epsilon(k^{-1}))$ has the effect:

$$\tilde{\tilde{g}}(\mathbf{t}) = g(\mathbf{t} - \epsilon(k^{-1}) - \epsilon((-k)^{-1}));$$

thus all the odd variables return to the original values,

$$\tilde{\tilde{g}}(\mathbf{t}) = -2\left(t_2 - \frac{1}{k^2}\right)^{-3}\left(\left(t_2 - \frac{1}{k^2}\right)^{-2} - \left(xt_3 - \tfrac{1}{12}x^4\right)^{-1}\right)^{-1} = \frac{\wp'(t_2 - k^{-2})}{\wp(t_2 - k^{-2}) - \wp(c)},$$

as predicted by 1.5 since the point P is $(\wp(-k^{-2}), \tfrac{1}{2}\wp'(-k^{-2}))$.

We conclude with a sequence of examples in the spirit of 3.1 above; conjugation by a divisor (of second order, this time) of L but not B is used to add a singular point to the curve at each step. More precisely, we work out an analog of the [SSY] correspondence between the Adler–Moser family of rational KdV solutions (cf. [AM]) and a fake rank 2 family of rational KN solutions (cf. 2.6).

3.2 Sequence of fake rank 2 Darboux transformations. Denote by N_I the partition $(n_1, \cdots, n_I)$ of the positive integer N, i.e. the $n_i \geq 1$ are such that $\sum n_i = N$. Define the corresponding KP τ-function τ_{N_I} by $\tau_{N_I} = \det P_{N_I}$, where P_{N_I} is the $I \times I$ matrix with elements $p_{n_i - i + j}$, and where in turn, the p_ν are the Schur polynomials defined by $\exp \sum_1^\infty k^i t_i = \sum_0^\infty k^\nu p_\nu$. If the variable t_j is given a degree of j then the p_ν are homogeneous polynomials of degree ν. Define the Adler–Moser [AM] polynomials ϑ_n by, $\vartheta_n = \tau_{n(n+1)/2}$ where the partition of $\tfrac{1}{2}n(n+1)$ is given by $(n, n-1, \cdots, 3, 2, 1)$, i.e. $n_i = n - i + 1$. It is well known that the ϑ_n are independent of the even KP variables and depend only on the odd variables up to and including t_{2n-1}. Furthermore, these polynomials generate rational solutions of the KdV equation,

$$V_{t_3} - \tfrac{3}{4}VV_x - \tfrac{1}{4}V_{xxx} = 0,$$

via the formula,

$$V_n = 4\partial_x^2 \log \vartheta_n, \qquad n \geq 0,$$

and satisfy the differential recursion,

$$\vartheta'_{n+1}\vartheta_{n-1} - \vartheta_{n+1}\vartheta'_{n-1} = \vartheta_n^2. \qquad (3.2.1)$$

The ϑ_n's can be generated from (3.2.1) by a suitable choice of integration constants, starting from $\vartheta_1 = x$ and $\vartheta_0 = 1$, but note that we take a different (nonmonic in x) normalization to others (cf. [AM], [DG]).

Consider for $n > 4$, the partition of length $I = n$ of the integer $N = \frac{1}{2}(n^2 - 3n + 10)$ given by: $(n, n-3, n-4, \cdots, 2, 1, 1, 1)$. For $n = 2, 3, 4$, we consider the special partitions $(2,2)$, $(3,1,1)$, $(4,1,1,1)$ respectively. These partitions may be considered as being "created" from the above KdV partitions of $\frac{1}{2}n(n+1)$ since $n_1 = n$, $n_i = n-i+1-2$ for $2 \leq i \leq n-2$, $n_{n-1} = 1$, $n_n = 1$. Let τ_n denote the τ-function for the given partitions of $N = \frac{1}{2}(n^2 - 3n + 10)$ with $n_1 = n$ (we have departed slightly from the initially defined notation for τ-functions). Our peculiar choice of partitions was guided by the fact that τ_n is always quadratic in t_2,

$$\tau_{n+1} = t_2^2 \vartheta_{n-1} - \Phi_n,$$

where Φ_n is a polynomial in only the odd KP variables up to and including t_{2n+1}. These τ-functions give a series of KP solutions for $n \geq 1$,

$$U_{n+1} = (c_2)_{n+1} = 4\partial_x^2 \log(t_2^2 \vartheta_{n-1} - \Phi_n),$$
$$g_{n+1} = \partial_{t_2} \log\left(\frac{1}{t_2^2} - \frac{\vartheta_{n-1}}{\Phi_n}\right). \qquad (3.2.2)$$

Hence the functions,

$$v_n = \frac{\vartheta_{n-1}}{\Phi_n}, \qquad n \geq 1, \qquad (3.2.3)$$

are solutions of the singular KN equations. Applying transference to the operator L whose coefficients are given according to (3.2.2) and using the point (t_2^{-2}, t_2^{-3}), we obtain the self adjoint operator L_n which is the Lax operator for the KN equation and whose v is v_n (cf. 2.1, 2.2). Applying transference to L_n using the cusp point $(0,0)$, gives the square of the KdV Lax operator with KdV solution (cf. 2.2),

$$\varphi_n = -\frac{v_n'''}{v_n'} + \frac{1}{2}\frac{v_n''^2}{v_n'^2} - \frac{2v_n^3}{v_n'^2} + 4\partial_x^2 \log v_n. \qquad (3.2.4)$$

Substituting (3.2.3) into (3.2.4), we find that,

$$\varphi_n = 4\partial_x^2 \log \vartheta_{n-1} = V_{n-1}.$$

Adler and Moser showed that the sequence V_n of KdV solutions can be obtained via rational Darboux transformations starting from $V_0 = 0$, each time using eigenvalue zero. We therefore obtain the following picture of transference for these solutions of the KP, singular KN and KdV equations;

$$
\begin{array}{ccccccccc}
\text{KP} & U_2 & U_3 & U_4 & \cdots & U_{n+1} & \cdots \\
 & \updownarrow & \updownarrow & \updownarrow & & \updownarrow & \\
\text{s}-\text{KN} & v_1 & v_2 & v_3 & \cdots & v_n & \cdots, \\
 & \downarrow & \downarrow & \downarrow & & \downarrow & \\
\text{KdV} & V_0 \longrightarrow & V_1 \longrightarrow & V_2 \longrightarrow & \cdots \longrightarrow & V_{n-1} \longrightarrow & \cdots
\end{array}
\qquad (3.2.5)
$$

Applying transference using the cusp point directly to the KP operator L for U_{n+1} takes it into the square of the KdV operator with potential V_{n-1}, thus bypassing the intermediate stage v_n of singular KN. We would like to return to the geometry of this correspondence in a future publication; in particular, since the arithmetic genus of the singular curve is n for V_n, we would like to investigate a suitably modified KN equation.

The action of single transference by common eigenfunctions on the v_n and U_n at the cusp point can be computed explicitly, however we omit this here. To further this example we give the first few Φ_n as differential polynomials in the ϑ_j's,

$$\Phi_1 = -\vartheta_1\vartheta_2 + \tfrac{1}{4}\vartheta_1^4,$$
$$\Phi_2 = \vartheta_3' - \vartheta_1^2\vartheta_2 + \tfrac{1}{4}\vartheta_1^5,$$
$$\Phi_3 = \tfrac{1}{2}\vartheta_4''' - \vartheta_1\vartheta_3 - \vartheta_1^2\vartheta_3' + \tfrac{1}{4}\vartheta_1^4\vartheta_2,$$
$$\Phi_4 = \tfrac{1}{16}\vartheta_5^{(v)} - \tfrac{3}{4}\vartheta_1^3\vartheta_4''' + \tfrac{1}{2}\vartheta_4 + \tfrac{3}{2}\vartheta_1^2\vartheta_2\vartheta_3' + \tfrac{3}{2}\vartheta_1^5\vartheta_3' + \tfrac{1}{2}\vartheta_1\vartheta_2\vartheta_3 + \tfrac{1}{4}\vartheta_1^4\vartheta_3 - \tfrac{3}{2}\vartheta_1^4\vartheta_2^2.$$

It is not known whether the Φ_n satisfy a recursion analogous to (3.2.1), however, it may be possible to establish such a recursion by using the scheme given by Drinfeld and Sokolov [DS] which relates the singular KN equation to the KdV equation. Finally, because ϑ_{n-1} contains only odd variables up to and including t_{2n-3} and Φ_n contains only odd variables up to and including t_{2n+1}, it is clear that for arbitrary $n \geq 1$, we have $[(L_n^{m/4})_+, L_n] = 0$ for any odd $m \geq 2n + 3$. In all the examples computed so far, and we think it a general fact, Φ_n is independent of t_{2n-1} for $n > 1$, so in these cases we can take $m = 2n - 1$. Example (a) in [G2] corresponds to v_2 here.

Acknowledgements: The second author as a member of MSRI is supported by NSF Grant DMS 8505550. Both authors are grateful for the use of the computer resources of the Centre for Information Science Research at the Australian National University and the Central Computing Services at the University of California, Berkeley, access to the latter being generously provided by Alberto Grünbaum.

References

[AvM] M. Adler and P. van Moerbeke, Limits of operators and the geometry of Grassmannians, in preparation.

[AM] M. Adler and J. Moser, On a class of polynomials connected with the Korteweg de Vries equations, *Comm. Math. Phys.* **61** (1978), 1-30.

[BC] J.L. Burchnall and T.W. Chaundy, a) Commutative ordinary differential operators, *Proc. London Math. Soc.* **211** (1923), 420-440; b) —, *Proc. Royal Soc. London (A)* **118** (1928), 557-583; c) — II. The identity $P^n = Q^m$, *ibid.* **134** (1932), 471-485.

[D] P. Dehornoy, Opérateurs différentiels et courbes elliptiques, *Compositio Math.* **43** (1981), 71-99.

[DG] J.J. Duistermaat and F.A. Grünbaum, Differential equations in the spectral parameter, *Comm. Math. Phys.* **103** (1986), 177-240.

[DS] V.G. Drinfeld and V.V. Sokolov, On equations related to the Korteweg de Vries equation, *Soviet Math. Dokl.* **32** (1985), 361-365.

[EK] F. Ehlers and H. Knörrer, An algebro-geometric interpretation of the Bäcklund transformation for the Korteweg-de Vries equation, *Comment. Math. Helvetici* **57** (1982), 1-10.

[EMcK] N.M. Ercolani and H.P. McKean, a) A quick proof of Fay's secant identities, in *Analysis, et cetera*, Moser's Birthday Volume, eds. P.H. Rabinowitz and E. Zhender, Academic Press Inc., Boston 1990, pp. 301–307; b) Geometry of KdV (4): Abel sums, Jacobi variety, and theta functions in the scattering case, *Invent. Math.* **99** (1990), 483–544.

[G1] F.A. Grünbaum, Commuting pairs of linear ordinary differential operators of orders four and six, *Physica* **31D** (1988), 424–433.

[G2] F.A. Grünbaum, The Kadomtsev-Petviashvili equation: an elementary approach to the "rank 2" solutions of Krichever and Novikov, *Phys. Lett. A* **139** (1989), 146–150.

[G3] F.A. Grünbaum, Darboux's method and some "rank two" explicit solutions of the KP equation, in *Nonlinear evolution equations: integrability and spectral methods*, eds. A Degasperis, A.P. Fordy and M. Lakshmanan, Manchester University Press, Manchester, UK, 1990, 271–277.

[JM] M. Jimbo and T. Miwa, Solitons and infinite dimensional Lie algebras, *Pub. RIMS Kyoto Univ.* **19** (1983), 943–1001.

[K] I.M. Krichever, Commutative rings of ordinary differential operators, *Functional Anal. Appl.* **12** (1978), 175–185.

[KN] I.M. Krichever and S.P. Novikov, Holomorphic fiberings and nonlinear equations, Finite zone solutions of rank 2, *Sov. Math. Dokl.* **20** (1979), 650–654.

[L] G.A. Latham, Solutions of the Kadomtsev-Petviashvili equation associated to higher rank commuting ordinary differential operators, PhD Thesis, Berkeley, 1989.

[PW] E. Previato and G. Wilson, Vector bundles over curves and solutions of the KP equations, *Proc. Sympos. Pure Math.* **49** (1989), 553–569.

[SW] G.B. Segal and G. Wilson, Loop groups and equations of KdV type, *Publ. Math. I.H.E.S.* **61** (1985), 5–65.

[SSY] S.I. Svinolupov, V. Sokolov and R. Yamilov, On Bäcklund transformations for integrable evolution equations, *Soviet Math. Dokl.* **28** (1983), 165–168.

[W] G. Wilson, Algebraic curves and soliton equations, in *Geometry Today*, eds. E. Arbarello *et al.*, Birkhäuser, Boston, 1985, 303–329.

ON THE INITIAL VALUE PROBLEM OF
THE WHITHAM AVERAGED SYSTEM

Fei Ran Tian [1]

Courant Institute
251 Mercer Street
New York, NY 10012

INTRODUCTION

The initial value problem in question is concerning the Whitham averaged system:

$$\beta_{it} + \lambda_i(\beta_1, \beta_2, \beta_3)\beta_{ix} = 0 \qquad i = 1, 2, 3 \tag{1}$$

where

$$\lambda_1(\beta_1, \beta_2, \beta_3) = 2(\beta_1 + \beta_2 + \beta_3) + 4(\beta_1 - \beta_2)\frac{K(s)}{E(s)} \tag{2}$$

$$\lambda_2(\beta_1, \beta_2, \beta_3) = 2(\beta_1 + \beta_2 + \beta_3) + 4(\beta_2 - \beta_1)\frac{s^2 K(s)}{E(s) - (1 - s^2)K(s)} \tag{3}$$

$$\lambda_3(\beta_1, \beta_2, \beta_3) = 2(\beta_1 + \beta_2 + \beta_3) + 4(\beta_2 - \beta_3)\frac{K(s)}{E(s) - K(s)} \tag{4}$$

$$s^2 = \frac{\beta_2 - \beta_3}{\beta_1 - \beta_3}$$

$K(s)$ and $E(s)$ are complete elliptic integrals of first and second kind.

Lax - Levermore - Venakides [3, 7, 8] use the inverse scattering method to transform the problem of determining the zero dispersion limit of the KdV equation into the initial value problem of multi-phase averaged systems. However, this initial value problem which is important to the understanding of the oscillations and phase transition in the KdV zero dispersion limit has not been shown to have physical solutions. As a result, the existence of physical solutions to the multi-phase averaged systems is vital to Lax-Levermore-Venakides theory. The difficulty lies in the fact that multi-phase averaged systems become singular at the phase transition boundaries.

The purpose of this research is to attack the existence problem of the multi-phase averaged systems. In this paper, we consider the most important case – the single phase situation. This case is governed by the Whitham averaged system (1). We describe its initial value problem as follows:

Consider a horizontal motion of the initial curve. Each point on the curve has a different speed. Initially, the curve is expressed by a single valued function $u = u_0(x)$, and it moves

---

[1] New address: Department of Mathematics, University of Chicago, Chicago, IL 60637

Singular Limits of Dispersive Waves, Edited by
N.M. Ercolani et al., Plenum Press, New York, 1994

according to the Burgers equation (zero-phase averaged equation):

$$u_t + 6uu_x = 0 \qquad (5)$$

At a later time after the breaking of the Burgers solution, the evolving curve can only, in general, be given by a multi-valued function with odd number of branches. In this paper, we only deal with the case when the number of branches is three (see **Figure 1**).

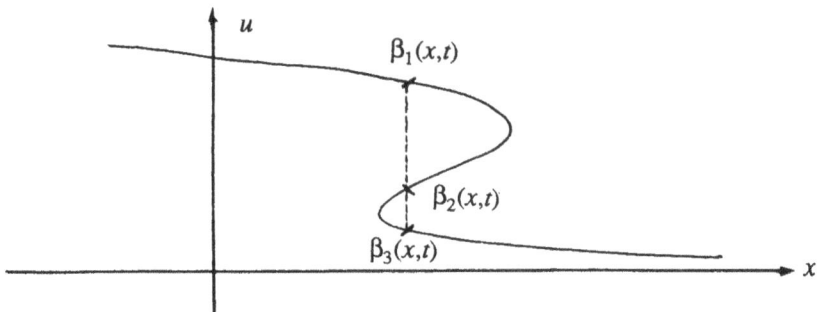

Figure 1

Let β_1, β_2 and β_3 be the three branches. Their motions are given by (1). Intuitively, β_1, β_2 and β_3 should match the Burgers solution at the boundaries of the multi-valued region:

a) At the trailing edge:

$$\beta_1 = the\ Burgers\ solution\ defined\ outside\ the\ region.$$
$$\beta_2 = \beta_3 \qquad (6)$$

b) At the leading edge:

$$\beta_1 = \beta_2$$
$$\beta_3 = the\ Burgers\ solution\ defined\ outside\ the\ region. \qquad (7)$$

The initial value problem of the Whitham averaged system is to determine the multi-branches β_1, β_2 and β_3 with boundary conditions (6) and (7) from the initial curve.

In this paper, we solve the initial value problem of the Whitham averaged system for very general initial data. Up to this point, only the initial value problems of system (1) for step-like initial data and initial data with cubic inverse functions were understood. The first one was straight forward [1, 3], while the latter was accomplished by Potemin [4] using Krichever [2]'s algebro-geometric method.

It is well known that the Burgers solution of (5) does not blow up only for increasing initial data. For convenience, we first consider the case when the initial data $u_0(x)$ is a decreasing function. Initial data with positive or negative humps will be considered later in this paper.

A LINEAR OVERDETERMINED SYSTEM OF EULER-POISSON-DARBOUX TYPE

Our strategy to solve the initial value problem of (1) is based on Tsarev [6]'s results about the Whitham averaged system: System (1) is locally integrable by a hodograph transform.

Theorem 1 (Tsarev) *If $w_i(\beta_1, \beta_2, \beta_3)$ solves the following linear overdetermined system:*

$$\frac{\partial w_i}{\partial \beta_j} = a_{ij}(\beta_1, \beta_2, \beta_3)[w_i - w_j] \qquad i, j = 1, 2, 3 \qquad i \neq j \tag{8}$$

where

$$a_{ij}(\beta_1, \beta_2, \beta_3) = \frac{\frac{\partial \lambda_i}{\partial \beta_j}}{\lambda_i - \lambda_j} \qquad i, j = 1, 2, 3 \qquad i \neq j \tag{9}$$

then the solution $(\beta_1(x, t), \beta_2(x, t), \beta_3(x, t))$ of the hodograph transformation:

$$x = \lambda_i(\beta_1, \beta_2, \beta_3)t + w_i(\beta_1, \beta_2, \beta_3) \qquad i = 1, 2, 3 \tag{10}$$

satisfies system (1). Conversely, any solution $(\beta_1, \beta_2, \beta_3)$ of system (1) can be obtained in this way in the neighborhood of (x_0, t_0) at which β_{ix}'s are not vanishing.

We shall use hodograph transform (10) to construct the Whitham solution satisfying boundary conditions (6) and (7). First, System (8) needs to be solved for $w_i(\beta_1, \beta_2, \beta_3)$'s. In this respect, we want to understand what kinds of boundary conditions should be imposed on $w_i(\beta_1, \beta_2, \beta_3)$'s.

Clearly, the Burgers solution outside the multi-valued region satisfies the characteristics equation:

$$x = 6ut + f(u) \tag{11}$$

where $x = f(u)$ is the inverse function of the initial data $u = u_0(x)$.

By (2), (3), (4), (10) and (11), we see:
At the trailing edge:

$$\begin{aligned} w_1(\beta_1, \beta_3, \beta_3) &= f(\beta_1) \\ w_2(\beta_1, \beta_3, \beta_3) &= w_3(\beta_1, \beta_3, \beta_3) \end{aligned} \tag{12}$$

Similar conditions hold at the leading edge:

$$\begin{aligned} w_1(\beta_1, \beta_1, \beta_3) &= w_2(\beta_1, \beta_1, \beta_3) \\ w_3(\beta_1, \beta_1, \beta_3) &= f(\beta_3) \end{aligned} \tag{13}$$

Therefore, it is natural to consider system (8) with boundary conditions (12) and (13). We shall explicitly construct all the solutions to this boundary value problem. Before that, a better understanding of (9) is necessary:

Lemma 2 *For $i, j = 1, 2, 3$ and $i \neq j$,*

$$a_{ij}(\beta_1, \beta_2, \beta_3) = \frac{1}{2} \frac{[\lambda_i - 2(\beta_1 + \beta_2 + \beta_3)] - 4(\beta_i - \beta_j)}{[\lambda_j - 2(\beta_1 + \beta_2 + \beta_3)](\beta_i - \beta_j)} \tag{14}$$

The proof is straight forwards, and involves only the complete elliptic integrals.

Using Lemma 2, we can construct solutions to system (8) with boundary conditions (12) and (13). This is done in the following theorem[2]:

Theorem 3 *If $q(\beta_1, \beta_2, \beta_3)$ is a symmetric solution of:*

$$2(\beta_i - \beta_j)\frac{\partial^2 q}{\partial \beta_i \partial \beta_j} = \frac{\partial q}{\partial \beta_i} - \frac{\partial q}{\partial \beta_j} \qquad i \neq j \tag{15}$$

$$q(\beta, \beta, \beta) = f(\beta) \qquad i, j = 1, 2, 3 \tag{16}$$

then (w_1, w_2, w_3) defined by:

$$w_i(\beta_1, \beta_2, \beta_3) = \frac{1}{2}[\lambda_i - 2(\beta_1 + \beta_2 + \beta_3)]\frac{\partial q}{\partial \beta_i} + q \qquad i = 1, 2, 3 \tag{17}$$

solves the boundary value problem (8), (12) and (13). Conversely, every solution of (8), (12) and (13) can be obtained in this way.

---

[2]The author was informed by the referee that (15) and (17) were also obtained independently by V. R. Kudashev and S. E. Sharapov.

Proof: By (17), we have:

$$\frac{\partial w_i}{\partial \beta_j} = \frac{1}{2}[\frac{\partial \lambda_i}{\partial \beta_j} - 2]\frac{\partial q}{\partial \beta_i} + \frac{1}{2}[\lambda_i - 2(\beta_1 + \beta_2 + \beta_3)]\frac{\partial^2 q}{\partial \beta_i \partial \beta_j} + \frac{\partial q}{\partial \beta_j}$$

$$w_i - w_j = \frac{1}{2}[\lambda_i - 2(\beta_1 + \beta_2 + \beta_3)]\frac{\partial q}{\partial \beta_i} - \frac{1}{2}[\lambda_j - 2(\beta_1 + \beta_2 + \beta_3)]\frac{\partial q}{\partial \beta_j}$$

Therefore, (8) is equivalent to:

$$\frac{1}{2}[\lambda_j - 2(\beta_1 + \beta_2 + \beta_3)][\frac{\partial q}{\partial \beta_i} - \frac{\partial q}{\partial \beta_j}]\frac{\frac{\partial \lambda_i}{\partial \beta_j}}{\lambda_i - \lambda_j}$$
$$= (\frac{\partial q}{\partial \beta_j} - \frac{\partial q}{\partial \beta_i}) + \frac{1}{2}[\lambda_i - 2(\beta_1 + \beta_2 + \beta_3)]\frac{\partial^2 q}{\partial \beta_i \partial \beta_j} \qquad (18)$$

Substituting (14) into (18), we finally obtain (15). Hence (8) is equivalent to (15).

Boundary conditions (12) and (13) can be checked as follows. We only consider the trailing edge, and the leading edge can be handled in the same way.

Since $q(\beta_1, \beta_2, \beta_3)$ is symmetric, we immediately deduce:

$$w_2(\beta_1, \beta_3, \beta_3) = w_3(\beta_1, \beta_3, \beta_3) \qquad (19)$$

As to the first condition of (12), it follows from (2) and (17) that

$$w_1(\beta_1, \beta_3, \beta_3) = 2(\beta_1 - \beta_3)\frac{\partial q}{\partial \beta_1} + q \qquad (20)$$

Differentiating (20) with respect to β_3, we have:

$$\frac{dw_1(\beta_1, \beta_3, \beta_3)}{d\beta_3} = -2\frac{\partial q}{\partial \beta_1} + 2(\beta_1 - \beta_3)[\frac{\partial^2 q}{\partial \beta_1 \partial \beta_2} + \frac{\partial^2 q}{\partial \beta_1 \partial \beta_3}] + \frac{\partial q}{\partial \beta_2} + \frac{\partial q}{\partial \beta_3}$$
$$= 0$$

where in the last equality we have used (15).

$w_(\beta_1, \beta_3, \beta_3)$ is independant of β_3, and therefore, the first condition of (12) follows by substituting $\beta_3 = \beta_1$ into (20). This proves the first part of Theorem 3. The second part follows from the uniqueness of the boundary value problem (8), (12) and (13).

System (15) and (16) can be solved explicitly:

Theorem 4 *The Cauchy problem (15) and (16) has one and only one solution. The solution is symmetric, and given by:*

$$q(\beta_1, \beta_2, \beta_3) = \frac{1}{2\sqrt{2\pi}} \int_{-1}^{1} \int_{-1}^{1} \frac{f(\frac{1+\mu}{2}\frac{1+\nu}{2}\beta_1 + \frac{1+\mu}{2}\frac{1-\nu}{2}\beta_2 + \frac{1-\mu}{2}\beta_3)}{\sqrt{(1-\mu)(1-\nu^2)}} d\mu d\nu \qquad (21)$$

The proof can be found in [5].

Remark *System:*

$$2(\beta_i - \beta_j)\frac{\partial^2 q}{\partial \beta_i \partial \beta_j} = \frac{\partial q}{\partial \beta_i} - \frac{\partial q}{\partial \beta_j} \qquad i \neq j \qquad (22)$$

$$q(\beta, \cdots, \beta) = f(\beta) \qquad i, j = 1, 2, \cdots, n \qquad (23)$$

can be solved in the same way. When $n = 2$, system (22) and (23) becomes the classical Euler-Poisson-Darboux equation.

Formula (21) gives all the smooth solutions to the Cauchy problem (15) and (16). We shall give a different approach to construct all the analytic solutions of the same Cauchy problem.

Consider the following Laurant expansion at $\xi = \infty$:

$$\frac{1}{\sqrt{(\xi - \beta_1)(\xi - \beta_2)(\xi - \beta_3)}} = \xi^{-\frac{3}{2}}[\Gamma_0 + \Gamma_1 \xi^{-1} + \cdots + \Gamma_k \xi^{-k} + \cdots] \qquad (24)$$

The following indentity is easily checked: for $i, j = 1, 2, 3, \ i \neq j$

$$2(\beta_i - \beta_j)\frac{\partial^2}{\partial\beta_i\partial\beta_j}[\frac{1}{\sqrt{(\xi - \beta_1)(\xi - \beta_2)(\xi - \beta_3)}}]$$
$$= \frac{\partial}{\partial\beta_i}[\frac{1}{\sqrt{(\xi - \beta_1)(\xi - \beta_2)(\xi - \beta_3)}}] - \frac{\partial}{\partial\beta_j}[\frac{1}{\sqrt{(\xi - \beta_1)(\xi - \beta_2)(\xi - \beta_3)}}]$$

which when combined with (24) establishes:

Theorem 5 $\Gamma_k(\beta_1, \beta_2, \beta_3)$ *of (24) is the unique solution to the Cauchy problem (15) and (16) with* $f(\beta) = \Gamma_k(1, 1, 1)\beta^k$. *All the analytic solutions can be constructed by superposition.*

Remark *It is interesting to note that (17) with* $q = -\Gamma_3(\beta_1, \beta_2, \beta_3)$ *gives Potemin [4]'s algebro-geometric solution. Later, we will see that all* $\Gamma_k(\beta_1, \beta_2, \beta_3)$ *(k = 3, 5, 7, $\cdots$) produce self-similar solutions of the Whitham averaged system.*

THE HODOGRAPH TRANSFORM

We are ready to use Tsarev's theorem to construct the solution to the initial value problem of (1).

Hodograph transform (10) with $w_i(\beta_1, \beta_2, \beta_3)$ given by (17) and (21) has to be solved for β_1, β_2 and β_3 as functions of x and t. In Section 1, we understand that, at the beginning, we have the Burgers solution, and that, later after the breaking of the Burgers solution, we can expect to have the Whitham solution.

Since $t_b = -[6minu_0'(x)]^{-1}$ is the breaking time of the Burgers solution of (5), the breaking is caused by an inflection point in the initial data. Without loss of generality, we may assume the breaking point to be at the origin of the u-x-t space. It means that we are starting at the breaking time, and that the evolving curve blows up at the origin of the x-u plane. It immediately follows that:

$$f(0) = f'(0) = f''(0) = 0 \tag{25}$$

where, as in the last section, $x = f(u)$ is the inverse function of the initial data $u = u_0(x)$. On the assumption that $x = f(u)$ has only one inflection point, it follows from the monotonicity of $f(u)$ that:

$$f''(u) = \begin{cases} < 0 & u > 0 \\ = 0 & u = 0 \\ > 0 & u < 0 \end{cases} \tag{26}$$

The following theorem says that, under a little bit stronger condition than (26), the hodograph transform can be solved for β_1, β_2 and β_3 as functions of x and t.

Theorem 6 *If, in addition to (25), $f'''(u) < 0$ for $u \neq 0$, then transform (10) with w_i's given by (17) and (21) can be solved for β_1, β_2 and β_3 within a cusp in the x-t plane for all t > 0. Furthermore, these β_1, β_2, β_3 satisfy boundary conditions (6) and (7) on the cusp.*

The proof is long, and can be found in [5]. The main mathematical tool is the Implicit Function Theorem. In this paper, we can only sketch the proof:

We eliminate x from transform (10) to obtain two equations involving β_1, β_2, β_3 and t. These two equations can be shown, for each fixed time after the breaking, to determine β_1 and β_3 as decreasing functions of β_2 within an interval whose end points depend on t. Substituting β_1 and β_3 as funtions of β_2 into the hodograph transform, we find that, within a cusp in the x-t plane, β_2 is a function of (x, t), and so, therefore, are β_1 and β_3.

Theorem 1 and Theorem 6 immediately give:

139

Theorem 7 *If, in addition to (25), $f'''(u) < 0$ for all $u \neq 0$, then system (1) has a solution $(\beta_1, \beta_2, \beta_3)$ within a cusp in the x-t plane for all $t > 0$. Furthermore, this solution satisfies boundary conditions (6) and (7) on the cusp.*

If, in particular, $f(u) = -u^k$ $(k = 3, 5, 7, \cdots)$, then by Theorem 5, $[-\frac{\Gamma_k(\beta_1, \beta_2, \beta_3)}{\Gamma_k(1,1,1)}]$ solves (15) and (16). It is clear from (2), (3) and (4) that $\lambda_i(\beta_1, \beta_2, \beta_3)$'s are homogeneous of order 1. $[-\frac{\Gamma_k(\beta_1, \beta_2, \beta_3)}{\Gamma_k(1,1,1)}]$ is homogeneous of order k, and so, therefore by (17), are the resulting $w_i(\beta_1, \beta_2, \beta_3)$'s.

Let

$$X = \frac{x}{t^{\frac{k}{k-1}}}$$

$$\Omega_i = \frac{\beta_i}{t^{\frac{1}{k-1}}} \qquad i = 1, 2, 3$$

Then (10) becomes a time-free hodograph transform:

$$X = \lambda_i(\Omega_1, \Omega_2, \Omega_3) + w_i(\Omega_1, \Omega_2, \Omega_3) \qquad i = 1, 2, 3$$

which by Theorem 6 determines Ω_i's as functions of X. Therefore, we have:

Corollary 8 *For $f(u) = -u^k$, $k = 3, 5, 7, \cdots$, the Whitham averaged system has a global self-similar solution $(\beta_1, \beta_2, \beta_3)$:*

$$\beta_i(x, t) = t^{\frac{1}{k-1}} \Omega_i(\frac{x}{t^{\frac{k}{k-1}}}) \qquad i = 1, 2, 3$$

within a cusp in the x-t plane: $bt^{\frac{k}{k-1}} < x < at^{\frac{k}{k-1}}$, $t > 0$ where $a > b$ are two real constants. Furthermore, this solution satisfies boundary conditions (6) and (7) on the cusp.

Local results also hold if we assume local conditions:

Theorem 9 *If, in addition to (25), $f'''(u) < 0$ locally holds in a deleted neighborhood of $u = 0$, then system (1) has a solution $(\beta_1, \beta_2, \beta_3)$ within a cusp in the x-t plane for a positive short time. Furthermore, this solution satisfies boundary conditions (6) and (7) on the cusp.*

INITIAL DATA WITH HUMPS

In Section 1, we have assumed for convenience that the initial data is a decreasing function. Here we indicate how arbitrary initial data can be handled by our method.

A hump-like initial data can be decomposed into two kinds of data: increasing and decreasing ones. These monotonic data will not interact with one another until a later time after the breaking. Our previous results say that for a decreasing initial data with an isolated inflection point, the Whitham averaged system has a solution within a cusp in the x-t plane for a positive short time after the breaking. Furthermore, this solution matches the Burgers solution on the cusp. A similar short time result clearly holds for a hump-like initial data.

We conclude this paper by an important remark. The uniqueness of solution to the initial value problem of the Whitham averaged system is guaranteed by Lax-Levermore-Venakides theory. The detail can be found in [3, 7, 8].

Acknowledgment. I am grateful to my thesis adviser, Peter D. Lax, for suggesting this problem and discussing it with me frequently. Above all, I thank him for giving me the opportunity to watch him doing mathematics. I would also like to thank Percy A. Deift, C. David Levermore, David W. McLaughlin and Stephanos Venakides for fruitful discussion and encouragement.

This research was supported in part by the NYU GSAS Dean's Dissertation Award for the 1990-1991 academic year.

References

[1] A. V. Gurevich and L. P. Pitaevskii, "Non-stationary Structure of a Collisionless Shock Wave", Soviet Phys. JETP, 38 (1974), 291-297.

[2] I. M. Krichever, "The Method of Averaging for Two-dimensional 'Integrable' Equations", Functional Anal. App., 22 (1988), 200-213.

[3] P. D. Lax and C. D. Levermore, "The Small Dispersion Limit for the Korteweg-de Vries Equation I, II, and III", CPAM, 36 (1983), 253-290, 571-593, 809-830.

[4] G. V. Potemin, "Algebro-geometric Construction of Self-similar Solutions of the Whitham Equations", Russian Math. Surveys, 43:5 (1988), 252-253.

[5] F. R. Tian, "Oscillations of the Zero Dispersion Limit of the Korteweg-de Vries Equation", to appear in CPAM.

[6] S. P. Tsarev, "Poisson Brackets and One-dimensional Hamiltonian Systems of Hydrodynamic Type", Soviet Math. Dokl., 31 (1985), 488-491.

[7] S. Venakides, "The Zero Dispersion Limit of the KdV Equation with Nontrivial Reflection Coefficient", CPAM, 38 (1985), 125-155.

[8] S. Venakides, "Higher Order Lax-Levermore Theory", CPAM, 43 (1990), 335-362.

ON THE INTEGRABILITY OF THE AVERAGED KDV AND BENNEY EQUATIONS

Serguei P. Tsarev

Steklov Mathematical Institute
117966,   Vavilova, 42
Moscow,   GSP-1, USSR

The general theory of hamiltonian systems of hydrodynamic type was developed by B. A. Dubrovin and S. P. Novikov [7], [8]. Here we study only one-dimensional hamiltonian systems possessing a complete set of Riemann invariants arising in the theory of multiphase averaging of completely integrable equations (see [2], [4], [13], [20], [32]), for example the Whitham equations (the averaged 1-phase KdV equation):

$$
(1) \quad
\begin{aligned}
u^1_t &= v_1(u) \cdot u^1_x \;, \quad v_1(u) = \frac{u^1 + u^2 + u^3}{3} - \frac{2 \cdot (u^2 - u^1) \cdot K(s)}{3 \cdot (K(s) - E(s))} \;, \\[2mm]
u^2_t &= v_2(u) \cdot u^2_x \;, \quad v_2(u) = \frac{u^1 + u^2 + u^3}{3} - \frac{2 \cdot (u^2 - u^1) \cdot (1 - s^2) \cdot K(s)}{3 \cdot (E(s) - (1 - s^2) \cdot K(s))} \;, \\[2mm]
u^3_t &= v_3(u) \cdot u^3_x \;, \quad v_3(u) = \frac{u^1 + u^2 + u^3}{3} + \frac{2 \cdot (u^3 - u^1) \cdot (1 - s^2) \cdot K(s)}{3 \cdot E(s)} \;,
\end{aligned}
$$

(here $s^2 = (u^2 - u^1)/(u^3 - u^1)$; $E(s)$, $K(s)$ are the complete elliptic integrals) and Benney equations:

$$
(2) \quad
\begin{aligned}
&h^i_t + (q^i \cdot h^i)_x = 0, \qquad i = 1, \ldots, N, \\[2mm]
&q^i_t + q^i \cdot q^i_x + S_x = 0, \quad S = h^1 + \ldots + h^N.
\end{aligned}
$$

We use in (2) Zakharov's [33] form for N-layer flows, its diagonalizability was proved by J. Gibbons [14]. Riemann invariants $\lambda^k, k = 1, \ldots, 2N$, for (2) and the diagonal coefficiens of the system in the diagonal form are implicitely defined as the branching points (λ^k, μ^k) of a rational Riemannian surface $\Gamma(\lambda, \mu)$:

$$
(3) \qquad \lambda = \mu + \sum_i \frac{h^i}{q^i + \mu}
$$

Let us recall briefly the main results of [7], [8]. A (generally non-diagonal) system $u^i_t = v^i_j(u) \cdot u^j_x$ is hamiltonian if there exist a hamilto nian $H = \int h(u)dx$ and a hamiltonian operator

$$\hat{A}^{ij} = g^{ij}(u)d/dx + b^{ij}_k(u) \cdot u^k_x$$

which defines a skew-symmetric Poisson bracket on functionals

$$\{I,J\} = \int \frac{\delta I}{\delta u^i(x)} \hat{A}^{ij} \frac{\delta J}{\delta u^j(x)} dx$$

satisfying the Jacobi identity and generates the system

$$(4) \qquad u^i_t = \{u^i, H\} = \hat{A}^{ij}\frac{\delta H}{\delta u^j(x)} = (g^{ij}\partial_k\partial_j h + b^{ij}_k\partial_j h)u^k_x = v^i_k(u) \cdot u^k_x$$

where $\partial_s = \partial/\partial u^s$. B. A. Dubrovin and S. P. Novikov [7] proved that the necessary and sufficient conditions for (4) to be a hamiltonian operator in the case of non-degeneracy of the matrix g^{ij} are:

   a) $g^{ij} = g^{ji}$, i.e. the inverse matrix g^{-1} defines a Riemannian metric.

   b) $b^{ij}_k = -g^{is}\Gamma^j_{sk}$ for the standard Christoffel symbols Γ^j_{sk} generated by g_{ij}.

   c) the metric g_{ij} has identically vanishing curvature tensor.

In such case we have $v^i_j(u) = \nabla^i\nabla_j h = g^{is}\nabla_s\nabla_j h$ with the covariant derivatives defined by g_{ij}. Consequently $g_{ik}v^k_j = g_{jk}v^k_i$, $\nabla_j v^i_k = \nabla_k v^i_j$ in view of the zero-curvature property of g_{ij}. For a *diagonal* matrix $v^i_j(u) = v_j(u)\delta^i_j$ this implies that (see [29], [30])

   a) g_{ij} is also diagonal;

   b) $$\frac{\partial_i v_k}{v_i - v_k} = \Gamma^k_{ki} = 1/2 \cdot \partial_i \ln g_{kk} \ , \quad \partial_s = \partial/\partial u^s,$$

(hereafter we do not imply the summation on repeated indices!).
From b) we deduce

$$(5) \qquad \partial_j \frac{\partial_i v_k}{v_i - v_k} = \partial_i \frac{\partial_j v_k}{v_j - v_k} \ , \quad \text{for all different } i, j, k.$$

Following [30] we call a diagonal system

$$(6) \qquad u^i_t = v_i(u) \cdot u^i_x \ , \quad i = 1, \ldots, n,$$

semihamiltonian if $n=2$ or if $n>2$ and $v_i(u)$ satisfy (5). As a physical example of a semihamiltonian (but non-hamiltonian for $n>3$) system one can mention the ideal Langmuir chromatography and electrophoresis systems ([23], [26], [30]).

   Theorem [29]. Any smooth solution $u^i(x,t)$ of a semihamiltonian system (6) in a neighbourhood of a generic point (x_0, t_0) (where u^i_x are non-zero for all i) can be found as the solution of the system

(7) $w_k(u) = v_k(u) \cdot t + x$, $i = 1, \ldots, n$,

where $w_k(u)$ are solutions of a linear system

(8) $\partial_i w_k = \Gamma_{ki}^k (w_i - w_k)$, $i \neq k$, $\Gamma_{ki}^k = \partial_i v_k / (v_i - v_k) = 1/2 \cdot \partial_i \ln g_{kk}$

We recall that $w_k(u)$ are the coefficients of commuting with (6) flows
$u_\tau^i = w_i(u) \cdot u_x^i$. System (8) is compatible for semihamiltonian systems, the Cauchy data are n functions of one variable – values of w_k on a non-characteristic curve $u^i = \gamma^i(y)$ (i.e. $\gamma_y^i \neq 0$ for all i and y).

The problem of integration of a semihamiltonian system is therefore reduced to the problem of integration of the *linear* system (8). Here we list the known semihamiltonian systems for which (8) is solved:

1) Numerous systems with $n=2$ equations ([5], [15], [27], [31], see also [16] for $n>2$).
2) Chromatography and electrophoresis system ([23], [30]).
3) Weakly nonlinear semihamiltonian systems (i.e. semihamiltonian systems (6) with $\partial_i v_i = 0$) ([12], [25]).
4) The Temple class of systems ([26]).
5) Hamiltonian systems (6) invariant with respect to:

a) Galilei transformations $(x, t) \longrightarrow (x - V \cdot t, t)$, $u^i \longrightarrow u^i \pm V$;

b) Scaling transformations $x \longrightarrow c \cdot x$, $u^i \longrightarrow c^i \cdot u^i$.

In this paper we will concentrate ourselves on the last class of systems. Evident examples of such systems are Whitham (1) and Benney (2) systems. We will construct a complete basis of solutions of (8) for them. As a byproduct a complete set of (infinitely many) concervation laws of hydrodynamic type will be given. In § 5, § 6, § 7 additional structures (other hamiltonian structures, higher-order symmetries and conservation laws) for (1), (2) are discussed.

Recently ([10], [17]) alternative procedures for the integration of (1), (2) were given.

The author thanks Professor B. A. Dubrovin, Professor S. P. Novikov and Professor E. V. Ferapontov for many helpful discussions. I would also like to thank Professor D. Serre, Professor N. Ercolani and Professor C. D. Levermore for an invitation to the Workshop and their hospitality in Lyon.

1. GALILEI INVARIANCE AND THE EGOROV CLASS OF ORTHOGONAL CURVILINEAR COORDINATE SYSTEMS IN $\mathbb{R}^n$

The theory of diagonal zero-curvature metrics is equivalent to the classical theory of orthogonal curvilinear coordinate systems in (pseudo) euclidean space $\mathbb{R}^n$ (cf. [30]). Moreover many formulas of the theory of diagonal hamiltonian systems have their counterparts in the classical tracts of G. Darboux [6], L. Bianchi [1] and D. Th. Egorov [11]. For example the solution of (8) is equivalent to the construction of all orthogonal coordinate systems in the same space, possessing the following property: one can find a one-to-one correspondence between such a coordinate system and the original orthogonal coordinate system (defined by the diagonal metric related to the diagonal hamiltonian system in question) in such a way that the hyperplanes tangent to the coordinate surfaces for both systems are parallel (see [6], [30]).

Another equivalence: if a hamiltonian system (6) is Galilei-invariant its diagonal zero-curvature metric is "typically" potential (i.e. $\partial_i g_{kk} = \partial_k g_{ii}$; such metrics were called in [6] "Egorov type metrics") and translation-invariant (i.e. $\hat{T} g_{ii} = 0$, $\hat{T} = \partial_1 + \ldots + \partial_n$). For the correct statement of the result see [30]. One can easily check both properties for the metric

$$g_{11} = \frac{(K - E)^2}{s^2 K^2}, \quad g_{22} = \frac{-(E - (1-s^2)K)^2}{s^2 (1-s^2) K^2}, \quad g_{33} = \frac{E^2}{(1-s^2)K^2}, -$$

for the Whitham equations (1) and the metric for Benney system (2) in diagonal form (g_{ii} coincides with the residues of the holomorphic 1-form $(\partial\mu/\partial\lambda)d\mu$ in the branching point (λ^1, μ^1) of the Riemannian surface (3); an analogous residue formula for Whitham system see in [9]). The consequences of these invariances are the existence of a recursion operator generating higher-order symmetries for (1), (2) and the existence of higher-order conservation laws for these systems (see [30] and § 6 below).

Established in [9] equivalence of the equations describing potential (Egorov) metrics in $\mathbb{R}^3$ and the well-known integrable system (pure imaginary reduction of 3-wave resonant interactions) is mirrored by practically identical formulas for their Bäcklund transformations (cf. [1], [3]).

Instead of the direct study of (8) we would prefer to introduce an equivalent but simpler system. Let $g_{ii}(u)$ be the diagonal zero-curvature metric connected with a diagonal hamiltonian system (6), $H_i = \sqrt{g_{ii}}$, $\beta_{ik}(u) = \partial_i H_k / H_i$. For the metrics connected with semihamiltonian systems one can easily find $\partial_j \beta_{ik} = \beta_{ij} \beta_{jk}$ for $i \neq j \neq k$.

Proposition 1. Let $\bar{H}_i(u)$ be a solution of equations

(9)
$$\partial_i \bar{H}_k = \beta_{ik} \cdot \bar{H}_k , \qquad i \neq k,$$

compatible for any semihamiltonian system. Then $w_i = \bar{H}_i / H_i$ satisfy (8). Conversely, for any solution $w_i(u)$ of (8) $\bar{H}_i = w_i H_i$ satisfy (9).

This proposition actually dates to [6] or earlier. As we have stated above the diagonal metrics for Whitham and Benney equations are translation-invariant, consequently $\hat{T} \beta_{ik} = 0$. Therefore the system

(10)
$$\partial_i \bar{H}_k = \beta_{ik} \cdot \bar{H}_k , \qquad i \neq k ,$$
$$\hat{T} \bar{H}_i = 0 ,$$

is compatible and has n linearly independent solutions denoted here by $H_i^{(0,s)}$, $s = 1, \ldots, n$. For the Whitham system $n=3$ and the basis is

$$H_1^{(0,1)} = H_1 = \frac{K - E}{s\,K} \quad, \quad H_2^{(0,1)} = H_2 = \frac{i\,(E - (1-s^2)K)}{s\,\sqrt{(1-s^2)}\,K} \quad,$$

$$H_3^{(0,1)} = H_3 = \frac{E}{\sqrt{(1-s^2)}\,K} \quad, \quad H_1^{(0,2)} = \frac{-\sqrt{(1-s^2)}}{s\,\sqrt{(u^3-u^2)}\,K} \quad,$$

$$H_2^{(0,2)} = \frac{i}{s\,\sqrt{(u^3-u^2)}\,K} \quad, \quad H_3^{(0,2)} = \frac{1}{\sqrt{(u^3-u^2)}\,K} \quad,$$

$$H_1^{(0,3)} = \frac{\sqrt{(u^3-u^2)}\,(E^2 - 2EK + (1-s^2)K^2)}{2\,s\,\sqrt{(1-s^2)}\,K} \quad,$$

$$H_2^{(0,3)} = \frac{-i\,\sqrt{(u^3-u^2)}\,(E^2 - 2(1-s^2)EK + (1-s^2)K^2)}{2\,s\,(1-s^2)\,K} \quad,$$

$$H_3^{(0,3)} = \frac{\sqrt{(u^3-u^2)}\,(-E^2 + (1-s^2)K^2)}{2\,(1-s^2)\,K} \quad,$$

Let us define recursively $H_i^{(p,s)}$, $s = 1,\ldots,n$, $p>0$ as solutions of

(11)
$$\partial_i H_k^{(p,s)} = \beta_{ik} \cdot H_k^{(p,s)} \quad, \qquad i \neq k,$$

$$\hat{T}\,H_i^{(p,s)} = H_i^{(p-1,s)} \quad,$$

<u>Theorem 1</u>. The constructed $H_i^{(p,s)}$, $p = 0, 1, \ldots$ are linearly independent and generate a basis in the space of all solution of (9).

<u>Proof</u>. Let $\gamma^i(y) = u_0^i + y$ be a curve containing a predefinite point u_0^i. We can suppose that $H_i^{(0,s)} = \delta_i^s$. Then $H_i^{(p,s)}(\gamma(y))$ are polynomials of degree n with the leading term $\delta_i^s y^p/p!$, so any Cauchy data can be approximated by linear combinations of $H_i^{(p,s)}$. Since one can prove the continuous dependence of solutions of (9) on the Cauchy data (see the proofs of analogous propositions in [6]) this completes the proof.

In principle solutions of the inhomogeneous system (11) may be found by quadratures using the known solutions of the homogeneous system (10). But for the systems possessing *the scaling symmetry* such as (1), (2) these quadratures can be avoided.

2. AN ALGEBRAIC ALGORITHM FOR CONSTRUCTION OF $H_i^{(p,s)}$ FOR WHITHAM EQUATIONS

The scaling invariance of (1) is mirrored by homogeneity of $H_i^{(0,k)}$:

$$\hat{S} \, H_i^{(0,k)} = s_k \, H_i^{(0,k)}, \text{ and } \beta_{ik} : \hat{S} \, \beta_{ik} = - \beta_{ik} \, , \text{ where}$$

$$\hat{S} = u^1 \partial_1 + \ldots + u^n \partial_n \text{ and } s_1 = 0, \; s_2 = -1/2, \; s_3 = 1/2.$$

Proposition 2. For the system (9) related to the Whitham equations we can find the homogeneous basis $H_i^{(p,k)}$ of solutions of (11):

$$\hat{S} \, H_i^{(p,k)} = (s_k + p) \, H_i^{(p,k)}.$$

Proof. Since $H_i^{(0,k)}$ are homogeneous we shall prove that for a homogeneous $H_i^{(p-1,k)}$, $\hat{S} \, H_i^{(p-1,k)} = \lambda \, H_i^{(p-1,k)}$, one can find a homogeneous solution of (11). If $\bar{H}_i$ is arbitrary solution of (11), the commutation relation $[\hat{T}, \hat{S}] = \hat{T}$ implies $\hat{T} \, (\hat{S} \, \bar{H}_i - (\lambda+1)\bar{H}_i) = 0$, i.e. $\hat{S} \, \bar{H}_i = (\lambda+1)\bar{H}_i +$

$+ c_1 H_i^{(0,1)} + c_2 H_i^{(0,2)} + c_3 H_i^{(0,3)}$. Then $H_i^{(p,k)} = \bar{H}_i + x_1 H_i^{(0,1)} +$

$+ x_2 H_i^{(0,2)} + x_3 H_i^{(0,3)}$ with $x_1 = c_1/(\lambda+1)$, $x_2 = c_2/(\lambda+3/2)$, $x_3 = c_3/(\lambda+1/2)$, are homogeneous. The only zero-denominator case for x_i is $p=1$, $k=2$, $\lambda=-1/2$. In this case we can find $H_i^{(1,2)}$ solving (11) via the standard variable coefficiets method (see below the proof of Theorem 2).

$$H_1^{(1,2)} = \frac{i((u^1-u^3)(2E^2K' - 4EKK' - \pi E) + 2K^2K'(u^2-u^3) + \pi K(2u^1-u^3))}{\pi\sqrt{(u^1-u^2)} \, K^2} \, ,$$

$$H_2^{(1,2)} = \frac{(u^1-u^3)E(2EK' - \pi) + 2(u^2-u^3)KK'(K - 2E) + \pi K(2u^2-u^3)}{\pi\sqrt{(u^2-u^3)} \, sK^2} \, ,$$

$$H_3^{(1,2)} = \frac{(u^1-u^3)E(2EK' - \pi) - 2(u^2-u^3)K^2K' + \pi K u^3}{i\pi\sqrt{(u^2-u^3)} \, sK^2} \, ,$$

$$K' = K(\sqrt{(1-s^2)}).$$

Theorem 2. All homogeneous $H_i^{(p,k)}$ can be found recursively with algebraic operations from $H_i^{(p-1,k)}$ for $(p,k) \neq (1,2)$.

<u>Proof</u>. Substituting the standard variable coefficients decomposition $H_i^{(p,k)} = \sum c_q(u) \cdot H_i^{(0,q)}$ in (11) we can find all derivatives $\partial_i c_q$ via $H_i^{(p-1,k)}$ and $H_i^{(0,q)}$. For $H_i^{(p,k)}$ to be homogeneous $c_q(u)$ shall satisfy $\hat{S} c_q = (s_k + p - s_q) \cdot c_q$. Then $c_q = \hat{S} c_q / (s_k + p - s_q) = \sum_i u^i \partial_i c_q / (s_k + p - s_q)$, where $\partial_i c_q$ are known. The only zero-denominator case is $p=1$, $k=2$, $q=3$. This case requires a quadrature resulting in $c_3 = -4K' / (\pi K)$.

The found $H_i^{(p,1)}$ series corresponds to the averaged Kruskal symmetries of the original KdV equation. Consequently these averaged symmetries give us *incomplete* family of hydrodynamic type symmetries for Whitham equations. This is also true for the averaged multiphase KdV equations: in this case $n=2g+1$ but only 1 of n series $H_i^{(p,k)}$ is the averaged Kruskal series. For Benney equations (2) which can be represented as a 0-phase averaged coupled Schrödinger equations ([33]) only N of $2N$ series can be found as the averaged ones. The complete basis was found in [24], see also § 4. Moreover any Galilei-invariant semihamiltonian system has higher-order symmetries and conservation laws in addition to the hydrodinamic ones (§ 6).

Generally speaking one can add the homogeneity condition $\hat{S} \bar{H}_i = s \bar{H}_i$ (with arbitrary s) to (9) and find n linearly independent solution . For Whitham equations we have found two of them for $s=1/2$, $3/2$, ... and only one for $s=0$, 1, 2, ... To find the additional (redundant) homogeneous $\bar{H}_i$ one can use a useful substitution proposed by V. R. Kudashev, S. E. Sharapov [18], [19] and independently by F. R. Tian: introducing operators

$$\hat{Q} = 1 + q_i(u) \partial_i \ , \qquad i=1,2,3, \qquad q_1 = 2K \cdot (u^1 - u^2) / (K - E),$$

$q_2 = 2K \cdot (u^1 - u^2)(1-s^2) / (E - (1-s^2)K)$, $q_3 = 2K \cdot (u^3 - u^2) / E$, one can easily check that the diagonal coefficients of (1) are $v_i(u) = \hat{Q} U$, $U = (u^1 + u^2 + u^3)/3$.

<u>Theorem 3</u> [18]. Every solution of (8) for Whitham equations (1) has the form $w_i(u) = \hat{Q} V$, $V(u)$ being any soliton of

$$(12) \qquad 2 \cdot (u^i - u^j) \partial_i \partial_j V = \partial_i V - \partial_j V \ , \qquad i \neq j.$$

Analogous substitutions give *the same system* (12) also for the averaged NLS ([19]) and sine-Gordon equations (V. R. Kudashev, S. E. Sharapov, in press). Another remarkable feature of (12) (see [18]): besides the symmetries with generators $\hat{T}$, $\hat{S}$ it has also a symmetry generator $\hat{R} = \sum (2(u^i)^2 \partial_i + u^i)$. Since (12) is linear, $\hat{R}$ is its recursion operator and for any homogeneous $V(u)$, $\hat{S} V = sV$, $V_1 = \hat{R} V$ we have $\hat{S} V_1 = (s+1) V_1$, that gives a differential recursion scheme for generation of $H_i^{(p,k)}$ in contrast with the algebraic procedure given above. We can receive a rich class of homogeneous solutions of (12) using the following simple result.

   Proposition 3 (Tsarev). If $w_i(u)$ are homogeneous: $\hat{S}\,w_i = s\,w_i$ and $s \neq -1/2$, their "potential" V can be found algebraically:
$V = (s+1/2)^{-1} \cdot \sum (u^i w_i(u)/q_i(u))$.

   By the way, $w_i = H_i^{(p,1)}/H_i^{(0,1)}$ (the averaged Kruskal symmetries) correspond to polynomial $V(u)$.

   To receive addditional homogeneous solution of (12) we can use arbitrary permutation $\hat{P}$ of coordinates (u^1,u^2,u^2) and the following discrete symmetry: $\hat{I} : V(u^1,u^2,u^2) \longrightarrow (u^1u^2u^2)^{-1/2}V(1/u^1,1/u^2,1/u^2)$. Applying operators $\hat{T}, \hat{L}, \hat{I}, \hat{P}$ we receive a wider (redundant) class of solutions for (12). That gives us additional homogeneous solutions to (9).

§ 3. THE COMPLETE SET OF HYDRODYNAMIC TYPE CONSERVED DENSITIES FOR WHITHAM
      EQUATIONS

   A functional $I=\int P(u)dx$ gives a conservation law for a semihamiltonian system (6) if and only if

(13) $\partial_i\partial_k P = \Gamma_{ki}^k \partial_k P + \Gamma_{ik}^i \partial_i P$, $i \neq k$, $\Gamma_{ki}^k = \partial_i v_k /(v_i - v_k) = 1/2 \cdot \partial_i \log g_{kk}$.

The system (13) is compatible and has solutions parametrized (up to a trivial constant term) with n functions of 1 variable - the Cauchy data $z_i = \partial_i P$ on a noncharacteristic curve $\gamma^i(y)$. We can rewrite (13) in a simpler form using $C_i = \partial_i P/H_i$:

(14) $\partial_i C_k = \beta_{ki} C_i$, $i \neq k$.

For potential metrics $\beta_{ki} = \beta_{ik}$ and (14) coincide with (9).

   Theorem 4. If the metric $g_{ii} = H_i^2$ of a hamiltonian system (6) is potential and translation-invariant then for any solution $\bar{H}_i$ of (9) the density $\bar{P} = \sum H_i \bar{H}_i$ satisfy (13). If $H_i^{(p,k)}$ is the basis built in Theorem 1, $P^{(s,k)} = \sum H_i H_i^{(s,k)}$, $s>0$, is the complete basis of solutions for (13) . $P^{(1,k)}$ are the flat coordinates for the metric tensor g_{ii}.

   Proof. $\partial_k(\sum H_i \bar{H}_i) = \sum_{i \neq k} (\beta_{ik} H_k \bar{H}_i + \beta_{ik} \bar{H}_k H_i) + \partial_k(H_k \bar{H}_k) =$

$= \sum_{i \neq k} (\beta_{ik} H_k \bar{H}_i + \beta_{ik} \bar{H}_k H_i) + (\hat{T}H_k)\bar{H}_k + (\hat{T}\bar{H}_k)H_k - \sum_{i \neq k} (\partial_i H_k \bar{H}_k + \partial_i \bar{H}_k H_k) =$

$= (\hat{T}H_k)\bar{H}_k + (\hat{T}\bar{H}_k)H_k = (\hat{T}\bar{H}_k)H_k$. Since $\hat{T}$ is a recursion operator for (9),
$C_i = \partial_i \bar{P}/H_i$ satisfy (9) and (14). The completeness of $P^{(s,k)}$ results from

Theorem 1. $P^{(0,k)}$ are constants: $\partial_i(P^{(0,k)}) = \hat{T}\overline{H}_i^{(0,k)} H_i = 0$. Let us compute now the hamiltonian system generated by $\overline{P}$ in (4):

$$w_i(u) = g^{ii}(\partial_i\partial_i\overline{P} - \sum \Gamma_{ii}^k\partial_k\overline{P}) = g^{ii}(\partial_i(\hat{T}\overline{H}_i H_i) - \sum_{i\neq k}(H_i\partial_k H_i\hat{T}\overline{H}_k)/H_k - $$
$$ - \partial_i H_i\hat{T}\overline{H}_i) = (H_i\hat{T}^2\overline{H}_i - H_i\hat{T}(\sum_{i\neq k}\partial_k\overline{H}_k H_i) + \sum_{i\neq k}\beta_{ki}H_i\hat{T}\overline{H}_k)/H_i^2 = \hat{T}^2\overline{H}_i/H_i .$$

So $P^{(1,k)}$ generate *zero* w_i. This is characteristic only for the flat coordinates p^i: in this coordinates $g^{ij}=\delta^{ij}$ and (4) gives $w_j^i(u) = \partial^2\overline{P}/\partial p^i\partial p^j =$

$ = 0$ only for linear combinations of p^i.

<u>Remark</u>. Formulas (13), (14) as well as (8), (9) can be found in tracts of G.Darboux [6] and L.Bianchi [1] about orthogonal curvilinear coordinates in euclidean space together with detailed investigations on potential and translation-invariant metrics and other related (to our interests in quasilinear systems ...) topics. It may seem a miracle since no "evident reasons" for such extensive studies existed in the beginning of the XX century.

4. THE COMPLETE BASES FOR SYMMETRIES AND CONSERVATION LAWS
   OF BENNEY EQUATIONS

The proved above Theorem 1 is also applicable in this case since (2) is a hamiltonian diagonalizable system with Galilei and scaling symmetries. But the implicit diagonal form does not permit to find $H_i^{(p,k)}$ explicitly. Moreover for p>0 $H_i^{(p,k)}$ are *nonhomogeneous* in this case (see $P^{(s,k)}$ below). Instead of $H_i^{(p,k)}$ we will find explicitly the complete basis for the space of all conserved quantities of hydrodynamic type $P^{(s,k)}$ in physical coordinates $(u^i) = (q^1,\ldots,q^N,h^1,\ldots,h^N)$. Applying (4) we find (nondiagonal) commuting with (2) flows $u_t^i = w_j^i(u)u_x^i$, $i=1,\ldots,2N$. The generalized hodograph formula

(15) $$w_j^i(u) = v_j^i(u)\cdot t + x\,\delta_j^i , \qquad i, j=1,\ldots,2N,$$

gives us the solution of (2). Since the matrices v and w commute (13) contains only $2N$ independent equations (instead of $2N\cdot2N$) for $2N$ variables h^i, q^i. V.E.Zakharov [33] found N series of conserved densities $P^{(s,k)}$, $k=1,\ldots,N$, $s=0,1,2,\ldots$. They are homogeneous and rational: $P^{(0,k)}= -h^k$, $P^{(1,k)}=h^k(q^k+\sum_{m\neq k} h^m/(q^k-q^m))),\ldots$ The additional N series were found in [24]:

$$P^{(0,k+N)} = q^k, \quad P^{(1,k+N)} = (q^k)^2/2 + \sum_{m\neq k}h^m\log(q^k-q^m) + h^k\log(h^k) - h^k,\ldots$$

In contrast to $P^{(s,k)}$ they are nonalgebraic and nonhomogeneous:

(16) $$\hat{S}P^{(s,k+N)} = (s+1)P^{(s,k+N)} + P^{(s,k)} + \sum_{m=1}^{N}P^{(s,m)} .$$

Using (16) and the standard variable coefficient method one can give an algebraic recursion procedure for construction of $Z_{ij} = \partial_i \partial_j P^{(s,k+N)}$, $s>1$ and after restore $P^{(s,k+N)}$ from (16). This procedure is similar to that for the Whitham system (§ 2). The completeness theorem for $P^{(s,k)}$, $k=1,\ldots,2N$, is easily proved along the lines of Theorem 1.

Now we would prefer to concentrate our attention on some additional properties of Whitham and Benney equations.

5 THE SECOND HAMILTONIAN STRUCTURE OF WHITHAM AND BENNEY EQUATIONS

The bihamiltonian nature of the averaged KdV equations was discovered in [7]. The Benney system (2) for $N=1$ is known to have three hamiltonian structures of hydrodynamic type and some higher-order ones ([15]). In fact the bihamiltonian nature of (1) and (2) (for arbitrary N) is a simple consequence of Galilei+scaling invariance.

Theorem 5 [24]. If the coefficients $v_i(u)$ of a hamiltonian diagonal system (6) with zero-curvature potential (Egorov) metric $g_{ii}(u)$ are homogeneous, the system possesses a second hamiltonian structure of hydrodynamic type with zero-curvature diagonal metric $\bar{g}_{ii} = g_{ii}/u^i$.

The hamiltonian operators defined by both metrics are compatible (in the sense of [21]). The standard Lenard scheme [21] for Whitham and Benney equations gives the evident recursion operator $\hat{T}$ and generates the averaged Kruskal symmetries from the original system.

One can check that Whitham equations have only two hamiltonian structures of hydrodynamic type, the same seems to be true also for (2) with $N>1$. But we can conjecture the existence of higher-order hamiltonian structures similar to those found in [15] for some 2x2 systems. The hypothetical hamiltonian operators shall have the form $\hat{R}^{2k} \cdot \hat{A}$ where $\hat{A}$ is the hydrodynamic type hamiltonian operator and $\hat{R}$ is the Teshukov's recursion operator (see § 6).

The second hamiltonian operator for Benney equations can be transformed to the "physical" coordinates q^i, h^i. For $N=2$ it has coefficients

$$\bar{g}^{11} = \bar{g}^{22} = 2\bar{g}^{12} = 2, \quad \bar{g}^{13} = q^1 + c^2, \quad \bar{g}^{14} = -c^2, \quad \bar{g}^{23} = c^1, \quad \bar{g}^{24} = q^2 - c^1, \quad \bar{g}^{33} = 2h^1 - 2c^1 c^2,$$

$$\bar{g}^{34} = 2c^1 c^2, \quad \bar{g}^{44} = 2h^2 - 2c^1 c^2, \quad c^1 = h^1/(q^1 - q^2), \quad c^2 = h^2/(q^1 - q^2),$$

$$b^{ij}_k = 0 \text{ for } j<3, \quad b^{33}_k = \partial_k g^{33}/2, \quad b^{44}_k = \partial_k g^{44}/2, \quad b^{34}_k = b^{43}_k = \partial_k g^{34}/2,$$

all other $b^{ij}_k = \partial_k g^{ij}$.

6. RECURSION OPERATORS, HIGHER-ORDER SYMMETRIES AND CONSERVATION LAWS

Theorem 4 [30]. A semihamiltonian system (6) with $\partial_i v_i \neq 0$ has a conserved density $P_1 = P(u, u_x)$ iff there exist n functions of one variable $f^i(u^i)$ such that $\sum f^i g_{ii} \partial_i v_i = 0$, here $g_{ii}(u)$ is the diagonal metric associated with (6) through (8). If so,

(17) $$P_1 = \sum_i f^i g_{ii} / u^i_x + \text{hydrodynamic conserved density.}$$

This is a direct generalization of a result by J. Verosky [31] for 2×2 system of one-dimensional polytropic gas flow. In fact the density (15) exists for any one-dimensional barotropic gas flow as one can easily check using Theorem 4.

For a Galilei-invariant hamiltonian diagonal system "typically" $\sum g_{ii} \partial_i v_i = const.$ Indeed, its metric shall be potential (§ 1) and translation-invariant. This is true for Whitham and Benney systems. Hence

$$\partial_i (\sum_k g_{kk}) = \sum_k \partial_i g_{kk} = \sum_k \partial_k g_{ii} = \hat{T} g_{ii} = 0, \text{ i.e. } \sum_k g_{kk} = c_g = const.$$

Galilei invariance gives $\hat{T} v_i = 1$, and $\sum g_{ii} \partial_i v_i = \sum g_{ii} (1 - \sum_{k \neq i} \partial_k v_i) =$

$$= \sum_{i,k} g_{ii} - \sum_{i,k} g_{ii} \Gamma^i_{ik} (v_k - v_i) = c_g - \sum_{i,k} \partial_k g_{ii} v_k + \sum_{i,k} \partial_i g_{kk} v_i = c_g = const.$$

For Whitham equations $c_g = 1$, for Benney equations $c_g = 0$, since g_{ii} in this case are residues of a holomorphic 1-form on a Riemannian surface (3). Consequently Benney equations possess a conserved density (17) and Whitham equations have $P'_1 = P_1 - t$.

V. M. Teshukov [28] proved the existence of a recursion operator for any semihamiltonian system (6) with translation-invariant (not necessary zero-curvature) metric. If the system is Galilei-invariant this is always true. The Teshukov operator in this case has the form

(18) $$\hat{R}^i_k = (\delta^i_k d/dx + \Gamma^i_{ik}(u^i_x - u^k_x) + \delta^i_k \sum_s u^s_x \Gamma^i_{is})/u^k_x .$$

Applying $(\hat{R})^2$, $(\hat{R})^3$, ... to the symmetry $u^i_\tau = (v_i(u) \cdot t + x)u^i_x$ we find symmetries $u^i_\tau = f(u, u_x, u_{xx}, \dots)$ of arbitrary order. Since the conserved densities of hidrodynamic type $P = P(u)$ are complete in Liouville sense for any hamiltonian diagonal system (6) (see [30]) those higher-order symmetries give rise to "superintegrability" of Whitham and Benney equations.

7. ADDITIONAL TOPICS AND UNSOLVED PROBLEMS

The "superintegrable" nature of (1) and (2) is characteristic for diagonalizable (semi)hamiltonian systems arising in physics and mechanics. Numerous systems of $n=2$ equations have practically all properties stated above for Whitham and Benney equations. Another interesting nontrivial system with $n>2$ is that of chromatography and electrophoresis [23], [26]. It has three different Teshukov operators hence higher-order symmetries. The physical nature of such additional symmetries for the systems in question is unclear.

Recently O. I. Mokhov and E. V. Ferapontov [22] found a nonlocal generalization of the hamiltonian formalism of hydrodynamic type (4). V. E. Ferapontov courteously communicated to the author about the further generalization resulting in the following beautiful theorem: any semihamiltonian system (6) has a nonlocal hamiltonian structure with a hydrodynamic hamiltonian and a hamiltonian operator with (possibly infinitely many) nonlocal terms similar to those in [22]. The technique used is closely connected with some remarcable constructions in the classical differential geometry and uncovers a parallellism between semihamiltonian systems, nonlocal hamiltonian operators and the theory of triply conjugate coordinate systems in euclidean space developed by G. Darboux, L. P. Eisenhart and others. These topics will be covered in subsequent publications.

Another remarcable subclass of semihamiltonian class of systems is the Temple class studied by D. Serre [26]. For such systems D. Serre gave an alternative to the generalized hodograph formula (7) method of solution. It would be interesting to connect both methods.

Weakly nonlinear semihamiltonian systems (i. e. systems (6) with $\partial_i v_i = 0$, such systems are also called "linearly degenerate") were studied in [12], [25]. The theory of such systems is connected to the theory of n-webs on euclidean plane, Dupin cyclids and Stäckel metrics (E. V. Ferapontov). Among the results are: quasiperiodic behaviour of their solutions ([25]), complete description of such systems and complete sets of their hydrodynamic symmetries ([12]). E. V. Ferapontov communicated to the author the following fact: any n-phase (n-zone) quasiperiodic (or a n-soliton) solution of the KdV equation can be represented with a solution of a weakly nonlinear semihamiltonian system $R_t^i = (\sum_{k \neq i} R^k) R_x^i$, $i=1,\ldots,n$. These results shall be compared with Curro and Fusco's results [5] on the soliton-like interactions of Riemann simple waves for some 2x2 systems.

The most challenging problem is certainly the integrability of the system describing the potential diagonal zero-curvature metrics [9] (i. e. the 3-wave system) since it can be transformed (via nonlocal changes of variables) to a *nondiagonalizable* quasilinear 3x3 system of hydrodynamic type.

REFERENCES

[1] L. Bianchi, Opere, v. 3: Sisteme tripli ortogonali, Ed. Cremonese, Roma (1955).

[2] R. F. Bikbaev, V. Yu. Novokshenov, Self-similar solutions of the Whitham equations and the Korteweg-de vries equation with finiite-gap boundary condition, Proc 3 Intern. Workshop "Nonlinear and turbulent processes in physics", Kiev, v. 1, p. 32 (1987).

[3] K. M. Case, S. C. Chiu, Bäcklund transformation for thr resonant three-wave process, Phys. Fluids, 20, p. 768 (1977).

[4] L. Chierchia, N. Ercolani, D. W. McLauhglin, On the weak limit of rapidly oscillating waves, Duke Math. J., 55, p. 759 (1987).

[5] C. Curro, D. Fusco, On a class of quasilinear hyperbolic reducible systems allowing for special wave interactions, Zeitschr. Angew. Math. and Physik, 38, p. 580 (1987).

[6] G. Darboux, Leçons sur les systémes orthogonaux et les coordonnées curvilignes, Paris (1910).

[7] B. A. Dubrovin, S. P. Novikov, Hydrodynamical formalism of one-dimensional systems of hydrodynamic type and the Bogolyubov-Whitham averaging method, Soviet Math. Doklady, 27, p. 665 (1983).

[8] B. A. Dubrovin, S. P. Novikov, Hydrodynamics of weakly deformed soliton lattices. Differential geometry and hamiltonian theory, Russ. Math. Surveys, 44, p. 35 (1989).

[9] B. A. Dubrovin, On the differential geometry of strongly integrable systems of hydrodynamic type, Funk. Anal., 24 (1990).

[10] B.A.Dubrovin, The differential geometry of moduli space and its applications to soliton equations and to the topological conformal field theory, Preprint Univ. Napoli, 1991.

[11] D. Th. Egorov, Collected papers on differential geometry, Nauka, Moscow (1970).

[12] E.V.Ferapontov, Integration of the weakly nonlinear semihamiltonian systems of hydrodynamic type with the web theory methods, Mat. Sbornik, 181, p. 1220 (1990).

[13] H.Flaschka, M.G.Forest, D.W.McLaughlin, Multiphase averaging and the inverse spectral solution of the Korteweg-de Vries equation, Comm. Pure Appl. Math., 33, p. 739 (1980).

[14] J.Gibbons, Collisionless Boltzmann equations and integrable moment equations, Physica D, 3, p. 503 (1981).

[15] H.Gümral, Y.Nutku, Hamiltonian structure of equations of hydrodynamic type, J. Math. Phys., 31, p. 2606 (1990).

[16] Y.Kodama, J.Gibbons, Integrability of the dispersionless KP hierarchy, Proc Intern. Workshop "Nonlinear and turbulent processes in physics", Kiev, 1989.

[17] I.M.Krichever, Spectral theory of two-dimensional periodic operators and its applications, Russian Math. Surveys, 44, #2 (1989).

[18] V.R.Kudashev, S.E.Sharapov, The inheritance of KdV symmetries under Whitham averaging and hydrodynamic symmetries for the Whitham equations, Theor. Math. Phys., 87, p. 40 (1991).

[19] V.R.Kudashev, S.E.Sharapov, Hydrodynamic symmetries for the Whitham equations for the nonlinear Schrödinger equation (NSE), Phys. Lett A, 154, p.445 (1991).

[20] C.D.Levermore, The hyperbolic nature of the zero dispersion KdV limit, Comm. Partial Diff. Eq., 13, p. 495 (1988).

[21] F.Magri, A simple model of the integrable hamiltonian system, J. Math. Phys., 19, p. 1156 (1978).

[22] O.I.Mokhov, E.V.Ferapontov, On the nonlocal hamiltonian operators of hydrodynamic type connected with constant-curvature metrics, Russian Math. Surveys, 45 (1991).

[23] M.V.Pavlov, Hamiltonian formalism of electrophoresis equations, ITPh preprint, Chernogolovka, 1987.

[24] M.V.Pavlov, S.P.Tsarev, On the conservation laws for Benney equations, Russian Math. Surveys, 45 (1991).

[25] D.Serre, Systèmes d'EDO invariants sous l'action de systèmes hyperboliques d'EDP, Ann. Inst. Fourier, 39, p. 953 (1989).

[26] D.Serre, Intégrabilité d'une classe de systèmes de lois de conservation, Preprint ENS Lyon, #45 (1991).

[27] M.B.Sheftel, On the integration of the hamiltonian systems of hydrodynamic type with two dependent variables with the help of a Lie-Bäcklund group, Funk. Anal., 20 (1986).

[28] V.M.Teshukov, Hyperbolic systems admitting a nontrivial Lie-Bäcklund group, Preprint LIIAN, Leningrad, 1989.

[29] S.P.Tsarev, On Poisson brackets and one-dimensional systems of hydrodynamic type, Soviet Math. Doklady, 31, p. 488 (1985).

[30] S.P.Tsarev, Geometry of hamiltonian systems of hydrodynamic type. Generalized hodograph method, Math. in USSR Izvestiya, 36 (1991).

[31] J.Verosky, Higher-order symmetries of the compressible one-dimensional isentropic fluid equations, J. Math. Phys., 25, p. 884 (1984).

[32] G.B.Whitham, Linear and nonlinear waves, N.Y., John Wiley (1974).

[33] V.E.Zakharov, Benney equations and the quasiclassical approximation in the inverse spectral problem method, Funk. Anal., 14 (1980).

Explicit Construction of the Lax-Levermore Minimizer for the KdV Zero Dispersion Limit

Otis C. Wright

Program in Applied and Computational Mathematics
Princeton University
New Jersey, 08544

1 Introduction

There has been a great deal of interest recently (Krichever [3], Potemin [5], Kudashev and Sharapov [2]) in constructing solutions of the Whitham [9] averaged system of modulation equations through an implicit construction proposed by Tsarev [7]. For a certain class of ramp-like initial data, Tian [6] has constructed the global solution of the Whitham equations which matches at the phase transition boundaries to the characteristic solution of the inviscid Burgers' equation for the given initial data.

In this article the relationship between the Tsarev solution of the Whitham equations and the Lax-Levermore [4] zero dispersion limit of the KdV equation is examined (see also Wright [11]). In general, any solution of the KdV modulation equations (derived for higher phases by Flaschka, Forest, and McLaughlin [1]), which is connected to initial data through a sequence of phase transition boundaries where the matching to the lower phase solution is at least continuous, is unique because Lax and Levermore showed that two such different solutions would produce two different solutions to a certain quadratic functional minimization problem. However it is known that this quadratic functional minimization problem has only one solution. The uniqueness of the global solution of the modulation equations is non-trivial because the speed function is not Lipschitz continuous.

The Lax-Levermore construction of the minimizer requires integration of the solution of the modulation equations across all phase transitions. This is sufficient to show the uniqueness of global solutions of the modulation equations, however it does not reveal the local character of the minimizer and its explicit dependence on the

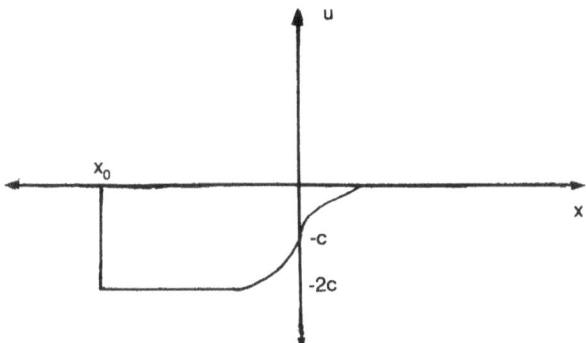

Figure 1. The Initial Data

initial data. In this article we show how the minimizer can be constructed explicitly in such a way to yield the leading order phase of the asymptotic form of the small dispersion KdV wave form through first breaking, as defined by Venakides [8].

2 The Minimization Problem

Consider the Korteweg-de Vries equation in the following form,

$$\frac{\partial u}{\partial t} - 6u\frac{\partial u}{\partial x} + \epsilon^2\frac{\partial^3 u}{\partial x^3} = 0 \tag{1}$$

with initial data which is analytic in the neighbourhood of an inflection point at $u = -c$ where c is a positive shift parameter chosen to be sufficiently small:

$$x_{\text{in}} = (c + u)^3 + \sum_{j=4}^{\infty} \alpha_j(c + u)^j. \tag{2}$$

The shift parameter c and a truncation of the initial data is introduced to create a well whose right-hand side has one inflection point given in the analytic form above which meets the $u = 0$ axis at some point and is identically $u = 0$ for all larger x values. To the left the well becomes identically $u = -2c$ at some point and then truncates to identically $u = 0$ for all $x < x_-$ where x_- is chosen to be sufficiently large and negative. With this adjustment the initial data is now amenable to the Lax-Levermore analysis. See Figure 1.

The zero dispersion limit exists in the weak L^2 sense, uniformly over compact x and t sets, and is given by

$$\bar{u}(x,t) = \frac{\partial^2}{\partial x^2}Q(\psi^*(\eta; x, t), x, t). \tag{3}$$

Here Q is a strictly convex quadratic functional

$$Q(\psi, x, t) = \frac{4}{\pi}[(a(\eta; x, t), \psi(\eta)) - \frac{1}{2}(L\psi, \psi)] \tag{4}$$

where

$$(f, g) = \int_0^{\sqrt{2c}} fg\,d\eta \tag{5}$$

and

$$L\psi(\eta) = \frac{1}{2\pi} \int_0^{\sqrt{2c}} \log\left(\frac{\eta - \mu}{\eta + \mu}\right)^2 \psi(\mu)\,d\mu \tag{6}$$

and $a(\eta; x, t)$ is determined by the initial condition:

$$a(\eta; x, t) = \eta x - 4\eta^3 t - \eta x_+(\eta) - \int_{x_+(\eta)}^{\infty} (\eta - (\eta^2 + u(y))^{\frac{1}{2}})\,dy \tag{7}$$

with $x_+(\eta)$ being determined by the right hand side of the well

$$-\eta^2 = u(x_+(\eta)) \tag{8}$$

for $0 < \eta < \sqrt{2c}$.

The minimizer ψ^* is the unique function of η on $[0, \sqrt{2c}]$ such that

$$0 \le \psi^* \le \phi$$

and the following variational conditions are satisfied,

$$\psi^* = 0 \quad \text{where} \quad L\psi^* - a < 0, \tag{9}$$

$$\psi^* = \phi \text{ where } L\psi^* - a > 0. \tag{10}$$

The function ϕ is determined by the initial conditions:

$$\phi(\eta) = \int_{x_-}^{x_+(\eta)} \frac{\eta\,dy}{(-u(y) - \eta^2)^{\frac{1}{2}}} \tag{11}$$

where x_- is the constant location of the left hand side truncation of the initial data.

3   Whitham's Equations

Lax and Levermore [4] showed that in a region of one phase oscillations the support of the minimizer evolves according to the Whitham equations. In particular if the support is given by

$$[0, \beta_2] \quad \text{and} \quad [\beta_1, \beta_0] \tag{12}$$

then $\lambda_k(X = x - 6ct, t) = c - \beta_k^2(x, t)$ evolve according to

$$\frac{\partial \lambda_k}{\partial t} + s_k \frac{\partial \lambda_k}{\partial X} = 0 \quad k = 0, 1, 2, \quad (NS), \tag{13}$$

where the speeds are non-local functions of the Riemann variables given by

$$s_k(\lambda_0, \lambda_1, \lambda_2) = -12 \frac{\lambda_k^2 - \frac{1}{2}(\lambda_0 + \lambda_1 + \lambda_2)\lambda_k - a_3}{\lambda_k - a_1} \tag{14}$$

with

$$a_1 = \lambda_0 + (\lambda_2 - \lambda_0)\frac{E(m)}{K(m)}$$

and

$$a_3 = \frac{1}{6}(\lambda_0 + \lambda_1 + \lambda_2)a_1 - \frac{1}{3}(\lambda_0\lambda_1 + \lambda_0\lambda_2 + \lambda_1\lambda_2)$$

and

$$m^2 = \frac{\lambda_2 - \lambda_1}{\lambda_2 - \lambda_0}$$

and $K(m)$ and $E(m)$ are the complete elliptic integrals of the first and second kind respectively.

A local solution of the Whitham equations, valid in an open wedge with tip at the origin in $x - t$ space and matching continuously to the characteristic solution of the inviscid Burgers' equation at the edges, can be defined using the implicit prescription of Tsarev, viz. the $\lambda_k(X, t)$ are defined implicitly by

$$X - ts_k(\vec{\lambda}) + w_k(\vec{\lambda}) = 0 \quad k = 0, 1, 2 \quad \text{(NS)} \tag{15}$$

where $\vec{w}$ satisfies the overdetermined system

$$\frac{\frac{\partial w_k}{\partial \lambda_j}}{w_j - w_k} = \frac{\frac{\partial s_k}{\partial \lambda_j}}{s_j - s_k} \tag{16}$$

for $j \neq k$, $j, k = 0, 1, 2$. In particular

$$w_k = \frac{4}{15}s_k^{(7)} - \sum_{j=4}^{\infty} \frac{\alpha_j}{12\delta_{j+1}}s_k^{(2j+1)} \tag{17}$$

where $s_k^{(2j+1)}$ are the speeds of the modulation equations of the $(j+1)$-th flow in the KdV hierarchy.

$$s_k^{(2j+1)} = -12\frac{P^{(2j+1)}(\lambda_k)}{P^{(1)}(\lambda_k)} \tag{18}$$

where $P^{(2j+1)}(\lambda)$ is a polynomial in λ uniquely determined by the following two conditions.

- For $\lambda = 1/\kappa^2 \to \infty$

$$\frac{P^{(2j+1)}(\lambda)d\lambda}{\sqrt{(\lambda - \lambda_0)(\lambda - \lambda_1)(\lambda - \lambda_2)}} \sim \left(\frac{1}{\kappa^{2j}} + \text{holomorphic}\right)d\kappa. \tag{19}$$

- The loop condition,

$$\int_{\lambda_1}^{\lambda_2} \frac{P^{(2j+1)}(\lambda)d\lambda}{\sqrt{(\lambda - \lambda_0)(\lambda - \lambda_1)(\lambda - \lambda_2)}} = 0. \tag{20}$$

Also

$$\delta_{j+1} = -\frac{1}{2^{2j}}\binom{2j}{j}, \tag{21}$$

and the α_j are given as part of the initial data.

160

		$(-\beta_0, -\beta_1)$	$(-\beta_1, -\beta_2)$	$(-\beta_2, \beta_2)$		(β_2, β_1)	(β_1, β_0)	
Im $R(c-\eta^2)$	-	0	+	0	0	-	0	+
Re $R(c-\eta^2)$	0	-	0	+	+	0	-	0
$\eta\, T(c-\eta^2)$	+	-	+/-	+	-	+/-	$\cdot$+	-
- Re$(\frac{\eta T}{R})$	0	-	0	-	+	0	+	0
Im $\frac{\eta T}{R}$	+	0	-/+	0	0	+/-	0	+
$-\int_{\beta_0}^{\eta}$ Im$\frac{\eta T}{R}d\eta$	+	0	+	0	0	-	0	-
				by loop condition				

Figure 2. Oscillation Diagram

4 Construction of the Minimizer

The Tsarev system can be written in the following way, first let

$$T(\lambda; \vec{\lambda}, x, t) = x P^{(1)}(\lambda) + 12t P^{(3)}(\lambda) - \frac{16}{5} P^{(7)}(\lambda) + \sum_{j=4}^{\infty} \frac{\alpha_j}{\delta_{j+1}} P^{(2j+1)}(\lambda) \qquad (22)$$

then require the λ_k to evolve such that this analytic function of λ (for $\vec{\lambda}$ sufficiently small) has zeros at the three λ_k, viz.

$$T(\lambda_k; \vec{\lambda}, x, t) = 0 \text{ for } k = 0, 1, 2. \qquad (23)$$

Theorem 1 *For all sufficiently small $c > 0$ there is an $x - t$ wedge with cusp at the inflection point of the initial data such that the unique minimizer ψ^* is given by*

$$- Re \left(\frac{\eta T(c - \eta^2; c - \vec{\beta^2}, x - 6ct, t)}{\sqrt{(\eta^2 - \beta_0^2)(\eta^2 - \beta_1^2)(\eta^2 - \beta_2^2)}} \right) \qquad (24)$$

where

$$\vec{\lambda}(x - 6ct, t) = c - \vec{\beta^2}(x, t)$$

and $\vec{\lambda}(x, t)$ is the previously defined Tsarev solution of Whitham's equations.

Proof

The first step in the proof is to show that the Tsarev system does in fact define a three sheeted local solution of Whitham's equations matching to the solution of the initial value problem of the inviscid Burgers' equation at the phase transition boundary. Then it is sufficient to verify that the proposed minimizer satisfies the variational conditions and lies in the admissible set of functions. This is accomplished by examining the signature patterns of

$$\psi^* = -\text{Re} \left(\frac{\eta T(c - \eta^2)}{R(c - \eta^2)} \right) \qquad (25)$$

and

$$L\psi^* - a = - \int_{\beta_0}^{\eta} \text{Im} \left(\frac{\eta T}{R} \right) d\eta \qquad (26)$$

which can be done because the λ_k evolve precisely so as to fix all the zeros of the function T. See Figure 2. The details can be found in Wright [10].

5   Higher Order Lax-Levermore Theory

Venakides [8] introduces an additional constraint in the Lax-Levermore minimization problem, the "quantum condition", which leads to an asymptotic representation of the KdV solution itself, uniformly valid on compact sets of $x - t$ within regions of constant phase. Since the explicit construction of the minimizer given above is essentially an infinite sum of Abelian differentials of the second kind, it is possible to determine a phase quantity in the oscillatory part of the KdV waveform through the use of a Riemann bilinear identity. Venakides shows that

$$u(x + \epsilon x_1, t + \epsilon t_1, \epsilon) \sim \bar{u}(x,t) - 2\frac{\partial^2}{\partial x_1^2} \log \theta(A(x,t); \tau(x,t)) + O(\epsilon^2) \qquad (27)$$

where

$$A(x,t) = \frac{\kappa x_1}{\pi} + \frac{\omega t_1}{\pi} - \frac{\alpha(x,t)}{\epsilon} + O(1) \qquad (28)$$

and

$$\alpha(x,t) = \frac{1}{\pi} \int_{\beta_1}^{\beta_0} \psi^*. \qquad (29)$$

Using the explicit form of the minimizer, the above equation, and a Riemann bilinear identity yields

$$\alpha(x,t) = 2x\sigma_0 - 8t\sigma_2 - 2\sum_{k=0}^{\infty} \theta_k \sigma_{2k}. \qquad (30)$$

The above quantities are defined by

$$\theta_k = \frac{2^k(-1)^k}{(2k+1)!!} \sum_{j=k}^{\infty} \frac{\alpha_j j! c^{j-k}}{(j-k)!} \qquad (31)$$

and

$$\sum_{j=0}^{\infty} \sigma_j z^j = \frac{i(a+bz)}{\sqrt{(1-\beta_0^2 z^2)(1-\beta_1^2 z^2)(1-\beta_2^2 z^2)}} \qquad (32)$$

where

$$a = -\frac{1}{4}\left(\int_{\beta_2}^{\beta_1} \frac{\eta \, d\eta}{\sqrt{(\beta_0^2 - \eta^2)(\beta_1^2 - \eta^2)(\beta_2^2 - \eta^2)}}\right)^{-1}$$

and

$$b = -\frac{1}{4}\left(\int_{\beta_2}^{\beta_1} \frac{d\eta}{\sqrt{(\beta_0^2 - \eta^2)(\beta_1^2 - \eta^2)(\beta_2^2 - \eta^2)}}\right)^{-1}.$$

6   Conclusion

The initial data must be shifted and truncated into a negative well so that the Lax-Levermore analysis can be applied. The shift is merely a change of origin and, as long as it is sufficiently small, may be regarded as arbitrary but fixed. Once the shift is fixed, the zero dispersion limit of the KdV equation determined by the explicitly constructed minimizer in the region of the Tsarev solution of the Whitham equations is independent of the truncation of the initial data as long as that truncation is outside the region of evolution of the Tsarev solution. Using the same shift and truncation of the initial data the uniqueness of global solutions to the KdV modulation equations, which are connected to initial data through continous matching at the phase transition boundaries, can be proven using the Lax-Levermore global construction of the minimizer.

The explicit construction of the minimizer relies on the assumption of analytic initial data since this enables the equivalent formulation of the Tsarev system as the requirement that a certain analytic function have zeros at the locations of the Riemann variables, see Equation (23). This explicit form of the minimizer uses only local function evaluations. Without necessarily knowing the phase transition properties of the solution, the explicit form of the minimizer identifies the Tsarev solution of the Whitham equations as the unique one corresponding to the zero dispersion limit of the KdV equation for the given initial data and it enables the calculation of the phase of the oscillatory part of the small dispersion KdV wave form through first breaking.

7 Acknowledgements

The author is indebted to Professor Nicholas Ercolani for his help in the development of these results. Part of this work was completed while the author was supported by a National Science Foundation Graduate Fellowship.

References

[1] Flaschka, H., M. G. Forest, and D. W. McLaughlin, "Multiphase Averaging and the Inverse Spectral Solution of the Korteweg-de Vries Equation", C. P. A. M., 33:739-784 (1980).

[2] Kudashev, V. R., and S. E. Sharapov, "Generalized Hodograph Method from the Group Theoretical Point of View," Theoretical and Math. Physics, 85(2):1155ff.

[3] Krichever, I. M., "Spectral theory of two-dimensional periodic operators and its applications," Russian Math. Surveys, 44(2):145-225 (1989).

[4] Lax, P. D., and Levermore, C. D., "The Small Dispersion Limit for the Korteweg-de Vries Equation I, II, and III," C. P. A. M., 36: 253-290, 571-593, 809-830 (1989).

[5] Potemin, G. V., "Algebro-geometric construction of self-similar solutions of the Whitham equations", Russian Math. Surveys 43(5): 252-253 (1988).

[6] Tian, F. R., Ph.D. dissertation, New York University (1991).

[7] Tsarev, S. P., "Poisson brackets and one-dimensional Hamiltonian systems of hydrodynamic type," Soviet Math. Dokl., 31:488-491 (1985).

[8] Venakides, S., "Higher Order Lax-Levermore Theory", C. P. A. M., **43**:335-361 (1990).

[9] Whitham, G. B., "Non-linear dispersive waves", Royal Society of London, Proceedings, Ser. A <u>283</u>: 238-261 (1965).

[10] Wright, O. C., Ph. D. dissertation, Princeton University (1991)

[11] Wright, O. C., "Korteweg-de Vries Zero Dispersion Limit: Through First Breaking for Cubic-Like Analytic Initial Data," C. P. A. M., **46**:423-440 (1993).

DISPERSIONLESS LIMIT OF INTEGRABLE SYSTEMS IN $2+1$ DIMENSIONS

V.E. Zakharov[1,2]

[1]Department of Mathematics, University of Arizona
Tucson, Arizona 85721, USA

[2] Landau Institute for Theoretical Physics
Moscow 117 334, Russia

1. General construction

A general scheme for construction of dispersionless limits of $2+1$ dimensional integrable systems was described first in the article [1]. Now we give its description in more details. Let us consider the following overdetermined system of two first–order nonlinear partial differential equations on a function $\chi = \chi(x, y, t)$:

$$
\begin{aligned}
\chi_y &= A(\chi_x), \\
\chi_t &= B(\chi_x).
\end{aligned}
\tag{1}
$$

Here $A(\xi)$ and $B(\xi)$ are rational in ξ. A compatibility condition for the system (1) has a form:

$$
A_t - B_y + A_\xi\, B_x - B_\xi\, A_x = 0.
\tag{2}
$$

In the simplest case of general position functions A, B can be decomposed in sum of elementary fractions:

$$
\begin{aligned}
A &= a_0 + \sum_{n=1}^{N_1} \frac{a_n}{\xi - u_n}\,, \\
B &= b_0 + \sum_{m=1}^{N_2} \frac{b_m}{\xi - v_m}\,.
\end{aligned}
\tag{3}
$$

Singular Limits of Dispersive Waves, Edited by
N.M. Ercolani et al., Plenum Press, New York, 1994

Here a_i, b_i, u_n, v_m depend on x, y, t. Substituting (3) into (2), one can get the following system of equations:

$$a_{0t} = b_{0y},$$

$$u_{nt} = \frac{\partial}{\partial x}\left(b_0 + \sum_{m=1}^{N_2} \frac{b_m}{u_n - v_m}\right),$$

$$v_{my} = \frac{\partial}{\partial x}\left(a_0 + \sum_{n=1}^{N_1} \frac{a_n}{v_m - u_n}\right),$$

$$a_{nt} = -\frac{\partial}{\partial x} a_n \sum \frac{b_m}{(u_n - v_m)^2},$$

$$b_{my} = \frac{\partial}{\partial x} b_m \sum_{n=1}^{N_1} \frac{a_n}{(u_n - v_m)^2} \qquad (4)$$

Resolving the first equation in (4):

$$a_0 = \phi_t, \qquad b_0 = \phi_y.$$

one can eliminate ϕ by substitution:

$$u_n - \phi_{0x} = \tilde{u}_n, \qquad v_m - \phi_{0x} = \tilde{v}_m.$$

We have now:

$$\tilde{u}_{nt} = \frac{\partial}{\partial x} \sum_{m=1}^{N_2} \frac{b_m}{\tilde{u}_n - \tilde{v}_m},$$

$$\tilde{v}_{my} = \frac{\partial}{\partial x} \sum_{n=1}^{N_1} \frac{a_n}{\tilde{v}_m - \tilde{u}_n},$$

$$a_{nt} = -\frac{\partial}{\partial x} a_n \sum_{m=1}^{N_2} \frac{b_m}{(u_n - v_m)^2}, \qquad (5)$$

$$b_{my} = -\frac{\partial}{\partial x} b_m \sum_{n=1}^{N_1} \frac{a_n}{(u_n - v_m)^2}.$$

The system (5) can be rewritten in a form:

$$u_{nt} = \frac{\partial}{\partial x} \frac{\delta H}{\delta a_n}, \qquad a_{nt} = \frac{\partial}{\partial x} \frac{\delta H}{\delta u_n}, \qquad (6)$$

$$v_{my} = -\frac{\partial}{\partial x} \frac{\delta H}{\delta b_m}, \qquad b_{my} = -\frac{\partial}{\partial x} \frac{\delta H}{\delta a_n}, \qquad (7)$$

$$H = \sum_{n=1}^{N_1} \sum_{m=1}^{N_2} \frac{a_n b_m}{\tilde{u}_n - \tilde{v}_m}. \qquad (8)$$

The system (6),(7) realizes an extremum of the action functional S ($\delta S = 0$):

$$S = \int dx\, dy\, dt \left(\sum_{n=1}^{N_1} \partial^{-1} u_{nt} a_n - \sum_{m=1}^{N_2} \partial^{-1} v_{my} b_m - H\right). \qquad (9)$$

It means that (6), (7) is a hamiltonian system of a very interesting type — each of variables t and y could be treated as a "time".

The system (4) contains $2(N_1 + N_2) + 1$ equations imposed on $2(N_1 + N_2) + 2$ unknown functions u_n, a_n, v_m, b_m. Lack of equations is explained by existence of the obvious transformation:

$$\chi \to \chi - \phi \,. \tag{10}$$

2. DISPERSIONLESS KP–EQUATION

Let us suppose that $A(\xi)$, $B(\xi)$ are as follows:

$$\begin{aligned} A(\xi) &= \xi^2 + u \,, \\ B(\xi) &= \xi^3 + v\xi + w \,. \end{aligned} \tag{11}$$

Substituting (11) to (2) yields :

$$v = 3/2u \,, \qquad w_x = 3/4u_y \,, \qquad \frac{\partial}{\partial x}(u_t - 3/2u\,u_x) = 3/4u_{yy} \,. \tag{12}$$

The equation (12) is a dispersionless limit of the Kadomtzev–Petviashvili equation (KP–2):

$$\frac{\partial}{\partial x}(u_t - 3/2u\,u_x - 1/4u_{xxx}) = 3/4u_{yy} \,. \tag{13}$$

and can be obtained from (13) by the limiting procedure:

$$\frac{\partial}{\partial x} \to \epsilon\frac{\partial}{\partial x} \,, \qquad \frac{\partial}{\partial t} \to \epsilon\frac{\partial}{\partial t} \,, \qquad \frac{\partial}{\partial y} \to \epsilon\frac{\partial}{\partial y} \,, \qquad \epsilon \to 0 \,. \tag{14}$$

This equation is a compatibility condition for the linear system:

$$\psi_y = \psi_{xx} + u\,\psi \,. \tag{15}$$

$$\psi_t = \psi_{xxx} + 3/2u\,\psi_x + 3/4u_x\,\psi + w\psi \,. \tag{16}$$

A substitution :

$$\psi = e^{\frac{1}{\epsilon}\chi} \,. \tag{17}$$

together with transformation (14) gives the system:

$$\begin{aligned} \chi_y &= \chi_x^2 + u \,, \\ \chi_t &= \chi_x^3 + 3/2\,u\chi_x + w \,. \end{aligned} \tag{18}$$

in accordance with (11). This example is a key for understanding of a general situation. Any equation is obtained from (2) by any specific choice of A and B is dispersionless limit of a certain 2+1 dimensional system , integrable by the Inverse Scattering Method. We will call it a "solitonic system". The equations (1) are quasiclassical limits for corresponding linear Lax operators. It is especially clear if $A(\xi)$ and $B(\xi)$ are polynomial.

Let us consider the overdetermined linear system:

$$\psi_y = A(\frac{\partial}{\partial x})\psi \,,$$
$$\psi_t = B(\frac{\partial}{\partial x})\psi \,. \tag{19}$$

together with its compatibility condition:

$$A_t - B_y + [A,\, B] = 0 \tag{20}$$

Here A and B are linear differential operators of arbitrary orders:

$$A = \frac{\partial^n}{\partial x^n} + \cdots \,, \qquad B = \frac{\partial^m}{\partial x^m} + \cdots \,. \tag{21}$$

It is obvious that the substitution (17) and following limiting procedure (14) transform the system (19) to the system (1) as well as they transform the equation (20) to the equation (2). On the other hand the equations (20) is the simpliest example of $2 + 1$ dimensional integrable by the Inverse Scattering transform "solitonic" systems.

Both conditions (2) and (20) impose $n + m + 1$ equations on $n + m + 2$ unknown coefficients of operators (or polynomials) A , B. This ambiguity for the system (19) is a result of a "gauge invariance" — a possibility to do the transformation:

$$\psi \rightarrow \psi g \,.$$

Here g is an arbitrary given function. A substitution $g = e^{-\frac{1}{\epsilon}\phi}$ gives the transformation (10).

3. Generalized Benney system

The equation (13) has a clear physical meaning . Another important example of the dispersionless integrable equation (from viewpoint of applications) is the *Benney equation*. Let us choose:

$$A = \sum_{n=1}^{N} \frac{a_n}{\xi - u_n} \,, \qquad B = -\frac{1}{2}\xi^2 - w \,. \tag{22}$$

Substituting to (2) gives the system:

$$a_{nt} + (a_n u_n)_x = 0 \,,$$
$$u_{nt} + u_n u_{nx} + w_x = 0 \,,$$
$$w_y + \frac{\partial}{\partial x} \sum_{n=1}^{N} a_n = 0 \,. \tag{23}$$

In the special case $\partial/\partial y = -\partial/\partial x$ this system is equivalent to the "n-layer" Benney system (see[2]) . The whole system (23) is one of possible generalizations of the Benney system to 2+1 dimensions. (It is not a unique possibility to do this generalization, but we will not discuss here a very interesting question about other similar possibilities.) Let us study instead the simplest of the systems (23) putting $N = 1$:

$$a_t + (au)_t = 0\,,$$
$$u_t + uu_x + w_x = 0\,,$$
$$w_y + a_x = 0\,.$$

(24)

It is a certain two-dimensional generalization of one-dimensional gas dynamic equation at $\gamma = 2$. Probably it can be used in applications. A question arises – what "solitonic" $2 + 1$ integrable equation has this system as its dispersionless limit? The following system:

$$u_t + uu_x + w_x = \frac{1}{2}u_{xx}\,,$$
$$a_t + (au)_x = -\frac{1}{2}a_{xx}\,,$$
$$w_y + a_x = 0\,.$$

(25)

obviously goes to (24) after the limiting transition (14). One can check that this system is a compatibility condition for linear equations:

$$\psi_{xy} = u\psi_y + a\psi\,,$$
$$\psi_t = -\frac{1}{2}\psi_{xx} - w\psi\,.$$

(26)

(It is enough to compare the partial derivative ψ_{xyt} computed by two different ways.) The substitution (17) together with the limiting process (14) leads from the system (26) to the system:

$$\chi_y = \frac{a}{\chi_x - u}\,, \qquad \chi_t = -\frac{1}{2}\chi_x^2 - w\,.$$

(27)

in according with (1) and (22).

We realized that the system (25) is a "solitonic prototype" of the dispersionless equations (24). The Cauchy problem for (25) is ill-posed, and this system could hardly be used for physical applications. It is important to mention, that system (24) has at least one more "solitonic prototype". Let us consider the following two-dimensional generalization of the NLS-equation:

$$i\psi_t + \psi_{xx} - \frac{1}{2}w\psi = 0\,,$$
$$w_y + \frac{\partial}{\partial x}|\psi|^2 = 0\,.$$

(28)

The substitution:

$$\psi = \sqrt{\frac{a}{z}}e^{\frac{i}{2\epsilon}\int u\,dx}$$

(29)

together with limiting process (14) allows to transform the system (28) to the system (24). But the system (24) is integrable. It is a limiting case of the well-known Davey-Stewarson equation and can be represented as a compatibility condition for the following linear system imposed on a two-component function $\Psi = \begin{pmatrix} \psi_1 \\ \psi_2 \end{pmatrix}$:

$$\frac{\partial \Psi}{\partial \xi} = I\frac{\partial \Psi}{\partial \eta} + [I, Q]\Psi \, ,$$

$$i\frac{\partial \Psi}{\partial t} = J\frac{\partial^2 \Psi}{\partial \eta^2} + [J, Q]\frac{\partial \Psi}{\partial \eta} + R\Psi \, . \tag{30}$$

Here:

$$\eta = \frac{\partial}{\partial y} + \frac{\partial}{\partial x} \, , \qquad \xi = \frac{\partial}{\partial y} - \frac{\partial}{\partial x} \, ,$$

and:

$$I = \begin{bmatrix} 1 & 0 \\ 0 & -1 \end{bmatrix} \, , \qquad J = \begin{bmatrix} 0 & 0 \\ 0 & 1 \end{bmatrix} \, , \qquad Q = \begin{bmatrix} 0 & \psi \\ \psi^* & 0 \end{bmatrix} \, .$$

This example shows that reconstruction of a "solitonic prototype" for a given dispersionless system in a general case is not an unique procedure.

4. Soliton prototype

There is a regular way to construct a "solitonic prototype" for any system (1,2) (see [3]). Let us have:

$$A(\xi) = \frac{l_1(\xi)}{m_1(\xi)} \, , \qquad B(\xi) = \frac{l_2(\xi)}{m_2(\xi)} \tag{31}$$

where $l_i(\xi)$, $m_i(\xi)$ are polynomial:

$$l_1 = c_1\xi^{p_1} + \cdots , \qquad l_2 = c_2\xi^{p_2} + \cdots , $$

$$m_1 = \xi^{q_1} + \cdots , \qquad m_2 = \xi^{q_2} + \cdots . \tag{32}$$

Here $p_2 \geq p_1$, $q_2 \geq q_1$ and c_1, c_2 are constants. The system (1) can be rewritten in a form:

$$m_1(\chi_x)\chi_y = l_1(\chi_x) \, , \tag{33}$$

$$m_2(\chi_x)\chi_t = l_2(\chi_x) \, . \tag{34}$$

Let us consider the following system of two linear differential equations:

$$M_1(\frac{\partial}{\partial x})\Psi_y = L_1(\frac{\partial}{\partial x})\Psi \, . \tag{35}$$

$$M_2(\frac{\partial}{\partial x})\Psi_t = L_2(\frac{\partial}{\partial x})\Psi \, . \tag{36}$$

Here:

$$L_1 = c_1\frac{\partial^{p_1}}{\partial x^{p_1}} + \cdots , \qquad L_1 = c_2\frac{\partial^{p_2}}{\partial x^{p_2}} + \cdots , $$

$$M_1 = \frac{\partial^{q_1}}{\partial x^{q_1}} + \cdots , \qquad M_2 = \frac{\partial^{q_2}}{\partial x^{q_2}} + \cdots . \tag{37}$$

It is clear that equations (33), (34) are quasiclassical limits for (35), (36) correspondingly. It means that a compatibility condition for (35), (36) must be a "solitonic prototype" for the system (2), (31). To find this condition one must solve first the equation:

$$R_2 M_1 - R_1 M_2 = 0.$$

(38)

Here:

$$R_1 = \frac{\partial^{a_1}}{\partial x_{q_1}} + \cdots, \qquad R_2 = \frac{\partial^{q_2}}{\partial x_{q_2}} + \cdots,$$

are unknown differential operators.

The equation (38) imposes $q_1 + q_2$ linear algebraic condition on $q_1 + q_2$ unknown coefficient s of operators R_1, R_2 and can be solved uniquely in a sense that the coefficients of operators R_1, R_2 can be expressed explicitly through coefficients of operators M_1, M_2. For instance, if $q_1 = q_2 = 1$:

$$M_1 = \frac{\partial}{\partial x} + u, \qquad M_2 = \frac{\partial}{\partial x} + v,$$

$$R_1 = \frac{\partial}{\partial x} + a, \qquad R_2 = \frac{\partial}{\partial x} + b,$$

(39)

$$a = u - \frac{u_x - v_x}{u - v}, \qquad b = v - \frac{u_x - v_x}{u - v}.$$

Assuming that R_i are known, we introduce now an operator S:

$$S = R_2 \frac{\partial}{\partial t} \left(M_1 \frac{\partial}{\partial y} - L_1 \right) - R_1 \frac{\partial}{\partial y} \left(M_2 \frac{\partial}{\partial t} - L_2 \right).$$

In virtue of (38):

$$S = \left(R_2 \frac{\partial M_1}{\partial t} + R_1 L_2 \right) \frac{\partial}{\partial y} - \left(R_1 \frac{\partial M_2}{\partial y} + R_2 L_1 \right) \frac{\partial}{\partial t} + R_1 \frac{\partial L_1}{\partial y} - R_2 \frac{\partial L_1}{\partial t}$$

(40)

As for as $S\Psi \equiv 0$ on solutions of (35), (36) one can demand:

$$S = P_2 \left(M_1 \frac{\partial}{\partial y} - L_1 \right) - P_1 \left(M_2 \frac{\partial}{\partial t} - L_2 \right).$$

(41)

(Here P_1, P_2 - differential by x operators). Comparing (40) and (4) one can get:

$$R_2 \frac{\partial M_1}{\partial t} + R_1 L_2 = P_2 M_1.$$

(42)

$$R_1 \frac{\partial M_2}{\partial y} + R_2 L_1 = P_1 M_2.$$

(43)

$$R_1 \frac{\partial L_2}{\partial y} - R_2 \frac{\partial L_1}{\partial t} = P_1 L_2 - P_2 L_1.$$

(44)

Operators P_1, P_2 can be found in a form:

$$P_1 = C_1 \frac{\partial^{P_1}}{\partial x^{P_1}} + \cdots, \qquad P_2 = C_2 \frac{\partial^{P_2}}{\partial x^{P_2}} + \cdots$$

(45)

In the trivial case:

$$M_1 = M_2 = 1, \qquad R_1 = R_2 = 1, \qquad P_1 = L_1, \qquad P_2 = L_2.$$

and

$$\frac{\partial L_1}{\partial t} - \frac{\partial L_2}{\partial y} + [L_x, L_z] = 0. \tag{46}$$

In this case operators L_1, L_2 contain $P_1 + P_2$ unknown coefficients . The condition (46) impose on them $P_1 + P_2 - 1$ equations . Lack of one equation is explained by gauge invariance . In the general case operators M_i, L_i, P_i contain $q_1 + q_2 + 2(p_1 + p_2)$ unknown coefficients. A dedicated analysis shows that conditions (42-44) impose on them $q_1 + q_2 + 2(p_1 + p_2) - 1$ independant equations , in according with possibility of a gauge transform $\psi \to \Psi g$ in the linear system (35) , (36).

It is possible to do the dispersionless limit directly in equations (38), (42-44) . We just must follow two simple rules of correspondence . In the process of dispersionless transition :

1. An operator $A(\frac{\partial}{\partial x})$ is replaced by a polynomial $A(\chi_x)$, where $A(\xi)$ is a symbol of the operator .

2. Commutator of two operators A, B goes to a Jacobian of their symbols:

$$\left[A(\frac{\partial}{\partial x}), B(\frac{\partial}{\partial x}) \right] \to \{A(\chi_x), B(\chi_x)\}, \tag{47}$$
$$\{A(\xi), B(\xi)\} = A_\xi B_x - B_\xi A_x.$$

Introducing:

$$R_i = M_i + S_i \qquad P_i = L_i + T_i \qquad (i = 1, 2),$$

and using formulated rules one can get for l_i, m_i, s_i, q_i — symbols of corresponding operators:

$$s_2 m_1 - s_1 m_2 = \{m_1, m_2\},$$
$$m_2 \frac{\partial m_1}{\partial t} + \{m_1, l_2\} = q_2 m_1 - s_1 l_2,$$
$$m_1 \frac{\partial m_2}{\partial y} + \{m_2, l_1\} = q_1 m_2 - s_2 l_1, \tag{48}$$
$$m_1 \frac{\partial m_2}{\partial y} - m_2 \frac{\partial l_1}{\partial t} = q_1 l_2 - q_2 l_1 + \{l_1, l_2\}.$$

Equations (48) form a complete system.

5. Symmetric systems

All equations studied above do not allow any nontrivial Lie group of symmetries. It does not mean that such symmetric systems do not exist. Let us consider the following example:

$$\chi_x = A(\chi_t), \qquad A(\xi) = u + e^\phi \cos \xi, \tag{49}$$
$$\chi_y = B(\chi_t), \qquad B(\xi) = v + e^\phi \sin \xi.$$

The compatibility conditions for (49) gives:

$$v_t = \phi_x, \qquad u_t = -\phi_y, \qquad u_x - u_y = \frac{\partial}{\partial t} e^{2\phi},$$

or

$$\frac{\partial^2}{\partial t^2}e^{2\psi} = \triangle\phi\,, \qquad \triangle = \frac{\partial^2}{\partial x^2} + \frac{\partial^2}{\partial y^2}\,. \tag{50}$$

The equation (50) has obvious rotational symmetry. Introducing $\phi = w_t$, one can rewrite the equation (50) in a form:

$$\frac{\partial}{\partial t}e^{2w_t} = \triangle w\,. \tag{51}$$

This equation together with the compatibility representation (49), was derived by author in 1982 [1]. It is Lagrangian and minimizes the functional of action:

$$S = \frac{1}{2}\int(e^{2w_t} - (\nabla w)^2)dxdydt\,. \tag{52}$$

The equation (50) can be used as a model for description of weak nonlinear phenomena in the two-dimensional gas dynamics.

The "solitonic prototype" of the system (50) is a two-dimensional generalization of the "Toda lattice" [4]:

$$\frac{1}{2}e^{2\phi(t+1)} - 2e^{2\phi(t)} + e^{2\phi(t-1)} = \triangle\phi\,. \tag{53}$$

This system is a compatibility condition for the linear differential–difference system:

$$\begin{aligned}
\psi_z &= u\psi(t) + e^{\phi(t+1)}\psi(t+1)\,, \\
\psi_{\bar{z}} &= v\psi(t) + e^{\phi(t)}\psi(t-1)\,.
\end{aligned} \tag{54}$$

Compatibility conditions for (54) include identities:

$$\begin{aligned}
u(t) - u(t-1) &= \frac{\partial\phi}{\partial z}\,, \\
v(t) - v(t-1) &= -\frac{\partial\phi}{\partial\bar{z}}\,.
\end{aligned} \tag{55}$$

Here:

$$\frac{\partial}{\partial z} = \frac{\partial}{\partial x} + i\frac{\partial}{\partial y}\,.$$

This example shows that the scheme, described in Section 1, can be extended to a case when A and B are trigonometric polynomials on ξ. Corresponding solitonic prototypes are differential-difference linear systems.

6. Mechanical interpretation

The equation:

$$\chi_t = B(\chi_x)\,, \qquad B = B(\xi, x, y, t)\,, \qquad \xi = \chi_x\,,$$

can be interpretated as a Hamilton-Jacobi equation for a Hamiltonian system of one degree of freedom with Hamiltonian $A = A(p, x, y, t)$.

173

The condition (2) is a condition of compatibility of two Hamilton-Jacobi equations with Hamiltonians $A(p, x, y, t), B(p, x, y, t)$ and different "times" t and y. It can be rewritten in form:

$$A_t - B_y + \{A, B\} = 0. \tag{56}$$

Here

$$\{A, B\} = A_p B_x - B_p A_x$$

is a Poisson bracket.

It is a sufficient condition for commutativity of two Hamiltonian fluxes:

$$
\begin{aligned}
p_y &= -\frac{\partial A}{\partial x}, & x_y &= \frac{\partial A}{\partial p}, \\
p_t &= -\frac{\partial B}{\partial x}, & x_t &= \frac{\partial B}{\partial p}.
\end{aligned}
\tag{57}
$$

Let us remind that A and B are rational functions on momentum P. In the simplest case:

$$A = \frac{1}{2}p^2 + u(x, y, t)$$

one of these systems is just a one-dimensional motion of a particle in a nonstationary field with a potencial $u(x, t)$. In other cases auxiliary Hamiltonian systems have no such simple physical interpretations.

We didn't touch in this section the most interesting problem — how to solve the nonlinear dispersionless equations we have found? There is a hope that a kind of a general procedure for solving mentioned dispersionless equations can exist. This hope stems from the fact that for all their "solitonic prototypes" such procedures exists .

This is so called "dressing method" (see[3]). And the urgent problem is to perform a non-local $\bar{\partial}$ problem used for dressing of the equations (35),(36).

REFERENCES

1. V.E.Zakharov, *Integrable systems in multidimensional spaces*, Lecture Notes in Physics , Springer-Verlag, Berlin **153** (1982), 190–216.
2. V.E.Zakharov, *Benney equation and quasiclassical approximation in the inverse transform model*, Funct. Anal. Appl. **14** (1980), 15–24.
3. V.E.Zakharov, *Multidimensional integrable systems*, Proceedings of the International Congress of Mathematicians, August 1983, Warsaw **2** (1983), 1225–1244.
4. A.V.Mikhailov, JETP Letters **30** (1979), 414–420.

KDV EQUATION WITH NONTRIVIAL BOUNDARY CONDITIONS AT $x \to \pm\infty$

R. F. Bikbaev

Leningrad Branch of V. A. Steklov Mathematical Institute
Academy of Science USSR
191011, Leningrad, Fontanka 27

I. INTRODUCTION

Let us consider KdV equation

$$u_t - 6uu_x + u_{xxx} = 0 , \tag{1}$$

with initial condition $u(x,0)$, which is a smooth function of x, rapidly (for simplicity - in the sense of Shwartz class) approaching two different limits as $x \to \pm\infty$:

$$u(x,0) \to v_\pm(x,0|\Gamma_\pm, D_\pm), x \to \pm\infty . \tag{2}$$

Here $v(x,t|\Gamma, D)$ is a finite-gap (or algebro-geometrical, or theta-functional) solution of KdV equation, given by the Its-Matveev formula (see [1]) and depending on free parameters Γ, D:

$$\Gamma : z^2(\lambda) = \prod_{i-1}^{2g+1} (\lambda - e_i), \qquad e_i \neq e_j$$

$$e_i \in \mathbf{R}, \qquad i = 1, \ldots, 2g+1$$

where Γ-is a nonsingular hyperelliptic curve of genus $g \geq 0$, and D is some real divisor of degree g.

The Cauchy problem (1), (2) looks very exotic at a first glance and a natural question arises: "why consider such a special problem?"

I want to present here two examples, that come from physics and motivate our investigation.

Example 1. $\Gamma_+ = \Gamma_- \equiv \Gamma; D_+ = D_- \equiv D$

This degenerate case is nevertheless very interesting. Considering it, we come to the well-known question about stability of the finite-gap solution $v(x,t|\Gamma, D)$. Let's

Singular Limits of Dispersive Waves, Edited by
N.M. Ercolani et al., Plenum Press, New York, 1994

formulate this problem more accurately. Consider for (1) an initial condition of the type

$$u(x,0) = v(x,0|\Gamma, D) + \epsilon v_1(x) , \qquad \epsilon \ll 1 ;$$

$$v_1(x) \to 0, \qquad x \to \pm\infty . \tag{3}$$

The question about stability means the following: "Is it true that as $t \to \infty$ uniformly in $x \in R$ the solution $u(x,t)$ of KdV equation (1), (3) has limit of the type

$$u(x,t) = v(x,t|\Gamma, D) + O(1) ; \tag{4}$$

or physically speaking, is the finite-gap solution $v(x,t|\Gamma, D)$ stable to small perturbation?".

An attempt to prove the stability result in the simplest case $g = 1$ was done in 1974 [2]. From our general results (see below §1) it follows that this proof is incorrect, and the finite-gap solution $v(x,t)$ is unstable in the sense (3), (4). However, this instability is not very serious and is connected with modulation of phase shift in the leading term of asymptotics at large t.

Example 2. $\Gamma_+ = z^2(\lambda) = \lambda; \Gamma_- : z^2(\lambda) = \lambda + 1$.
Initial problem (2) in this case is very simple

$$u(x,0) \to \begin{cases} 0, & x \to +\infty, \\ -1, & x \to -\infty. \end{cases} \tag{5}$$

It is a shock problem for the KdV equation, which was first investigated (in purely step-like form) by heuristical and numerical considerations in 1973 [3]. It should be noted that problem (1), (5) is the simplest physical model for describing collisionless shock waves is dispersive media.

So the Cauchy problem (1), (2) is a natural generalization of interesting physical problems. Here we shall not investigate general problem (1), (2). Let us note however that in [4], [5] inverse scattering method (ISM) in the modern form of Riemann-Hilbert problem was applied to the problem (1), (2). Interesting connections with Whitham hyperbolic systems were found at this route [6], [7]. In some interesting cases long-time asymptotics of the solution of (1), (2) was constructed [8], [9].

In this paper we restrict ourselves to description of the results concerning physically interesting Examples 1 and 2.

I.1. The case $\Gamma_+ = \Gamma_-$

This case was investigated in [4], [8]; here let's recall the main results. Consider scattering problem on the axis $x \in]-\infty, +\infty[$ for the Schrödinger operator

$$\mathcal{L} = -\partial_x^2 + u(x) , \tag{6}$$

with potential $u(x)$ of the type (2). Analogously to the usual case $u(x) \to 0, |x| \to \infty$ it is possible to define a set of "scattering data"

$$S(\lambda) = (z(\lambda); D_-; \lambda_1, \dots, \lambda_N; \beta_1, \dots, \beta_N) . \tag{7}$$

Here $2(\lambda)$ is a reflection coefficient, defined on the continuous spectrum E

$$E = [e_1, e_2] \cup [e_3, e_4] \cup \ldots \cup [e_{2g+1}, +\infty[,$$

$\lambda_1, \ldots, \lambda_N$ - points of discrete spectrum of $\mathcal{L}$, $\beta_1, \ldots, \beta_N$ - corresponding normalizing constants, D_- - divisor of the finite-gap potential $v_-(x, t)$. (One can take D_+ instead of D_- in (7)).

In the usual manner (for example, using Gelfand-Levitan integral equations [8]) it can be proved that the map

$$u(x) \longrightarrow S(x)$$

may be inverted and we obtain in this manner potential $u(x)$, which is rapidly decreasing to finite-gap ones $v_\pm(x)$ at $x \to \pm\infty$.

So, the general ISM scheme works in our situation, but we are interested in some concrete results. Let's investigate the behavior of the solution $u(x, t)$ of the Cauchy problem (1), (2) as $t \to +\infty$. First suppose (for simplicity only) that we have no discrete spectrum $\{\lambda_1 \ldots, \lambda_N\} = \emptyset$

Theorem 1. In the absence of the discrete spectrum, the solution $u(x, t)$ of the problem (1) in the case $\Gamma_+ = \Gamma_- = \Gamma$ behaves for $t \to +\infty$ as a finite-gap potential $v(x, t)$ with modulated divisor $\tilde{D} = D(\xi), \equiv \frac{x}{t}$:

$$u(x, t) = v(x, t | \Gamma, D(\xi)) + \delta(\xi, t),$$

$$\delta(\xi, t) = O(t^{-\epsilon}), \qquad \epsilon > 0. \tag{8}$$

The dependence $D(\xi)$ can be obtained in the closed form

$$\hat{A}(D(\xi)) = \hat{A}(D_-) + \frac{1}{2\pi i} \int_{\gamma(\xi)} ln(1 - |z(P)|^2)\,\omega(P). \tag{9}$$

Here P-point on the curve Γ; $\omega(P)$ - vector of normalized holomorphic differentials on Γ, $\hat{A}(D)$-Abel-transform of D; $\gamma(\xi)$ - some contour on Γ (details see in [8]).

Remark 1. There are finite number of sectors $A_i, i = 1, 2, \ldots, g + 1$ on the (x, t) plane such that

$$D(\xi) = const, \qquad \xi \in A = \bigcup_{i=1}^{g+1} A_i.$$

In this A-region the $\delta(\xi, t)$ term of asymptotics (8) is very small

$$|\delta(\xi, t)| \le t^{-N}, \qquad \forall N > 0, \qquad \xi \in A. \tag{10}$$

These sectors A_i are analogs of soliton region $\xi > \epsilon_1 > 0$ in the rapidly decreasing case $u(x) \to 0, |x| \to \infty$.

Remark 2. From (8), (9) it follows that in general situation (i.e. $z(P) \not\equiv 0$) there are regions $B_i, i = 1, \ldots, g + 1$ on the (x, t) plane, where divisor $D(\xi)$ really changes. These B-sectors are analogs of radiation section $\xi < -\epsilon_2 < 0$ in the rapidly decreasing case.

Suppose that $D_+ = D_-$. Nevertheless, in the "intermediate" B-sectors the asymptotic divisor $\tilde{D} = D(\xi)$ changes. This means that finite-gap solution $v(x, t|\Gamma, D)$ is unstable in the sense (3), (4). (Of course you can, as physicists usually do, change the initial definition of stability. For an example from the theory of flame propagation, see [10]).

Remark 3. It is possible to say more about the structure of the $\delta(\xi, t)$ term in (8) not only in the A-regions (see (10)), but in the B-regions as well. More precisely, if $\xi \in B_i$, then

$$\delta(\xi, t) = \frac{1}{\sqrt{t}} \left(y(\xi, t) \partial_x e^2(x, t, P_0(\xi) + \text{c. c.)}) \right) + O\left(\frac{1}{t}\right). \tag{11}$$

Here $e(x, t, P)$ is the Baker-Akhiezer function, corresponding to the potential $v(x, t)$ in (8); $P_0(\xi)$-stationary point, which is moving to the left with increasing of ξ; $\zeta(\xi, t)$ is some bounded in t, ξ function, given by explicit, but rather complicated formulae (see [8]).

Let us comment on the first term in (11). The function $w = \partial_x e^2(x, t, P)$ is an exact solution of the linearized KdV equation

$$w_t - \partial(vw)_x - w_{xxx} = 0.$$

That is why it looks quite natural to find the first correction term of the asymptotics (8) in the form of the Ansatz (11).

Remark 4. Influence of the discrete spectrum on the character of the solution $u(x, t)$ can be taken into account by the "dressing" method (see example [11]), or by some modification of equations of ISM [4].

The final result is the following. Suppose we have N points of discrete spectrum $\lambda_1, \dots, \lambda_N$. It is possible to find different numbers (velocities of solitons) $\xi_i, i = 1, \dots, N$ such that $\xi_i \in A, \forall i$ (see Remark 1). Asymptotics of $u(x, t)$ at $t \to +\infty, \xi = \xi_i$ is not given by the formula (8) but represents a soliton (or dislocation) on the background of the finite-gap solution.

Out of soliton rays, i.e. at $\xi \neq \xi_i$, asymptotical formula (8) is qualitatively correct at $t \to +\infty$, but it is necessary to add to the right side of (9) some extra term, equal to

$$2 \cdot \sum_{j=1}^{n} \hat{A}(\lambda_j),$$

where $n = n(\xi)$ is a number of solitons, whose velocities $\xi_j = f(\lambda_j)$ are less fixed than fixed value of ξ, and the summation is expanded only on such solitons.

We see that the picture of interaction of the modes of continuous and discrete spectrum in the case $\Gamma_+ = \Gamma_-$ is in general features the same, as in trivial case $u(x) \to 0, |x| \to \infty$

I.2. Shock problem for KdV equation

In this section we present a solution of the shock problem (3) for KdV equation. This solution [9] is based on some heuristic predictions of [3] and modern analytical advantages of the ISM (the so called Riemann-Hilbert method).

The shock problem (3) is one of the simplest Cauchy problems of the type (2) with $\Gamma_+ \neq \Gamma_-$.

Asymptotical Ansatz for $u(x,t)$ at $t \to +\infty$ from [3] in spectral terms can be interpreted as a one-gap (elliptic) solution of KdV equation, corresponding to a very special (depending on $\xi = \frac{x}{t}$ only) deformation of spectrum $\tilde{\Gamma}(\xi)$, such that $\tilde{\Gamma}(\xi) \to \Gamma_\pm$, when $\xi \to \pm\infty$.

$$\tilde{\Gamma}(\xi) = \begin{cases} \Gamma_- & \xi < -6 , \\ \Gamma(\xi) & \xi \in [-6, 4] , \\ \Gamma_+ & \xi > 4 . \end{cases} \tag{12}$$

Here $\Gamma_- : z^2(\lambda) = (\lambda + 1); \Gamma_+ : z^2 = \lambda; \Gamma(\xi) : z^2(\lambda) = \lambda(\lambda+1)(\lambda - \alpha), 0 < \alpha < -1$. The branching point $\alpha = \alpha(\xi)$ is monotonously moving from $\alpha(-6) = 0$ to $\alpha(4) = -1$ according to Whitham equation (see [3], [9]). The hypothesis of [3] predicted, that at $t \to +\infty$

$$u(x,t) = \tilde{v}\left(x, t | \tilde{\Gamma}(\xi), \tilde{D}(\xi)\right) + \delta(\xi, t),$$

$$\delta(\xi, t) = 0(t^{-\epsilon}), \qquad \epsilon > 0, \tag{13}$$

where

$$\tilde{v}(x,t) = \begin{cases} 0 & \xi > 4, \\ -1 & \xi < -6, \end{cases}$$

and $\tilde{v}(x,t)$ is a modulated one-gap solution with some divisor $\tilde{D}(\xi)$ in the oscillation region $\xi \in [-6, 4]$.

Mathematical problems which arise here are: 1) to prove this hypothesis, 2) to calculate the asymptotic divisor (phase shift) $\tilde{D}(\xi)$, 3) to say more about correction term $\delta(\xi, t)$, etc.

For investigation of these problems it is necessary to consider scattering theory for $\mathcal{L}$-operator (6) with step-like potential (3). It is easy to define the continuous spectrum $E = [-1, +\infty[$, reflection coefficient $2(\lambda), \lambda \in E$, discrete spectrum $\lambda_1, \ldots, \lambda_N$ and normalizing constants $\beta_1, \ldots, \beta_N$.

Let us formulate now the main asymptotical result. Suppose for simplicity that we have no discrete spectrum. (In particular, it is so in the case of monotonous potential $u_x(x) \geq 0$ of the type (3)).

Theorem 2. In the absence of discrete spectrum solution $u(x,t)$ of the KdV shock problem (1), (3) behaves at $t \to +\infty$ as modulated one-gap potential (13). Phase shift $\tilde{D}(\xi)$ can be obtained by explicit formulae in terms of scattering data [9]. For the second term in (13) the following estimates are valid

$$\delta(\xi, t) = \begin{cases} 0(t^{-N}) \forall N > 0 & \text{if } \xi > 4 + \epsilon_1, \\ 0(t^{-\frac{1}{2}}) & \text{if } \xi < -6 - \epsilon_2. \end{cases} \tag{14}$$

Remark 5. Of course, more detailed information about in (14) can be obtained in the same manner as in Remark 3.

A much more interesting and unsolved question is to describe the structure of $\delta(\xi, t)$ in the oscillatory region $\xi \in [-6, 4]$.

Remark 6. In the presence of discrete spectrum $\lambda_1, \ldots, \lambda_N$, qualitative character of asymptotics (13) as $t \to +\infty$ remains valid. The only serious difference is that in the region $\xi > 4N$ different soliton rays $\xi = \xi_i, i = 1, \ldots, N$ appear, along which soliton excitations $S(x - \xi_i t)$ are expanding in the usual way.

Remark 7. We have described asymptotics of the shock problem (1), (3) as $t \to +\infty$. It is interesting that the character of the solution $u(x, t)$ at $t \to -\infty$ is completely different from one as $t \to +\infty$. More precisely, asymptotics at $t \to -\infty$ is a "rarefaction wave" $u(\xi)$-self-similar solution of dispersionless KdV equation $u_t - 6uu_x = 0$. This problem is more simple than the one considered earlier, but the proof can be obtained by the same method.

CONCLUSION

The ideas described in this paper can be easily applied to many other integrable equations with real spectrum (see, for example, attractive nonlinear Schrödinger equation [12]; Toda lattice [13], modified KdV equation [13], etc.). However, for the problems with complex spectrum some new ideas are necessary. The simplest example is "finite-density" Cauchy problem for NS equation with repulsion

$$
\begin{cases}
iu_t + u_{xx} + 2|u|^2 u = 0, \\
|u(x, 0)| \to \rho & x \to \pm\infty \\
\rho = \text{const} > 0.
\end{cases} \tag{15}
$$

It is known that NS solution with constant modulus $|u(x, t)|$ is unstable. But nevertheless how one can describe the behavior of the solution of (15) at $t \to +\infty$? So the problem, which seems to me just intriguing, is to give effective asymptotical description of instability.

Another interesting approach to the problem of instability of finte-gap solutions is an application of the modified infinite dimensional KAM theory. But this is a subject of another report.

References

[1.] Theory of Solitons. S. P. Nikov ed. 1980. Moskva. Nauka.

[2.] Kuznetsov, E. A., Mikhailov, A. V. 1974. *JETP*. V. **67**, N 11, pp. 1717-1727.

[3.] Gurevich, A. V., Pitaevskii, L. P. 1973. *JETP* V. **65**, N 2, pp. 590-604.

[4.] Bikbaev, R. F. 1988. KdV equation with finite-gap boundary conditions. Preprint of Baskhir Scientific Center.

[5.] Bikbaev, R. F. 1989. *Funct. analis i ego pril.* V. **23**, N4, pp. 1-10

[6.] Bikbaev, R. F., Novokshenov. V. Ju. In Proc. III Intern. Workshop on Nonlinear Process, 1987; *Kiev. Naukova Dumka*, 1988, V **1**, pp. 32-35.

[7.] Bikbaev, R. F. 1989. *Zapiski nauch. semin. LOMI* V. **180**, pp. 23-32.

[8.] Bikbaev, R. F., Sharipov, R. A. 1989. *Theor. i matem. phys.* V. **78**, N 3, pp. 345-356.

[9.] Bikbaev, R. F. 1989. *Physics Letters A* V. **141**, N 5-6, pp. 289-293.

[10.] Zeldovich, Ja. B., Barenblatt, G. I. 1957. *Prikl. Matem. i meh.* V. **21**, pp. 856-859.

[11.] Bikbaev, R. F. 1988. *Theor. i matem. phys.* V. **77**, N 2, pp. 163-170.

[12.] Bikbaev, R. F. 1990. *Algebra i analis* V. **2**, N 3, pp. 131-144.

[13.] Bikbaev, R. F. 1991. *Physics Letters A*, to appear.

LONG-TIME ASYMPTOTICS FOR THE AUTOCORRELATION FUNCTION OF THE TRANSVERSE ISING CHAIN AT THE CRITICAL MAGNETIC FIELD

P. Deift - Courant Institute

X. Zhou - Yale University

New York, NY, New Haven, Connecticut

INTRODUCTION

We consider the particular case of the spin $-\frac{1}{2} XY$ model in a magnetic field with Hamiltonian

$$H = -\frac{1}{2} \sum_{\ell \in \mathbb{Z}} (\sigma_\ell^x \sigma_{\ell+1}^x + \sigma_\ell^z) \,, \tag{1.1}$$

where σ_ℓ^x, σ_ℓ^z are the standard Pauli matrices at the ℓth site of a one-dimensional lattice. As is well known, the Hamiltonain H can clearly be identified with the transverse Ising model at the critical transverse magnetic field ([LSM]). We will study the long-time behavior of the autocorrelation function $\chi(t)$ of the first spin component

$$\begin{aligned} \chi(t) &= \langle \sigma_0^x(t) \sigma_0^x \rangle_T \\ &= \frac{\text{Tr}(e^{-\beta H}(e^{-iHt}\sigma_0^x e^{iHt})\sigma_0^x)}{\text{Tr}(e^{-\beta H})} \,, \end{aligned} \tag{1.2}$$

where $\beta = 1/T$ is the inverse temperature.

As explained in [IIKN], the methods of [MPS], together with the methods in [IIK], [IIKS], can be used to compute $\chi(t)$ in terms of a Riemann-Hilbert (RH) problem, as follows. Set

$$\sigma(t) = \log(e^{t^2/2} \chi(t)) \,, \tag{1.3}$$

and set

$$g(z) = \tanh \beta \sqrt{1 - z^2} > 0 \,, \qquad -1 < z < 1 \,, \tag{1.4}$$

and also define

$$v^{(1)}(z,t) = \begin{pmatrix} 1 + g(z) & -g(z)e^{2zt} \\ g(z)e^{-2zt} & 1 - g(z) \end{pmatrix} \,, \qquad -1 < z < 1 \,. \tag{1.5}$$

If $m^{(1)}(z,t)$ solves the RH problem

$$m_+^{(1)}(z,t) = m_-^{(1)}(z,t)\,v^{(1)}(z,t) \,, \qquad -1 < z < 1 \,, \tag{1.6}$$

$$m^{(1)}(\cdot, t) \text{ analytic in } \mathbb{C} \setminus [-1, 1] , \tag{1.7}$$

$$m^{(1)}(z, t) \to I \qquad \text{as } z \to \infty , \tag{1.8}$$

where

$$m_\pm(z, t) = \lim_{\epsilon \downarrow 0} m(z \pm i\epsilon, t) , \qquad -1 < z < 1 , \tag{1.9}$$

then

$$\frac{d\sigma}{dt} = -2(m_1^{(1)})_{11} , \tag{1.10}$$

where

$$m^{(1)}(z, t) = I + \frac{m_1^{(1)}}{z} + O(\frac{1}{z^2}) \qquad \text{as } z \to \infty . \tag{1.11}$$

Our main result is the following.

Theorem 1.12. As $t \to \infty$,

$$\chi(t) = e^{[(t/\pi) \int_{-1}^{1} \log|\tanh \beta\zeta| d\zeta + O(\log t)]} . \ \square \tag{1.13}$$

Note that as the temperature increases, the decay rate $\frac{1}{\pi} \int_{-1}^{1} \log|\tanh \beta\zeta| \, d\zeta$ for the correlation function $\chi(t)$ also increases, as it should.

As noted in [IIKN]

$$\langle \sigma_0^y(t) \sigma_0^y \rangle_T = -\frac{\partial^2 \chi}{\partial t^2} \tag{1.14}$$

and

$$\langle \sigma_0^x(t) \sigma_0^x \rangle_T^{(XY)} = \chi^2(t/2) , \tag{1.15}$$

where the LHS of (1.15) denotes the autocorrelator for the isotropic XY model, $H_{XY} = -\frac{1}{4} \sum_{\ell \in \mathbb{Z}} (\sigma_\ell^x \sigma_{\ell+1}^x + \sigma_\ell^y \sigma_{\ell+1}^y)$. It follows that our methods also give the long-time asymptotics for the above two autocorrelation functions.

Remark 1.16. For finite t, it is not clear whether the RH problem (1.5)–(1.8) is solvable. One of the results of this paper, however, is that the RH problem is indeed solvable for large t.

Remark 1.17. It will be clear from the calculations that follow that the method of the paper applies not only to the specific function $g(z)$ in (1.4), but to a far broader class of functions. Indeed, using the approximation methods of [DZ], one can also consider functions which are not analytic in $-1 < z < 1$.

In [DZ], the authors have developed a general method for evaluating the long-time behavior of oscillatory RH problems. In (1.5), however, the factors $e^{\pm 2zt}$ lead to exponential growth and are not oscillatory. We will see, however, tht the RH problem (1.5)–(1.8) can be converted into a problem with oscillatory factors, and the methods of [DZ] then apply.

2. CONVERSION TO AN OSCILLATORY RIEMANN-HILBERT PROBLEM

The map

$$\omega \mapsto z(\omega) \equiv (\omega - \omega^{-1})/2i \tag{2.1}$$

is one-to-one from $\{|\omega| < 1\}$ onto $\mathcal{C} \setminus [-1,1]$ and also one-to-one from $\{|\omega| > 1\}$ onto $\mathcal{C} \setminus [-1,1]$. Also

$$z(\omega) = z(-1/\omega) . \tag{2.2}$$

Under the map $\omega \mapsto z$, the RH problem $m^{(1)}$ pulls back to a problem on the contour $\Sigma^{(2)} \equiv \{|\omega| = 1\}$.

i

$+$ $-$ $+$ $-$ $\Sigma^{(2)}$

$-i$

Figure 2.1.

$$m_+^{(2)}(\omega) = m_-^{(2)}(\omega) \, v^{(2)}(\omega) , \quad |\omega| = 1 , \tag{2.3}$$

$$m^{(2)}(\omega) \to I , \quad \text{as} \quad \omega \to \infty . \tag{2.4}$$

(Notational remark: we use the convention that the +-side of a contour lies to the *left* of the indicated orientation.)

Here

$$m^{(2)}(\omega) \equiv m^{(1)}(z(\omega)) = m^{(1)}\left(\frac{\omega - \omega^{-1}}{2i}\right) , \quad |\omega| \neq 1 , \tag{2.5}$$

and

$$v^{(2)}(\omega) = v^{(1)}(z(\omega)) = v^{(1)}\left(\frac{\omega - \omega^{-1}}{2i}\right) , \quad |\omega| = 1 . \tag{2.6}$$

Observe that

$$m^{(2)}(\omega) = m^{(2)}\left(-\frac{1}{\omega}\right) . \tag{2.7}$$

Reversing the orientation on the part $\{e^{i\theta} : \frac{\pi}{2} < \theta < \frac{3\pi}{2}\}$ of $\Sigma^{(2)}$, we obtain a Riemann-Hilbert problem on $\Sigma^{(3)} \equiv \{|\omega| = 1\}$

i

$-$ $+$ $+$ $-$ $\Sigma^{(3)}$

$-i$

Figure 2.2.

$$m_+^{(3)}(\omega) = m_-^{(3)}(\omega) \, v^{(3)}(\omega) , \quad |\omega| = 1 , \tag{2.8}$$

$$m^{(3)}(\omega) \to I \quad \text{as} \quad \omega \to \infty . \tag{2.9}$$

Again

$$m^{(3)}(\omega) = m^{(2)}(\omega) = m^{(2)}(-\frac{1}{\omega}) = m^{(3)}(-\frac{1}{\omega}) , \qquad |\omega| \neq 1 . \tag{2.10}$$

Observing that

$$\tanh \beta \sqrt{1 - z^2} = \tan \beta \frac{\omega + \omega^{-1}}{2} , \qquad \omega = e^{i\theta} , \ -\frac{\pi}{2} < \theta < \frac{\pi}{2}, \tag{2.11}$$

$$= -\tanh \beta \left(\frac{\omega + \omega^{-1}}{2} \right) , \qquad \omega = e^{i\theta} , \ \frac{\pi}{2} < \theta < \frac{3\pi}{2} , \tag{2.12}$$

we obtain

$$v^{(3)}(\omega, t) = \begin{pmatrix} 1 + h(\omega) & -h(\omega) e^{-i(\omega - \omega^{-1})t} \\ h(\omega) e^{i(\omega - \omega^{-1})t} & 1 - h(\omega) \end{pmatrix} , \qquad |\omega| = 1 , \tag{2.13}$$

where

$$h(\omega) \equiv \tanh \beta \frac{\omega + \omega^{-1}}{2} . \tag{2.14}$$

We see explicitly that

$$v^{(3)}(\omega, t) = (v^{(3)}(-1/\omega, t))^{-1} . \tag{2.15}$$

As $v^{(3)}(\cdot, t)$ is clearly smooth on $\Sigma^{(3)}$, the factorization $m^{(3)}(\cdot, t)$, provided it exists, will also be smooth up to the boundary. Also, as $\det v^{(3)} = 1$, $\det m_+^{(3)} = \det m_-^{(3)}$ on $\Sigma^{(3)}$, and hence

$$\det m^{(3)}(\omega, t) \equiv 1 \tag{2.16}$$

by Liouville's theorem.

Lemma 2.17. The RH problem for $v^{(1)}$ has a solution if and only if the RH problem for $v^{(3)}$ has a solution. Moreover

$$m^{(3)}(\omega) = m^{(1)} \left(\frac{\omega - \omega^{-1}}{2i} \right) . \tag{2.18}$$

Proof: The foregoing computations show that if $m^{(1)}$ solves the first RH problem, then $m^{(3)}$ given by (2.18) solves the problem for $v^{(3)}$. To prove the converse it is enough to show that if $m^{(3)}$ solves the RH problem for $v^{(3)}$ then

$$m^{(3)}(\omega) = m^{(3)}(-1/\omega) , \qquad |\omega| \neq 1 \tag{2.19}$$

and hence (2.18) can be used to define $m^{(1)}$ unambiguously. But one checks easily that $m^{(3)}(\omega)(m^{(3)}(-1/\omega))^{-1}$ is analytic in $\mathbb{C} \setminus \Sigma^{(3)}$, has no jump across Σ^3, and converges to $(m^{(3)}(0))^{-1}$ as $\omega \to \infty$. By Liouville's theorem, $m^{(3)}(\omega) = (m^{(3)}(0))^{-1} m^{(3)}(-1/\omega)$ for all $|\omega| \neq 1$. Setting $\omega = i$, we find $m_+^{(3)}(i) = (m^{(3)}(0))^{-1} m_-^{(3)}(i)$. But $v^{(3)}(i) = I$, and hence $m_+^{(3)}(i) = m_-^{(3)}(i)$ and (2.19) follows. $\square$

To remove the exponentially growing terms we proceed as follows. Define

$$m^{(4)}(\omega) = m^{(3)}(\omega) e^{i\omega^{-1}t\sigma_3} , \qquad |\omega| > 1 , \tag{2.20}$$

$$= m^{(3)}(\omega) e^{-i\omega t\sigma_3} , \qquad |\omega| < 1 , \tag{2.21}$$

where $\sigma_3 = \begin{pmatrix} 1 & 0 \\ 0 & -1 \end{pmatrix}$. This leads to a RH problem on $\Sigma^{(4)} = \Sigma^{(3)}$,

$$m_+^{(4)} = m_-^{(4)} v^{(4)} \quad \text{on} \quad \Sigma^{(4)} \tag{2.22}$$

$$m^{(4)}(\omega) \to I \quad \text{as} \quad \omega \to \infty \tag{2.23}$$

with

$$
\begin{aligned}
v^{(m)} &= e^{-i\omega^{-1}t\sigma_3} v^{(3)}(\omega) e^{-i\omega t\sigma_3} \\
&= \begin{pmatrix} (1 + h(\omega)) e^{-i(\omega+\omega^{-1})t} & -h(\omega) \\ h(\omega) & (1 - h(\omega)) e^{i(\omega+\omega^{-1})t} \end{pmatrix}, \quad |\omega| = 1 .
\end{aligned} \tag{2.24}
$$

The problem for $m^{(4)}$ is clearly equivalent to the problem for $m^{(3)}$.

Observe that $e^{\pm i(\omega+\omega^{-1})t}$ is now oscillatory on $\Sigma^{(4)}$. As in [DZ], we now deform the contour to obtain a RH problem which converges to an explicitly solvable problem as $t \to \infty$.

The deformation which follows is guided by the following signature table for Re $i(\omega + \omega^{-1}) = (r^{-1} - r)\sin\theta$, $\omega = r\, e^{i\theta}$.

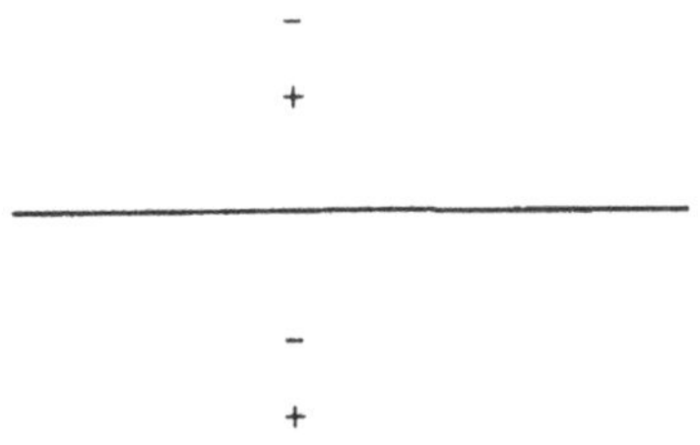

Figure 2.3.

As $v^{(4)}(\omega)$ is analytic near the unit circle, we may deform $\Sigma^{(4)}$ to an indented contour $\Sigma^{(5)}$

Figure 2.4

to obtain by analytic continuation an equivalent RH problem

$$m_+^{(5)} = m_-^{(5)} v^{(5)} \quad \text{on} \quad \Sigma^{(5)} , \tag{2.25}$$

$$m^{(5)} \to I \quad \text{as} \quad \omega \to I . \tag{2.26}$$

The precise shape of the indentation is clearly immaterial, but we do require that $\Sigma^{(5)}$ lies in some small annular region

$$A_\eta = \{ 1 - \eta < |\omega| < 1 + \eta \} \tag{2.27}$$

on which h vanishes only at $\pm i$. Note in particular that $h(\omega) \neq 0$ on $\Sigma^{(5)}$. Hence $v^{(5)}(\omega)$ admits the following smooth factorizations

$$v^{(5)}(\omega) =$$
$$\begin{pmatrix} 1 & 0 \\ \frac{h(\omega)-1}{h(\omega)} e^{(i(\omega+\omega^{-1})t} & 1 \end{pmatrix} \begin{pmatrix} 0 & -h(\omega) \\ h(\omega)^{-1} & 0 \end{pmatrix} \begin{pmatrix} 1 & 0 \\ \frac{h(\omega)+1}{-h(\omega)} e^{-i(\omega+\omega^{-1})t} & 1 \end{pmatrix} \quad (2.28)$$
$$v^{(5)}(\omega) =$$
$$\begin{pmatrix} 1 & \frac{1+h(\omega)}{h(\omega)} e^{-i(\omega+\omega^{-1})t} \\ 0 & 1 \end{pmatrix} \begin{pmatrix} 0 & -h(\omega)^{-1} \\ h(\omega) & 0 \end{pmatrix} \begin{pmatrix} 1 & \frac{1-h(\omega)}{h(\omega)} e^{i(\omega+\omega^{-1})t} \\ 0 & 1 \end{pmatrix} \quad (2.29)$$

Extend the contour $\Sigma^{(5)}$ to $\Sigma^{(6)}$

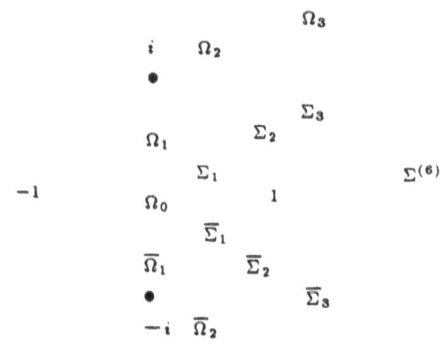

Figure 2.5

where $\Sigma^{(6)}$ remains within the annulus A_η. Also note that $\Sigma^{(5)} = \Sigma_2 \cup \overline{\Sigma}_2$ and, for later convenience, we require that $\Sigma_1, \overline{\Sigma}_1, \Sigma_3, \overline{\Sigma}_3$ form straight lines near 1 and -1, making an angle of $45°$ with the horizontal.

Define

$$m^{(6)}(\omega) \equiv m^{(5)}(\omega) , \quad \omega \in \Omega_0 \cup \Omega_3 , \quad (2.30)$$

$$\equiv m^{(5)}(\omega) \begin{pmatrix} 1 & 0 \\ \frac{h(\omega)+1}{h(\omega)} e^{-i(\omega+\omega^{-1})t} & 1 \end{pmatrix} , \quad \omega \in \Omega_1 , \quad (2.31)$$

$$\equiv m^{(5)}(\omega) \begin{pmatrix} 1 & \frac{h(\omega)-1}{h(\omega)} e^{i(\omega+\omega^{-1})t} \\ 0 & 1 \end{pmatrix} , \quad \omega \in \overline{\Omega}_1 , \quad (2.32)$$

$$\equiv m^{(5)}(\omega) \begin{pmatrix} 1 & 0 \\ \frac{h(\omega)-1}{h(\omega)} e^{i(\omega+\omega^{-1})t} & 1 \end{pmatrix} , \quad \omega \in \Omega_2 , \quad (2.33)$$

$$\equiv m^{(5)}(\omega) \begin{pmatrix} 1 & \frac{h(\omega)+1}{h(\omega)} e^{-i(\omega+\omega^{-1})t} \\ 0 & 1 \end{pmatrix} , \quad \omega \in \overline{\Omega}_2 . \quad (2.34)$$

Observe that $m^{(6)}(\omega)$ is meromorphic in $\mathbb{C} \setminus \Sigma^{(6)}$ with simple poles at $i \in \Omega_2$ and $-i \in \overline{\Omega}_2$. A direct computation using (2.28) (on $\Sigma^{(5)} \cap \{\mathrm{Im}\, \omega > 0\}$) and (2.29) (on $\Sigma^{(5)} \cap \{\mathrm{Im}\, \omega < 0\}$) shows that

$$m_+^{(6)}(\omega) = m_-^{(6)}(\omega) v^{(6)}(\omega) , \quad \omega \in \Sigma^{(6)} , \quad (2.35)$$

where

$$v^{(6)}\,|\,\Sigma_3 = \begin{pmatrix} 1 & 0 \\ \frac{h(\omega)-1}{h(\omega)}\,e^{i(\omega+\omega^{-1})t} & 1 \end{pmatrix} \tag{2.36}$$

$$v^{(6)}\,|\,\Sigma_2 = \begin{pmatrix} 0 & -h(\omega) \\ h^{-1}(\omega) & 0 \end{pmatrix} \tag{2.37}$$

$$v^{(6)}\,|\,\Sigma_1 = \begin{pmatrix} 1 & 0 \\ -\frac{h(\omega)+1}{h(\omega)}\,e^{-i(\omega+\omega^{-1})t} & 1 \end{pmatrix} \tag{2.38}$$

$$v^{(6)}\,|\,\overline{\Sigma}_1 = \begin{pmatrix} 1 & -\frac{(h(\omega)-1)}{h(\omega)}\,e^{i(\omega+\omega^{-1})t} \\ 0 & 1 \end{pmatrix} \tag{2.39}$$

$$v^{(6)}\,|\,\overline{\Sigma}_2 = \begin{pmatrix} 0 & -h(\omega)^{-1} \\ h(\omega) & 0 \end{pmatrix} \tag{2.40}$$

$$v^{(6)}\,|\,\overline{\Sigma}_3 = \begin{pmatrix} 1 & \frac{h(\omega)+1}{h(\omega)}\,e^{-i(\omega+\omega^{-1})t} \\ 0 & 1 \end{pmatrix}. \tag{2.41}$$

Also

$$\operatorname*{Res}_{\pm i} m^{(6)}(\cdot) = \lim_{\omega\to\pm i} m^{(6)}(\omega)\,v^{(6)}(\pm i)\,, \tag{2.42}_{\pm}$$

where

$$v^{(6)}(i) = \begin{pmatrix} 0 & 0 \\ -\beta^{-1} & 0 \end{pmatrix}, \qquad v^{(6)}(-i) = \begin{pmatrix} 0 & \beta^{-1} \\ 0 & 0 \end{pmatrix}. \tag{2.43}_{\pm}$$

Together with

$$m^{(6)}(\omega) \to I \quad \text{as} \quad \omega \to \infty \tag{2.44}$$

we obtain a RH problem (2.35)–(2.44) for $m^{(6)}(\omega)$ which is clearly equivalent to the RH problem for $v^{(5)}$.

Using the signature table Figure 2.3, we note that the exponential factors in $v^{(6)}$ are all exponentially decreasing as $t \to \infty$. Thus the RH problem "converges" to a time independent RH problem on $\Sigma^{(5)} \cup \{i\} \cup \{-i\}$ with jump matrices given by (2.37), (2.40), (2.43) and (2.44). This limiting problem will be solved explicitly in the next section, and the error estimates will be derived in Section 4.

3. THE MODEL PROBLEM

Here we consider the model RH problem $v^{(7)}$ on $\Sigma^{(7)} = \Sigma^{(5)} \cup \{i\} \cup \{-i\} = \Sigma_2 \cup \overline{\Sigma}_2 \cup \{i\} \cup \{-i\}$,

$$m_+^{(7)} = m_-^{(7)}\,v^{(7)} \quad \text{on} \quad \Sigma^{(5)}\,, \tag{3.1}$$

$$\operatorname*{Res}_{\pm i} m^{(7)}(\cdot) = \lim_{\omega\to\pm i} m^{(7)}(\omega)\,v^{(7)}(\pm i)\,, \tag{3.2}$$

$$m^{(7)}(\omega) \to I \quad \text{as} \quad \omega \to \infty\,, \tag{3.3}$$

where

$$v^{(7)}\,|\,\Sigma_2 = v^{(6)}\,|\,\Sigma_2 = \begin{pmatrix} 0 & -h(\omega) \\ h(\omega)^{-1} & 0 \end{pmatrix}\,, \tag{3.4}$$

$$v^{(7)}\,|\,\overline{\Sigma}_2 = v^{(6)}\,|\,\overline{\Sigma}_2 = \begin{pmatrix} 0 & -h(\omega)^{-1} \\ h(\omega) & 0 \end{pmatrix}\,, \tag{3.5}$$

and

$$v^{(7)}(i) = v^{(6)}(i) = \begin{pmatrix} 0 & 0 \\ -\beta^{-1} & 0 \end{pmatrix}, \quad v^{(7)}(-i) = v^{(6)}(-i) = \begin{pmatrix} 0 & \beta^{-1} \\ 0 & 0 \end{pmatrix}. \tag{3.6}_{\pm}$$

Define

$$m^{(8)} \equiv m^{(7)} \delta^{\sigma_3} , \quad \omega \text{ outside } \Sigma^{(5)} , \tag{3.7}$$

$$\equiv m^{(7)} \delta^{-\sigma_3} , \quad \omega \text{ inside } \Sigma^{(5)} , \tag{3.8}$$

where δ solves the following scalar RH problem,

$$\left.\begin{array}{ll} \delta_+ = h(\omega)^{-1}\delta_- & \text{on} \quad \Sigma_2 \\ \quad = h(\omega)\delta_- & \text{on} \quad \overline{\Sigma}_2 \end{array}\right\} \tag{3.9}$$

$$\delta(\omega) \to 1 \quad \text{as} \quad \omega \to \infty . \tag{3.10}$$

As we will see below, $\delta(\omega)$ and $\delta(\omega)^{-1}$ are both bounded in $\overline{\mathbb{C}} \setminus \Sigma^{(5)}$.

We find

$$m_+^{(8)} = m_-^{(8)} \begin{pmatrix} 0 & -1 \\ 1 & 0 \end{pmatrix} , \quad \omega \in \Sigma^{(5)} , \tag{3.11}$$

$$\operatorname*{Res}_{\pm i} m^{(8)}(\cdot) = \lim_{\omega \to \pm i} m^{(8)}(\omega) v^{(8)}(\pm i) , \tag{3.12}$$

where

$$v^{(8)}(i) = \delta(i)^{-\sigma_3} v(i) \delta(i)^{\sigma_3} = \begin{pmatrix} 0 & 0 \\ -\beta^{-1}(\delta(i))^2 & 0 \end{pmatrix} ,$$

$$v^{(8)}(-i) = \delta(-i)^{-\sigma_3} v(-i) \delta(-i)^{\sigma_3} = \begin{pmatrix} 0 & \beta^{-1}\delta(-i)^{-2} \\ 0 & 0 \end{pmatrix} , \tag{3.13}_\pm$$

and

$$m^{(8)}(\omega) \to I \quad \text{as} \quad \omega \to \infty . \tag{3.14}$$

The RH problem (3.9), (3.10) is solved by the following standard formula,

$$\delta(\omega) = \exp\left\{ -\frac{1}{2\pi i} \int_{\Sigma_2} \frac{\log h(s)}{s-\omega} \, ds + \frac{1}{2\pi i} \int_{\overline{\Sigma}_2} \frac{\log h(s)}{s-\omega} \, ds \right\} \tag{3.15}$$

Here the integrals are taken with the orientation of $\Sigma^{(5)}$, i.e. counterclockwise, and $\log h(s)$ is defined by continuation of the principal branch $\log h(1) \in \mathbb{R}$ at $s = 1$. For convenience in the calculations that follow we assume that $\Sigma^{(5)}$ has the shape

$$i \qquad |\omega - i| = \rho(\epsilon) = 2\sin\epsilon/2$$

$$|\omega| = 1$$

$$\cdot \, 0 \qquad\qquad \Sigma^{(5)}$$

$$-i \qquad |\omega + i| = \rho(\epsilon) = 2\sin\epsilon/2$$

Figure 3.1.

We find

$$\log h(\omega) = \log \left| \tanh \frac{\beta(\omega + \omega^{-1})}{2} \right| , \quad |\omega| = 1 , \quad \mathrm{Re}\,\omega > 0 , \tag{3.16}$$

$$= \log \left| \tanh \frac{\beta(\omega + \omega^{-1})}{2} \right| - i\pi , \quad |\omega| = 1 , \quad \mathrm{Re}\,\omega < 0, \ \mathrm{Im}\,\omega > 0 , \tag{3.17}$$

$$= \log \left| \tanh \frac{\beta(\omega + \omega^{-1})}{2} \right| + i\pi , \quad |\omega| = 1 , \quad \mathrm{Re}\,\omega < 0, \ \mathrm{Im}\,\omega < 0 . \tag{3.18}$$

To compute $\delta(i)$, note that $\delta(i)$ is clearly independent of ϵ. Evaluating (3.15) at $\omega = i$, we obtain as $\epsilon \downarrow 0$,

$$\frac{1}{2\pi i} \int_{\Sigma_2} \frac{\log h(s)}{s - i} \, ds = \frac{1}{2\pi} \int_0^{\pi/2 - \epsilon} \log |\tanh \beta \cos \theta| \left(\frac{1}{2} + \frac{i \cos \theta}{2(1 - \sin \theta)} \right) d\theta$$

$$+ \frac{1}{2\pi} \int_{\pi/2 + \epsilon}^{\pi} (\log |\tanh \beta \cos \theta| - i\pi) \left(\frac{1}{2} + \frac{i \cos \theta}{2(1 - \sin \theta)} \right) d\theta$$

$$+ \frac{1}{2\pi i} \int_{\substack{|\omega - i| = \rho(\epsilon) \\ \omega \in \Sigma_2}} \frac{\log h(s)}{s - i} \, ds$$

$$= \left(\frac{1}{4\pi} \int_0^{\pi} \log |\tanh \beta \cos \theta| \, d\theta - \frac{i\pi}{8} + \frac{1}{4} \log(1 - \cos \epsilon) \right)$$

$$+ \left(\frac{i\pi}{4} - \frac{1}{2} \log \beta - \frac{1}{2} \log \epsilon \right) + o(1)$$

$$= \frac{1}{4\pi} \int_0^{\pi} \log |\tanh \beta \cos \theta| \, d\theta + \frac{i\pi}{8} - \frac{1}{2} \log \beta - \frac{1}{4} \log 2 + o(1) ,$$

and simiarly

$$\frac{1}{2\pi i} \int_{\overline{\Sigma}_2} \frac{\log h(s)}{s - i} \, ds = \frac{1}{4\pi} \int_{\pi}^{2\pi} \log |\tanh \beta \cos \theta| \, d\theta + \frac{i\pi}{8} + \frac{\log 2}{4} + o(1) .$$

Thus

$$\delta(i) = e^{(1/2)\log(2\beta)} = \sqrt{2\beta} . \tag{3.19}$$

On the other hand the $\omega \to \bar{\omega}$ symmetry of (3.9) shows that

$$\delta(\omega) = (\overline{\delta(\bar{\omega})})^{-1} \tag{3.20}$$

and hence

$$\delta(-i) = 1/\sqrt{2\beta} . \tag{3.21}$$

Thus

$$v^{(8)}(i) = \begin{pmatrix} 0 & 0 \\ -2 & 0 \end{pmatrix} , \quad v^{(8)}(-i) = \begin{pmatrix} 0 & 2 \\ 0 & 0 \end{pmatrix} . \tag{3.22}$$

The RH problem for $v^{(8)}$ can be solved explicitly by rational functions. One checks easily that

$$m^{(8)}(\omega) = \begin{pmatrix} \frac{\omega}{\omega - i} & \frac{1}{\omega + i} \\ \frac{-1}{\omega - i} & \frac{\omega}{\omega + i} \end{pmatrix} , \quad \omega \text{ outside } \Sigma^{(5)} , \tag{3.23}$$

$$= \begin{pmatrix} \frac{\omega}{\omega - i} & \frac{1}{\omega + i} \\ \frac{-1}{\omega - i} & \frac{\omega}{\omega + i} \end{pmatrix} \begin{pmatrix} 0 & -1 \\ 1 & 0 \end{pmatrix} , \quad \omega \text{ inside } \Sigma^{(5)} . \tag{3.24}$$

We have proved the following result.

Proposition 3.25.

$$m^{(7)}(\omega) = \begin{pmatrix} \frac{\omega}{\omega-i} & \frac{1}{\omega+i} \\ \frac{-1}{\omega-i} & \frac{\omega}{\omega+i} \end{pmatrix} \delta^{-\sigma_3}, \quad \omega \text{ outside } \Sigma^{(5)}, \tag{3.26}$$

$$= \begin{pmatrix} \frac{\omega}{\omega-i} & \frac{1}{\omega+i} \\ \frac{-1}{\omega-i} & \frac{\omega}{\omega+i} \end{pmatrix} \begin{pmatrix} 0 & -1 \\ 1 & 0 \end{pmatrix} \delta^{\sigma_3}, \quad \omega \text{ inside } \Sigma^{(5)}. \tag{3.27}$$

$\square$

As $\omega \to \infty$,

$$\delta(\omega) = 1 + \frac{\delta_1}{\omega} + O(\frac{1}{\omega^2}) \tag{3.28}$$

where

$$\delta_1 = \frac{1}{2\pi i} \int_{\Sigma_2} \log h(s)\, ds - \frac{1}{2\pi i} \int_{\overline{\Sigma}_2} \log h(s)\, ds \tag{3.29}$$

$$= -\frac{1}{i\pi} \int_{-1}^{1} \log|\tanh \beta s|\, ds - \frac{1}{i} \tag{3.30}$$

by a similar calculation to that of $\delta(i)$. This leads to the expansion,

$$m^{(7)}(\omega) = I + \frac{m_1^{(7)}}{\omega} + O(\frac{1}{\omega^2}), \quad \omega \to \infty,$$

where

$$m_1^{(7)} = \left(\frac{1}{i\pi} \int_{-1}^{1} \log|\tanh \beta s|\, ds\right) \sigma_3 + \begin{pmatrix} 0 & -1 \\ 1 & 0 \end{pmatrix}. \tag{3.31}$$

By the constructions of Section 2,

$$m^{(6)}(\omega) = m^{(1)}\left(\frac{\omega - \omega^{-1}}{2i}\right) e^{i\omega^{-1} t\sigma_3} \tag{3.32}$$

for $|\omega|$ large, and hence

$$m_1^{(1)} = \frac{m_1^{(6)}}{2i} - \frac{t}{2}\sigma_3 \tag{3.33}$$

so that

$$\frac{d\sigma}{dt} = i(m_1^{(6)})_{11} + t \tag{3.34}$$

by (1.10).

Approximating $(m_1^{(6)})$ by $m_1^{(7)}$, we obtain

$$\sigma_{\text{as}}(t) \equiv \int^{t} (i(m_1^{(7)})_{11} + t')\, dt'$$

$$= \frac{t}{\pi} \int_{-1}^{1} \log|\tanh \beta s|\, ds + \frac{t^2}{2} + \text{const.} \tag{3.35}$$

In the next section we will prove the following result.

192

Proposition 3.36. As $t \to \infty$,

$$\sigma(t) - \sigma_{as}(t) = O(\log t) . \tag{3.37}$$

$\square$

This proves (1.13).

4. ERROR ESTIMATES AND THE SOLVABILITY OF THE RH PROBLEM FOR LARGE t

On $\mathbb{C} \setminus \Sigma^{(6)}$, set

$$m^{(10)}(\omega) = m^{(6)}(\omega) \left(m^{(7)}(\omega) \right)^{-1} .$$

One verifies directly that $m^{(10)}(\omega)$ has no poles at $\pm i$ and no jump across $\Sigma^{(5)} = \Sigma_2 \cup \bar{\Sigma}_2$. We obtain a RH problem on $\Sigma^{(10)} = \Sigma_1 \cup \bar{\Sigma}_1 \cup \Sigma_3 \cup \bar{\Sigma}_3$

$$m_+^{(10)} = m_-^{(10)} v^{(10)} , \quad \omega \in \Sigma^{(10)} , \tag{4.1}$$

$$m^{(10)} \to I \quad \text{as} \quad \omega \to \infty , \tag{4.2}$$

where

$$v^{(10)} = m^{(7)} v^{(6)} (m^{(7)})^{-1} \quad \text{on} \quad \Sigma^{(10)} . \tag{4.3}$$

As $m_1^{(6)} = m_1^{(7)} + m_1^{(10)}$, $\int^t (m_1^{(10)})_{11}$ is precisely the error in Proposition 3.36.

For the convenience of the reader, we recall the solution procedure for a RH problem on an oriented contour Σ (see, for example, [BC]). Suppose the jump matrix v has a factorization

$$v = (I - W_-)^{-1} (I + W_+) . \tag{4.4}$$

Then define

$$C_W f = C_+(f W_-) + C_-(f W_+) \tag{4.5}$$

where $C_\pm$ are the Cauchy integral operators

$$C_\pm g(\omega) = \lim_{\substack{\omega' - \omega \\ \omega' \in \pm \text{ side of} \Sigma}} \int_\Sigma \frac{g(\omega')}{\omega' - \omega} \frac{d\omega'}{2\pi i} , \quad \omega \in \Sigma . \tag{4.6}$$

If $W_\pm \in L^\infty(\Sigma)$, then C_W is bounded from $L^2(\Sigma) \to L^2(\Sigma)$, and if $W_\pm \in L^2(\Sigma)$, then C_W is bounded from $L^\infty(\Sigma) \to L^2(\Sigma)$.

Suppose $\mu \in L^2(\Sigma) + \mathbb{C} I$ is a solution of

$$(1 - C_W)\mu = I . \tag{4.7}$$

Then

$$m(\omega) \equiv I + \int_\Sigma \frac{\mu(\omega')(W_+(\omega') + W_-(\omega'))}{\omega' - \omega} \frac{d\omega'}{2\pi i} , \quad \omega \in \mathbb{C} \setminus \Sigma , \tag{4.8}$$

solves the RH problem

$$m_+ = m_- v \quad \text{on} \quad \Sigma , \tag{4.9}$$

$$m \to I \quad \text{as} \quad \omega \to \infty . \tag{4.10}$$

In particular, if we write

$$m = I + \frac{m_1}{\omega} + O\left(\frac{1}{\omega^2}\right), \tag{4.11}$$

then

$$m_1 = -\frac{1}{2\pi i} \int_\Sigma \mu(\omega) \left(W_+(\omega) + W_-(\omega)\right) d\omega. \tag{4.12}$$

For $\epsilon > 0$ small, let $\Sigma^{(11)} = \Sigma_A^{(11)} \cup \Sigma_B^{(11)}$ denote the union of two crosses

$$\Sigma^{(11)}$$

$$-1 \qquad\qquad\qquad 1$$

$$\Sigma_A^{(11)} \qquad\qquad\qquad \Sigma_B^{(11)}$$

Figure 4.1

where $\Sigma_A^{(11)} = \Sigma^{(10)} \cap \{|\omega - 1| < \epsilon\}$, $\Sigma_B^{(11)} = \Sigma^{(10)} \cap \{|\omega + 1| < \epsilon\}$. \tag{4.13}

The lines in $\Sigma^{(11)}$ are straight and make an angle of $45°$ with the horizontal. Let

$$m^{(11)} = m^{(11)} v^{(11)}, \quad \omega \in \Sigma^{(11)}, \tag{4.14}$$
$$m^{(11)} \to I \quad \text{as} \quad \omega \to \infty, \tag{4.15}$$

where $v^{(11)} \equiv v^{(10)}|\Sigma^{(11)}$, be the assocaited RH problem. From (2.36), (2.38), (2.39) and (2.41), we see that

$$\|v^{(10)}|(\Sigma^{(10} \setminus \Sigma^{(11)})\|_{L^\infty} \le c\, e^{-\gamma t} \tag{4.16}$$

for some $c, \gamma > 0$.

We will show below that $(1 - C_{W^{(11)}})^{-1}$ exists for large t and

$$\|(1 - C_{W^{(11)}})^{-1}\|_{L^2 \to L^2} \text{ is bounded as } t \to \infty. \tag{4.17}$$

By the resolvent formula together with (4.16), this implies that $(1 - C_{W^{(10)}})^{-1}$ exists for large t and

$$\|(1 - C_{W^{(10)}})^{-1}\|_{L^2 \to L^2} \text{ is bounded as } t \to \infty, \tag{4.18}$$

and hence the RH problems for $v^{(11)}$ and $v^{(10)}$ are solvable for large t. Also by the proof of Lemma 2.44 in [DZ], we obtain from (4.16)

$$(m_1^{(10)})_{11} = (m_1^{(11)})_{11} + O(e^{-\gamma t}). \tag{4.19}$$

Of course the existence of a solution for the RH problem $v^{(10)}$ for large t, immediately implies the existence of a solution for the equivalent RH problems $v^{(1)}, v^{(2)}, \ldots$, for large t.

Let $m_A^{(11)}$, $m_B^{(11)}$ be the solutions of the RH problems associated with

$$v_A^{(11)} \equiv v^{(11)} \,|\, \Sigma_A^{(11)} \;, \qquad v_B^{(11)} \equiv v^{(11)} \,|\, \Sigma_B^{(11)} \tag{4.20}$$

on $\Sigma_A^{(11)}$, $\Sigma_B^{(11)}$ respectively. We will show below that $(1 - C_{W_A^{(11)}})^{-1}$, $(1 - C_{W_B^{(11)}})^{-1}$ exist for large t and that

$$\|(1 - C_{W_A^{(11)}})^{-1}\| \;,\; \|(1 - C_{W_B^{(11)}})^{-1}\| \text{ are bounded as } t \to \infty \;, \tag{4.21}$$

As in [DZ], this implies in turn that (4.17) is true, and hence that the RH problems associated with $v^{(11)}$, $v_A^{(11)}$, $v_B^{(11)}$ are solvable for large t. Also the proof of Proposition 3.66 in [DZ], now shows that the contributions of the two crosses are additive to highest order,

$$(m_1^{(11)})_{11} = (m_{A,1}^{(11)})_{11} + (m_{B,1}^{(11)})_{11} + O(\tfrac{1}{t}) \;. \tag{4.22}$$

We consider the RH problem on one of the crosses, say $\Sigma_B^{(11)}$. The RH problem on $\Sigma_A^{(11)}$ is similar and can in fact be obtained by symmetry from $\Sigma_B^{(11)}$ (see (4.58) below, et seq.).

Extend the lines in $\Sigma_B^{(11)}$ to infinity to obtain the infinite contour $\Sigma_B^{(12)}$

1                          $\Sigma_B^{(12)}$

Figure 4.2.

and set v

$$v_B^{(12)}(\omega) \equiv v_B^{(11)}(\omega) \;, \quad \omega \in \Sigma_B^{(11)} \;, \tag{4.23}$$

$$\equiv I \;, \quad \omega \in \Sigma_B^{(12)} \setminus \Sigma_B^{(11)} \;, \tag{4.24}$$

Then

$$m_B^{(12)}(\omega) = m_B^{(11)}(\omega) \tag{4.25}$$

solves the RH problem for $v_B^{(12)}$. Introduce the scaling

$$\omega \mapsto \frac{\omega}{\sqrt{2t}} + 1$$

taking $\Sigma_B^{(12)} \to \Sigma_B^{(13)} \equiv \Sigma_B^{(12)} - 1$. On $\Sigma_B^{(13)}$ we have the RH problem

$$(m_B^{(13)})_+(\omega) = (m_B^{(13)})_-(\omega) v_B^{(13)}(\omega) \;, \quad \omega \in \Sigma_B^{(13)} \;, \tag{4.26}$$

$$m_B^{(13)} \to I \quad \text{as} \quad \omega \to \infty \;, \tag{4.27}$$

where

$$v_B^{(13)}(\omega) = v_B^{(11)}\left(\frac{\omega}{\sqrt{2t}} + 1\right)$$

$$= m^{(7)}\left(\frac{\omega}{\sqrt{2t}} + 1\right) v^{(6)}\left(\frac{\omega}{\sqrt{2t}} + 1\right) \left(m^{(7)}\left(\frac{\omega}{\sqrt{2t}} + 1\right)\right)^{-1} \tag{4.28}$$

for $\omega \in \Sigma_B^{(13)}$, $|\omega| < \epsilon\sqrt{2t}$, and

$$v_B^{(13)}(\omega) = I \tag{4.29}$$

for $\omega \in \Sigma_B^{(13)}$, $|\omega| > \epsilon\sqrt{2t}$.

Also clearly

$$m_B^{(13)}(\omega) = m_B^{(11)}\left(\frac{\omega}{\sqrt{2t}} + 1\right) , \tag{4.30}$$

so that

$$m_{B,1}^{(11)} = (m_{B,1}^{(13)})/\sqrt{2t} . \tag{4.31}$$

Inserting this relation with its A-analog into (4.22), we obtain from (4.19)

$$(m_1^{(10)})_{11} = (m_{A,1}^{(13)})_{11}/\sqrt{2t} + (m_{B,1}^{(13)})_{11}/\sqrt{2t} + O\left(\frac{1}{t}\right) . \tag{4.32}$$

Remark 4.33. By the dilation-translation invariance of the Cauchy operator,

$$\|(1 - C_{W_B^{(13)}})^{-1}\|_{L^2(\Sigma_B^{(13)}) \to L^2(\Sigma_B^{(13)})} \tag{4.34}$$

is independent of t.

To proceed we need more precise information about δ, the solution of the scalar factorization problem (3.9), (3.10). Define

$$\hat{h}(\omega) = \left(h(\omega)/h(1)\right)^{-1} , \quad \omega \in \Sigma_2 , \tag{4.35}$$

$$= h(\omega)/h(1) , \quad \omega \in \overline{\Sigma}_2 . \tag{4.36}$$

Then one checks that

$$\delta(\omega) = \left(\frac{\omega - 1}{\omega + 1}\right)_{i\mathbb{R}_+}^{i\nu} \left(\frac{\omega - 1}{\omega + 1}\right)_{i\mathbb{R}_-}^{i\nu} \exp\left\{\frac{1}{2\pi i} \int_{\Sigma_2 \cup \overline{\Sigma}_2} \frac{\log \hat{h}(s)}{s - \omega} \, ds\right\} \text{ for } \omega \in \mathbb{C} \setminus (\Sigma_2 \cup \overline{\Sigma}_2) , \tag{4.37}$$

where

$$\nu = -\frac{1}{2\pi} \log h(1) = -\frac{1}{2\pi} \log(\tanh \beta) > 0 , \tag{4.38}$$

$$(z)_{i\mathbb{R}_+}^{i\nu} = \exp\left\{i\nu(\log|z| + i \arg_{i\mathbb{R}_+}(z))\right\} , \quad -3\pi/2 < \arg_{i\mathbb{R}_+}(z) < \pi/2 , \tag{4.39}$$

and

$$(z)_{i\mathbb{R}_-}^{i\nu} = \exp\left\{i\nu(\log(z) + i \arg_{i\mathbb{R}_-}(z))\right\} , \quad -\pi/2 < \arg_{i\mathbb{R}_-}(z) < \frac{3\pi}{2} . \tag{4.40}$$

Also a simple calculation shows that $\log \hat{h}$ is in the Sobolev space $H^2(\Sigma^{(5)}) = H^2(\Sigma_2 \cup \overline{\Sigma}_2)$, and hence, in particular, $\exp\{\frac{1}{2\pi i} \int_{\Sigma_2 \cup \overline{\Sigma}_2} \frac{\log \hat{h}(s)}{s-\cdot} \, ds\}$ is Lipschitz in $\mathbb{C} \setminus \Sigma^{(5)}$ with a global bound. Also, this implies that δ and δ^{-1} are bounded in $\overline{\mathbb{C}} \setminus \Sigma^{(5)}$, as indicated earlier.

We conclude that

$$\delta(\frac{\omega}{\sqrt{2t}} + 1) = (8t)^{-i\nu}(\omega)_{\mathbb{R}_+}^{i\nu}(\omega)_{\mathbb{R}_-}^{i\nu}\,\hat{\delta}(1)(1 + O(\frac{\omega}{\sqrt{t}})) \qquad (4.41)$$

for $\omega \in \Sigma_B^{(13)} \cap \{|\omega| < \epsilon\sqrt{2t}\}$, where

$$\hat{\delta}(1) = \exp\{\frac{1}{2\pi i}\int_{\Sigma_2 \cup \overline{\Sigma}_2}\frac{\log \hat{h}(s)}{s-1}\,ds\}\ . \qquad (4.42)$$

From (3.26), we obtain

$$m^{(7)}(\frac{\omega}{\sqrt{2t}} + 1) = \begin{pmatrix} \frac{1}{1-i} & \frac{1}{1+i} \\ \frac{-1}{1-i} & \frac{1}{1+i} \end{pmatrix}(8t)^{i\nu\sigma_3}(\omega)_{\mathbb{R}_+}^{-i\nu\sigma_3}(\omega)_{\mathbb{R}_-}^{-i\nu\sigma_3}\hat{\delta}(1)^{-\sigma_3} + O(\frac{\omega}{\sqrt{t}}) \quad (4.43)$$

for $\omega \in \Sigma_B^{(13)} \cap \{|\omega| < \epsilon\sqrt{2t}\} \cap \{\mathrm{Re}\,\omega > 0\}$, and

$$m^{(7)}(\frac{\omega}{\sqrt{2t}} + 1) = \begin{pmatrix} \frac{1}{1-i} & \frac{1}{1+i} \\ \frac{-1}{1-i} & \frac{1}{1+i} \end{pmatrix}\begin{pmatrix} 0 & -1 \\ 1 & 0 \end{pmatrix}(8t)^{-i\nu\sigma_3}(\omega)_{\mathbb{R}_+}^{i\nu\sigma_3}(\omega)_{\mathbb{R}_-}^{i\nu\sigma_3}\hat{\delta}(1)^{\sigma_3} + O(\frac{\omega}{\sqrt{t}})\ ,$$
$$(4.44)$$

for $\omega \in \Sigma_B^{(13)} \cap \{|\omega| < \epsilon\sqrt{2t}\} \cap \{Re\,\omega < 0\}$. Define $v_B^{(14)}$ on $\Sigma_B^{(14)} = \Sigma_B^{(13)}$ as follows,

$$\begin{pmatrix} 1 & 0 \\ -\frac{(h(1)+1)}{h(1)}e^{-2it}e^{-i\omega^2/2} & 1 \end{pmatrix} \qquad \begin{pmatrix} 1 & 0 \\ \frac{h(1)-1}{h(1)}e^{2it}e^{i\omega^2/2} & 1 \end{pmatrix}$$

$$\Sigma_B^{(14)} = \Sigma_B^{(13)}$$

$$\begin{pmatrix} 1 & \frac{1-h(1)}{h(1)}e^{2it}e^{i\omega^2/2} \\ 0 & 1 \end{pmatrix} \qquad \begin{pmatrix} 1 & \frac{h(1)+1}{h(1)}e^{-2it}e^{-i\omega^2/2} \\ 0 & 1 \end{pmatrix}\ .$$

Figure 4.3

A simple computation shows that

$$v^{(6)}|_{\Sigma_B^{(11)}}(\frac{\omega}{\sqrt{2t}} + 1) = v_B^{(14)}(\omega) + O(\frac{e^{-|\omega|^2/4}}{\sqrt{t}})\ . \qquad (4.45)$$

Let $v_B^{(15)}$ be the constant jump matrix on $\Sigma_B^{(15)} = \Sigma_B^{(14)} = \Sigma_B^{(13)}$ given by

$$\begin{pmatrix} 1 & \frac{h(1)+1}{h(1)} \\ 0 & 1 \end{pmatrix} \qquad \begin{pmatrix} 1 & 0 \\ \frac{h(1)-1}{h(1)} & 1 \end{pmatrix}$$

$$\Sigma_B^{(15)}$$

$$\begin{pmatrix} 1 & 0 \\ \frac{h(1)-1}{h(1)} & 1 \end{pmatrix} \qquad \begin{pmatrix} 1 & \frac{h(1)+1}{h(1)} \\ 0 & 1 \end{pmatrix}\ .$$

Figure 4.4

For convenience we use the notation

$$A^{\text{ad } \sigma_3} B = A^{\sigma_3} B A^{-\sigma_3} \ . \tag{4.46}$$

Inserting the above computations and definitions in (4.28), (4.29) we find that on $\Sigma_B^{(15)}$,

$$v_B^{(13)}(\omega) = v_B^{(17)}(\omega) + O(e^{-|\omega|^2/4}/\sqrt{t}) \ , \tag{4.47}$$

where

$$v_B^{(17)}(\omega) = D_B \, v_B^{(16)}(\omega) \, D_B \ , \tag{4.48}$$

$$D_B = \begin{pmatrix} \frac{1}{1-i} & \frac{1}{1+i} \\ \frac{-1}{1-i} & \frac{1}{1+i} \end{pmatrix} (8t)^{i\nu\sigma_3} \, e^{-it\sigma_3} \, \hat{\delta}(1)^{-\sigma_3}, \tag{4.49}$$

and

$$v_B^{(16)}(\omega) = (\omega)_{\mathbb{R}_+}^{-i\nu \text{ ad } \sigma_3} (\omega)_{\mathbb{R}_-}^{-i\nu \text{ ad } \sigma_3} \, e^{-i(\omega^2/4) \text{ ad } \sigma_3} \, v_B^{(15)} \ . \tag{4.50}$$

We will show shortly that the RH problem associated with $v_B^{(17)}$ on $\Sigma_B^{(15)}$ has a solution for all t with

$$\|(1 - C_{W_B^{(17)}})^{-1}\| \text{ bounded as } t \to \infty \ . \tag{4.51}$$

As before, (4.47) then implies that the RH problem associated with $v_B^{(13)}$ on $\Sigma_B^{(13)}$ (and hence all the RH problem, $v^{(1)}$, $v^{(2)}$, ...) exists for large t, and

$$(m_{B,1}^{(13)})_{11} = (m_{B,1}^{(17)})_{11} + O(\frac{1}{\sqrt{t}}) \ . \tag{4.52}$$

Inserting this estimate and its A-analog into (4.32), we obtain

$$(m_1^{(10)})_{11} = (m_{A,1}^{(17)})_{11}/\sqrt{2t} + (m_{B,1}^{(17)})_{11}/\sqrt{2t} + O(\frac{1}{t}) \ . \tag{4.53}$$

To prove that the RH problem associated with $v_B^{(17)}$ has a solution and that (4.51) holds true, it is sufficient to show that $(1 - C_{W_B^{(16)}})^{-1}$ exists in L^2 for the (time-independent) RH problem associated with $v_B^{(16)}$ on $\Sigma_B^{(16)} = \Sigma_B^{(15)} = \Sigma_B^{(13)}$.

Now the problem $v_B^{(16)}$ on $\Sigma_B^{(16)}$ can be "folded" onto the imaginary axis, oriented from $-i\infty$ to $i\infty$, as follows. In the right half-plane, let $m_B^{(18)}$ denote the analytic continuation (which exists) of $m_B^{(16)} \, | \, \{\omega : -\frac{\pi}{4} < \arg \omega < \pi/4\}$, and in the left half-plane, let $m_B^{(18)}$ denote the analytic continuation of $m_B^{(16)} \, | \, \{\omega : \frac{3\pi}{4} < \arg \omega < \frac{5\pi}{4}\}$. Let $v^{(18)} = (m_-^{(18)})^{-1} m_+^{(18)}$ on $i\mathbb{R}$.

On $i\mathbb{R}_+$,

$$v^{(18)} = \left((\omega)_{\mathbb{R}_+}^{-i\nu\sigma_3}\right)_- \left((\omega)_{\mathbb{R}_-}^{-i\nu\sigma_3}\right)_- e^{-i(\omega^2/4)\sigma_3} \begin{pmatrix} 1 & 0 \\ \frac{h(1)-1}{h(1)} & 1 \end{pmatrix} \left((\omega)_{\mathbb{R}_+}^{i\nu\sigma_3}\right)_-$$

$$\times \left((\omega)_{\mathbb{R}_-}^{i\nu\sigma_3}\right)_- e^{i(\omega^2/4)\sigma_3} \left((\omega)_{\mathbb{R}_+}^{-i\nu\sigma_3}\right)_+ \left((\omega)_{\mathbb{R}_-}^{-i\nu\sigma_3}\right)_+ e^{-i(\omega^2/4)\sigma_3}$$

$$\times \begin{pmatrix} 1 & \frac{h(1)+1}{h(1)} \\ 0 & 1 \end{pmatrix} \left((\omega)_{\mathbb{R}_+}^{i\nu\sigma_3}\right)_- \left((\omega)_{\mathbb{R}_-}^{i\nu\sigma_3}\right)_+ e^{i(\omega^2/4)\sigma_3} \ .$$

Clearly $\left((\omega)_{i\mathbb{R}_-}^{-i\nu\sigma_3} \right)_+ = \left((\omega)_{i\mathbb{R}_-}^{-i\nu\sigma_3} \right)_-$ on $i\mathbb{R}_+$, but

$$
\begin{aligned}
\left((\omega)_{i\mathbb{R}_+}^{i\nu\sigma_3} \right)_- / \left((\omega)_{i\mathbb{R}_+}^{i\nu\sigma_3} \right)_+ &= e^{i\nu(i\pi/2 - (-3i\pi/2))\sigma_3} \\
&= e^{-2\pi\nu\sigma_3} \\
&= h(1)^{\sigma_3} ,
\end{aligned}
\tag{4.54}
$$

by (4.38). Inserting (4.54) we obtain on $i\mathbb{R}_+$,

$$
\begin{aligned}
v^{(18)} &= \left((\omega)_{i\mathbb{R}_+}^{-i\nu\sigma_3} \right)_- \left((\omega)_{i\mathbb{R}_-}^{-i\nu\sigma_3} \right)_- e^{-i(\omega^2/4)\sigma_3} \begin{pmatrix} h(1) & h(1)+1 \\ h(1)-1 & h(1) \end{pmatrix} \\
&\quad \times \left((\omega)_{i\mathbb{R}_+}^{i\nu\sigma_3} \right)_+ \left((\omega)_{i\mathbb{R}_-}^{i\nu\sigma_3} \right)_+ e^{i(\omega^2/4)\sigma_3} \\
&= \left((\omega)_{i\mathbb{R}_-}^{-2i\nu\sigma_3} \right)_- e^{-i(\omega^2/4)\sigma_3} \begin{pmatrix} 1 & (h(1)+1)h(1) \\ \frac{h(1)-1}{h(1)} & h(1)^2 \end{pmatrix} \\
&\quad \times e^{i(\omega^2/4)\sigma_3} \left((\omega)_{i\mathbb{R}_-}^{2i\nu\sigma_3} \right)_+ .
\end{aligned}
\tag{4.55}
$$

A similar computation leads to the identical form (4.55) for $v^{(18)}$ on $i\mathbb{R}_-$.

Define

$$
r = \sqrt{1 - h(1)^2} = \sqrt{1 - (\tanh\beta)^2} < 1 \quad \text{and} \quad a = \sqrt{\frac{r}{h(1)(1 + h(1))}} .
\tag{4.56}
$$

Then

$$
\begin{aligned}
v^{(19)} &\equiv a^{\operatorname{ad}\sigma_3} v^{(18)} \\
&= \left((\omega)_{i\mathbb{R}_-}^{-2i\nu\sigma_3} \right)_- e^{-i(\omega^2/4)\sigma_3} \begin{pmatrix} 1 & r \\ -r & 1-r^2 \end{pmatrix} e^{i(\omega^2/4)\sigma_3} \left((\omega)_{i\mathbb{R}_-}^{2i\nu\sigma_3} \right)_+
\end{aligned}
\tag{4.57}
$$

on $i\mathbb{R}$. But as in [DZ] (see (3.100) et seq.)), the RH problem with $v^{(19)}$ is equivalent to the small norm problem associated with $e^{-i(\omega^2/4)\operatorname{ad}\sigma_3} \begin{pmatrix} 1 & r \\ -r & 1-r^2 \end{pmatrix}$, $0 \le r < 1$. This proves that $(1 - C_B^{(16)})^{-1}$ exists and hence (4.51) is true, etc. (Furthermore the RH problem for $v^{(19)}$ (and hence for $v^{(18)}$) can be solved explicitly in terms of parabolic cylinder functions (see [I] or [DZ]), but the exact form of the solution is not needed in this paper.) This establishes the error estimate (4.53).

Finally we observe directly from (2.36)–(2.41),that

$$
v^{(6)}(\omega) = \sigma_1 v^{(6)}(-\omega)\sigma_1
\tag{4.58}
$$

on $\Sigma^{(6)}$ (we can always ensure that $\Sigma^{(6)} = -\Sigma^{(6)}$), where $\sigma_1 = \begin{pmatrix} 0 & 1 \\ 1 & 0 \end{pmatrix}$. The succeeding problems inherit this symmetry, and in particular one finds that

$$
v_A^{(17)}(\omega) = \sigma_1 v_B^{(17)}(-\omega)\sigma_1 .
\tag{4.59}
$$

By uniqueness

$$
m_A^{(17)}(\omega) = \sigma_1 m_B^{(17)}(-\omega)\sigma_1
\tag{4.60}
$$

so that

$$
m_{A,1}^{(17)} = -\sigma_1 m_{B,1}^{(17)} \sigma_1
\tag{4.61}
$$

and in particular
$$(m^{(17)}_{A,1})_{11} = -(m^{(17)}_{B,1})_{22} \ . \tag{4.62}$$

Inserting (4.62) in (4.53), we obtain

$$(m^{(10)}_1)_{11} = \left((m^{(17)}_{B,1})_{11} - (m^{(17)}_{B,1})_{22} \right) / \sqrt{2t} + O(\tfrac{1}{t}) \ . \tag{4.63}$$

But from (4.48),
$$m^{(17)}_B = D_B \, m^{(16)}_B \, D_B^{-1} \tag{4.64}$$

and inserting (4.49), (4.64) into (4.63) we obtain

$$
\begin{aligned}
(m^{(10)}_1)_{11} = &(1-i)^{-2}\hat{\delta}(1)^{-2}(m^{(16)}_{B,1})_{12}(8t)^{2i\nu}e^{-2it}t^{-1/2} \\
&+ (1+i)^{-2}\hat{\delta}(1)^{2}(m^{(16)}_{B,1})_{21}(8t)^{-2i\nu}e^{2it}t^{-1/2} + O(\tfrac{1}{t})
\end{aligned} \tag{4.65}
$$

(recall that $m^{(16)}_B$ is time-independent). As the integrals $\int^{\infty}(16t)^{\pm 2i\nu}\,e^{\mp 2it}\,t^{-1/2}\,dt$ are convergent, we obtain

$$\int^t (m^{(10)}_1)_{11}(t')\,dt' = O(\log t) \quad \text{as} \quad t \to \infty \ ,$$

which completes the proof of Theorem 1.12.

Remark 4.66. It is possible to evaluate the asymptotics of $m^{(1)}_1$ to higher order in t and so obtain an asymptotic expansion for $\chi(t)$ as $t \to \infty$ (up to the integration constant for $\sigma = \int^t \frac{d\sigma}{dt'}\,dt'$; this constant is of course not determined by the large t behavior of the RH problem).

Remark 4.67. We have shown that the RH problem for $v^{(1)}$ has a solution for sufficiently large t. There is no physical reason to expect that the RH problem exists for all t. Indeed it is possible that $\chi(t) = 0$ for some (small) $t = t_0$, and hence $\sigma(t) = \log e^{t^2/2}\,\chi(t)$ has a singularity at t_0. Thus the RH problem may not have a solution for all t.

Acknowledgments

The authors would like to thank A. R. Its for bringing this problem to their attention. The work of the authors was supported in part by NSF Grants DMS-9001857 and DMS-919 6033 respectively. The first author would also like to thank NATO for support in attending the conference at Lyons.

REFERENCES

[BC] R. Beals and R. Coifman, Scattering and inverse scattering for first order systems, Comm. Pure Appl. Math. 37, 39–90 (1984).

[DZ] P. Deift and X. Zhou, A steepest descent method for oscillatory Riemann-Hilbert problems, Asymptotics for MKdV, to appear in Annals of Math. Announcement in Bull. Amer. Math. Soc., (New Series), 26, 119–123 (1992).

[I] A. R. Its, Asymptotics of solutions of the nonlinear Schrödinger equation and isomonodromic deformations of systems of linear differential equations, Sov. Math. Dokl. 24(3), 452–456 (1981).

[IIK] A. R. Its, A. G. Izergin, V. E. Korepin, Temperature correlators of the impenetrable Bose gas as an integrable system, ICTP preprint IC/89/120, Trieste, Italy, June 1989.

[IIKN] A. R. Its, A. G. Izergin, V. E. Korepin and V. Yu. Novokshenov, Temperature autocorrelations of the transverse Ising chain at the critical magnetic field, preprint, 1989.

[IIKS] A. R. Its, A. G. Izergin, V. E. Korepin and N. A. Slavnov, Differential equations for quantum correlation functions, preprint CMA-R26-89, Australian Nat. Univ., 1989.

[MPS] B. M. McCoy, J. H. H. Perk and R. E. Shrock, Time-dependent correlation functions of the transverse Ising chain at the critical magnetic field, Nucl. Phys. B220 [FS8], 35–47 (1983).

[LSM] E. Lieb, T. Schultz and D. Mattis, Ann. of Phys. 16, 406 (1961).

RESONANCES IN MULTIFREQUENCY AVERAGING THEORY

S. YU. DOBROKHOTOV

1. INTRODUCTION

The Whitham method allows to obtain rapidly oscillating asymptotic solutions of nonlinear equations with a small parameter ε which characterizes the dispersion. The principal term of such asymptotic solutions can be represented in the form

$$(1.1) \qquad u = f\left(\frac{S(x,t)}{\varepsilon}, x, t \right),$$

where $f(\tau, x, t)$ and $S(x,t)$ are smooth functions and f is 2π-periodic in the argument τ. The function (1.1) describes the distribution of wave packets. Such solutions were used in different physical and mechanical problems (see, for example, the bibliography in [1–16]); they satisfy the special Cauchy data

$$(1.2) \qquad u\,\big|_{t=0} = f\left(\frac{S^0(x)}{\varepsilon}, x, 0 \right).$$

The functions of such type also appear when we look for asymptotics (as $t \to \infty$) of the Cauchy problem solution for nonlinear equations with localized initial data or derivatives (the problem about decay of steps)

$$u\,\bigg|_{t=0} = v^0\left(\frac{x}{\varepsilon} \right),$$

Moreover, they also appear in the Cauchy problem with initial data which yield the overturning of the solution of limit ($\varepsilon = 0$) equation (Gurevich–Pitaevsky problem and Lax–Levermore problem). Here we consider only the problem (1.2) and all the investigations are carried out for the Korteweg de Vries equation

$$(1.3) \qquad K[u] \overset{\text{def}}{=} u_t - 6uu_x + \varepsilon^2 u_{xxx} = 0.$$

(A large but noncomplete bibliography concerning these problems can be found, for example, in [1–8, 12], as well as in other papers of this collection).

The solutions of the form (1.1) as well as any other asymptotic solutions are based on certain exact solutions. Such exact solutions are conoidal waves which are the solutions of the following equations (without a parameter)

$$(1.4) \qquad v_\eta - 6vv_\xi + v_{\xi\xi\xi} = 0.$$

Singular Limits of Dispersive Waves, Edited by
N.M. Ercolani et al., Plenum Press, New York, 1994

They are defined by the formula

$$(1.5) \qquad\qquad v = f(v(E)\xi + V(E)\eta + P|E) .$$

Here $E = (E_1, E_2, E_3)$, P are parameters,

$$(1.6) \qquad\qquad F(\tau|E) = C(E) - 2\left(U \cdot \frac{\partial}{\partial \tau}\right)^2 \ln \theta(\tau|E) ,$$

$$(1.7) \qquad\qquad \theta = \sum_k \exp\left(i\frac{Bk \cdot k}{2}\right) \exp(ik \cdot \tau) ,$$

where θ is the Jacobi θ-function (the sum is taken over all integer k), U, V, C, B ($\mathrm{Im}\, B > 0$) are expressed in terms of E by means of elliptic integrals. The solution (1.1) is related to (1.5)–(1.7) as follows. The parameters E and P become functions in (x, t), i.e., $E = E(x, t)$, $P = P(x)$ and the phase $S(x, t)$ is defined by the relations

$$(1.8) \qquad\qquad \frac{\partial S}{\partial x} = U(E(x, t)), \qquad \frac{\partial S}{\partial t} = V(E(x, t)),$$

so that

$$(1.9) \qquad\qquad u_0 = f\left(\frac{S(x, t)}{\varepsilon} + P(x, t)|E(x, t)\right).$$

The functions $E(x,t)$ are obtained from the Whitham equation (the explicit form $s_j(E)$ is given in [4,10])

$$(1.10) \qquad\qquad E_{jt} = s_j(E)E_{jx} .$$

The shift of the phase $P(x, t)$ is defined by a system of equations of the second order (see [2,8,9]).

The function (1.9) describes the propagation of a single wave packet and is called an one-phase solution. Naturally, in problems about the wave packets interaction, the multiphase asymptotics appear which have the same form (1.8), but already with vector-functions $S = (S_1, ..., S_l)$ and $P = (P_1, ..., P_l)$ and a $(2l+1)$-dimensional vector $E = (E_1(x, t),...,E_{2l+1}(x, t))$.

The multiphase asymptotics are based on exact l-valued almost periodic solutions of the equation (1.4) which are also defined by the Matveev–Its formulas (1.5), but where U, V, P are assumed to be l-dimensional vectors, B is assumed to be a $l \times l$-matrix; $\theta(\tau|E)$ is the Riemann θ-function. The components of the vectors U, V, the elements of the matrix B and C are expressed in terms of E already by means of hyperelliptic integrals. The phases S are related to E by the formulas (1.8), and the functions E are defined by the Witham–Flaschka–Forest–McLaughlin equations (1.10) [10]. The multiphase asymptotics first appeared in the work [11], however it was possible to use them in the constructive form [2,3,10,16,17] only after the almost-periodic finite-gap solutions were discovered. The explicit formulas of finite-gap integration, in particular, the formulas for U, V, B, C, θ, etc., can be found, for example, in [12,13].

Here we only recall that functions f are called l-gap or l-phase (almost periodic) solutions, and that the parameters $(E_1, E_2, ..., E_s)$ are the ends of prohibited gaps $(\infty, E_1), (E_2, E_3), (E_4, E_5)$ of the spectrum of the Sturm–Liouville operator

$$-\frac{d^2}{d\xi^2} + f(U\xi + V\xi + P|E).$$

They are situated on the real axis E as follows

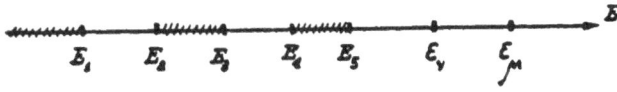

Figure 1

The formula (1.9) defines only the principal term of the asymptotic solution. The substitution of this solution in the equation yields the correction $O(1)$, and in order to obtain such corrections it is sufficient to demand that only the relations (1.6)–(1.9) hold. Equation (1.10) can be obtained only in the process of constructing a correction to (1.9). In the one-phase case, the procedure of constructing corrections is well developed [14]. In this case the following asymptotic series can be constructed

$$u = u_0 + u_1 + \cdots + u_N,$$

(1.11)
$$u_j(x, t, s) = \varepsilon^j f_j\left(\frac{S(x,t)}{\varepsilon} + P(x,t), x, t\right),$$

each term of which has the same structure as the principal term U_0 and satifies the equation with the same accuracy $O(\varepsilon^N)$.

The multiphase case is esssntially different from the one-phase case. Here there are resonance points which are everywhere dense on the x-axis. Since these points are placed in the configuration space, it is impossible to throw them away together with their small neighbourhoods as was done in the KAM-theory. Because of these points, the expansion (1.11) does not hold even in the simplest case of two phases. Already the first correction u_1 to u_0 is of different structure and of order at least $O(\sqrt{\varepsilon})$. A similar estimate holds for the derivative $\varepsilon \partial u_1/\partial x$. This circumstance yields the necessity to find the correction u_1 not from a linearized equation as is done in the standard theory of perturbations, but from a nonlinear equation. Actually, if we neglect the summand $u_1 \partial u_1/\partial x$ when looking for u_1, the correction obtained after the substitution of $u_0 + u_1$ into the initial equation (1.3) yields the correction $-u \partial u_1/\partial x = O(1)$ of the same order as the result of substitution of u_0 in the equation. Thus, in order to find u_1, we must, actually, solve a more complicated equation, namely, the same KdV-equation, but inhomogeneous and with additional oscillating coefficient $(\varepsilon \partial/\partial x(u_0 u_1))$. Nevertheless, it turns out that it is possible to construct the desired solution which yields a correction the order of smallness of which is higher than $O(1)$. Just this problem is the central point of this report. Though we consider a two-phase case only, the complete proof of this problem occupies too large place, and here restrict ourselves only to the ideological side of the problem and state the central ideas. The complete version based on [2,3,15,18,19,26] will be published in Russian Journal of Pure and Applied Mathematics.

2. Zero Approximation

Here and everywhere below the brackets $\langle \, , \, \rangle$ denote the real scalar product.

Lemma 2.1. *Suppose smooth functions $S_j(x,t)$, $P_j(x,t)$, $j = 1, ..., l$, $E_j(x,t)$ $E_j < E_{j+1}$, $j = 1, 2, ..., 2l+1$, are related by (1.2) for $0 \leqslant t \leqslant T$. Then the function u_0 of form (1.9) satisfies the relation*

$$K[u_0] = R\left(\frac{S}{\varepsilon} + P, x, t\right) + \varepsilon R_1\left(\frac{S}{\varepsilon} + P, x, t\right) + \varepsilon^2 R_2\left(\frac{S}{\varepsilon} + P, x, t\right).$$

Here

$$(2.1) \quad R(\tau, x, t) = -\left(\langle P_t, \frac{\partial}{\partial t} \rangle f - 3\langle P_x, \frac{\partial}{\partial \tau} \rangle (f^2 - \langle U, \frac{\partial}{\partial \tau} \rangle^2 f \right.$$

$$\left. + 3(\frac{\partial f}{\partial x} - \langle U, \frac{\partial}{\partial \tau} \rangle^2 \frac{\partial F}{\partial x} - \langle U, \frac{\partial}{\partial \tau} \rangle \langle U, \frac{\partial}{\partial \tau} \rangle f) - \frac{\partial f}{\partial t} \right),$$

$R_{1,2}(\tau, x, t)$ are polynomials in the second power of f, S, D, E, U, V and their derivatives.

Proof. As the result of the substitution of u_0 in the left-hand side of equation (1.3) and the succeeding differentiation, we obtain that in order to prove the Theorem, it is sufficient to prove the inequality

$$(2.2) \qquad \langle V, \frac{\partial}{\partial \tau} \rangle f - 6f\langle U, \frac{\partial}{\partial \tau} \rangle f + \langle U, \frac{\partial}{\partial \tau} \rangle^3 f = 0.$$

We introduce new coordinates $\eta, \xi, \zeta_1, ..., \zeta_{l-2}$ on the torus $T^l = \{\tau_j \in [0, 2\pi], j = 1, ..., l\}$ in the following way. Suppose $W_1, ..., W_{l-2}$ are l-dimensional vectors completing the vectors V, U to a basis in R^l and orthogonal to V, U. The coordinates $\tau = (\tau_1, ..., \tau_l)$ are related to the coordinates by formulas

$$\tau = V\eta + U\xi + W_1\zeta_1 + \cdots + W_{l-2}\zeta_{l-2},$$

where τ is a vector-column.

In these coordinates the relation (2.2) has the form of equation (1.4). Since the function (1.5) is a solution of the equation (1.9) for any real vector P, then the function $f(\tau|E)$ is also a solution of (1.4) in coordinates $(\eta, \xi, \zeta_1, ..., \zeta_{l-2})$.

Thus the corrections obtained by the substitution of u_0 in equation (1.3) are equal to $O(1)$ and have the same structure as u_0, i.e., they can be expressed in terms of a 2π-periodic function in each of the arguments $(\tau = \tau_1, .., \tau_l)$.

3. Linear equation in variations

Now we consider the two-phase case $l = 2$. First we make an assumption that is untrue for this problem, namely, we assume that, in order to obtain the first correction u_1 to u_0, we must consider, just as in the standard theory of perturbations, a linearized KdV-equation

$$(3.1) \qquad Lu_1 \overset{\text{def}}{=} u_{1t} - 6(u_0u_1)_x + \varepsilon^2 u_{xxx} = R\left(\frac{S}{\varepsilon} + P, x, t\right) + O(\varepsilon).$$

We shall need most of the formulas obtained when this equation is investigated in order to construct a correction u_1. Equation (3.1) is an equation with rapidly oscillating coefficients. For equations of this type, there is a well developed theory of averaging [20]. However, this theory does not suit for us because of the following circumstances. First, the principal term of asymptotic solution oscillates as well as the coefficients, second, the coefficients here will be functions, in a certain sense, almost periodic in rapid variables $S_1/\varepsilon, S/\varepsilon$. Thus we shall use here the other scheme [21].

First, we regularize the coefficients of the problem by representing the solution of (3.1), without loss of generality, in the form

$$(3.2) \qquad u_1 = w\left(\frac{S}{\varepsilon} + P, x, t, \varepsilon\right),$$

without assuming the regular dependence of the function $w(\tau, x, t, \varepsilon)$ on ε. By substituting (3.2) in the left-hand side of (3.1) and after differentiating, we obtain the following Lemma.

Lemma 3.1. *Suppose the functions $w(\tau, x, t, \varepsilon)$ satisfy the equation up to $O(\varepsilon^\alpha)$,* $0 \leqslant \alpha \leqslant 1$,

$$(3.3) \qquad \begin{aligned} \varepsilon\frac{\partial w}{\partial \tau} + (\langle V, \frac{\partial}{\partial \tau}\rangle)w - 6(\varepsilon\frac{\partial}{\partial x} + (\langle U, \frac{\partial}{\partial \tau}\rangle)(fw)+ \\ + (\varepsilon\frac{\partial}{\partial \tau} + (\langle U, \frac{\partial}{\partial \tau}\rangle))^3 w = \varepsilon R(\tau, x, t) + O(\varepsilon^{1+\alpha}), \quad 1 \geqslant \alpha \geqslant 0 \, . \end{aligned}$$

Then the function u_1 (3.2) satisfies (3.1) up to $O(\varepsilon^\alpha)$.

Thus we obtain for w an equation where the variables t, x and τ do not play equal roles. There is a parameter at the derivatives in (t, x), and there is no parameter at the derivatives in τ_1, τ_2. In the theory of qiasi-classical asymptotics, such equations are called equations with operator-valued symbol [9]. Namely, we can write the equation (3.3) in the form

$$(3.4) \qquad -\varepsilon\frac{\partial}{\partial t}w + (\mathcal{L}(-i\varepsilon\frac{\partial}{\partial x}, x, t) + \sum_{j=1}^{3} \varepsilon^j \mathcal{L}_j(-i\varepsilon\frac{\partial}{\partial x}, x, t)w = \varepsilon R,$$

where $\mathcal{L}(p, x, t), \mathcal{L}_j(p, x, t)$ are families of derivative operators (depending on p, x, t, $p \in R$) on a two-dimensional torus $Y^2 = \{(\tau_1, \tau_2) \in [0.2\pi]^2\}$. We assume that in (3.4) the operators $-i\varepsilon\frac{\partial}{\partial x}$ act first, and x act second. We shall need only the symbol operator $\mathcal{L}$

$$(3.5) \qquad \mathcal{L} = i(\langle V, \frac{\partial}{\partial \tau}\rangle - 6(ip + \langle U, \frac{\partial}{\partial \tau}\rangle)(f\cdot) + (ip + \langle U, \frac{\partial}{\partial \tau}\rangle)^3).$$

We note that the dependence on x, t in V, U and f, and hence on $\mathcal{L}$, is expressed in terms of the parameters $E = (E_1, \ldots, E_s)$, thus we write further $\mathcal{L}(p, x, t)$ instead of $\mathcal{L}(p, E)$.

Obviously, the spectral properties of the operator $\mathcal{L}$ and the operator

$$\mathcal{L}^+ = i(\langle V, \frac{\partial}{\partial \tau}\rangle - 6f(ip + \langle U, \frac{\partial}{\partial \tau}\rangle)) + (ip + \langle U, \frac{\partial}{\partial \tau}\rangle)^3)$$

adjoint to the operator $\mathcal{L}$ play an important role in solving the equation (3.4). Both standard methods of spectral theory and the theory of solitons [27] cannot be applied to the investigation of the spectra of $\mathcal{L}$ and $\mathcal{L}^+$. Nevertheless, by the fact that KdV equation is integrable, the spectra of $\mathcal{L}$ and $\mathcal{L}^+$ can be calculated. It is real and purely pointwise, i.e., the eigenfunctions and adjoint functions of the operators $\mathcal{L}$ and $\mathcal{L}^+$ form biorthogonal basis systems, but in dependence on p and E it can either cover the real axis everywhere dense, or some of its values turn out to be infinitely degenerate. The whole description of the spectrum [18] is rather complicated, so we restrict ourselves only to certain facts.

First, we present the central ideas of calculating the spectra of the operators $\mathcal{L}$ and $\mathcal{L}^+$. Suppose $w_\nu(\tau, p, t)$, $\lambda_\nu(p, E)$ and $w_\nu^+(\tau, p, t)$, $\lambda_\nu(p, E)$ are the functions and eigenvalues of operators $\mathcal{L}$ and $\mathcal{L}^+$ on the torus T^2. They depend on p, E as well as the operators $\mathcal{L}$. The following propositions play an important role in the investigation of spectra of $\mathcal{L}$ and $\mathcal{L}^+$:

1) If $\mathcal{L}^+ w_\nu^+ = \lambda_\nu^+ w_\nu^+$, then the function $(ip + \langle U, \partial/\partial \tau\rangle)w_\nu^+$ is an eigenfunction of the operator $\mathcal{L}$ related to the same eigenvalue $\lambda_\nu^+ = \lambda_\nu$. This fact implies, in particular, that the spectra of $\mathcal{L}^+$ and $\mathcal{L}^-$ coincide. Thus we further denote the eigenvalues of $\mathcal{L}^+$ also by λ_ν.

2) If $\mathcal{L}w_\nu = \lambda_\nu w_\nu$, and $\mathcal{L}^+ w_\nu^+ = \lambda_\nu w_\nu^+$, then the functions $w(\xi, \eta) = \exp(i(p\xi - \lambda_\nu \eta))w_\nu(U\xi + V\xi + P, E, p)$ and $w^+(\xi, \eta) = \exp(i(p\xi - \lambda_\nu \eta))w_\nu^+(U\xi + V\xi + P, E, p)$ are solutions of the linearized KdV-equation

(3.6) $$w_\eta - 6(f(U\xi + V\eta + P|E)w)_\xi + w_{\xi\xi\xi} = 0$$

and the adjoint equation

(3.7) $$w_\eta^+ - 6f(U\xi + V\eta + P|E)w_\xi^+ + w_{\xi\xi\xi}^+ = 0$$

Equation (3.6) implies that the function w can be obtained by linearizing the exact three-gap solution of the KdV-equation (1.4) on the background of the two-gap solution $f(U\xi + V\eta + P|E)$. On the other hand, it is known (for example, see [3, 5, 12, 27]) that the second powers of eigenfunctions $\psi(\xi, \eta)$ of the operator $-\partial^2/\partial \xi^2 + f$ are solutions of equation (3.7).

3) If $w_0(\tau, p, E)$, $\lambda_0(p, E)$ are the eigenfunction and the eigenvalue of the operator $\mathcal{L}(p, E)$, then the function

(3.8) $$w_\nu = w_0(\tau, p + \langle U, \nu\rangle, E)\exp(i\langle \tau, \nu\rangle), \quad \lambda_\nu = \lambda_0(p + \langle U, \nu\rangle, E) - \langle V, \nu\rangle,$$

are also the eigenfunctions and eigenvalues of the operator $\mathcal{L}(p, E)$. Here $\nu = (\nu_1, \nu_2) \in \mathbb{Z}^2$ is a multi-index. By this fact, we can point out several "vacuum" eigenfunctions and eigenvalues, in terms of which one can express all the other eigenfunctions and eigenvalues according to formulas (3.8). And, as a matter of fact, one must pay attention to the fact that the operation of choosing the multi-index ν does not give the eigenfunctions already obtained. The same operation can be performed with eigenfunctions of the operator $\mathcal{L}^+$.

4) For a certain $p = p_k$, in particular, for $p = 0$, adjoint functions appear in the spectra of $\mathcal{L}$ and $\mathcal{L}^+$. The calculation of these functions can be performed by passing to the limit as $p \to p_k$ and having in mind the following: there is a family of problems depending on the parameters p and E.

5) After the systems of eigenfunctions and adjoint functions is found, we have the problem how to prove whether they form a basis. A "bad" arrangement of spectra of operators $\mathcal{L}$ and $\mathcal{L}^+$ does not allow to apply standard methods of spectral theory. The main idea of proof that they form a basis (just as in [22]) is based on the fact that there is a family of operators $\mathcal{L}(p, E)$ smoothly depending on the parameters (p, E). And there exists a family (everywhere dense) of parameters E for which the eigenfunctions (as well as the adjoint functions) of the operator $\mathcal{L}$ can be represented in the form of pairwise products of functions from two systems such that one of them obviously forms a basis, and the other one forms a basis by the work [24].

These considerations allow to calculate the eigenfunctions and eigenvalues w_ν, w_ν^+ and λ_ν and investigate their properties. They can be expressed, as well as l-gap solutions of the KdV-equation, in terms of hyperelliptic integrals.

It is convenient to show the spectrum of the operators $\mathcal{L}$ and $\mathcal{L}^+$ for fixed E on the plane (p, λ). It is convenient to choose five "vacuum" eigenvalues, i.e., the functions $\lambda_0^{(j)}(p, E)$ with the following properties (see Figure 3.1).

$$\lambda_0^{(1,2)}(p, E) = (2U_1 - \text{periodic}) - V_1 p / U_1,$$

$$\lambda_0^{(3,4)}(p, E) = (2(U_2 - U_1) - \text{periodic}) - p(V_2 - V_1)/(U_2 - U_1),$$

$$\lambda^{(5)}(p, E) = ((p \pm U_2)^2 + \gamma p + O(|p|^{-1}) \quad \text{for } p \to \pm\infty,$$

$\gamma = \gamma(E)$ are certain constants.

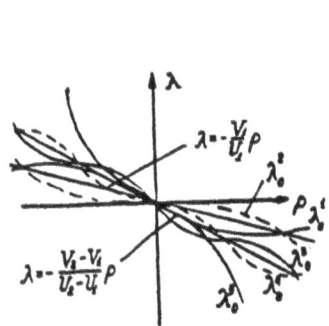

Figure 2

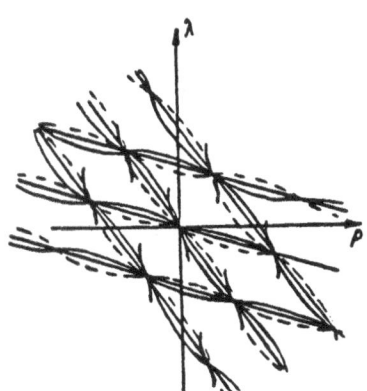

Figure 3

The other eigenvalues $\lambda_\nu^{(j)}(p, E)$ are obtained by formulas (3.8), and one must take $\nu = (0, \nu_2)$, $\nu_2 \in \mathbb{Z}$ for $j = 1, 2, 3, 4$, and $\nu \in \mathbb{Z}^2$ for $j = 5$.

The spectrum of the operators $\mathcal{L}$ and $\mathcal{L}^+$ is the intersection of these curves with vertical lines $p = \text{const}$. Another interpretation of the spectrum is possible. On the plane (λ, p), we have a skew lattice N formed by the vectors $(V_1 - V_2, U_2 - U_1)$, $(V_1, -U_1)$. We issue the curves λ_ν^j from each knot of our lattice as from the origin, and thus we obtain the curvilinear projections of each knot on each line $p = \text{const}$. The union of all such "projections" of knots just yields the spectrum of the operators $\mathcal{L}$ and $\mathcal{L}^+$ (see Figure 3.1).

Each of the curves λ_ν^j is associated with the eigenfunctions of the operators $\mathcal{L}$ and $\mathcal{L}^+$. These eigenfunctions depend smoothly both on the parameters (τ_1, τ_2) on the torus T^2, and on the parameters $p \in R$ and E under the condition $E_j < E_{j+1}$, $j = 1, 2, 3, 4$. However, at the lattice knots M, the eigenfunctions corresponding to four curves — "snakes" and a curve of parabolic type issued from this knot become linearly dependent, and we have here only three eigenfunctions and two adjoint functions appear. (The "parabolas" issued form other knots may, generally speaking, enter these knots, but the eigenfunctions corresponding to them are linearly independent of those).

Everything said above concerns to the same extent the spectrum of the operators $\mathcal{L}$ and $\mathcal{L}^+$ for $p = 0$.

Suppose $\mathcal{D}$ is a certain compact in the space of parameters $E = (E_1, \ldots, E_s)$, $E_j < E_{j+1}$, $\delta > 0$ is a certain small positive number. The curves $\lambda_\nu^{(j)}$ will be called nonresonance curves if $E \in \mathcal{D}$, $|\lambda_\nu^{(j)}| \geqslant \delta$ and resonance curves otherwise.

Obviously, the eigenvalues $\lambda_0^{(j)}$, $j = 1, \ldots, 5$ are resonance values, and $\lambda_0^{(j)}(0, E) = \blacksquare$ 0 for all E. Thus these values will be called strongly resonance values in contrast to weakly resonance values $\lambda_\nu^{(j)}$ which vanish for $p = 0$ only for certain $E \in \mathcal{D}$, We denote

$$(\psi_1, \psi_2) = \frac{1}{(2\pi)^2} \int_0^{2\pi} \int_0^{2\pi} \psi_1(\tau) \psi_2(\tau) d\tau_1 .$$

Suppose $\epsilon(s)$ is a smooth cutting function: $e = 1$ for $|s| < \delta$ and $e = 0$ for $|s| > 2\delta$.

Lemma 3.2. *(About the nonresonance component of correction.) Suppose the following condition holds: for all $E \in \mathcal{D}$, the curves $\lambda = \lambda_0^{1,2}(p, E)$ on the plane (p, λ) do not intersect the lines $\lambda = -V_1/U_1(p \pm (V_1 - V_2)) \pm (V_1 - V_2)$ for $|p| < U_1$, and the curves $\lambda = \lambda_0^{3,4}(p, E)$ do not intersect the lines $\lambda = -(V_2 - V_1)/(U_2 - U_1)(p \pm U_1) \pm V_1$ for $|p| < |U_1 - U_1|$, (or otherwise, they lie between these straight lines). Then*

1) The values $\lambda_\nu^{(j)}$, $j = 1, 2, 3, 4$, $\nu \neq 0$ are nonresonance.

2) the curves $\lambda_0^{(5)}(p, E)$ and $\lambda_0^{(2n-1)}$, $n = 1, 2, 3$, correspond to the eigenfunctions (smoothly depending on the parameters $\tau \in T^2$, $p \in R$. $E \in \mathcal{D}$) $w_\nu(\tau, p, E)$. $z_n(\tau, p, E)$ and $w_n^+(\tau, p, E)$, $z_n^+(\tau, p, E)$ of the operators $\mathcal{L}$ and $\mathcal{L}^+$. These functions are mutually biorthogonal in $L_2(T^2)$: $(w_\nu^+, w_\mu) \overset{\text{def}}{=} \delta_{\nu,\mu}$; $(z_k^+, z_m) = \delta_{k,m}$; $(z_k^+, w_\mu) = (w_\nu^+, z_m) = 0$, $\mu \neq 0$, $\nu \neq 0$.

3) Any smooth function $R(\tau)$ on the torus T^2 can be represented in the form

(3.9)
$$R = R_{nr} = \sum_{k=1}^{3} b_k z_{2k-1} + R_{res},$$

(3.10)
$$R_{res} = \sum_{\nu \in N} c_\nu w_\nu(\tau, 0E).$$

210

Here $b_k = (z^+_{2k-1}, \varphi)|_{p=0}$, $c_\nu = e(\lambda^{(j)}_\nu(0, E))(w^+_\nu, \varphi)|_{p=0}$ are the Fourier coefficients of the expansion of R in the eigenfunctions of the operator $\mathcal{L}(0, E)$, N is the set of two-dimensional indices for which the curves $\lambda^{(5)}_\nu$ are the resonance curves. And for $|\nu| \to \infty$ the Fourier coefficients c_ν decrease quicker than any power of c_ν.

4) If $E(x, t)$ are smooth functions for $x \in R$ and $E \in \mathcal{D}$, then there exists the single solution of the equation

$$\mathcal{L}(0, E(x, t))g = R_{nr}(\tau, x, t)$$

orthogonal to the kernel of the operator $\mathcal{L}^+(0, E(x, t))$. This solution is a smooth function of the arguments $\tau \in T^2$, $x \in R$, $t \in [0, T]$.

Proof. We can proof the statements of items 1) and 2) by analysing the explicit formulas for the values $\lambda^{(j)}_\nu$. The statement 3) is a consequence of the fact that the system of eigenfunctions and adjoint functions of the operators $\mathcal{L}(0, E)$, $\mathcal{L}^+(0, E)$ is a basis (see [18]). Then it is rather difficult to verify whether the solution y is smooth since, for certain $E(x, t)$, the Jordan cells appear in the spectrum of the operators $\mathcal{L}(0, E(x, t))$, $\mathcal{L}^+(0, E(x, t))$. Therefore, by inverting $\mathcal{L}(0, E)$ by means of the Fourier method, we use the basis smoothly depending on (τ, x, t) and such that $\mathcal{L}(0, E)$ can be represented in blocks. Each block in this representation corresponds to less than five-dimensional invariant subspace of the operator $\mathcal{L}$, and the matrix elements in them are smooth functions $E \in \mathcal{D}$ decreasing sufficiently quickly when the block number increases.

We present here the formulas explaining the structure of functions w_0 and w^+_0 (see [18])

$$w_0 = a\left(\langle U, \frac{\partial}{\partial \tau}\rangle - ip\right)^2 \frac{\theta(\tau + A(p, E)|E)}{\theta(\tau|E)},$$

$$w^+_0 = e^{i\tau_2}\left(\frac{\theta(\tau + \frac{1}{2}A(p, E) + \frac{1}{2}B + \bar{\pi}|E)}{\theta(\tau|E)}\right)^2,$$

where θ is a two-dimensional Riemann θ-function, $A(p, E)$ is a complex vector-function with asymptotics $-\frac{1}{2}\bar{\pi} + 2iU_2/p + O(p^{-2})$ for $|p| \to \infty$, where b is a vector-column $\begin{pmatrix} B_{12} \\ B_{22} \end{pmatrix}$ of the matrix B- periods (see item 1), $\bar{\pi} = \begin{pmatrix} \pi \\ \pi \end{pmatrix}$, $a = a(p, E)$ is the coefficient of normalization; A and B are expressed in terms of E by means of hyperelliptic integrals.

We note that the value $p/2$ is a quasi-momentum in the gap theory of the Sturm-Liouville operator with potential p. If we introduce the energy $\mathcal{E}(p/2)$ (a spectral parameter) of the operator, then $\mathcal{E}_\nu = \mathcal{E}((p + V + U)/2)$ will characterize the "contracted" boundaries of prohibited gaps of a three-gap solution which w_ν was obtained by linearization on the background of the two-gap solution f. The parameter of linearization is just the width of a closed prohibited gap with end $\mathcal{E}_\nu$. The functions w_ν in the expansion (3.10) correspond to $\mathcal{E}_\nu$ in the allowed gap (E_5, ∞) (see Figure 1.1).

We denote $u_{1nr} = \varepsilon g(S/\varepsilon + P, x, t)$. The Lemma yields immediately

Corollary. *The following relations hold*

(3.11) $$Lu_{1nr} = R_{nr}(S/\varepsilon + P, x, t) + O(\varepsilon),$$

(3.12) $$u_{1nr} = O(\varepsilon), \quad \varepsilon \frac{\partial}{\partial x} u_{1nr} = O(\varepsilon).$$

We now consider the resonance summands in the expansion (3.9) of the function R. It is easy to show that the terms of strong resonance $c_k z_k$ in this expansion into u_1 are of the same order $O(1)$ as U_0. Thus the demand that the correction u_1 be small yields immediately (see also [23]) the following relations

(3.13) $$(z_k^+(\tau, 0, E(x,t)), R(\tau, x, t)) \equiv 0, \quad k = 1, 2, 3.$$

Lemma 3.3. *The relations (3.12) together with (1.8) yield a closed system of equations for $E(x,t)$ which are equivalent to the Whitham- Flaschka-Forest-McLaughlin system (1.10).*

The proof (see [3,10,15]) is based on the fact that the eigenfunctions $z_k^+(\tau, 0, E)$ are linear combinations of elementary symmetric polynomials well known in the theory of KdV functions γ_j (see [12]).

Remark. In the one-phase situation, this does not complete the process of constructing the correction u_1 since the spectrum of the operators $\mathcal{L}$ and $\mathcal{L}^+$ is discrete, all eigenvalues are separated, and the kernels of the operators $\mathcal{L}$ and $\mathcal{L}^+$ are of equal dimensions for $E_1 < E_2 < E_3$.

Further we assume that

i_1) the functions $E_1(x,t)$ are solutions of system (1.10) and satisfy the conditions of Lemma 3.2 for $x \in R$ and $t \in [0, T]$, and the functions $S_j(x, t)$ are related to E by formulas (1.8).

Thus after separating the function u_{1nr} from u_1, we obtain in the right-hand side of equation (2.4) the function

(3.14)
$$R_{\text{res}} = \sum_{\nu \in N} c_\nu(x, t)) w_\nu(\tau, 0, E(x,t)),$$
$$c_\nu = e(\lambda_\nu^{(5)}(0, E(x, t))(w_\nu^+(\tau, 0, E(x, t)), R(\tau, x, t))$$

In order to invert the operator $i\varepsilon\partial/\partial t + \mathcal{L}(-i\varepsilon\partial/\partial x, E(x, t))$, we use the Duhamel integral [3].

Here we follow the method of [25] applied to the simplest equations with operator-valued symbol, i.e., a system of equations with partial derivatives, i.e., with characteristics of variable multiplicity. As the result we obtain the formula

$$u_{1\text{res}} = w_{\text{res}}(S/\varepsilon, x, t, \varepsilon),$$

(3.15)
$$w_{\text{res}}(\tau, x, t, \varepsilon) = \sum_{\nu \in N} \int_0^t dq \, \exp \frac{i\Phi_\nu(x, t, q)}{\varepsilon} \varphi_\nu(x, t, q) \times$$
$$\times w_\nu(\tau, \frac{\partial \Phi_\nu(x, t, q)}{\partial x}, E(x, t)).$$

Here the new phases Φ_ν are solutions of the Hamilton-Jacobi equations

(3.16) $$\frac{\partial \Phi_\nu}{\partial t} + \lambda_\nu^{(5)}\left(\frac{\partial \Phi_\nu}{\partial x}, E(x,t)\right) = 0, \quad \Phi_\nu\bigg|_{t=q} = 0.$$

The functions $\varphi_\nu(x,t,q)$ saisfy the transport equations corresponding to (3.16)

(3.17) $$\varphi_\nu(x,t,q)\bigg|_{t=q} = c_\nu(x,q).$$

We obtain the answer for $u_{1\mathrm{res}}$ in the form of a sum of integrals from rapidly oscillating exponents. The further analysis for $u_{1\mathrm{res}}$ is based on the method of stationary phase. Easy considerations show that the existence of stationary points in these integrals is equivalent to vanishing of the value $\lambda_\nu(0, E(x, t))$ at certain points (x, t) (i.e., to appearing of resonances). The order (in ε) of certain summands in (3.15) is determined by the degree of degeneration of stationary points. We restrict ourselves to the case when stationary points are nondegenerate.

The sum in (3.15) contains, generally speaking, an infinite number of summands, therefore it is necessary not only to estimate each separate summand, but also to prove that the estimates hold after the summing. Of course, theis fact is the main difficulty in estimating the function (3.15).

In order to obtain the stimate $O(\sqrt{\varepsilon})$ for the weakly resonating part of correction, it is sufficient to obtain the following conditions for $E(x, t)$ (together with item i_1). They, in particular, imply that the stationary points in (3.15) are nondegenerate, (i_2). The equation for x

$$U(E(x,t)) \cdot n = 0$$

possesses not more than one solution $x = x_0(t, u)$ for each fixed $n = \binom{n_1}{n_2}$, $|n| = 1$. and at these points we have

$$\frac{\partial}{\partial x}(U \cdot n) \neq 0.$$

i_3) The equation for x

$$\lambda_0^{(5)}(\langle U, \nu\rangle, x, t) + \langle V, \nu\rangle = 0$$

possesses not more than one solution $x_\nu^R(t)$ for each fixed $U \in N$. and at these points

$$\frac{\partial}{\partial x}(\lambda_0^{(5)}(\langle U, \nu\rangle, x, t) + \langle V, \nu\rangle \neq 0.$$

We note that the condition i_2) implies that i_1) holds for sufficiently large $|\nu| = \sqrt{\nu_1^2 + \nu_2^2}$.

i_4) At the points $x_0(t, n)$, we have the inequalities

$$V \cdot n \cdot \frac{\partial(U \cdot n)}{\partial x} < 0; \quad \frac{\partial(U \cdot n)}{\partial x} \cdot \frac{\partial(V \cdot n)}{\partial x} < 0,$$

i_5) The vectors $U(E(x,t))$ and $V(E(x,t))$ cannot be parallel to each other for any $x \in R$. $t \in [0, T]$.

213

i_6) At the points x_ν^R we have

$$\left(\frac{\partial}{\partial t} + \frac{\partial \lambda_0^{(5)}}{\partial p}(\langle U, \nu \rangle, E(x,t))\frac{\partial}{\partial x}\right)\left(\lambda_0(\langle U, \nu \rangle, E(x,t)) + \langle V, \nu \rangle\right) \neq 0.$$

(These inequalities hold beforehand for sufficiently large $|\nu| = \sqrt{\nu_1^2 + \nu_2^2}$). From a geometrical point of view, this means that the characteristics of (3.6) corresponding to the resonance sets

$$\{(x,t) : \lambda_\nu^{(5)}(0, E(x,t)) = 0\}$$

are transversal.

Lemma 3.4. *Under the conditions $i_1) - i_6$) we have*

1) the phases $\Phi_\nu(x,t,q)$ are defined (there are no focal points in WKB-asymptotics (3.15)) and the formula (3.15) has the sense;

2)

$$u_{1\mathrm{res}} = O(\sqrt{\varepsilon}), \quad \varepsilon\frac{\partial u_{1\mathrm{res}}}{\partial x} = O(\sqrt{\varepsilon});$$

$$Lu_{1\mathrm{res}} = \varepsilon R_{1\mathrm{res}}(\frac{S}{\varepsilon}, x, t) + O(\varepsilon^{3/2}).$$

Lemma 3.3 implies that the substitution of the function $u_0 = u_{1nr} + u_{1\mathrm{res}}$ in the equation (1.3) yields the correction of the same order $O(1)$ as the substitution of the principal term of asymptotics u_0, and if we want to construct a right correction, we must take into account the summand $-6\varepsilon u_1 \partial u_1 / \partial x$ in the equation in variations (3.3).

4. SECOND-ORDER SUMMANDS IN THE FIRST CORRECTION

We add to the function $u_1 = u_{1nr} + u_{1\mathrm{res}}$ obtained in the previous section a correction u_2 which allows us to kill the summand $\varepsilon u_1 \partial u_1 / \partial x$ up to $o(1)$.

First we present heuristic considerations. We preserve the summands of order $O(1)$ and $O(\sqrt{\varepsilon})$ in formulas for $u_0 + u_1$, and in formula (3.15), we preserve a certain (sufficiently large but finite) number of summands, and then we replace the integral with respect to dq by Riemannian sums. As the result, we obtain

(4.1) $$u = F(\frac{S}{\varepsilon} + P, \frac{\Phi}{\varepsilon}, x, t, \Delta),$$

(4.2) $$F(\tau, \tau', x, t, \Delta) = f(\frac{\tau}{\varepsilon}, E(x,t)) +$$

$$+ \sum_{\nu \in N, |\nu| \leqslant L} \sum_{k=0}^{K} \Delta \varphi_\nu(\Delta k, x)\exp(i\tau_{\nu k})w_\nu(\tau, \frac{\partial \Phi_\nu}{\partial x}(x, t, k\Delta), E(x,t))$$

where Δ and k are the step and the index of summing in the Riemannian sum, $k = [t/\Delta]$, L is a sufficiently large number, τ' is a set of variables $\{\tau_{\nu k}\}$, Φ is a set of phases $\{\Phi_\nu(x,t,k\Delta)\}$. Thus the function (4.1) is a multiphase asymptotic solution with phases S_1, S_2 and $\{\Phi_\nu(x,t,k\Delta)\}$, $\nu \in N$, $|\nu| \leqslant L$ and $k = 0, 1, \ldots, K$.

By substituting the function (4.1) into the multiphase equation (2.2), we see that it satisfies this equation up to $O(\Delta^2)$. This fact is obvious since the eigenfunctions in formula (3.15) were obtained by linearization of multigap solutions on the background of two-gap solutions. One can construct a multiphase function which exactly satisfies (2.2) and the first two elements of the expansion of which with respect to the parameter Δ coincide with (4.1). By taking more terms in the corresponding expansion into a Taylor series (for example, those containing Δ^2), we obtain a function satisfying (2.2) with a greater than $O(\Delta^2)$ accuracy. In this new (still approximate) formula, we can pass to the limit as $\Delta \to 0$ and $h \to 0$ and obtain a more precise resonance correction.

To perform this procedure precisely with a certain procedure of regularization yields a correction $u_{2\text{res}}$ and $u_{1\text{res}}$ such that the function $u_0 + u_{1nr} + u_{1\text{res}} + u_{2\text{res}}$ will satisfy equation (1.3) up to $O(\varepsilon^{1/4})$. A curious and important fact (not obvious apriori) is the fact that the phases $\Phi_\nu(x,tq)$ do not vary when the correction becomes more precisely (in contract to the linear case of the theory of perturbations).

We present here the formula which shows the structure of this addition

$$
u_{2\text{res}} = \sum_{\nu,\mu \in N} \int_0^t \int_0^t \exp(\frac{i}{\varepsilon}(\Phi_\nu(x,t,q) + \Phi_\mu(x,l,y))\varphi_\nu(x,t.q) \times
$$

(4.3)
$$
\times \varphi_\mu(x,t,y) W_{\nu\mu}(\tau, \Phi_{\nu x}(x,t,q), \Phi_{\nu x}(x,t,y), E(x,t))\, dq\, dy +
$$

$$
+ \sum_{j=1}^{5} \rho_j(x,t,\varepsilon)\frac{\partial f_0}{\partial \tau_j}(\tau, E(x,t)) + \sum_{j=1}^{5} \tilde{\rho}_j(x,t,\varepsilon)\frac{\partial f_0}{\partial E_j}(\tau, E(x,t)).
$$

Here $\tau = S/\varepsilon + P$, Φ_ν and φ_ν are the same as before,

$$
W_{\nu,\mu}(\tau, p, p'E) =
$$
$$
= -((\langle U, \frac{\partial}{\partial \tau}\rangle + i(p + p'))^2 \{\exp(i\langle \tau_1(\mu + \nu)\rangle)[\exp(iQ_{\mu\nu}) \times
$$
$$
\times \frac{\theta(\tau + A_\mu(p, E) + A_\nu(p', E)|E)}{\theta(\tau, E)} - \frac{\theta(\tau + A_\nu(p, E)|E)\theta(\tau + A_\mu(p', E)|E)}{\theta^2(\tau, E)}]\},
$$

where $A_\nu(p, E) = A(p + \langle V, \nu\rangle, E)$, and A is the same as in section 3, $Q_{\mu\nu} = Q(p + \langle V, \nu\rangle, p' + \langle V, \mu\rangle, E)$ is a certain function expressed in terms of hyperelleptic integrals from E and $\mathcal{E}_\nu = \mathcal{E}(p + \langle V, \nu\rangle/2)$, $\mathcal{E}_\mu = \mathcal{E}(p + \langle V, \mu\rangle/2)$, $\rho_j(x,t,\varepsilon)$ and $\tilde{\rho}_j(x,t,\varepsilon)$ are certain functions depending on ε irregularly, $\rho_j = O(\varepsilon^{3/4})$, $\tilde{\rho}_j = O(\varepsilon^{3/4})$, (see [19, 26]).

Theorem 4.1. *We assume that, for $x \in R$ and $t \in [0,T]$, the functions $E(x,t)$ satisfy the Cauchy problem $E|_{t=o} = E^0(x)$ for system (1.10) and the conditions $i_1) - -i_6$), and suppose $S^0 = \int U(E^0(x))\, dx$. Then the function*

(4.4)
$$
u = u_0 + u_{1nr} + u_{1\text{res}} + u_{2\text{res}},
$$

satisfies equation (1.3) up to $O(\varepsilon^{1/4})$. In this case, $u - u_0 = O(\varepsilon^{1/2})$ and satisfies the following Cauchy data

$$u \Big|_{t=0} = f\left(\frac{S^0}{\varepsilon}, E^0(x)\right) + O(\varepsilon).$$

To prove this statement we must estimate the integrals in (4.2) and in the correction obtained after the substitution of the function u in (1.3).

We present the geometric interpretation of formula (4.4). As known, the KdV equation is a completely integrable Hamiltonian system in a certain phase space of infinite dimension. The finite-gap KdV solutions define a finite-dimensional torus in this space. The function u_0 describes an "adiabatically" perturbed torus, namely, it describes its "constants" E (similar to variables of action) which are slowly varying functions in x and t. The resonances imply that this torus becomes infinite-dimensional already in the next approximation. And almost all generating curves of this torus except two initial ones corresponding to the principal term of the solution are contracted to a point. The basic cycles (circles) of torus are divided in groups, so that the parameters characterising them (the variables of action) are close to each other. This object is replaced by another torus which is a squared germ of the first one. After this, we pass to the limit continuously with respect to the variable of action close to each other which yields the integrals in the formulas for resonance corrections. The "actual" torus turns out to be invariant with respect to the KdV equation in rapid variables (1.4). The germ of this torus is already "almost" invariant, but it is sufficient to obtain an correction small with respect to ε, and to write clear formulas for a torus of "continual" dimension the generating curves of which are "numbered" by continuously varying parameters of integration. We note that similar considerations in other situations are rather effective [28–31].

From the point of view of finite-gap integrating, we see that the obtained almost invariant limit torus is associated with a Riemannian surface of infinite (continual) order characterized by the main ends of gaps $E_1, \ldots, E_s$ and by traces of infinitely small gaps $\mathcal{E}_\nu$. The summand u_{1res} describes the interaction of the gap ends $E_1, \ldots, E_s$ with $\mathcal{E}_\mu$, and u_{2res} describes the interaction of $\mathcal{E}_\mu$ and $\mathcal{E}_\nu$ (see Figure 3.1). We note that the extension of gaps in spectrum of the Sturm-Liouville operator does not always yield changes of the dimension of corresponding KdV torus. The nonresonance correction u_{1nr} in (4.3) corresponds just to such extension of gaps. The resonance summands correspond to opening of gaps which changes the torus dimension.

We can also interpret the summands with functions ρ_j and $\hat{\rho}_j$ in (4.2) as small (but irregular) corrections to phases S and the ends of gaps E. However, these corrections do not change equations (1.10), (1.3). It should be noted that "large" correction P to phases remains undetermined in this approximation.

REFERENCES

1. Whitham G.B., *Linear and Nonlinear Waves*, Wiley Interscience, New York, 1974.
2. Scott A.C., Chu F.Y., McLaughlin D.W., *The soliton: a new concept in applied science*, Proc. IEEE 1 (1973), 1443.
3. Dobrokhotov S.Yu., Maslov V.P., *Finite-gap almost periodic solutions in WKB-approximations*, Itogi Nauki i Tehniki. Modern Problems of Mathematics 15 (1980). Moscow, VINITI AN SSSR, 3 – 94.

4. Dobrokhotov S.Yu., Maslov V.P., *Multiphase asimptotics of nonlinear partial differential equations with a small parameter*, Sov. Sci. Rev.– Math. Phys. Rev. **3** (1982), 221 – 311.

5. Dubrovin B.A., Novikov S.P., Uspekhi Mat.Nauk **XLIV** (1989), 29.

6. Bikbaev R.F., *KdV equation with finite-zone boundary conditions and Whitham's deformations of Riemann surfaces*, Func. anal. i priloj. **23:4** (1989), 1 – 10.

7. Krichever I.M., *Spectral theory of two-dimensional periodic operators and its applications*, Uspekhi Mat. Nauk **XLIV:2** (1989), 121 – 184.

8. Haberman R., *The modulated phase shift for weakly dissipated non-linear oscillatory waves of the KdV type*, Stud. in Appl. Math. **LXXVIII:1** (1988), 73 – 90.

9. Maslov V.P., *Asymptotic methods and perturbation theory*, Nauka, Moscow, 1988.

10. Flashka H., Forest M.G., McLaughlin D.W., *Multiphase averaging and the inverse spectral solution of the KdV equation*, Comm. Pure and Appl. Math. **33:6** (1980), 739 – 784.

11. Ablowitz M.A., Benny D.Y., *The evolution of multiphase modes for non-linear dispersive waves*, Stud. Appl. Math. **49:3** (1970).

12. Zakharov V.E., Manakov S.V., Novikov S.P., Pitaevsky L.P., *The soliton theory*, Nauka, Moscow, 1979.

13. Dubrovin B.A., Matveev V.B., Novikov S.P., *Non-linear KdV-equations, finite-gap linear operators and Abelian manifolds*, Uspekhi Mat. Nauk **31:1** (1976), 55 – 136.

14. Luke J.C., *A perturbation method for non-linear dispersive wave problems*, Proc. Roy. Soc. **A292:1430** (1966), 403 – 412.

15. Dobrokhotov S.Yu., Maslov V.P., *Finite-gap almost periodic solutions in asymptotic expansions*, Math. Studies **47** (1981), Noth Holland Publ. Comp., Amsterdam, 1 – 27.

16. Dobrokhotov S.Yu., Maslov V.P., *Solution asymptotics of mixed problem for the non-linear wave equation $h^2 \Box u + a \sinh u = 0$*, Uspekhi Mat. Nauk **34:3** (1979), 225 – 226.

17. Dobrokhotov S.Yu., Maslov V.P., *Boundary reflection problem for the equation $h^2 \Box u + a \sinh u = 0$ and finite-gap quasiperiodic solutions*, Func. Anal. i Priloj. **13:3** (1973), 79 – 80.

18. Dobrokhotov S.Yu., Vorob'ev Yu.M., *Basic systems on the torus generated by finite-zone integration of the KdV quation*, Matem. Zametki **47:1** (1990), 47 – 61.

19. Dobrokhotov S.Yu., *Resonance correction for adiabatically perturbed finite-zone almost- periodic solution of KdV equation*, Matem. Zametki **44:4** (1988), 551 – 554.

20. Bensousan A., Lions J.L., Papanicolaou G., *Asymptotic analysis for periodic structures*, Noth-Holl. Publ. Comp., Amsterdam, 1978.

21. Dobrokhotov S.Yu., *Resonances in asymptotic solutions of Cauchy problem for the Schrödinger equation with rapidly oscillating finite-zone potential*, Matem. Zametki **44:3** (1988), 319 – 340.

22. Dobrokhotov S.Yu., Vorob'ev Yu.M., *Completness of the system of eigenfunctions of a nonelliptic operator on the torus generated by the Hill operator with finite-zone potential*, Func. Anal. i Priloj. **22:2** (1988), 65 – 66.

23. Dobrokhotov S.Yu., Krichever I.M., *Multiphase solutions of Benjamin-Ono equation and its averaging*, Matem. Zametki **49:6** (1991).

24. Krichever I.M., *Hessians of integrals for KdV-equations and perturbations of finite-zone solutions*, Dokl. AN SSSR **270:6** (1983), 1312–1317.

25. Kucherenko V.V., *Asymptotic solution of system $A(x, -ih(\partial/\partial x))u = 0$ case of characteristic of varying multiphcity*, Izv. AN SSSR, ser. Math. **38:3** (1974), 378 – 383.

26. Dobrokhotov S.Yu., *Doctor thesis* (1989), LOMI AN SSSR, Leningrad.

27. Kaup D.J., *A perturbation theory for inverse scattering transforms*, SIAM J. Appl. Math. **31:1** (1976), 121 – 132.

28. Maslov V.P., *Operational methods*, MIR, Moscow, 1976.

29. Babich V.M., Buldyrev V.S., *Asymptotic method in short wave problems*, Nauka, Moscow, 1972.

30. Lazutkin V.E., *Convex hilliard and eigenfunctions of Laplace operator*, LGU, Leningrad, 1981.

31. Krakhnov A.D., *On asymptotics of eigenvalues of pseudodifferential operators and invariant tori*, Uspekhi Mat.Nauk **XXXI:3** (1976), 217 – 218.

BILLIARDS SYSTEMS AND THE TRANSPORT EQUATION

François Golse

Université Paris VII
Département de Mathématiques
F75251 Paris Cédex 05

1. INTRODUCTION

Consider a system of like particles that do not interact between themselves but collide with a periodic array of strictly convex obstacles. This type of dynamical system is referred to as a dispersive billiards systems in this work. The most genuine example of such systems is the so-called Lorentz gas model of hard spheres[7]: the interaction between the particles and the obstacles is the pure (Descartes) reflection law. In large time, due to multiple collisions and the strict convexity of the obstacles, the distribution of the particles tends to forget its velocity dependence[7]. Bunimovich and Sinai[7] have proved that the density of particles approaches in some sense the solution of a diffusion (heat) equation for large time and on a large space scale. Their proof relies on symbolic dynamics *via* the construction of a Markov partition[8,9]. The role of the strict convexity of the scatterers is reminiscent of the ergodic properties of the geodesic flow on compact manifolds with negative curvature: see Arnold [1] (although the explanation provided there is not fully rigorous). However, the method used by Bunimovich-Sinai[7] does not provide the value of the diffusion coefficient, except through the use of the ubiquitous Kubo series. See also Gaspard-Nicolis[11] for an interesting expression of the diffusion coefficient. None of these expressions clearly provides the dependence of the diffusion coefficient in terms of the geometry of the array of scatterers. Numerical experiments have been conducted by Machta-Zwanzig, as well as some asymptotics in the case where the distance between neighboring scatterers goes to zero (see Machta[14]).

The purpose of the present work is to analyze models like the Lorentz gas of hard spheres by PDE methods. This type of methods works best when the reflection law is of the same type as Maxwell's accomodation condition[10]. In this case, the same scaling as in Bunimovich-Sinai[7] also leads to a diffusion limit: see Bardos-Dumas-Golse[2]. However, the result by Bunimovich-Sinai contains a particularly interesting feature: the Lorentz gas of hard spheres with pure reflections on the scatterers is naturally a reversible system (an ODE with global solution in time) whereas the diffusion limit is

Singular Limits of Dispersive Waves, Edited by
N.M. Ercolani et al., Plenum Press, New York, 1994

irreversible — it is known that the backward heat equation is an ill-posed system. In the case treated by Bardos-Dumas-Golse, the billiards system is already irreversible due to the accomodation condition at the boundary of the scatterers. This makes the long time, large scale asymptotics towards diffusion a lot easier to establish. However, PDE methods can also be used on the Lorentz gas of hard spheres: they result in a computation of the diffusion coefficient (Golse[12]), based on an approximation of the number density a priori different from that in Bunimovitch-Sinai[7]. The basis for this computation consists in deriving a Boltzmann type equation which is intermediate between the billiards system and its diffusion limit. Whether such a procedure is valid and consistent with the result by Bunimovich-Sinai[7] still remains an open question.

The rest of the paper is organized as follows. Section 2 is devoted to a general presentation of the various models considered and their scalings. Section 3 gives the main results that have been obtained. The tools and methods used in the proofs are discussed in Section 4. Conclusive remarks belong to Section 5.

Finally, it is my pleasure to thank the organizing committee of this NATO Conference on Oscillations and Dispersive Waves for giving me the opportunity to present these results in Lyon. I also wish to thank C. Bardos for helpful discussions about this problem.

2. PRESENTATION OF THE MODELS

We shall confine our investigations in two particular cases of billiards systems for the sake of clarity in the exposition. However, our analysis can be carried through in other cases, provided that the main ingredients remain present: the periodic structure of the array of scatterers and their dispersive character (based on the assumption of strictly convex obstacles in the case of the Lorentz gas of hard spheres). The two models discussed here are the Lorentz gas of hards spheres with spheric scatterers distributed at the nodes of a square lattice in arbitrary dimension, and the case of circular scatterers distributed at the nodes of a regular triangular lattice in dimension two with accomodation reflection law[10].

The Lorentz Gas of Hard Spheres

The geometry of the array of scatterers is given as follows. The scatterers are centered at the nodes of a square lattice of $\mathbf{R}^D$

$$\mathcal{L} = \mathbf{Z}(a, 0, \ldots, 0) \oplus \ldots \oplus \mathbf{Z}(0, \ldots, a) \tag{2.1}$$

and are balls of radius $r < a/2$ so that they do not overlap. The domain left free for particle flight is

$$X = \{x \in \mathbf{R}^D \,/\, \text{dist}(x, \mathcal{L}) > r\}. \tag{2.2}$$

Particles considered here have velocities of modulus denoted by $c > 0$. The phase space is therefore $X \times \mathbf{S}^{D-1}$, the set of positions in X and directions of the velocities of the particles. Their number density is denoted by $f \equiv f(t, x, \Omega)$: this notation means that the number of particles at time t in a volume $dx d\Omega$ of the phase space $X \times \mathbf{S}^{D-1}$ centered around (x, Ω) is $f(t, x, \Omega) dx d\Omega$. Since the particles do not interact between themselves but only with the scatterers, and travel freely between those, the number density solves the transport equation:

$$\partial_t f + c\Omega \cdot \nabla_x f = 0, \quad t > 0, \ x \in X, \ \Omega \in \mathbf{S}^{D-1}. \tag{2.3}$$

The particles bounce off the surface of the scatterers according to the pure reflection law

$$f(t, ak + r\omega, \Omega) = f(t, ak + r\omega, R(\omega)\Omega), \quad k \in \mathbf{Z}^D, \ \omega \in \mathbf{S}^{D-1}, \ \Omega \in \mathbf{S}^{D-1}, \quad (2.4)$$

where $R(\omega)$ denotes the reflection defined by the vector ω; in other words,

$$R(\omega)\Omega = \Omega - 2(\omega \cdot \Omega)\omega.$$

Finally the initial density of particles is prescribed:

$$f(0, x, \Omega) = \phi(x, \Omega), \quad x \in X, \ \Omega \in \mathbf{S}^{D-1}. \quad (2.5)$$

The solution of the problem (2.3)-(2.4)-(2.5) is given by

$$f(t, x, \Omega) = \phi(X_{-t}(x, \Omega), \Omega_{-t}(x, \Omega)), \quad t > 0, \ x \in X, \ \Omega \in \mathbf{S}^{D-1}, \quad (2.6)$$

where (X, Ω) is the flow generated by the motion of the particles between the scatterers. In other words,

$$\frac{d}{dt}\Omega_t(x, \Omega) = 0, \quad \frac{d}{dt}X_t(x, \Omega) = \Omega_t(x, \Omega), \quad (2.7)$$

whenever $X_t(x, \Omega) \in X$, and

$$\Omega_{t+0}(x, \Omega) = R(\omega)\Omega_{t-0}(x, \Omega), \quad X_{t+0}(x, \Omega) = X_{t-0}(x, \Omega) \quad (2.8)$$

if $X_{t-0}(x, \Omega) = ak + r\omega$; moreover

$$\Omega_0(x, \Omega) = \Omega, \quad X_0(x, \Omega) = x. \quad (2.9)$$

The 2D Lorentz Gas with Accomodation Reflection

The next model discussed in this paper is very much similar to the previous one. Here, the scatterers are disks centered at the nodes of the regular triangular lattice

$$\mathcal{L} = \mathbf{Z}(a, 0) \oplus \mathbf{Z}(\frac{a}{2}, \frac{a\sqrt{3}}{2}). \quad (2.10)$$

The scatterers have radius $r < a/2$, so that they do not overlap. The domain left free for particle flight is

$$X = \{x \in \mathbf{R}^2 \ / \ \text{dist}(x, \mathcal{L}) > r\}. \quad (2.11)$$

The particles considered also have velocities of modulus $c > 0$. The phase space is therefore $X \times \mathbf{S}^1$, the set of positions in X and directions of the velocities of the particles. Their number density is denoted $f \equiv f(t, x, \Omega)$ with the same meaning as above. The analogues of equations (2.3)-(2.4)-(2.5) are as follows:

$$\partial_t f + c\Omega \cdot \nabla_x f = 0, \quad t > 0, \ x \in X, \ \Omega \in \mathbf{S}^1, \quad (2.12)$$

$$f(t, x, \Omega) = \tfrac{1}{2} \int_{\Omega' \cdot n_x < 0} f(t, x, \Omega')|\Omega' \cdot n_x| d\Omega', \quad x \in \partial X, \ \Omega \cdot n_x > 0, \quad (2.13)$$

where n_x is the unit normal vector at point $x \in \partial X$ directed towards X. Finally the initial density of particles is prescribed:

$$f(0, x, \Omega) = \phi(x, \Omega), \quad x \in X, \ \Omega \in \mathbf{S}^1. \tag{2.14}$$

The meaning of the reflection law (2.13) is as follows: particles impinging on the boundary ∂X are reflected in a random direction. More specifically, particles impinging on the scatterers with an angle θ relative to the normal are reflected isotropically with probability proportional to the cosine of θ.

A very important property of this second model is the one called "finite horizon". A billiards system is said to have finite horizon if and only if there exists a finite constant C such that

$$\sup\{|x - y| \ / \ [x, y] \subset X\} \leq C. \tag{2.15}$$

(where $[x, y]$ denotes the line segment connecting x and y). Elementary geometry shows that any billiards system defined by a triangular lattice as X in (2.11) with $\mathcal{L}$ given by (2.10) has finite horizon when

$$\frac{a\sqrt{3}}{4} < r. \tag{2.16}$$

The Scalings

Three length scales are present in the two models above.
— the characteristic scale of variation of the initial data, that can be defined as the order of magnitude of

$$L = \frac{\phi(x, \Omega)}{|\partial_x \phi(x, \Omega)|}.$$

To avoid ambiguities, we shall henceforth assume that ϕ is a smooth function defined on $\mathbf{R}^D \times \mathbf{S}^{D-1}$ and that the initial condition (2.5) holds only when $x \in X$;
— the length a (the distance between two neighboring scatterers);
— the length r (the radius of the scatterers).
The present work is based on the assumption that a and r are small compared to L. This suggests the introduction of a small parameter ϵ such that

$$a = \hat{a}\epsilon, \quad r = \hat{r}\epsilon^\gamma, \tag{2.17}$$

where the length scales L, $\hat{a}$ and $\hat{r}$ are of the same order of magnitude. The parameter $\gamma \geq 1$ is here to adjust the size of the scatterers to the distance between neighboring scatterers.

With the scalings defined above, the model for the Lorentz Gas of Hard Spheres can be rewritten as follows, once all hats have been dropped:

$$\partial_t f_\epsilon + c\Omega \cdot \nabla_x f_\epsilon = 0,$$

$$t > 0, \ x \in X_\epsilon, \ \Omega \in \mathbf{S}^{D-1}. \tag{2.18}$$

$$f_\epsilon(t, \epsilon a k + \epsilon^\gamma r \omega, \Omega) = f_\epsilon(t, \epsilon a k + \epsilon^\gamma r \omega, R(\omega)\Omega),$$

$$k \in \mathbf{Z}^D, \ \omega \in \mathbf{S}^{D-1}, \ \Omega \in \mathbf{S}^{D-1}, \tag{2.19}$$

$$f_\epsilon(0, x, \Omega) = \phi(x, \Omega), \quad x \in X_\epsilon, \ \Omega \in \mathbf{S}^{D-1}. \tag{2.20}$$

In the above equations, we have denoted

$$\mathcal{L}_\epsilon = \mathbf{Z}(\epsilon a, 0, \ldots, 0) \oplus \ldots \oplus \mathbf{Z}(0, \ldots, \epsilon a) \tag{2.21}$$

and

$$X_\epsilon = \{ x \in \mathbf{R}^D \,/\, \mathrm{dist}(x, \mathcal{L}_\epsilon) > \epsilon^\gamma r \}. \tag{2.22}$$

As for the second model introduced above, we shall only consider the scaling corresponding to the case where $\gamma = 1$. This restriction is consistent with the rationale for considering the 2D Lorentz Gas with Accomodation Reflection, that is, the property of finite horizon (see (2.16)), compatible only with $\gamma = 1$. Again, after dropping all hats, the rescaled model for the 2D Lorentz Gas with Accomodation Reflection is recast as

$$\partial_t f_\epsilon + c\Omega \cdot \nabla_x f_\epsilon = 0, \quad t > 0, \ x \in X_\epsilon, \ \Omega \in \mathbf{S}^1, \tag{2.23}$$

$$f_\epsilon(t, x, \Omega) = \tfrac{1}{2} \int_{\Omega' \cdot n_x < 0} f_\epsilon(t, x, \Omega') |\Omega' \cdot n_x| d\Omega', \quad x \in \partial X_\epsilon, \ \Omega \cdot n_x > 0, \tag{2.24}$$

where n_x is the unit normal vector at point $x \in \partial X_\epsilon$ directed towards X_ϵ. The initial density of particles is prescribed:

$$f(0, x, \Omega) = \phi(x), \quad x \in X_\epsilon, \ \Omega \in \mathbf{S}^1. \tag{2.25}$$

In the above equations

$$\mathcal{L}_\epsilon = \mathbf{Z}(\epsilon a, 0) \oplus \mathbf{Z}(\frac{\epsilon a}{2}, \frac{\epsilon a\sqrt{3}}{2}). \tag{2.26}$$

and

$$X_\epsilon = \{ x \in \mathbf{R}^2 \,/\, \mathrm{dist}(x, \mathcal{L}_\epsilon) > \epsilon r \}. \tag{2.27}$$

3. MAIN RESULTS

In the case of the Lorentz Gas of Hard Spheres, we shall present an asymptotic analysis of the different scalings corresponding to the different admissible values of $\gamma \geq 1$. It is based on an expansion of the number density in powers of ϵ, the so-called Hilbert expansion[10]. This will provide statements of approximation of the number density in the sense of strong consistency (a notion that will be defined more precisely in the sequel). The case of the 2D Lorentz Gas with Accomodation Reflection will be handled using a multiscale expansion of the number density in powers of ϵ with strongly oscillating coefficients [4], like the WKB expansion. The resulting approximation of the number density holds in a sense much stronger than the one that we give for the hard sphere case, viz. in the sense of uniform convergence.

We shall henceforth use the notation

$$\langle\!\langle F \rangle\!\rangle = \int_{\mathbf{R}^D} \int_{\mathbf{S}^{D-1}} F(x, \Omega) d\Omega dx$$

Main Results for the Lorentz Gas of Hard Spheres

The key notion to understand the difference between the various scalings introduced so far is the mean free path of particles. It is given by

$$\lambda_\epsilon = \frac{1}{\mathcal{N}_\epsilon \Sigma_\epsilon}$$

where $\mathcal{N}_\epsilon$ is the number of scatterers per unit volume and Σ_ϵ is the surface of the scatterers. Denoting by $\Sigma(D)$ the $D-1$-dimensional volume of $\mathbf{S}^{D-1}$, the scaling above leads to

$$\lambda_\epsilon = \frac{a^D}{\Sigma(D)r^{D-1}}\epsilon^{(\gamma_c - \gamma)(D-1)}, \tag{3.1}$$

with the following critical value of the parameter γ

$$\gamma_c = \frac{D}{D-1}. \tag{3.2}$$

It is only natural to extend the number density to the whole $\mathbf{R}^D$ space by setting its value to zero inside the scatterers; we shall abuse the notation f_ϵ for the number density extended in this way. Equation (2.18) is recast as

$$\partial_t f_\epsilon + c\Omega \cdot \nabla_x f_\epsilon - c\Omega \cdot n_x f_{\epsilon|\Gamma_\epsilon}\delta_{\Gamma_\epsilon} = 0, \quad t > 0, \ x \in \mathbf{R}^D, \ \Omega \in \mathbf{S}^{D-1}, \tag{3.3}$$

where $\Gamma_\epsilon = \partial X_\epsilon$ and δ_{Γ_ϵ} is the surface measure concentrated on Γ_ϵ. Then the collision term is split into absorption and scattering terms

$$-c\Omega \cdot n_x f_{\epsilon|\Gamma_\epsilon}\delta_{\Gamma_\epsilon} = c(\Omega \cdot n_x)_- f_{\epsilon|\Gamma_\epsilon}\delta_{\Gamma_\epsilon} - c(\Omega \cdot n_x)_+ f_{\epsilon|\Gamma_\epsilon}\delta_{\Gamma_\epsilon}.$$

Taking into account the boundary condition (2.19), and abusing the notation $f \circ R(n_x)(t, x, \Omega)$ for $f(t, x, R(n_x)\Omega)$,

$$\partial_t f_\epsilon + c\Omega \cdot \nabla_x f_\epsilon + c(\Omega \cdot n_x)_- f_{\epsilon|\Gamma_\epsilon}\delta_{\Gamma_\epsilon} - c(\Omega \cdot n_x)_+ f_{\epsilon|\Gamma_\epsilon} \circ R(n_x)\delta_{\Gamma_\epsilon} = 0,$$

$$t > 0, \ x \in \mathbf{R}^D, \ \Omega \in \mathbf{S}^{D-1}, \tag{3.4}$$

and the initial condition reads

$$f_\epsilon(0, x, \Omega) = \mathbf{1}_{X_\epsilon}(x)\phi(x, \Omega), \quad x \in \mathbf{R}^D, \ \Omega \in \mathbf{S}^{D-1}. \tag{3.5}$$

The collision terms can be recast with the help of a collision operator Λ_ϵ defined as

$$\Lambda_\epsilon \psi = c(\Omega \cdot n_x)_- f_{\epsilon|\Gamma_\epsilon}\delta_{\Gamma_\epsilon} - c(\Omega \cdot n_x)_+ f_{\epsilon|\Gamma_\epsilon} \circ R(n_x)\delta_{\Gamma_\epsilon}. \tag{3.6}$$

The key idea in the present work is the introduction of an asymptotic collision integral operator. This is done in the following proposition.

Proposition 1. Let Λ be the operator defined by

$$(\Lambda\psi)(x, \Omega) = \frac{cr^{D-1}}{a^D}\left(\frac{s(D)}{D-1}\psi(x, \Omega) - \int_{\mathbf{S}^{D-1}}\psi(x, \Omega')\frac{d\Omega'}{4|\Omega - \Omega'|^{D-3}}\right), \tag{3.7}$$

where $s(D) = |\mathbf{S}^{D-2}|$ if $D > 2$ and $s(2) = 2$. For all $\psi \in L^\infty(\mathbf{S}^{D-1}; C_0^\infty(\mathbf{R}^D))$

$$\Lambda_\epsilon \psi - \epsilon^{(D-1)(\gamma - \gamma_c)}\Lambda\psi = O(\epsilon^\infty) \tag{3.8}$$

for the topology of $L^\infty(\mathbf{S}^{D-1}; \mathcal{D}'(\mathbf{R}^D))$. More precisely, for all $\chi \in L^1(\mathbf{S}^{D-1}; C_0^\infty(\mathbf{R}^D))$

$$\left| \left\langle\!\!\left\langle \chi \left(\Lambda_\epsilon \psi - \epsilon^{(D-1)(\gamma-\gamma_c)} \Lambda \psi \right) \right\rangle\!\!\right\rangle \right|$$

$$\leq C_{D,q,\mathrm{supp}(\psi_\chi)} \epsilon^q \sup_{\Omega,\Omega'} ||\nabla_x{}^{\rho(q)}(\psi(.,\Omega)\chi(.,\Omega'))||_{L^\infty(\mathbf{R}^D)}, \qquad (3.9)$$

where

$$\rho(q) = 1 + \left[\frac{q-D}{\gamma} - D \right]^+ .$$

(In the above expression, $[x]$ is the largest integer less than or equal to x).

The subcritical case is defined as $\gamma > \gamma_c$. It follows from (3.1) that the mean free path of particles tends to infinity as ϵ tends to zero. It is only natural to expect that, in this case, the number density converges to the solution of a free flight problem.

Theorem 2. Assume that $\phi \in L^\infty(\mathbf{R}^D \times \mathbf{S}^{D-1})$ and that $\gamma > \gamma_c$. Then

$$f_\epsilon \to f, \quad \text{in } L^\infty([0,\infty) \times \mathbf{R}^D \times \mathbf{S}^{D-1}) \text{ weak } * \quad \text{as } \epsilon \to 0, \qquad (3.10)$$

where f is the solution of the following free flight problem:

$$\partial_t f + c\Omega \cdot \nabla_x f = 0, \quad t \geq 0, \ x \in \mathbf{R}^D, \ \Omega \in \mathbf{S}^{D-1}. \qquad (3.11)$$

$$f(0, x, \Omega) = \phi(x, \Omega), \quad x \in \mathbf{R}^D, \ \Omega \in \mathbf{S}^{D-1}. \qquad (3.12)$$

The critical case is defined by $\gamma = \gamma_c = \frac{D}{D-1}$. Again, the behavior of the number density can be surmised from (3.1) which shows that the mean free path of particles is of order one as $\epsilon \to 0$. This particular scaling corresponds to the so-called Boltzmann-Grad limit of kinetic theory[10]. This indicates that the asymptotic number density is going to be the solution of a kinetic equation like the linearized Boltzmann equation[10].

Theorem 3. Assume that $\phi \in C^\infty(\mathbf{R}^D \times \mathbf{S}^{D-1})$ and that $\gamma = \gamma_c$. Let $f \equiv f(t, x, \Omega)$ be the solution of

$$\partial_t f + c\Omega \cdot \nabla_x f + \Lambda f = 0, \quad t \geq 0, \ x \in \mathbf{R}^D, \ \Omega \in \mathbf{S}^{D-1}. \qquad (3.13)$$

$$f(0, x, \Omega) = \phi(x, \Omega), \quad x \in \mathbf{R}^D, \ \Omega \in \mathbf{S}^{D-1}. \qquad (3.14)$$

Let $R_\epsilon = f_\epsilon - \mathbf{1}_{X_\epsilon} f$, then

$$\partial_t R_\epsilon + c\Omega \cdot \nabla_x R_\epsilon + \Lambda R_\epsilon = O(\epsilon^\infty) + O(\epsilon)_{|L^\infty(\mathbf{R}^D \times \mathbf{S}^{D-1})}, \qquad (3.15)$$

and

$$R_\epsilon(0, x, \Omega) = O(\epsilon)_{|L^\infty(\mathbf{R}^D \times \mathbf{S}^{D-1})}, \qquad (3.16)$$

uniformly on compact t-sets. In the above equations, $O(\epsilon^\infty)$ is to be understood in the same sense as in Proposition 1, and $O(\epsilon)_{|L^\infty(\mathbf{R}^D \times \mathbf{S}^{D-1})}$ denotes the class of functions whose L^∞ norm is of order $O(\epsilon)$.

The subcritical case defined by $1 \leq \gamma < \gamma_c$ leads to the so-called hydrodynamic limit of the billiards system since the mean free path defined by (3.1) tends to zero as ϵ tends to zero. It is therefore expected that the asymptotic behavior of the number density is given by the diffusion approximation[3,5] of the linearized Boltzmann equation

(3.13). Before stating the corresponding result in the present case, some preparations are required.

First, diffusion occurs in long time; one should first introduce the time scale adapted to this diffusion. Let $\eta = \epsilon^{(D-1)(\gamma_c - \gamma)}$ and $\tau = \eta t$. Define $F_\epsilon \equiv F_\epsilon(\tau, x, \Omega)$ as

$$F_\epsilon(\tau, x, \Omega) = f_\epsilon(\frac{\tau}{\eta}, x, \Omega)$$

so that F_ϵ solves

$$\eta \partial_\tau F_\epsilon + c\Omega \cdot \nabla_x F_\epsilon + \Lambda_\epsilon F_\epsilon = 0, \quad \tau \geq 0, \ x \in \mathbf{R}^D, \ \Omega \in \mathbf{S}^{D-1}, \tag{3.17}$$

with initial condition

$$F_\epsilon(0, x, \Omega) = \mathbf{1}_{X_\epsilon}(x)\phi(x), \quad x \in \mathbf{R}^D, \ \Omega \in \mathbf{S}^{D-1}. \tag{3.18}$$

It is assumed that the initial data $\phi(x)$ does not depend on Ω so that there is no initial layer[13].

Second, we introduce the notation $\theta_\gamma = 1$ if $\gamma > 1$ and $\theta_1 = 1 - \left(\frac{r}{a}\right)^D |B^D|$ where B^D is the closed unit ball of $\mathbf{R}^D$ and $|B^D|$ its Lebesgue measure. In other word, θ_γ is the volume fraction left free for the diffusion motion.

Third, we shall denote $\beta \equiv \beta(\Omega)$ the vector field on $\mathbf{S}^{D-1}$ such that

$$(\Lambda \beta_j)(\Omega) = \Omega_j, \quad \int_{\mathbf{S}^{D-1}} \beta_j(\Omega) d\Omega = 0. \tag{3.19}$$

This definition of β makes sense thanks to the following lemma.

Lemma 4. Λ is a self-adjoint Fredholm operator on $L^2(\mathbf{S}^{D-1})$ with nullspace $N(\Lambda) = \mathbf{R}1$. Moreover,

$$\beta(\Omega) = \frac{a^D}{cr^{D-1}} \frac{1}{s(D)} \frac{D^2 - 1}{4} \Omega. \tag{3.20}$$

That being said, the asymptotics of the number density in the subcritical case is given by the following theorem.

Theorem 5. Assume that F_ϵ solves (3.17)-(3.18) with $1 \leq \gamma < \gamma_c$ and $\nabla_x{}^4\phi \in L^\infty(\mathbf{R}^D)$. Define $F_0 \equiv F_0(\tau, x)$ as the solution of the diffusion equation

$$\partial_\tau F_0 - \frac{\kappa^2}{2} \Delta_x F_0 = 0, \quad \tau \geq 0, \ x \in \mathbf{R}^D, \tag{3.21}$$

with initial condition

$$F_0(0, x) = \phi(x), \quad x \in \mathbf{R}^D, \tag{3.22}$$

where the difusion coefficient κ is given by

$$\kappa^2 = \tfrac{2}{D}c^2\theta_\gamma \int_{\mathbf{S}^{D-1}} tr(\Omega \otimes \beta(\Omega)) \frac{d\Omega}{|\mathbf{S}^{D-1}|}$$

$$= \theta_\gamma \frac{D^2 - 1}{2Ds(D)} \frac{ca^D}{r^{D-1}}. \tag{3.23}$$

Introduce the functions

$$F_1 = -c\theta_\gamma \beta(\Omega) \cdot \nabla_x F_0$$

and

$$F_2 = \tfrac{1}{2}\mu(\Omega) : \nabla_x{}^2 F_0$$

where $\mu(\Omega)$ is the unique tensor such that

$$(\Lambda\mu_{ij})(\Omega) = \theta_\gamma^2 c^2 \left[\Omega_i \beta_j(\Omega) + \Omega_j \beta_i(\Omega)\right] - \theta_\gamma \kappa^2 \delta_{ij} \tag{3.24}$$

and

$$\int_{\mathbf{S}^{D-1}} \mu_{ij}(\Omega) d\Omega = 0. \tag{3.25}$$

Define

$$R_\epsilon = F_\epsilon - \mathbf{1}_{X_\epsilon}(F_0 + \eta F_1 + \eta^2 F_2). \tag{3.26}$$

Then,

$$\partial_\tau R_\epsilon + \frac{1}{\eta} c\Omega \cdot \nabla_x R_\epsilon + \frac{1}{\eta} \Lambda R_\epsilon = O(\eta^\infty) + O(\eta)_{|L^\infty(\mathbf{R}^D \times \mathbf{S}^{D-1})}, \tag{3.27}$$

and

$$R_\epsilon(0, x, \Omega) = O(\eta)_{|L^\infty(\mathbf{R}^D \times \mathbf{S}^{D-1})}, \tag{3.28}$$

Another way to state the results (3.15) and (3.27) would be to say that $\mathbf{1}_{X_\epsilon} f$ in Theorem 3 or $\mathbf{1}_{X_\epsilon}(F_0 + \eta F_1 + \eta^2 F_2)$ in Theorem 5 are consistent approximations of the solutions of (3.13) and (3.17), where consistency is measured in the topology of distributions. Here, consistency is to be understood in the sense of Lax Equivalence Theorem. This is what we call "strong consistency" in the present article.

Main Results for the 2D Lorentz Gas with Accomodation Reflection

As for the subcritical case for the Lorentz Gas of Hard Spheres, the diffusion aproximation of the 2D Lorentz Gas with Accomodation Reflection first requires a rescaling of the time variable. As in the subsection above introduce $F_\epsilon \equiv F_\epsilon(\tau, x, \Omega)$ as

$$F_\epsilon(\tau, x, \Omega) = f_\epsilon(\frac{\tau}{\epsilon}, x, \Omega)$$

so that F_ϵ solves

$$\epsilon \partial_\tau F_\epsilon + c\Omega \cdot \nabla_x F_\epsilon = 0, \quad \tau \geq 0, \ x \in X_\epsilon, \ \Omega \in \mathbf{S}^{D-1}, \tag{3.29}$$

with boundary condition

$$F_\epsilon(\tau, x, \Omega) = \tfrac{1}{2} \int_{\Omega' \cdot n_x < 0} F_\epsilon(\tau, x, \Omega') |\Omega' \cdot n_x| d\Omega', \quad x \in \partial X, \ \Omega \cdot n_x > 0, \tag{3.30}$$

and with initial condition

$$F_\epsilon(0, x, \Omega) = \phi(x), \quad x \in X_\epsilon, \ \Omega \in \mathbf{S}^{D-1}. \tag{3.31}$$

As in the case of hard spheres, it is expected that the number density F_ϵ is approximated in some sense by the solution of a diffusion equation as $\epsilon \to 0$. In view of writing the Einstein-Kubo formula for the diffusion coefficient, one introduces for $j = 1, 2$ a solution $\chi_j \equiv \chi_j(y, \Omega)$ of the problem

$$\Omega \cdot \nabla_y \chi_j(y, \Omega) = \Omega_j , \quad y \in Y , \ \Omega \in \mathbf{S}^1 \tag{3.32}$$

with boundary condition

$$\chi_j(y, \Omega) = \tfrac{1}{2} \int_{\Omega' \cdot n_y < 0} \chi_j(y, \Omega') |\Omega' \cdot n_y| d\Omega' , \quad y \in \partial Y , \ \Omega \cdot n_y > 0 . \tag{3.33}$$

In the above equations, Y is the image of $\overline{X}$ by the projection of $\mathbf{R}^2$ onto $\mathbf{R}^2/\mathcal{L}$, with X and $\mathcal{L}$ being as in (2.10)-(2.11). That the problem (3.32)-(3.33) admits a solution is a nontrivial fact and the key to the diffusion approximation of the 2D Lorentz Gas with Accomodation Reflection. By analogy with the classical Einstein-Kubo formula, we surmise that the diffusion coefficient is given by

$$K'^2 = \int_Y \int_{\mathbf{S}^1} (c\Omega \otimes \chi(y, \Omega) + \chi(y, \Omega) \otimes c\Omega) \, d\Omega dy . \tag{3.34}$$

Moreover, it is easily deduced from the symmetries of the lattice $\mathcal{L}$ that the above matrix is invariant when conjugated with a rotation of angle multiple $\pi/3$. Hence this matrix is a pure dilation, that is

$$K'^2 = \kappa'^2 \mathbf{I} \tag{3.35}$$

where

$$\kappa'^2 = \int_Y \int_{\mathbf{S}^1} (c\Omega_1 \chi_1(y, \Omega) + c\Omega_2 \chi_2(y, \Omega)) \, d\Omega dy . \tag{3.36}$$

The asymptotic behavior of the number density as $\epsilon \to 0$ is explained in the following theorem, due to Bardos-Dumas-Golse[2].

Theorem 6. Assume that condition (2.16) of finite horizon holds. Let $\phi \equiv \phi(x)$ be a smooth function defined on $\mathbf{R}^2$ with bounded derivatives up to order 4, and let $F \equiv F(\tau, x)$ be the solution of the diffusion equation

$$\partial_\tau F - \frac{\kappa'^2}{2} \Delta_x F = 0 , \quad \tau \geq 0 , \ x \in \mathbf{R}^2 , \tag{3.37}$$

with initial condition

$$F(0, x) = \phi(x) , \quad x \in \mathbf{R}^2 , \tag{3.38}$$

where κ' is given by (3.36). Then for all $T > 0$,

$$\sup_{t \in [0,T]} \|F_\epsilon(t, \cdot, \cdot) - F(t, \cdot)\|_{L^\infty(X_\epsilon \times \mathbf{S}^1)} \leq C_T \epsilon , \tag{3.39}$$

where C_T is a positive constant depending on T.

The difference between this result and that of Theorem 5 lies in the sense in which the diffusion approximation holds in both cases. In the case of the accomodation reflection, the diffusion approximation holds in the sense of uniform convergence. The reason for this much stronger result than the one stated as Theorem 6 lies in the much stronger stability of equation (3.29) supplemented with boundary condition (3.30): we shall discuss it at length in the sequel.

4. METHODS OF PROOF

We shall begin with a sketch of the proof of the approximation in the sense of strong consistency for the Lorentz Gas of Hard Spheres.

Sketch of the Proofs in the Case of Hard Spheres

The key result here is Proposition 1. We shall discuss it in detail. Once Proposition 1 and Lemma 4 are proved, the results stated as Theorems 2, 3, and 5 are quite standard, and we shall say no more about them. On the other hand, the proof of Lemma 4 is straightforward enough so that there is little need to discuss it.

Proof of Proposition 1. The collision term is split into two parts: the absorption term

$$\mathcal{A}_\epsilon \psi = c(\Omega \cdot n_x)_- \psi_{|\Gamma_\epsilon} \delta_{|\Gamma_\epsilon} \tag{4.1}$$

and the scattering term

$$\mathcal{S}_\epsilon \psi = c(\Omega \cdot n_x)_+ \psi_{|\Gamma_\epsilon} \circ R(n_x) \delta_{|\Gamma_\epsilon} . \tag{4.2}$$

We shall discuss only the asymptotics of the absorption term; that of the scattering term is very smilar.

Let ψ and χ be two functions in $C_0^\infty(\mathbf{R}^D \times \mathbf{S}^{D-1})$. Then

$$\langle\!\langle (\mathcal{A}_\epsilon \psi)\chi \rangle\!\rangle =$$

$$\sum_{k \in \mathbf{Z}^D} \int_{\mathbf{S}^{D-1}} (c\Omega \cdot \omega)_- (\epsilon^\gamma r)^{D-1} d\omega \int_{\mathbf{S}^{D-1}} \psi\chi(\epsilon a k + \epsilon^\gamma r\omega, \Omega) d\Omega . \tag{4.3}$$

First Taylor expand $\psi\chi(\epsilon a k + \epsilon^\gamma r\omega, \Omega)$ at $\epsilon a k$ in (4.3). This operation yields

$$\langle\!\langle (\mathcal{A}_\epsilon \psi)\chi \rangle\!\rangle = \sum_{k \in \mathbf{Z}^D} \int_{\mathbf{S}^{D-1}} (c\Omega \cdot \omega)_- (\epsilon^\gamma r)^{D-1} d\omega$$

$$\cdot \Big[\int_{\mathbf{S}^{D-1}} \psi\chi(\epsilon a k, \Omega) d\Omega + \int_{\mathbf{S}^{D-1}} \sum_{m=1}^{M} \frac{1}{m!} \nabla_x{}^m \psi\chi(\epsilon a k, \Omega) d\Omega$$

$$+ \int_{\mathbf{S}^{D-1}} \int_0^1 \frac{(1-t)^M}{M!} \nabla_x{}^{M+1} \psi\chi(\epsilon a k + t\epsilon^\gamma r\omega, \Omega) d\Omega \Big] . \tag{4.4}$$

Proposition 1 results from two observations:
— First, according to a multidimensional analogue of the Euler-McLaurin formula,

$$\Big| (a\epsilon)^D \sum_{k \in \mathbf{Z}^D} \varphi(\epsilon a k) - \int_{\mathbf{R}^D} \varphi(x) dx \Big| \leq C_{D,q}(a\epsilon)^{q+1} \|\partial_{x_1}^{q+1}...\partial_{x_D}^{q+1}\varphi\|_{L^\infty(\mathbf{R}^D)} ; \tag{4.5}$$

for any $\varphi \in C_0^\infty(\mathbf{R}^D)$ and any integer $q \geq 0$.
— Second,

$$\Big| (a\epsilon)^D \sum_{k \in \mathbf{Z}^D} \nabla_x{}^m \varphi(\epsilon a k) \Big| \leq C_{D,q}(a\epsilon)^{q+1} \|\partial_{x_1}^{q+1}...\partial_{x_D}^{q+1} \nabla_x{}^m \varphi\|_{L^\infty(\mathbf{R}^D)}$$

for any integer $m > 0$ in view of (4.5) since

$$\int_{\mathbf{R}^D} \nabla_x{}^m \varphi(x) dx = 0\,,$$

φ having being with compact support.

With these two remarks, one easily sees that

$$(a\epsilon)^D \langle\!\langle (\mathcal{A}_\epsilon \psi) \chi \rangle\!\rangle = (\epsilon^\gamma r)^{D-1} c \left(\int_{\mathbf{S}^{D-1}} (\Omega \cdot \omega)_- d\omega \right) \langle\!\langle \psi \chi \rangle\!\rangle + O(\epsilon^\infty)\,. \qquad (4.6)$$

Then

$$\int_{\mathbf{S}^{D-1}} (\Omega \cdot \omega)_- d\omega = \frac{s(D)}{D-1}\,, \qquad (4.7)$$

so that

$$(\mathcal{A}_\epsilon \psi)(x, \Omega) = \epsilon^{(\gamma - \gamma_c)(D-1)} \frac{c r^{D-1}}{a^D} \frac{s(D)}{D-1} \psi(x, \Omega) + O(\epsilon^\infty) \qquad (4.8)$$

in the sense of distributions. The asymptotic scattering term is found in quite the same way. This completes the sketch of the proof of Proposition 1. //

The proof of Lemma 4 is done by inspection (for example, for $D < 6$, the operator

$$\theta \mapsto \int_{\mathbf{S}^{D-1}} \theta(\Omega') \frac{d\Omega'}{|\Omega - \Omega'|^{D-3}}$$

is Hilbert-Schmidt; for higher dimensions, some iterate of this self-adjoint operator is Hilbert-Schmidt, whence the conclusion of Lemma 4 follows).

Next, the proofs of Theorems 3 and 5 are done by introducing the Hilbert type expansion

$$f_\epsilon(t, x, \Omega) \sim \mathbf{1}_{X_\epsilon} \sum_{m \geq 0} \epsilon^m f_m(t, x, \Omega)$$

for the proof of Theorem 3, or

$$F_\epsilon(\tau, x, \Omega) \sim \mathbf{1}_{X_\epsilon} \sum_{m \geq 0} \epsilon^m F_m(\tau, x, \Omega)$$

for the proof of Theorem 5 and writing down the equations for the remainders at the proper order.

The detailed proofs of Proposition 1, Lemma 4 and Theorems 2, 3, and 5 can be found in Golse[12], along with an extended treatment of the Hard Sphere case.

Sketch of the Proofs in the Case of Accomodation Reflection

The proofs for this second model are quite different from those for the Hard Sphere case exposed in the preceding subsection.

The main result in the proof of Theorem 6 is the following statement, which is the Fredholm Alternative for the streaming operator with accomodation reflection in Y.

Proposition 7. Assume that the finite horizon condition (2.16) holds, and let $S \equiv S(y, \Omega) \in L^\infty(Y \times \mathbf{S}^1)$. Consider the problem

$$\Omega \cdot \nabla_y \theta = S\,, \quad y \in Y\,, \ \Omega \in \mathbf{S}^1\,, \qquad (4.9)$$

with boundary condition

$$\theta(y, \Omega) = \tfrac{1}{2} \int_{\Omega' \cdot n_y < 0} \theta(y, \Omega') |\Omega' \cdot n_y| d\Omega', \quad y \in \partial Y, \ \Omega \cdot n_y > 0. \tag{4.10}$$

The two following assertions are equivalent:
(i) $\iint_{Y \times \mathbf{S}^1} S(y, \Omega) dy d\Omega = 0$,
(ii) there exists a unique solution $\theta \in L^\infty(Y \times \mathbf{S}^1)$ of (4.9)-(4.10) such that

$$\iint_{Y \times \mathbf{S}^1} \theta(y, \Omega) dy d\Omega = 0.$$

Proposition 7 shows that the equations (3.32)-(3.33) have a unique bounded solution of zero average, whence the tensor defined in the right side of (3.34) exists. That it is of the form predicted in the right sides of (3.35)-(3.36) is proved by inspection since this tensor commutes with the matrix

$$\begin{pmatrix} \frac{1}{2} & -\frac{\sqrt{3}}{2} \\ \frac{\sqrt{3}}{2} & \frac{1}{2} \end{pmatrix}$$

since the lattice $\mathcal{L}$ defined in (2.10) is invariant under rotations of angle multiple of $\pi/3$. To prove the positivity of the tensor defined in the right side of (3.34), multiply (3.32) by $\xi_j \chi_j$ where ξ is a vector of $\mathbf{R}^2$, then apply Green's formula with boundary condition (3.33): one finds that

$$\xi^T \left(\iint_{Y \times \mathbf{S}^1} [\Omega \otimes \chi(y, \Omega) + \chi(y, \Omega) \otimes \Omega] \, dy d\Omega \right) \xi$$

$$= \int_{\partial Y} \left[\tfrac{1}{2} \int_{\Omega \cdot n_y < 0} (\xi \cdot \chi)^2 |\Omega \cdot n_y| d\Omega - \left(\tfrac{1}{2} \int_{\Omega' \cdot n_y < 0} \xi \cdot \chi |\Omega' \cdot n_y| d\Omega' \right)^2 \right] dy. \tag{4.11}$$

Positivity of the right side of the above equality follows from the Cauchy-Schwarz inequality. The definite character follows from a careful analysis of the equality case in the above Cauchy-Schwarz inequality. Indeed, it follows that

$$\tfrac{1}{2} \int_{\Omega \cdot n_y < 0} (\xi \cdot \chi)^2 |\Omega \cdot n_y| d\Omega - \left(\tfrac{1}{2} \int_{\Omega' \cdot n_y < 0} \xi \cdot \chi |\Omega' \cdot n_y| d\Omega' \right)^2 = 0$$

whence

$$\xi \cdot \chi(y, \Omega) = \varphi(y) \tag{4.12}$$

is independent of Ω on ∂Y. One easily deduces from (4.12) that

$$\xi \cdot \chi(y, \Omega) - \xi \cdot \chi(y', \Omega) = \xi \cdot (y - y'),$$

for $y, y' \in \partial Y$. This ensures that $\xi \cdot \chi = 0$ for all $y \in Y$, whence $\xi = 0$.

Now we turn to the

Proof of Proposition 7. That (ii) implies (i) follows from Green's formula. It remains to prove that (i) implies (ii).

As for the uniqueness part, let θ_0 be a solution of (4.9)-(4.10) with $S = 0$. Multiplying (4.9) by θ_0 and applying Green's formula and the Cauchy-Schwarz inequality as in (4.11)-(4.12), one can show that $\theta_{0|\partial Y}$ does not depend on Ω. This piece of information is propagated to Y by using (4.9) so that one easily conludes that θ_0 is a constant.

The existence part is more technical and relies essentially on the following

Lemma 8. Consider the map

$$\varphi \mapsto \tfrac{1}{2} \int_{\Omega' \cdot n_y < 0} \beta(y, \Omega') |\Omega' \cdot n_y| d\Omega' \qquad (4.13)$$

defined on $L^p(\partial Y)$ for $1 \le p < +\infty$, where β is the unique solution in $L^p(Y \times \mathbf{S}^1)$ of the problem

$$\Omega \cdot \nabla_x \beta = 0, \quad y \in Y, \ \Omega \in \mathbf{S}^1, \qquad (4.14)$$

$$\beta(y, \Omega) = \phi(y), \quad y \in \partial Y, \ \Omega \cdot n_y > 0. \qquad (4.15)$$

This map is compact.

Lemma 8 shows the existence of a solution of (4.9)-(4.10) in L^p with $1 \le p < \infty$ by the Schauder fixed point theorem. The existence in L^∞ is proved by showing that some iterate of the map defined above maps L^2 into L^∞. This concludes the proof of Proposition 7. //

Sketch of the Proof of Lemma 8. First write

$$\int_{\Omega \cdot n_y < 0} \beta(y, \Omega) |\Omega \cdot n_y| d\Omega = \int_{\Omega \cdot n_y < 0} \varphi(y - \tau_{y,\Omega} \Omega) |\Omega \cdot n_y| d\Omega \qquad (4.16)$$

where $\tau_{y,\Omega}$ denotes the backward exit time:

$$\tau_{y,\Omega} = \sup\{t \ge 0 / [y, y - t\Omega] \subset \overline{X}\}$$

Notice that the exit time $\tau \in L^\infty(Y \times \mathbf{S}^1)$ thanks to the assumption of finite horizon (2.16). In the formula (4.16) above, one performs the change of variables

$$\Omega' \mapsto y - \tau_{y,\Omega} \Omega.$$

The announced compactness follows from applying the dominated convergence theorem. //

Proof of Theorem 6. The number density F_ϵ is seeked in the form

$$F_\epsilon(\tau, x, \Omega) \sim \sum_{m \ge 0} \epsilon^m F_m(\tau, x, y, \Omega)_{|y = \frac{x}{\epsilon}}$$

where $F_m(\tau, x, \cdot, \Omega)$ is defined on Y for all (τ, x, Ω). The asymptotic expansion above is introduced in (3.29); identifying the coefficients of the powers of ϵ, one easily determines F_0, F_1 and F_2 thanks to the Fredholm Alternative (Proposition 7). In particular, one easily sees that

$$F_0(\tau, x, y, \Omega) = F(t, x), \qquad F_1(\tau, x, y, \Omega) = -\chi(y, \Omega) \cdot \nabla_x F(t, x)$$

where F is the solution of (3.37)-(3.38) and χ the mean zero solution of (3.32)-(3.33) (whose existence and uniqueness in L^∞ is asserted by Proposition 7). Introducing then

$$R_\epsilon(\tau, x, \Omega) = F_\epsilon(\tau, x, \Omega) - F(t, x) + \epsilon \chi(y, \Omega) \cdot \nabla_x F(t, x) - \epsilon^2 F_2(\tau, x, \frac{x}{\epsilon}, \Omega)$$

one computes

$$\partial_\tau R_\epsilon + \tfrac{1}{\epsilon}\Omega \cdot \nabla_x R_\epsilon . \tag{4.17}$$

One concludes as in the case of hard spheres, but in the topology of uniform convergence. In fact, it suffices to use the maximum principle for equation (4.17) with accomodation reflection to obtain the announced conclusion. //

5. CONCLUSIONS AND FINAL REMARKS

We shall finish with some remarks to compare the results above with other results on billiards systems existing in the literature. Various possible extensions of this work are also listed.

a) The result in Theorem 3 is known as the Boltzmann-Grad limit. The corresponding approximation has been proved in the case where the distribution of scatterers is not periodic but random (with a Poisson law): see Boldighrini-Bunimovich-Sinai[6] and Spohn[15] (for the case where the scatterers are replaced with central repulsive potentials). In both cases, the randomness smooths out the Dirac measures on the boundaries of the scatterers and gives more stability to the Liouville equation than in the purely periodic case. This extra stability shows that the number density which solves (2.18)-(2.19)-(2.20) stays close to the solution of the "linearized Boltzmann equation" (3.13)-(3.14) in L^p topologies. One could think that this kind of approximation results from the weak consistency proved in the present article plus the extra stability provided by the random character of the distribution of scatterers, very much in the same spirit as in the Lax Equivalence Theorem for numerical schemes.

b) Bunimovich-Sinai[7] have treated the problem posed in Theorem 5 in the case of finite horizon planar billiards, like for example (2.10)-(2.11). They proved the convergence of the process defined by the positions of particles to a Brownian motion with diffusion matrix given by Einstein-Kubo formula[7]:

$$\kappa''^2 = 2c^2 E(\tau(x_0,\Omega_0)^2 \Omega_0 \otimes \Omega_0) + \sum_{m=0}^{\infty} c^2 E(\tau(x_0,\Omega_0)\tau(x_m,\Omega_m)\Omega_0 \otimes \Omega_m)$$

where (x_m,Ω_m) are the coordinate of the particle immediatly after the mth impact knowing that the coordinate of the particle after the initial impact are (x_0,Ω_0), $\tau(x_m,\Omega_m)$ is the time lap between the mth and the $m+$1st impact and the expectation corresponds ot the probability measure on

$$\{(x_0,\Omega_0) \in \partial X \times \mathbf{S}^1 \ / \ \Omega_0 \cdot n_{x_0} > 0\}$$

proportional to

$$\Omega_0 \cdot n_{x_0} d\Omega_0 dx_0 .$$

The proof of Bunimovich-Sinai[7] is done in two steps: first, a Markov partition is constructed for these billiards systems, providing a good coding of the dynamics of the particles; second, the central limit theorem for processes with mixing properties is used to prove the convergence to the Brownian motion.

However, the Bunimovich-Sinai approach does not provide an approximation in the sense of weak consistency as in this article. Second, the diffusion coefficient given in the form of the Einstein-Kubo formula is not explicit in terms of the geometry of the lattice and of the scatterers, as (3.23) is. It should be emphasized that there is no rigorous proof that the diffusion coefficient given by (3.23) and the one given by Einstein-Kubo formula, since they correspond to approximations of the Liouville equation in two *a priori* different senses.

However, the diffusion coefficient given by (3.23) does not tend to zero as $r \to a/2$ in dimension 2 as would be expected. Indeed, in this case, the configuration of scatterers is a periodic arrangement of closed cells, and each particle stays in the cell where it initially is. For that reason, the mean square displacement of particles is a uniformly bounded function of time, whence the diffusion coefficient should be zero. A careful analysis of the proof of Theorem 5 seems to indicate that the prediction (3.23) for the diffusion coefficient defined as the mean square displacement per unit of time (in large time) is likely to be valid when $\gamma > 1$.

c) Theorem 6 could be extended to other finite horizon billiards systems. The choice (2.10)-(2.11) is the most elementary instance of such a billiards system. It can be seen that the diffusion coefficient given by (3.35)-(3.36) vanishes as $r = a/2$. The key observation in this case is that $\chi_j(y, \Omega) = y_j$ for $j = 1, 2$ is a solution of (3.32)-(3.33).

The problem of extending the proof of Theorem 6 to infinite horizon billiards systems will be dealt with in subsequent publications. Also, accomodation reflection laws more general than (2.13) can be treated by the same method: see Bardos-Dumas-Golse[2].

REFERENCES

1. V. Arnold: Mathematical Methods of Classical Mechanics; Springer Verlag.
2. C. Bardos, L. Dumas, F. Golse: Diffusion de particules par un réffiseau d'obstacles circulaires; to appear in C. R. Acad. Sci.
3. C. Bardos, R. Santos, R. Sentis: Diffusion Approximation and Computation of the Critical Size of a Transport Operator; Trans. of the A.M.S. 284, 617–649, (1984).
4. A. Bensoussan, J.-L. Lions, G. Papanicolaou: Asymptotic Study of Periodic Structures; North Holland (1978).
5. A. Bensoussan, J.-L. Lions, G. Papanicolaou: Boundary Layers and Homogenization of Transport Processes; Publ. R.I.M.S. Kyoto 15, 53–115, (1979).
6. C. Boldighrini, L. Bunimovich, Ya. Sinai: On the Boltzmann Equation for the Lorentz Gas 32, 477–502, (1983).
7. L. Bunimovich, Ya. Sinai: Statistical Properties of the Lorentz Gas with Periodic Configuration of Scatterers; Commun. Math. Phys. 78, 479–497, (1981).
8. L. Bunimovich, Ya. Sinai: Markov Partitions of Dispersed Billiards; Commun. Math. Phys. 73, 247–280, (1980).
9. L. Bunimovich, Ya. Sinai, N. Chernov: Markov Partitions for Two-Dimensional Hyperbolic Billiards; Russian. Math. Surveys 45, 105–152, (1990).
10. C. Cercignani: The Boltzmann Equation, Springer Verlag.
11. P. Gaspard, G. Nicolis: Transport Properties, Lyapunov Exponents, and Entropy per Unit Time; Phys. Rev; Letters 65, 1693–1696, (1990).
12. F. Golse: Transport dans les milieux composites fortement contrastés I: le modèle du billard; preprint C. E. A., Limeil (1991).
13. E. Larsen, G. Pomraning, V. Badham: Asymptotic Analysis of Radiative Transfer Problems; J. Quant. Spectrosc. and Radiative Transfer 29, 285–310, (1983).
14. J. Machta: Power Law Decay of Correlations in a Billiard Problem; J. of Stat. Phys. 32, 555–564, (1983).
15. H. Spohn: The Lorentz Process Converges to a Random Flight Process; Commun. in Math. Phys. 60, 277, (1978).

THE BEHAVIOR OF SOLUTIONS OF THE NLS EQUATION IN THE SEMICLASSICAL LIMIT

Shan Jin[1], C. David Levermore[2] and David W. McLaughlin[3]

[1]Program in Applied Mathematics
University of Arizona, Tucson, Arizona 85721, USA
[2]Department of Mathematics
University of Arizona, Tucson, Arizona 85721, USA
[3]Department of Mathematics
Program in Applied and Computational Mathematics
Princeton University, Princeton, NJ 08544, USA

Abstract

We report a numerical and theoretical study of the generation and propogation of oscillations in the semiclassical limit ($\hbar \to 0$) of the Nonlinear Schrödinger equation. In a general setting of both dimension and nonlinearity, we identify essential differences between the "defocusing" and "focusing" cases. Numerical comparisons of the oscillations are made between the linear ("free") and the cubic (defocusing and focusing) cases in one dimension. The integrability of the one-dimensional cubic NLS is exploited to give a complete global characterization of the weak limits of the oscillations in the defocusing case.

1. INTRODUCTION

1.1. The NLS Equation

One of the simplest nonlinear wave equations is the nonlinear Schrödinger (NLS) equation for a complex-valued field $\Psi(x, t)$ over a spatial domain $\Omega \subset \mathbf{R}^D$,

$$i\hbar\, \partial_t \Psi + \frac{\hbar^2}{2}\Delta \Psi - U'(|\Psi|^2)\, \Psi = 0 \,, \tag{1.1}$$

where U' is the first derivative of a twice differentiable nonlinear real-valued function and $\hbar$ is a positive parameter. This is just the usual Schrödinger equation of quantum mechanics with the potential $V(x)$ replaced by $U'(|\Psi|^2)$. The parameter $\hbar$ is analogous to Planck's constant, which in the quantum setting is usually very small when evaluated in the natural dimensional scales of the equation as determined by its initial and boundary conditions. For the moment, the precise specification of the

Singular Limits of Dispersive Waves, Edited by
N.M. Ercolani et al., Plenum Press, New York, 1994

domain Ω and the nature of the boundary conditions is left vague in order to make some general statements regarding the structure of (1.1). It will be assumed that they are consistent with all formal calculations.

That the nonlinear function $U : \mathbf{R}_+ \to \mathbf{R}$ is the potential energy density of the field is clearly seen when the NLS equation (1.1) is recast as a Hamiltonian system in the form

$$i\hbar \, \partial_t \Psi = \frac{\delta H}{\delta \bar{\Psi}} \, , \qquad H = \int_\Omega \frac{\hbar^2}{2} |\nabla \Psi|^2 + U(|\Psi|^2) \, d^D x \, . \tag{1.2}$$

The associated Poisson bracket of any two functionals F and G is given by

$$\{F, G\} \equiv \frac{1}{i\hbar} \int_\Omega \left(\frac{\delta F}{\delta \Psi} \frac{\delta G}{\delta \bar{\Psi}} - \frac{\delta F}{\delta \bar{\Psi}} \frac{\delta G}{\delta \Psi} \right) d^D x \, ; \tag{1.3}$$

the evolution of any functional F under the NLS flow (1.2) is then

$$\frac{dF}{dt} = \{F, H\} \, . \tag{1.4}$$

This Hamiltonian structure plays a major role in our analysis.

Also associated with the NLS equation (1.1) are $D + 2$ local conservation laws corresponding to mass, momentum, and energy conservation. Their densities, ρ, μ, and ϵ respectively, are given by

$$\rho = |\Psi|^2 \, , \qquad \mu = -i\frac{\hbar}{2} \left(\bar{\Psi} \nabla \Psi - \Psi \nabla \bar{\Psi} \right) \, , \qquad \epsilon = \frac{\hbar^2}{2} |\nabla \Psi|^2 + U(|\Psi|^2) \, . \tag{1.5}$$

The mass and momentum densities determine the field Ψ up to a constant phase; the energy density can be written in terms of them as

$$\epsilon = \frac{1}{2} \frac{|\mu|^2}{\rho} + \frac{\hbar^2}{8} \frac{|\nabla \rho|^2}{\rho} + U(\rho) \, . \tag{1.6}$$

The local conservation laws are then

$$\partial_t \rho + \nabla \cdot \mu = 0 \, ,$$

$$\partial_t \mu + \nabla \cdot \left(\frac{\mu \otimes \mu}{\rho} \right) + \nabla P(\rho) = \frac{\hbar^2}{4} \nabla \cdot \left[\rho \nabla^2 \log \rho \right] \, ,$$

$$\partial_t \epsilon + \nabla \cdot \left(\frac{\mu}{\rho} (\epsilon + P(\rho)) \right) = \frac{\hbar^2}{4} \nabla \cdot \left[\frac{\mu \Delta \rho}{\rho} - \frac{\nabla \cdot \mu \nabla \rho}{\rho} \right] \, , \tag{1.7}$$

where $P(\rho) \equiv \rho \, U'(\rho) - U(\rho)$. The first two of these are a closed system governing ρ and μ that has the form of a perturbation of the compressible Euler equations of fluid dynamics with the "pressure" given by $P(\rho)$. If the "Euler part" of these equations is to be hyperbolic, then the "pressure" $P(\rho)$ must be a strictly increasing function of ρ; in that case $P'(\rho) = \rho \, U''(\rho) > 0$. This means that U must be a strictly convex function of ρ and corresponds to a "defocusing" NLS equation. In this context a "focusing" NLS equation can be understood as a fluid whose pressure *decreases* when the mass density increases, a phenomemon leading to the development of mass concentrations.

1.2. Posing the Semiclassical Limit

The "semiclassical limit" of the NLS equation can be described as follows. Consider the family, parametrized by $\hbar > 0$, of solutions $\Psi^{(\hbar)}(x, t)$ to the Cauchy problems

$$i\hbar\, \partial_t \Psi^{(\hbar)} + \frac{\hbar^2}{2} \Delta \Psi^{(\hbar)} - U'(|\Psi^{(\hbar)}|^2)\, \Psi^{(\hbar)} = 0, \tag{1.8a}$$

$$\Psi^{(\hbar)}(x, 0) = A(x) \exp\left(\frac{i}{\hbar} S(x)\right), \tag{1.8b}$$

where the (nonnegative) amplitude $A(x)$ and (real) phase $S(x)$ are assumed to be smooth and independent of $\hbar$. The initial conserved densities are then

$$\rho^{(\hbar)}(x, 0) = |A(x)|^2, \qquad \mu^{(\hbar)}(x, 0) = |A(x)|^2 \nabla S(x), \tag{1.9a}$$

$$\epsilon^{(\hbar)}(x, 0) = \frac{1}{2}|A(x)|^2 |\nabla S(x)|^2 + \frac{\hbar^2}{2}|\nabla A(x)|^2 + U(|A(x)|^2). \tag{1.9b}$$

The general problem of the semiclassical limit is to determine the limiting behavior of any function of the field $\Psi^{(\hbar)}$ as $\hbar \to 0$. In particular, to ascertain the existence (in some sense) of the limits of the conserved densities

$$\rho = \lim_{\hbar \to 0} \rho^{(\hbar)}, \qquad \mu = \lim_{\hbar \to 0} \mu^{(\hbar)}, \qquad \epsilon = \lim_{\hbar \to 0} \epsilon^{(\hbar)},$$

and, if the limits exist, to determine their dynamics.

Arguing formally, it is natural to conjecture for the defocusing case that the $\mathcal{O}(\hbar^2)$ dispersive terms appearing in (1.7) are negligible as $\hbar \to 0$, and that the limiting densities ρ and μ satisfy the hyperbolic system (the Euler system)

$$\partial_t \rho + \nabla\cdot\mu = 0,$$
$$\partial_t \mu + \nabla\cdot\left(\frac{\mu \otimes \mu}{\rho}\right) + \nabla P(\rho) = 0, \tag{1.10a}$$

with initial conditions inferred from (1.9a) given by

$$\rho(x, 0) = |A(x)|^2, \qquad \mu(x, 0) = |A(x)|^2 \nabla S(x). \tag{1.10b}$$

This argument is self-consistent only so long as the solution of the Euler system (1.10) remains classical. In that case the limiting energy density will be given by

$$\epsilon = \frac{1}{2}\frac{|\mu|^2}{\rho} + U(\rho), \tag{1.11}$$

and will satisfy

$$\partial_t \epsilon + \nabla\cdot\left(\frac{\mu}{\rho}(\epsilon + P(\rho))\right) = 0, \tag{1.12}$$

hence, playing the role of a Lax entropy for the Euler system (1.10a). We have not tried to prove the above conjecture in this generality; however, a fairly general proof could certainly be carried out in any setting for which the local well-posedness of classical solutions of the Euler system (1.10) is known. In Section 3 of this paper we recover this result as a part of a much stronger result for a more restricted problem, that of the cubic NLS equation in one spatial dimension.

The genuinely nonlinear nature of the Euler system (1.10) will ensure that its classical solution will develop singular behavior (an infinite derivative) for all but rarefaction initial data. At the instant such a breaking occurs, the formally small dispersive terms on the right side of (1.7) will no longer be negligible, and the above characterization of the semiclassical limit will breakdown. Since this small regularizing term is dispersive, one expects that the impending singularity in ρ and μ will be regularized by the development of small wavelength oscillations. In the remainder of this paper we illuminate this phenomena, first through numerical experiments in Section 2, and then theoretically in Section 3 where we give a brief summary of our results in [9].

2. NUMERICAL EXPERIMENTS

2.1. The Onset of Breaking

Following the lead set by Fornberg and Whitham in their study [4] of solutions of the Korteweg-de Vries equation, we turn to numerics in order to illustrate the breakdown of the "Euler" description of the semiclassical limit for the defocusing NLS equation. We consider the problem cast in one spatial dimension with a cubic nonlinearity

$$i\hbar\, \partial_t \Psi = -\frac{\hbar^2}{2}\partial_{xx}\Psi + \gamma\,|\Psi|^2\,\Psi\,, \tag{2.1a}$$

$$\Psi(x,0) = A_{in}(x)\exp\left(\frac{i}{\hbar}S_{in}(x)\right), \tag{2.1b}$$

where γ here is a positive constant. For this one-dimensional case, the Euler system (1.10) describing the formal semiclassical limit reduces to the initial-value problem

$$\partial_t\rho + \partial_x\mu = 0\,,$$
$$\partial_t\mu + \partial_x\left(\frac{\mu^2}{\rho} + \gamma\,\frac{\rho^2}{2}\right) = 0\,, \tag{2.2a}$$

$$\rho(x,0) = A_{in}^2(x)\,, \qquad \mu(x,0) = A_{in}^2(x)\partial_x S_{in}(x)\,. \tag{2.2b}$$

Riemann invariants for the Euler system (2.2a) are given by

$$r_\pm = \frac{\mu}{2\rho} \pm \sqrt{\gamma\rho}\,, \tag{2.3}$$

and the system can be placed in the Riemann invariant form

$$\partial_t r_+ + \tfrac{1}{2}(3r_+ + r_-)\partial_x r_+ = 0\,,$$
$$\partial_t r_- + \tfrac{1}{2}(r_+ + 3r_-)\partial_x r_- = 0\,, \tag{2.4a}$$

with the initial conditions

$$r_\pm(x,0) = \tfrac{1}{2}\partial_x S_{in}(x) \pm \sqrt{\gamma} A_{in}(x) \,. \tag{2.4b}$$

The normalization of the Riemann invariants (2.3) has been so chosen that it is most natural for the theoretical study in the next section, which centers around the integrable structure of (2.1).

As our first example, consider the initial data given by

$$A_{in} = \exp(-x^2) \,, \qquad\qquad \partial_x S_{in}(x) = -\tanh x \,, \tag{2.5}$$

so that

$$r_\pm(x,0) = -\tfrac{1}{2}\tanh x \pm \sqrt{\gamma}\exp(-x^2) \,. \tag{2.6}$$

With the compressional initial data determined by (2.5), the solution of the reduced hyperbolic system (2.2a) will certainly breakdown in a finite time. To see the effect of this breakdown upon behavior in the solution of the original NLS equation (2.1a), we integrate the NLS equation numerically for a small value of $\hbar$. This was done by Indik and Wolfson [8] using a standard "split step" method in which during one half of the time step the term $\exp(i\tfrac{1}{2}\hbar\Delta t\partial_{xx})$ is implemented in Fourier space, while the nonlinearity is integrated in x-space during the other half of the time step. For these experiments, the values of the parameters γ and $\hbar$ are fixed (usually, $\hbar = 0.1$). The spatial scale is $x \in (-40, 40)$, which is discretized into 4096 points ($\Delta x \sim 0.02$). A sufficiently small time step is $\Delta t = 0.004$. Periodic boundary conditions are used, although the experiments are not run long enough for boundary effects to become important.

For $\hbar = 0.025$ and $\gamma = 1.0$, we calculate Ψ numerically, and then construct the Riemann invariants $r_\pm$ from this numerical data using definitions (1.5) and (2.3). This numerical data is depicted in Fig. 2.1. Note the substantial steepening for this small (0.025) value of $\hbar$. While the reduced system (2.2a) would develop an infinite spatial derivative in a finite time, the dispersive term $\hbar^2\partial_{xx}$ prevents the development of singularities in solutions of the NLS equation itself. Because of the dispersive nature of this small regularizing term, one expects that the impending singularity will be regularized by the development of small wavelength oscillations.

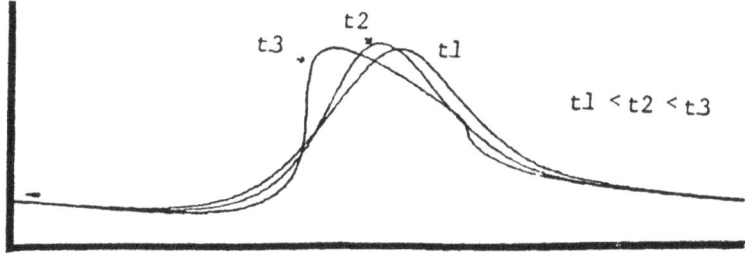

Fig. 2.1. The Riemann invariant r_+ as a function of x, for several times before "breaktime". These curves are computed from numerical data generated by direct simulation of the NLS equation, with the Riemann invariant constructed using equation (2.3) written directly in terms of $\Psi^\hbar$.

2.2. Post Breaking Phenomena

Another method for deriving the reduced system (2.2a) is the classical WKB method. It considers

$$i\hbar\, \partial_t \Psi^{(\hbar)} = -\frac{\hbar^2}{2} \partial_{xx} \Psi^{(\hbar)} + V(x)\, \Psi^{(\hbar)} \,, \tag{2.7}$$

and begins with the Ansatz that $\Psi^{(\hbar)}$ is in the form

$$\Psi^{(\hbar)}(x,t) = A(x,t) \exp\left(\frac{i}{\hbar} S(x,t)\right) + \mathcal{O}(\hbar)\,. \tag{2.8}$$

Inserting this Ansatz into the NLS equation (2.7), and balancing the leading two powers of $\hbar$ yields

$$\begin{aligned} \partial_t u + u\partial_x u + \partial_x V &= 0\,, \\ \partial_t A + u\partial_x A + \tfrac{1}{2}A\partial_x u &= 0\,, \end{aligned} \tag{2.9}$$

where $u = \partial_x S$, which is equivalent to the reduced system (2.2a) upon making the identifications

$$\rho = A^2\,, \qquad \mu = A^2 u\,. \tag{2.10}$$

The development of a singularity in this reduced system must then be interpreted as a breakdown in the Ansatz (2.8). In other words, after the breaktime the wave form no longer continues to resemble that of the Ansatz!

In linear theories such as quantum mechanics and classical electromagnetism, the presence of singularities in reduced systems and their consequences for the full system are well understood. For example, in quantum mechanics the characteristics of the reduced hyperbolic system (2.9) are the paths of classical particles in the conservative force field $F(x) = -\partial_x V(x)$, which is defined in terms of the prescribed potential energy function $V(x)$. In this manner classical mechanics arises as the semiclassical limit of quantum mechanics. (In the electromagnetic setting, the characteristics are the "rays" of "geometrical optics".) Singularities in the reduced semiclassical equations result from foci and envelopes of families of these classical paths. These envelopes separate regions in the (x,t)-plane that consist of points lying on only one classical path from regions consisting of points that lie on multiple paths. Along these envelopes (called caustics), neighboring rays coalesce and geometric conservation law properties of the transport equation for the amplitude $A(x,t)$ force $|A|$ to diverge along these caustics.

More mathematically, in the linear case of quantum mechanics, the reduced hyperbolic system is degenerate in the sense that the eikonal equation for u and the eikonal equation for A have identical characteristics. Since the eikonal equation does not depend upon the amplitude A, it can be solved first and its characteristics are the classical paths. The transport equation for the amplitude A is then integrated along these classical paths, and geometrical considerations force $|A|$ to diverge along caustic envelopes. This *divergence of the amplitude A* is a direct consequence of the degeneracy of system (2.9).

This caustic behavior can be present even in the trivial case of the free Schrödinger equation (with $V(x) = 0$) provided one begins with compressional initial data such as

$$u(x,0) = \partial_x S_{in}(x) = -\tanh(x). \qquad (2.11)$$

In this case the classical paths are straight lines, and the family for initial data (2.11) is depicted in Fig. 2.2, where the focus and envelopes of the classical paths are clearly visible. In this $V(x) = 0$ case the solution $\Psi^{(\hbar)}(x,t)$ of initial value problem (2.1) can be represented exactly as a Fourier integral,

$$\Psi^{(\hbar)}(x,t) = \frac{1}{\sqrt{2\pi i \hbar t}} \int_{-\infty}^{+\infty} \exp\left(\frac{i}{\hbar}\left(\frac{(x-y)^2}{2t} + S_{in}(y)\right)\right) A_{in}(y)\,dy. \qquad (2.12)$$

The behavior of $\Psi^{(\hbar)}(x,t)$ as $\hbar \to 0$, uniformly in (x,t), can be obtained from an asymptotic (stationary phase) evaluation of this integral. The result shows that, away from the characteristic envelopes, $\Psi^{(\hbar)}(x,t)$ behaves asymptotically as the linear superposition

$$\Psi^{(\hbar)}(x,t) \sim \sum_j \exp\left(\frac{i}{\hbar} S^{(j)}(x,t)\right) A^{(j)}(x,t), \qquad (2.13)$$

where the index j in the sum runs over the different classical paths through the point (x,t), $S^{(j)}$ is the classical action computed by integrating the eikonal equation along the jth classical path, and $A^{(j)}$ is computed by integrating the transport equation along the jth classical path, with adjustments by phase shifts of the form $\exp(in^{(j)}\pi/4)$, where the integers $n^{(j)}$ are computed from the number of times the jth path touches the caustic envelope.

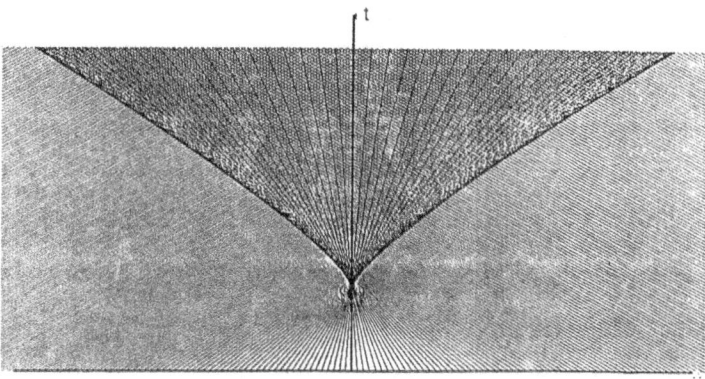

Fig. 2.2. The "free" particle paths governing the semiclassical behavior. Again, note the focus and the caustic envelope.

In the linear case for general $V(x)$, the classical paths are not straight lines and the asymptotic behavior cannot be calculated with Fourier theory; nevertheless, the asymptotic behavior is still given by the superposition of (2.13). Using the method of characteristics for the eikonal equation, one constructs a surface $u = u(x,t)$ over the (x,t)-plane. With this surface, one integrates the transport equation for $A(x,t)$

and assembles formula (2.13). Arguments from the theory of uniform asymptotic expansions then show that formula (2.13) is asymptotically valid except at the focus and along the characteristic envelopes, the only effect of which is phase shifts of integer multiples of $\pi/4$. These integers are related to the Morse (Maslov-Keller) indices for the surface $u = u(x,t)$.

In this linear case, the qualitative consequences of formula (2.13) are striking. Before the caustic, only one classical path passes through each space time point (x,t); only one term appears in the sum in (2.13); the intensity $|A(x,t)|^2$ is slowly varying on the $\hbar$ scale. After the caustic, three classical paths pass through each point (x,t); three terms appear in the sum; the intensity $|A(x,t)|^2$,

$$|A(x,t)|^2 = \left| \sum_j \exp\left(\frac{i}{\hbar} S^{(j)}(x,t) \right) A^{(j)}(x,t) \right|^2, \qquad (2.14)$$

has rapid oscillations on the $\hbar$ scale do to phase interference between the three terms in the sum. Mathematically, these rapid oscillations prevent strong convergence after the caustic. Moreover, the weak limit, which continues to exist, is independent of the Keller-Maslov phase shifts. In summary, in this linear case the weak limit can be constructed simply by summing over the contributions from each classical path, with interference between different terms in this sum causing rapid oscillations.

We now describe several simple experiments carried out by Indik and Wolfson [8] which illustrate and contrast the formation of oscillations in solutions of (2.1) for the linear ($\gamma = 0$), the nonlinear defocusing ($\gamma > 0$), and the nonlinear focusing ($\gamma < 0$) cases. In each experiment, the compressional initial data (2.5) is used.

In the first experiment, we consider the linear case ($\gamma = 0$) in order to illustrate the semiclassical linear theory just described. Figure 2.3a, which shows $|\Psi(x,t)|$ as a surface over the (x,t) plane, clearly contains one focus of that theory, from which two caustics emanate. Notice that $\mathcal{O}(\hbar)$ wavelength oscillations in the *intensity*

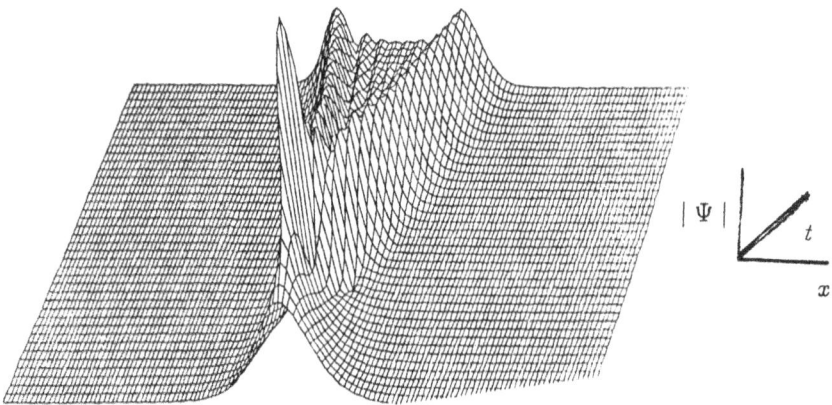

Fig. 2.3a. The amplitude $|\Psi^{\hbar}|$ for the linear ($\gamma = 0$) Schrödinger equation as a surface over the (x,t)-plane, for $\hbar = 0.1$.

242

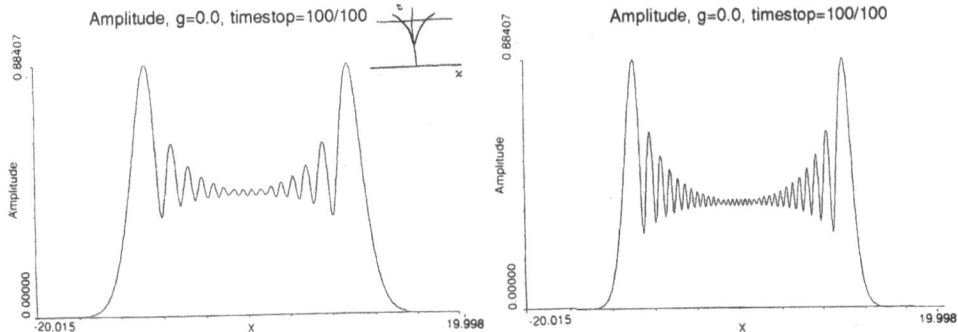

Fig. 2.3b. The amplitude $|\Psi^\hbar|$ as a function of x for a fixed time past breaktime, at two values of $\hbar = 0.05, 0.025$. Notice the regular structure of the oscillations in $|\Psi^\hbar|$ beyond breaktime.

form at the focus and persist for later times. The spatial region containing these oscillations is bounded by the two caustics. Figure 2.3b shows that the amplitude of these oscillations is $\mathcal{O}(1)$, while their wavelength is $\mathcal{O}(\hbar)$.

The next experiment treats the "defocusing" case with a repulsive nonlinearity ($\gamma > 0$). Figure 2.4a shows that, as in the linear case, oscillations in the intensity form at specific points in space and time, and then persist. However, when compared with the linear case, several distinguishing features arise from the nonlinearity. First, the amplitude $|\Psi(x,t)|$ is much less intense at the focus than in the linear case, which is depicted from the same perspective (but with a different vertical scale) in Fig. 2.4b. Second, while the oscillations are again $\mathcal{O}(1)$ in amplitude and $\mathcal{O}(\hbar)$ in wavelength as shown in Fig. 2.4c (compare with Fig. 2.3b), in the "defocusing" case these oscillations form two packets, one traveling to the right and the other to the left. The central region of the spatial profile that separates the two oscillatory regions is a quiescent plateau at times beyond the focus.

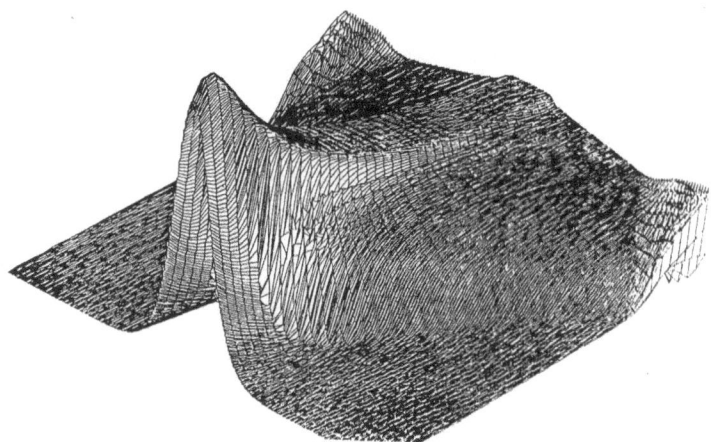

Fig. 2.4a. The amplitude $|\Psi^\hbar|$ for the defocusing ($\gamma > 0$) NLS equation as a surface over the (x,t)-plane, for $\hbar = 0.1$.

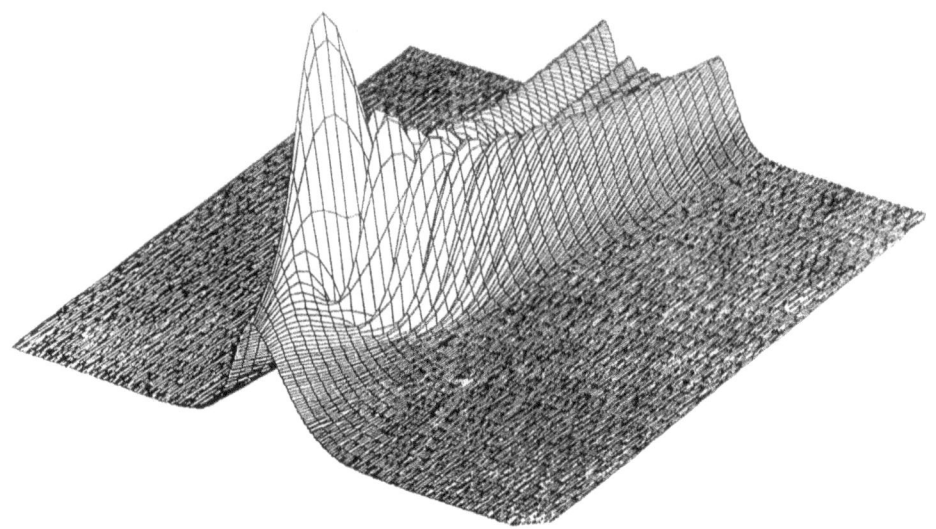

Fig. 2.4b. For comparison, the result for the linear case ($\gamma = 0$) for the identical initial data and h is depicted here. Note the change in vertical scale.

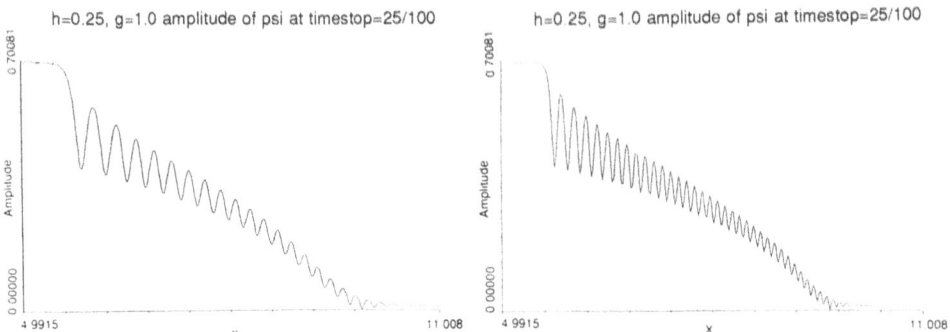

Fig. 2.4c. The amplitude $|\Psi^h|$ as a function of x for fixed time beyond breaktime, for two values of $h = 0.05, 0.025$.

The third experiment treats the "focusing" case with an attractive nonlinearity ($\gamma < 0$). The response in this focusing case (Fig. 2.5a) is the most violent of the three. Indeed, the defocusing case had the mildest reponse, followed the the linear case, with the most extreme behavior found with focusing nonlinearity. This ordering is not too surprising. It is the focusing nonlinearity which supports solitons in one spatial dimension and which blows up in finite time in two or more dimensions. In this focusing case instabilities in the equation itself make numerical integration difficult, and one should not trust the numerical accuracy for detailed information about the oscillations. Nevertheless, the blow up of a portion of the oscillatory region shown in Fig. 2.5b admits the intriguing interpretation of these oscillations as a dense "sea of solitons" located near the center of the spatial profile, with a sharp boundary separating the oscillatory from the quiescent regions of space. More information about this focusing case may be found in Bronski and McLaughlin [1].

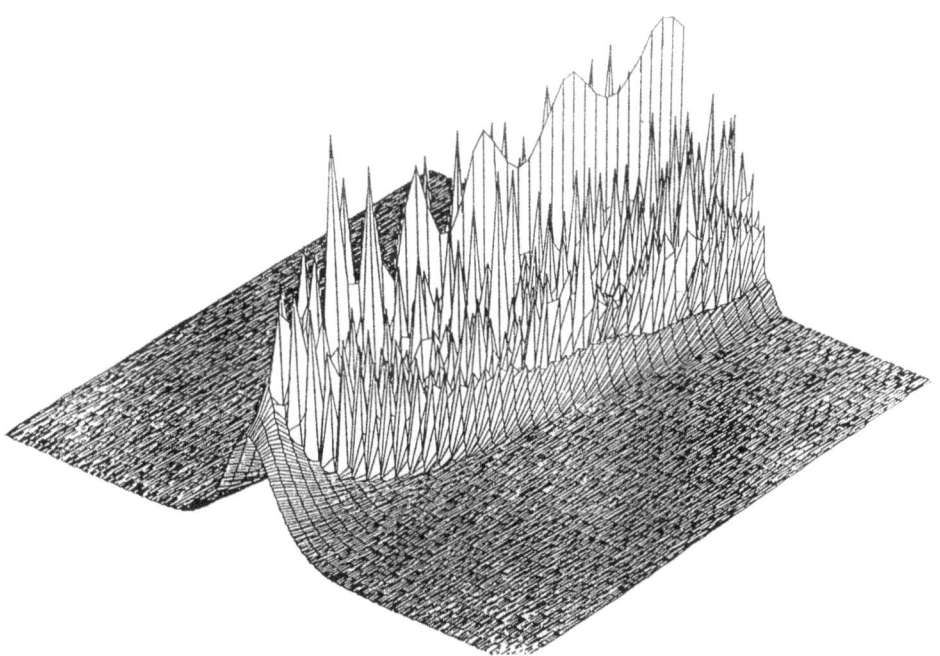

Fig. 2.5a. The amplitude $|\Psi^h|$ for the focusing ($\gamma < 0$) NLS equation as a surface over the (x,t)-plane, for $h = 0.1$.

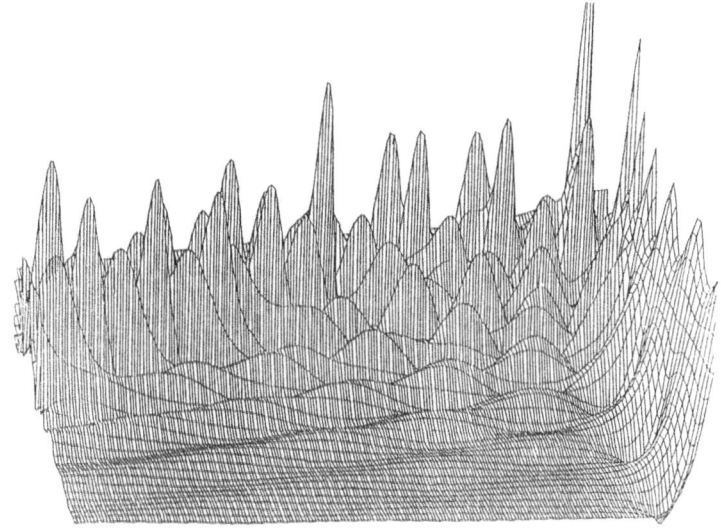

Fig. 2.5b. An enlargment of part of the oscillatory region from Fig. 2.5a.

3. THEORETICAL STUDY

3.1. The Defocusing NLS Hierarchy

We summarize a recent theoretical study [9] that established the semiclassical limit for the defocusing one-dimensional cubic Schrödinger equation given by

$$i\hbar\,\partial_t\Psi + \frac{\hbar^2}{2}\partial_{xx}\Psi + \left(1 - |\Psi|^2\right)\Psi = 0\,, \tag{3.1a}$$

with the far-field boundary conditions

$$\Psi(x,t) \sim \exp\left(\pm\frac{i}{\hbar}S_\infty\right) \qquad \text{as } x \to \pm\infty, \tag{3.1b}$$

for some $S_\infty \in \mathbf{R}$, and the initial condition

$$\Psi(x,0) = A_{in}(x)\exp\left(\frac{i}{\hbar}S_{in}(x)\right), \tag{3.1c}$$

for some smooth $A_{in}(x)$ and $S_{in}(x)$ that are independent of $\hbar$ and consistent with the far-field boundary conditions (3.1b).

Zakharov and Shabat [14] showed that this problem is completely integrable using the inverse scattering transform associated with the self-adjoint Dirac operator

$$\mathcal{L} = \begin{pmatrix} i\hbar\,\partial_x & \bar{\Psi} \\ \Psi & -i\hbar\,\partial_x \end{pmatrix}. \tag{3.2}$$

The strategy centers around the eigenvalue problem

$$\mathcal{L}f = \lambda f, \qquad \text{where } f = \begin{pmatrix} f^{(1)} \\ f^{(2)} \end{pmatrix}. \tag{3.3}$$

Given $\Psi = \Psi(x,0)$, the asymptotics of the eigenfunctions $f(\lambda, x, 0)$ as $|x| \to \infty$, referred to as the scattering data, can be calculated in principle. The evolution of the scattering data is then determined and the $\Psi(x,t)$ is then obtained from the knowledge of the large $|x|$ asymptotics of $f(\lambda, x, t)$ using the inverse scattering theory.

More specifically, the L^2 spectrum of $\mathcal{L}$ consists of two semi-infinite intevals, $(\infty, -1]$ and $[1, \infty)$, comprising the continuous spectrum, along with a finite set (possibly empty) of simple eigenvalues $\lambda_1, \cdots, \lambda_N$ in the interval $(-1, 1)$. The asymptotic behavior of one eigenfunction $f(\lambda, x)$ corresponding to a λ in the continuous spectrum is given by

$$f(\lambda, x) \sim \begin{cases} T(k)\bar{E}_-(k)\exp\left(\dfrac{-ikx}{\hbar}\right), & \text{for } x \to -\infty, \\[3mm] E_-(k)\exp\left(\dfrac{-ikx}{\hbar}\right) + R(k)E_+(k)\exp\left(\dfrac{ikx}{\hbar}\right), & \text{for } x \to +\infty, \end{cases} \tag{3.4}$$

where the vectors $E_\pm(k)$ are

$$E_\pm(k) \equiv \begin{pmatrix} \exp\left(\dfrac{-iS_\infty}{2\hbar}\right)(\lambda+k)^{\mp 1/2} \\[3mm] \exp\left(\dfrac{iS_\infty}{2\hbar}\right)(\lambda+k)^{\pm 1/2} \end{pmatrix},$$

and $k = \sqrt{\lambda^2 - 1}$. The complex-valued functions $R(k)$ and $T(k)$ are referred to as the reflection and transmission coefficients respectively [14]. For $|\lambda| > 1$, two independent eigenfunctions are then

$$\begin{pmatrix} f^{(1)} \\ f^{(2)} \end{pmatrix}, \qquad \begin{pmatrix} \bar{f}^{(2)} \\ \bar{f}^{(1)} \end{pmatrix}.$$

The asymptotic behavior of one eigenfunction $f = f_j(x)$ corresponding to a discrete eigenvalue $\lambda_j \in (-1, 1)$ is given by

$$f_j(x) \sim \begin{pmatrix} \exp\left(\dfrac{-iS_\infty}{2\hbar}\right)(\lambda_j - i\eta_j)^{1/2} \\ \exp\left(\dfrac{iS_\infty}{2\hbar}\right)(\lambda_j + i\eta_j)^{1/2} \end{pmatrix} \exp\left(\dfrac{-\eta_j x + \chi_j}{\hbar}\right), \qquad \text{for } x \to +\infty, \quad (3.5)$$

where $\eta_j = \sqrt{1 - \lambda_j^2}$ and the so-called norming exponents χ_j, which are real-valued, are determined by the normalization

$$\int \left| f_j^{(1)} f_j^{(2)} \right| dx = 1. \qquad (3.6)$$

Since the eigenfunction (3.5) satisfies the symmetry relation $f_j^{(1)} = \bar{f}_j^{(2)}$, the absolute values in (3.6) are redundant.

The inverse theory prescribes that the fundamental scattering data consists of the reflection coefficient $R(k)$, the eigenvalues λ_j and the norming exponents χ_j. The transmission coefficient $T(k)$, as well as all other asymptotic information, can be computed in terms of this fundamental set. The eigenvalues $\{\lambda_j\}$ are independent of time as Ψ evolves according to (3.1), while the time dependence of the other scattering data is

$$\chi_j(t) = \chi_j(0) + \eta_j \lambda_j t, \qquad R(k,t) = R(k,0) \exp\left(\dfrac{-i2k\lambda t}{\hbar}\right). \qquad (3.7)$$

Hence, given $R(k,0)$, λ_j, and $\chi_j(0)$ computed from the initial data $\Psi(x,0)$, the solution $\Psi(x,t)$ of the NLS equation (3.1) is then determined by inverse scattering from the $R(k,t)$, λ_j, and $\chi_j(t)$ as given by (3.7).

The complete integrability of the cubic Schrödinger equation implies the existence of an infinite family of independent conserved quantities [5],

$$H_m = \int_{-\infty}^{\infty} \rho_m \, dx, \qquad \text{for } m = -1, 0, 1, 2, \cdots, \qquad (3.8)$$

which are in involution with respect to the Poisson bracket (1.3),

$$0 = \{H_m, H_n\} = \frac{1}{i\hbar} \int_{-\infty}^{\infty} \left(\frac{\delta H_m}{\delta \Psi} \frac{\delta H_n}{\delta \bar{\Psi}} - \frac{\delta H_m}{\delta \bar{\Psi}} \frac{\delta H_n}{\delta \Psi} \right) dx. \qquad (3.9)$$

The first three of these densities correspond to the general conserved densities mentioned earlier (1.5) and are given by

$$\begin{aligned} \rho_{-1} &= |\Psi|^2 - 1, \\ \rho_0 &= -i\frac{\hbar}{2}\left(\bar{\Psi}\partial_x\Psi - \Psi\partial_x\bar{\Psi}\right), \\ \rho_1 &= \frac{\hbar^2}{2}|\partial_x\Psi|^2 + \frac{1}{2}\left(|\Psi|^2 - 1\right)^2, \end{aligned} \qquad (3.10)$$

Henceforth, the problem of the semiclassical limit is understood as the evaluation of the limiting behavior of all the conserved densities,

$$\rho_m = \lim_{\hbar \to 0} \rho_m^{(\hbar)}. \qquad (3.11)$$

Other limits can then be determined from these.

All of the H_m except H_{-1} are Hamiltonians that generate flows which commute with the cubic NLS flow (3.1a) and leave the boundary condition (3.1b) invariant, the so-called NLS hierarchy. Letting t_m denote the time variable associated with the mth flow, its evolution is then given by

$$i\hbar \partial_{t_m} \Psi = \frac{\delta H_m}{\delta \bar{\Psi}}, \qquad \text{for } m = 0, 1, 2, \cdots . \tag{3.12}$$

The t_0 flow is just spatial translation, the t_1 flow is given by the NLS equation (3.1a) with $t = t_1$, while the t_2 flow is that of the complex mKdV equation

$$\partial_{t_2} \Psi - \frac{3}{2} |\Psi|^2 \partial_x \Psi + \frac{\hbar^2}{4} \partial_{xxx} \Psi = 0 . \tag{3.13}$$

Every H_n is conserved by each flow; their densities satisfy the local conservation laws

$$\partial_{t_m} \rho_{n-1} + \partial_x \mu_{m,n} = 0, \qquad \text{for } m, n = 0, 1, \cdots . \tag{3.14}$$

Here $\mu_{m,n}$ is the flux for the $(n-1)^{th}$ conserved quantity under the mth flow.

Since all of these flows commute, they may be solved simultaneously for $\Psi^{(\hbar)}(x, \mathbf{t})$ satisfying the initial condition (3.1c), where $\mathbf{t} = (t_0, t_1, \cdots)$ such that all but finitely many t_m are zero. Associated with each $\mathbf{t}$ is a polynomial $p(\cdot, \mathbf{t})$ defined by

$$p(\lambda, \mathbf{t}) = \sum_{m=1}^{\infty} t_m \lambda^m . \tag{3.15}$$

The simultaneous evolution of the scattering data is then given by

$$\chi_j(\mathbf{t}) = \chi_j(0) + \eta_j p(\lambda_j, \mathbf{t}), \qquad R(k, \mathbf{t}) = R(k, 0) \exp\left(\frac{-i2kp(\lambda, \mathbf{t})}{\hbar}\right), \tag{3.16}$$

and $\Psi(x, \mathbf{t})$ is determined by inverse scattering.

The scope of the semiclassical limit for the defocusing NLS can then be enlarged to consider the solution $\Psi^{(\hbar)}(x, \mathbf{t})$ of the whole hierarchy that satisfies the initial condition

$$\Psi^{(\hbar)}(x, 0) = A_{in}(x) \exp\left(\frac{i}{\hbar} S_{in}(x)\right), \tag{3.17}$$

for some smooth $A_{in}(x)$ and $S_{in}(x)$ that are independent of $\hbar$ and consistent the far-field boundary conditions

$$\Psi^{(\hbar)}(x, t) \sim \exp\left(\pm \frac{i}{\hbar} S_\infty\right), \qquad \text{as } x \to \pm\infty . \tag{3.18}$$

The goal is then to determine the limiting behavior of all the conserved densities $\rho_n^{(\hbar)}$ and fluxes $\mu_{m,n}^{(\hbar)}$ associated with the entier NLS hierarchy of flows as $\hbar$ tends to zero.

3.2. The Analogy with the KdV Equation

This problem has many similarities with that of the zero-dispersion limit of the Korteweg-deVries (KdV) equation. There one studies the limit as $\varepsilon \to 0$ of the conserved densities for the scaled KdV equation

$$\partial_t u^{(\varepsilon)} - 6u^{(\varepsilon)}\partial_x u^{(\varepsilon)} + \varepsilon^2 \partial_{xxx} u^{(\varepsilon)} = 0 \,, \qquad (3.19a)$$

$$u^{(\varepsilon)}(x,0) = u_{in}(x) \,. \qquad (3.19b)$$

The limit is strong and given by the solution of the Hopf equation

$$\partial_t u - 6u \, \partial_x u = 0 \,, \qquad (3.20a)$$

$$u(x,0) = u_{in}(x) \,, \qquad (3.20b)$$

so long as its solution is classical. After the breaktime the limit is weak due to the development of regularizing small wavelength oscillations with an amplitude of order unity; thereafter its evolution is no longer governed by the Hopf equation (3.20a).

In their seminal paper, Gardner, Greene, Kruskal and Muria [7] showed that the KdV equation is completely integrable using the inverse scattering transform associated with the self-adjoint Schrödinger operator

$$\mathcal{L}_S = -\varepsilon^2 \partial_{xx} + u \,. \qquad (3.21)$$

Lax and Levermore [10] analyzed the limiting behavior of the scattering and inverse scattering transform using a WKB analysis of (3.21) and a kind of steepest descent argument to obtain a characterization of the (weak) limits in terms of the solution of a variational problem. The solution of this variational problem was then constructed through the solution of a Riemann-Hilbert problem. Venakides [12] has analyzed the microstructure of the limiting solutions, bridging the gap to the local approach of modulation theory developed by Flaschka, Forest and McLaughlin [3]. These results are surveyed in [11].

We have employed and improved the same strategy to analyze the semiclassical limit for the defocusing NLS hierarchy in [9]; some of these results are summerized below. More recently, a similar analysis has obtained the semiclassical limit for the odd flows of the focusing NLS hierarchy [2].

3.3. Asymptotic Analysis of the Initial Scattering Data

Here we outline the analysis in [9] for the case when the initial data $A_{in}(x)$ is a single up-side-down positive bump with a unique minimum and a horizontal asymptote as $|x| \to \infty$ which is also its upper bound. Moreover, we represent the initial data $A_{in}(x)$ and $S_{in}(x)$ in terms of the associated Riemann invariants $r_-(x,0)$ and $r_+(x,0)$, given by (2.4b) with $\gamma - 1$, as depicted in Fig. 3.1.

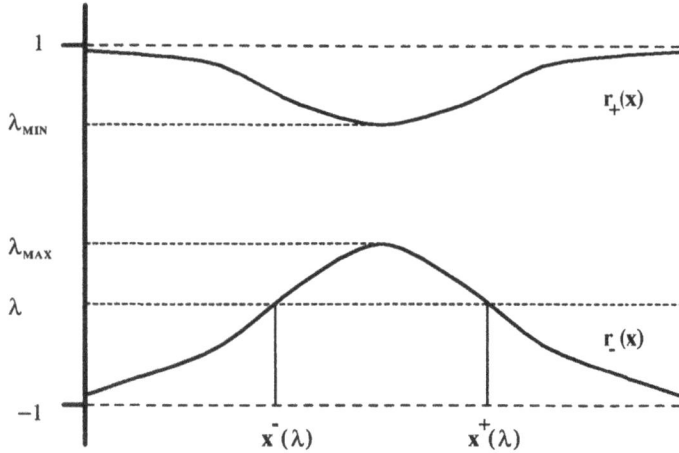

Figure 3.1. The initial data $r_\pm(x)$. Note their critical values λ_{min} and λ_{max}, and the indicated defining relations for the turning points $x^\pm(\lambda)$.

The scattering data for $A_{in}(x)$ and $S_{in}(x)$ through $r_\pm$ can then be computed asymptotically for small $\hbar$ by the semiclassical (WKB) method. The discrete eigenvalues are packed in the intervals $(-1, \lambda_{max})$ and $(\lambda_{min}, 1)$. In terms of the transformed spectral variable λ and $\eta(\lambda) = \sqrt{1 - \lambda^2}$, the asymptotic density of eigenvalues is given by the formula

$$\text{density of eigenvalues in } (-1, \lambda_{max}) \cup (\lambda_{min}, 1) \sim \frac{1}{\pi\hbar}\varphi(\lambda) \,,$$

where

$$\varphi(\lambda) = \int_{x^-(\lambda)}^{x^+(\lambda)} \frac{\left|\lambda - \frac{1}{2}\big(r_+(x) + r_-(x)\big)\right|}{\sqrt{\big(\lambda - r_+(x)\big)\big(\lambda - r_-(x)\big)}} \, dx \,. \tag{3.22}$$

The total number of eigenvalues $N = N_\hbar$ is given asymptotically by

$$N_\hbar \sim \frac{1}{\pi\hbar}\int_{-1}^{1} \varphi(\lambda) \, d\lambda \,. \tag{3.23}$$

The norming exponent obtained from the WKB analysis at $\mathbf{t} = 0$ is given by

$$\chi_j \sim \chi(\lambda_j) \,,$$

where

$$\chi(\lambda) = \eta(\lambda)\, x^+(\lambda) + \int_{x^+(\lambda)}^{\infty} \left(\eta(\lambda) - \sqrt{\big(r_+(x) - \lambda\big)\big(\lambda - r_-(x)\big)}\right) dx \,. \tag{3.24}$$

The reflection coefficient is calculated to be zero to all orders of the WKB expansion.

Based on the above calculation, we choose to neglect the scattering data related to the continuous spectrum. More precisely, we replace the initial data given by $A_{in}(x)$ and $S_{in}(x)$ in (3.17) with the reflectionless

$$\Psi^{(\hbar)}(x,0) = A_{in}^{(\hbar)}(x) \exp\left(\frac{i}{\hbar}S_{in}^{(\hbar)}(x)\right) \,, \tag{3.25}$$

corresponding to the WKB scattering data given by (3.22) and (3.24). This device sidesteps one aspect of the important question concerning the limiting behavior of the inverse scattering machinery, however is justified a posteriori by the fact that the resulting conserved densities and fluxes have the same strong limit as with the original initial data. In particular, $A_{in}^{(\hbar)}$ and $S_{in}^{(\hbar)}$ converge strongly to A_{in} and S_{in} respectively.

The solution $\Psi^{(\hbar)}(x, \mathbf{t})$ of the defocusing NLS hierarchy corresponding to this reflectionless initial data is given in terms of a determinant $\tau^{(\hbar)}(x, \mathbf{t})$ through the relations

$$|\Psi^{(\hbar)}(x, \mathbf{t})|^2 - 1 = -\hbar^2 \partial_{xx} \log \tau^{(\hbar)}(x, \mathbf{t}), \qquad (3.26a)$$

$$-i\frac{\hbar}{2}\left(\bar{\Psi}^{(\hbar)}(x, \mathbf{t})\partial_x \Psi^{(\hbar)}(x, \mathbf{t}) - \Psi^{(\hbar)}(x, \mathbf{t})\partial_x \bar{\Psi}^{(\hbar)}(x, \mathbf{t})\right) = \hbar^2 \partial_{xt} \log \tau^{(\hbar)}(x, \mathbf{t}), \qquad (3.26b)$$

where $t = t_1$. Relations (3.26) determine $\Psi^{(\hbar)}(x, \mathbf{t})$ up to a constant phase that is then fixed by the far-field boundary conditions (3.18). A formula for $\tau^{(\hbar)}(x, \mathbf{t})$ for the N-soliton solution was first derived by Zakharov and Shabat [14]. Their formula, which involves a $2N \times 2N$ matrix, was not suitable for our analysis, so we modified it into the more suitable formula

$$\tau^{(\hbar)}(x, \mathbf{t}) = \det(I + G), \qquad (3.27)$$

where the $N \times N$ matrix G is written as $G = \hbar DBD$ in terms of the trigonometric Cauchy matrix B and diagonal matrix D defined by

$$B = \left(\frac{1}{2\sin(\frac{1}{2}(\sigma_m + \sigma_n))}\right), \qquad D = \mathrm{diag}\left(\exp\left(\frac{a(\sigma_n, x, \mathbf{t})}{\hbar}\right)\right), \qquad (3.28)$$

where

$$a(\sigma, x, \mathbf{t}) = -\sin\sigma\, x + \sin\sigma\, p(\cos\sigma, \mathbf{t}) + \chi(\cos\sigma), \qquad (3.29)$$

All the associated conserved densities and fluxes are given in terms of $\tau^{(\hbar)}(x, \mathbf{t})$ by

$$\rho_{n-1}^{(\hbar)}(x, \mathbf{t}) = \hbar^2 \partial_{xt_n} \log \tau^{(\hbar)}(x, \mathbf{t}),$$
$$\mu_{m,n}^{(\hbar)}(x, \mathbf{t}) = -\hbar^2 \partial_{t_m t_n} \log \tau^{(\hbar)}(x, \mathbf{t}). \qquad (3.30)$$

An important observation is that for each $\hbar$ the function

$$(x, \mathbf{t}) \mapsto \log \tau^{(\hbar)}(x, \mathbf{t}) \qquad \text{is convex.} \qquad (3.31)$$

This follows from a direct computation of its Hessian from (3.27).

We write the expansion of the determinant (3.27) in an unusual form by replacing the summations over terms having the same number of factors from G by integrations over an atomic measures. Thus, setting $\sigma = \arccos\lambda$,

$$\tau^{(\hbar)}(x, \mathbf{t}) = 1 + \sum_{k=1}^{N_\hbar} \frac{1}{k! \pi^k} \tau_k^{(\hbar)}(x, \mathbf{t}), \qquad (3.32a)$$

$$\tau_k^{(\hbar)}(x, \mathbf{t}) = \int \cdots \int \exp\left(\frac{2}{\hbar} \sum_{j=1}^{k} a(\sigma_j, x, \mathbf{t})\right) \frac{\prod_{\substack{i,j=1 \\ i \neq j}}^{k} \left| \sin(\frac{1}{2}(\sigma_i - \sigma_j)) \right|}{\prod_{i,j=1}^{k} \left| \sin(\frac{1}{2}(\sigma_i + \sigma_j)) \right|} \, d\nu(\sigma_1) \cdots d\nu(\sigma_k),$$

(3.32b)

where the atomic measure $d\nu(\sigma)$ is given by

$$d\nu(\sigma) = \pi\hbar \sum_{i=1}^{N_\hbar} \delta(\sigma - \theta_i) \, d\sigma,$$

and the θ_i corresponds to the ith eigenvalue λ_i defined as $\lambda_i = \cos\theta_i$.

We recast $\tau_k^{(\hbar)}$ in the more transparent form

$$\tau_k^{(\hbar)}(x, \mathbf{t}) = \int \cdots \int \exp\left(\frac{2(a^{(\hbar)}, \psi_k) + (L\psi_k, \psi_k)}{\hbar^2}\right) d\nu(\sigma_1) \cdots d\nu(\sigma_k),$$

(3.33)

where the distribution ψ_k is defined by

$$\psi_k(\theta; \sigma_1, \cdots, \sigma_k) = \pi\hbar \sum_{i=1}^{k} \delta(\theta - \sigma_i),$$

(3.34)

and we have introduced the function $a^{(\hbar)}$ and operator L by

$$a^{(\hbar)}(\sigma, x, \mathbf{t}) = a(\sigma, x, \mathbf{t}) - \frac{\hbar}{2} \log(\sin\sigma),$$

(3.35a)

$$L\psi(\theta) = \frac{1}{\pi} \int_0^\pi \log \left| \frac{\sin(\frac{1}{2}(\theta - \sigma))}{\sin(\frac{1}{2}(\theta + \sigma))} \right| \psi(\sigma) \, d\sigma,$$

(3.35b)

where the kernel of L is defined to be zero on the diagonal and (3.35a) incorporates the diagonal term of the denominator of (3.32b). Here the inner product is defined by

$$(a, \psi) = \frac{1}{\pi} \int_0^\pi a(\theta) \psi(\theta) \, d\theta.$$

It is possible to write the functions $\tau_k^{(\hbar)}$ in real exponential form because they are sums of positive terms.

3.4. The Semiclassical Limit

The existence of the semiclassical limit rests upon establishing the existence of the limit

$$\lim_{\hbar \to 0} \hbar^2 \log \tau^{(\hbar)}(x, \mathbf{t}) = Q(x, \mathbf{t}),$$

(3.36)

where the limit is proven [9] to be uniform over compact subsets of $(x, \mathbf{t})$. Inserting this limit into (3.30) produces the weak limits

$$\lim_{\hbar \to 0} \rho_{n-1}^{(\hbar)}(x, \mathbf{t}) = \partial_{x t_n} Q(x, \mathbf{t}), \qquad \lim_{\hbar \to 0} \mu_{m,n}^{(\hbar)}(x, \mathbf{t}) = -\partial_{t_m t_n} Q(x, \mathbf{t}),$$

(3.37)

for the conserved densities and fluxes associated with the solution $\Psi^{(\hbar)}(x,\mathbf{t})$. In particular, this gives the weak limit

$$\lim_{\hbar \to 0} \left(|\Psi^{(\hbar)}(x,\mathbf{t})|^2 - 1 \right) = -\partial_{xx} Q(x,\mathbf{t}) . \tag{3.38}$$

Moreover, by analyzing the expansion (3.32) using an argument in the spirit of steepest-descents, we showed that the leading contribution to $\hbar^2 \log \tau^{(\hbar)}(x,\mathbf{t})$ comes from the largest term in that sum. Following the suggestive form of (3.33), $Q(x,\mathbf{t})$, which is the limit of these terms, is then characterized by the maximization problem

$$Q(x,\mathbf{t}) = \max \left\{ 2(a,\psi) + (L\psi,\psi) : \psi \in \mathcal{A} \right\} ,$$
$$\mathcal{A} \equiv \left\{ \psi \in L^1([0,\pi]) : 0 \le \psi \le \phi \right\} , \tag{3.39}$$

where $\phi(\sigma) = \varphi(\cos \sigma) \sin \sigma$, and the maximum being attained within the admissible set $\mathcal{A} \subset L^1$.

We observe that the maximization occurs over nonnegative L^1 functions, not over the atomic distributions ψ_k of (3.34). The functional maximizer $\psi^*(x,\mathbf{t})$ is the continuum limit of the multivariate maximizers ψ_k as $\hbar \to 0$. The variational problem is strictly concave and therefore has a unique solution that depends continuously on $(x,\mathbf{t})$ in the weak topology of measures. This in turn implies that $Q(x,\mathbf{t})$ is continuously differentiable with

$$\partial_x Q(x,\mathbf{t}) = -2 \left(\sin \theta, \psi^*(x,\mathbf{t}) \right) ,$$
$$\partial_t Q(x,\mathbf{t}) = 2 \left(\sin \theta \, p_t(\cos \theta), \psi^*(x,\mathbf{t}) \right) . \tag{3.40}$$

where $p_t(\lambda) \equiv \partial_t p(\lambda,\mathbf{t})$ is independent of $\mathbf{t}$ by (3.15). The convexity of $Q(x,\mathbf{t})$ follows from its characterization in (3.39) as the supremum of linear functions. The convexity (3.31) and regularity of the approximating functions $\log \tau^{(\hbar)}$ then implies the existence of the limits

$$\lim_{\hbar \to 0} \hbar^2 \partial_x \log \tau^{(\hbar)}(x,\mathbf{t}) = \partial_x Q(x,\mathbf{t}), \qquad \lim_{\hbar \to 0} \hbar^2 \partial_t \log \tau^{(\hbar)}(x,\mathbf{t}) = \partial_t Q(x,\mathbf{t}), \quad (3.41)$$

uniform over compact subsets of $(x,\mathbf{t})$.

The maximization problem (3.39) is attacked analytically by solving its variational conditions given by

$$L\psi(\theta) + a(\theta,x,\mathbf{t}) \le 0 , \qquad \text{when } \psi(\theta,x,\mathbf{t}) = 0 ,$$
$$L\psi(\theta) + a(\theta,x,\mathbf{t}) = 0 , \qquad \text{when } 0 < \psi(\theta,x,\mathbf{t}) < \phi(\theta) , \tag{3.42}$$
$$L\psi(\theta) + a(\theta,x,\mathbf{t}) \ge 0 , \qquad \text{when } \psi(\theta,x,\mathbf{t}) = \phi(\theta) .$$

Let I be the interior of the set of $(\theta,x,\mathbf{t})$ in which equality holds; let $\bar{I}$ be its closure. By direct differentiation we have

$$L\psi_x(\theta) = \sin \theta , \qquad L\psi_t(\theta) = -\sin \theta \, p_t(\cos \theta) , \qquad \text{when } (\theta,x,\mathbf{t}) \in I ,$$
$$\psi_x(\theta) = 0 , \qquad \psi_t(\theta) = 0 , \qquad \text{when } (\theta,x,\mathbf{t}) \notin \bar{I}. \tag{3.43}$$

These differentiated variational conditions do not have any explicit dependence on $(x, \mathbf{t})$, but rather their dynamics, as well as all their memory of the initial data, is contained in the set I. Consequently, these conditions have more general validity than the variational conditions (3.42). Indeed, the latter vary (although in a sense insignificantly) when the initial data are considered in different classes (such as periodic), while the differentiated conditions remain unchanged.

Making the Ansatz that at fixed $(x, \mathbf{t})$ the set $I(x, \mathbf{t})$, defined as

$$I(x, \mathbf{t}) \equiv \left\{ \theta \in (0, \pi) \ : \ (\theta, x, \mathbf{t}) \in I \right\}, \tag{3.44}$$

consists of a finite union of disjoint open intervals, one can uniquely determine the functions $\psi_x(\theta)$ and $\psi_{\mathbf{t}}(\theta)$ in terms of the endpoints of these intervals. More precisely, $I(x, \mathbf{t})$ is assumed to take the form

$$I(x, \mathbf{t}) = (0, \beta_1) \cup (\beta_2, \beta_3) \cup \cdots \cup (\beta_{2g+2}, \pi)$$

for some nonnegative integer g, where the β_i and g depend on $(x, \mathbf{t})$ and

$$0 \le \beta_1 < \beta_2 < \cdots < \beta_{2g+2} \le \pi \,.$$

Indeed, if any function $\psi(\theta)$ defined on $[0, \pi)$ is extended to be an odd function over the interval $(-\pi, \pi)$, thought of as the boundary of the unit disk, then the θ-derivative of L is the Hilbert transform:

$$\frac{d}{d\theta} L\psi(\theta) = \frac{1}{\pi} \int_{-\pi}^{\pi} \frac{\psi(\sigma)}{2\tan\left(\frac{1}{2}(\theta - \sigma)\right)} \, d\sigma \equiv H\psi(\theta) \,. \tag{3.45}$$

This observation is applied to ψ_x and $\psi_{\mathbf{t}}$ to transform the differentiated variational conditions (3.43) into Riemann-Hilbert problems through which they can be solved explicitly for ψ_x and $\psi_{\mathbf{t}}$ in terms of hyperelliptic functions involving the radical

$$R(\theta, x, \mathbf{t}) = \left(\prod_{i=1}^{2g+2} \frac{\cos\beta_i - \cos\theta}{2(1 + \cos\beta_i)(1 + \cos\theta)} \right)^{1/2} . \tag{3.46}$$

Note that the extended $I(x, \mathbf{t})$ is the set over which $R(\theta)$ is real-valued and g is just the genus of the associated Riemann surface.

The dependence of the β_i on $(x, \mathbf{t})$ is then derived from the compatibility constraint $\partial_t \psi_x = \partial_x \psi_{\mathbf{t}}$. This reduces to the first order system of hyperbolic equations in the Riemann invariant (diagonal) form

$$\partial_{t_k} \beta_i + S_{ki}(\beta_1, \cdots, \beta_{2g+2}) \, \partial_x \beta_i = 0, \qquad \text{for } i = 1, \cdots, 2g+2. \tag{3.47}$$

The weak limits are strong initially, and wherever the genus $g = 0$. In that case the equations (3.47) become equivalent to the Euler system (2.2a) upon identifying

$$r_+ = \cos\beta_2 \,, \qquad\qquad r_- = \cos\beta_1 \,, \tag{3.48}$$

where $r_\pm$ are the Riemann invariants defined by (2.3). Wherever $g = 1$, equations (3.47) are identical to the modulation equations for a family of periodic conidal waves that were first obtained by Forest and Lee [6] through the Whitham averaging method [13], the so-called Whitham equations. When $g > 1$, equations (3.47) are the NLS analogues of the generalized Whitham equations that describe the zero-dispersion limit of the Korteweg-deVries equation which were first derived in [10] and [3].

Acknowledgements. The authors thank Robert Indik and Michael Wolfson for carrying out the numerical experiments cited above. In addition they gratefully acknowledge the support provided to them: S.J. from the Air Force under grant AFOSR-90-021; C.D.L. from the NSF under grant DMS-8914420; D.W.M. from the Air Force under grant AFOSR-90-0161 and from the NSF under grant DMS-8922717.

References

[1] J.C. Bronski and D.W. McLaughlin, *Semiclassical Behavior in the NLS Equation: Optical Shocks–Focusing Instabilities*, in this volume.

[2] N. Ercolani, S. Jin, C.D. Levermore and W. MacEvoy, *The Zero Dispersion Limit of the NLS/mKdV Hierarchy for the Nonselfadjoint ZS Operator*, preprint (1992).

[3] H. Flaschka, M.G. Forest and D.W. McLaughlin, *Multiphase Averaging and the Inverse Spectral Solutions of the Korteweg-de Vries Equation*, Comm. Pure Appl. Math. **33** (1980), 739–784.

[4] B. Fornberg and G.B. Whitham, *A Numerical and Theoretical Study of Certain Nonlinear Wave Phenomena*, Phil. Trans. Roy. Soc. A **289** (1978), 373–404.

[5] H. Flaschka, A.C. Newell and T. Ratiu, *Kac-Moody Lie Algebras and Soliton Equations II: Lax Equations Associated with $A_1^{(1)}$*, Physica D **9** (1983), 300–323.

[6] M.G. Forest and J.E. Lee, *Geometry and Modulation Theory for Periodic Nonlinear Schrödinger Equation*, in "Oscillation Theory, Computation, and Methods of Compensated Compactness", C. Dafermos, J.L. Erickson, D. Kinderleher and M. Slemrod, eds, IMA Volumes on Mathematics and its Applications **2**, Springer-Verlag, New York (1986), 35-69.

[7] C.S. Gardner, J.M. Greene, M.D. Kruskal and R.M. Miura, *Method for Solving the Korteweg-deVries Equation*, Phys. Rev. Lett. **19** (1967), 1095–1097.

[8] R. Indik and M.A. Wolfson, (private communication).

[9] S. Jin, C.D. Levermore and D.W. McLaughlin, *The Semiclassical Limit for the Defocusing Nonlinear Schrödinger Hierarchy*, preprint (1991).

[10] P.D. Lax and C.D. Levermore, *The Zero Dispersion Limit of the Korteweg-deVries Equation*, Proc. Nat. Acad. Sci. USA **76** #8 (1979), 3602–3606.
    P.D. Lax and C.D. Levermore, *The Small Dispersion Limit of the Korteweg-deVries Equation I, II, III*, Comm. Pure Appl. Math. **36** (1983), 253–290, 571–593, 809–829.

[11] P.D. Lax, C.D. Levermore and S. Venakides *The Generation and Propagation of Oscillations in Dispersive IVPs and Their Limiting Behavior*, in "Important Developments in Soliton Theory 1980–1990", T. Fokas and V.E. Zakharov eds., Springer-Verlag, New York (to appear in 1992).

[12] S. Venakides, *The Generation of Modulated Wavetrains in the Solution of the Korteweg - deVries Equation*, Comm. Pure Appl. Math. **38** (1985), 883–909.
    S. Venakides, *The Zero Dispersion Limit of the Korteweg-deVries Equation with Periodic Initial Data*, AMS Trans. **301** (1987), 189–225.
    S. Venakides, *Higher Order Lax-Levermore Theory*, Comm. Pure Appl. Math. **43** (1990), 335–362.

[13] G.B. Whitham, *Non-Linear Dispersive Waves*, Proc. Royal Soc. London Ser. A **283** (1965), 238–261.
    G.B. Whitham, *Linear and Nonlinear Waves*, J. Wiley, New York (1974).

[14] V.E. Zakharov and A.B. Shabat, *Exact Theory of Two-Dimensional Self-Focusing and One-Dimensional Self-Modulation of Waves in Nonlinear Media*, Sov. Phys. JETP **34** (1973), 62–69.
    V.E. Zakharov and A.B. Shabat, *Interaction Between Solitons in a Stable Medium*, Sov. Phys. JETP **37(5)** (1973), 823–828.

CRITICAL AND SUBCRITICAL CASES OF THE TODA SHOCK PROBLEM

Spyridon Kamvissis

Courant Institute, New York

I. Introduction

1. Statement and Background of the Problem

In this paper we consider the Toda lattice of interacting particles. Physically, this is a one-dimensional lattice of particles interacting with exponential forces,

$$\dot{x}_n = y_n \tag{I.1.1a}$$

$$\dot{y}_n = e^{x_{n-1} - x_n} - e^{x_n - x_{n+1}}. \tag{I.1.1b}$$

Here x_n is the distance of the nth particle from the origin, and y_n is its velocity. The dot denotes differentiation with respect to time. As is well known ([F];see also [O],[VDO]), these equations are Hamiltonian, (formally) completely integrable, with a Lax-pair representation,

$$\dot{L} = [L, B(L)], \tag{I.1.2}$$

where L and $B(L)$ are the doubly infinite tridiagonal operators defined by

$$(Lf)_n = b_{n-1}f_{n-1} + a_n f_n + b_n f_{n+1},$$

$$(Bf)_n = -b_{n-1}f_{n-1} + b_n f_{n+1}.$$

The variables a_n and b_n are connected to the initial variables x_n and y_n as follows:

$$a_n = -y_n/2 \qquad b_n = \frac{1}{2} exp\left\{\frac{1}{2}(x_n - x_{n+1})\right\} \tag{I.1.3}$$

Writing out (I.1.2) explicitly, we have:

Singular Limits of Dispersive Waves, Edited by
N.M. Ercolani et al., Plenum Press, New York, 1994

$$\dot{a}_n = 2(b_n^2 - b_{n-1}^2) \tag{I.1.4a}$$

$$\dot{b}_n = b_n(a_{n+1} - a_n) \tag{I.1.4b}$$

In particular we consider equations (I.1.4) under initial conditions corresponding to the so-called Toda shock problem:

Suppose we have a distinguished particle driven at fixed velocity $2a$ (where $a > 0$) into a semi-infinite Toda lattice. Numerical experiments ([HoS], [HoFIMcL]) indicate that in the moving frame, the particles eventually settle down into a purely time-periodic motion (which is also asymptotically spatially periodic of period 2), provided $a > 1$, while they eventually come to rest if $0 < a < 1$. Our numerical calculations show that also in the case $a = 1$ the particles eventually come to rest in the moving frame. Recently, Venakides, Deift and Oba have given a complete derivation of the results of [HoFIMcL] for the case $a > 1$ in [VDO]. In this paper we consider the long time behavior of the Toda lattice in the cases $a = 1$ and $0 < a < 1$. Our results are described in section 2 below.

Doubling up the system, the above (non-autonomous) Toda shock problem reduces to an initial value problem for the autonomous doubly infinite Toda lattice, i.e. equations (I.1.4) above, with $n \in \mathbf{Z}$, and with initial conditions:

$$a_0(0) = 0 \tag{I.1.5a}$$

$$a_n(0) = sgn(n) \cdot a \qquad \text{for } n \neq 0 \tag{I.1.5b}$$

$$b_n(0) = \frac{1}{2} \qquad \text{for } n \in \mathbf{Z}. \tag{I.1.5c}$$

Note that from (I.1.4) and (I.1.5), the symmetry relations

$$a_n(t) = -a_{-n}(t) \qquad b_n(t) = b_{-n-1}(t) \tag{I.1.6}$$

follow easily. In terms of x_n and y_n the Toda shock problem is expressed by equations (I.1.1), for $n \in \mathbf{Z}$, together with

$$x_n = 0 \tag{I.1.7a}$$

$$y_n = -sgn(n) \cdot 2a \qquad \text{for } n \neq 0 \tag{I.1.7b}$$

$$y_0 = 0. \tag{I.1.7c}$$

Note that it follows from the above that

$$x_0 \equiv 0 \qquad \text{for } all \ t. \tag{I.1.8}$$

In the case $a > 1$ the spectrum of L (preserved by the flow) consists of two bands $[-a - 1, -a + 1]$ and $[a - 1, a + 1]$ together with an eigenvalue at zero. In the cases $a < 1$ and $a = 1$, the (continuous) spectrum of L consists of one band, the interval $[-a - 1, a + 1]$.

<u>Remark.</u> The initial condition $b_n(0) = \dfrac{1}{2}$ implies that all the particles are initially placed at the same position. It is easy to transform the problem with $b_n(0) = b_0$ for any b_0 into the one with $b_n(0) = \dfrac{1}{2}$.

2. Solution of the Problem

The basic model for the work that follows is the analysis of Lax and Levermore in [LL] of the small dispersion limit of the KdV equation.

(i)   The case $a = 1$.

In this case, the Toda lattice eventually comes to rest, with the a_n's approaching zero and the b_n's approaching the limiting value 1. Thus, under the Toda dynamics the initial lattice configuration is transformed eventually into a different configuration as the time goes to infinity. In fact, the complete picture for large times is the following: Define $N \equiv n/t$. For large values of N, both a_n and $b_n - 1/2$ are exponentially small: the shock is not yet felt. There is a critical value N^* for N (giving the shock front speed) at which the shock first appears. For $N < N^*$, the lattice undergoes decaying oscillations and an explicit formula describing these oscillations is derived. In particular, the rate of decay is of order $1/t$.

The proof proceeds as follows: first one develops the (direct and inverse) scattering theory for operators with nonstandard asymptotics,

$$a_n \to 1 \ \text{ as } \ n \to \infty$$

$$a_n \to -1 \ \text{ as } \ n \to -\infty.$$

The solution of the inverse problem is obtained via a Marchenko type equation, which is then solved to obtain an expression for the solutions of the flow $a_n(t)$ and $b_n(t)$, involving Fredholm determinants. Finally, after some manipulations one obtains a Bargman-Dyson-type formula for b_n as a ratio of infinite series of multiple integrals of real exponentials. These integrals are evaluated asymptotically by solving a maximization problem, with a certain quantum condition, as constraint, in analogy with [V] and [VD0]. Substituting the maximizers into the multiple integrals, we get an expression for the infinite series involving theta functions, which one recognizes as the solution of the 1-gap quasiperiodic (in space) Toda lattice. The moduli of these theta functions vary (slowly) with N and in fact as N approaches zero, the gap closes. Making use of identities involving theta functions we obtain an explicit (and simple when N is small) formula for the answer.

(ii)   The case $a < 1$.

In this case the lattice also eventually comes to rest. The limiting value of the a_n's is zero, while that of the b_n's is $(1 + a)/2$. One distinguishes three different regions:

There is a region (as for $a = 1$) where the shock is not felt yet ($N > N^*$). For intermediate N, the lattice undergoes decaying oscillations. Finally, for small N the oscillations are very small and the lattice settles to rest.

The analysis of this case is similar to the case above, at least in the first two regions. When N is small, the problem is more subtle because the "quantum condition" which usually enables us to capture the oscillations, is ineffective here. In this case, our method computes the limiting values of the a_n's and b_n's to leading order, but without the oscillations.

3. The Dyson Formula and the Variational Problem (for both $0 < a < 1$ and $a = 1$)

We only consider non-negative n. The results for negative n follow immediately from the symmetry relations (I.1.6).

The Dyson formula (I.3.2) is the starting point of our analysis (for a proof see [Ka]). We have assumed that the contribution to $D(\acute{n},t)$ due to the part $[a-1, a+1]$ of the spectrum is negligible. Simplifying assumptions of this type were also made in the calculations of [LL] and [VDO].

We thus have

$$b_n^2 = \frac{D(n+1,t)\, D(n-1,t)}{D(n,t)^2} \tag{I.3.1}$$

where

$$D(n,t) = 1 + \sum_{k=1}^{\infty} I_k \tag{I.3.2}$$

with

$$I_k = \int_{\Omega_k} exp\left\{ t^2 Q_k(\mu_1 \ \cdots \ ,\mu_k;n,t) \right\} d\mu_1 \ \cdots \ d\mu_k \tag{I.3.3}$$

and

$$\Omega_k = \{(\mu_1, \ \cdots \ ,\mu_k)\ \mu_j \in [-a-1, a-1],\ j = 1, \ \cdots \ ,k;\ \mu_1 < \mu_2 < \ \cdots \ < \mu_k\} \tag{I.3.4}$$

Here,

$$Q_k(\mu_1, \ \cdots \ ,\mu_k,n,t) = \sum_{j=1}^{k} \frac{1}{t}[\frac{n}{t}\ln(z^2(\mu_j)) + z(\mu_j) - \frac{1}{z(\mu_j)} + \frac{1}{t}y(\mu_j)]$$

$$+ \sum_{j=1}^{k} \sum_{\substack{i=1 \\ i \neq j}}^{k} \frac{1}{t^2} \ln \mid \frac{z(\mu_i) - z(\mu_j)}{1 - z(\mu_i)z(\mu_j)} \mid \tag{I.3.5}$$

Note that y is a function depending on the reflection coefficient for the particular shock initial data. It will not play any role in the asymptotic analysis. Also

$$z(\mu) = \mu - a - ((\mu - a)^2 - 1)^{\frac{1}{2}} \tag{I.3.6}$$

where the square root is chosen to be negative when $\mu < a - 1$.

We can write (I.3.1) in terms of the variables x_n. From (I.1.3) we can express x_n in terms of the b_n. Taking (I.1.8) into account, we eventually obtain

$$x_n = -2at + \ln \frac{D(n-1,t)}{D(n,t)}. \qquad (I.3.7)$$

We have

$$Q_k(\mu_1, \cdots, \mu_k, n, t) = (f, \psi_k) + (L\psi_k, \psi_k)$$

where

$$f(\mu) = \frac{n}{t}\ln z^2(\mu) + z(\mu) - \frac{1}{z(\mu)},$$

$$L\psi(\lambda) = \frac{1}{\pi}\int_{-a-1}^{a-1} \ln \left| \frac{z(\lambda) - z(\mu)}{1 - z(\lambda)z(\mu)} \right| \psi(\mu)d\mu, \qquad (I.3.8)$$

and

$$\psi_k(\mu) = \frac{\pi}{t}\sum_{j=1}^{k} \delta(\mu - \mu_j). \qquad (I.3.9)$$

Note that

$$\psi \geq 0 \qquad (I.3.10)$$

and

$$\int_{-a-1}^{a-1} \psi(\mu)\,d\mu = \frac{k\pi}{t} \quad \text{for some } k \in \mathbf{Z} \qquad (I.3.11)$$

Remarks:

1. It is proved in [VD0] that L is a negative definite operator.

2. It is clear that as $t \to \infty$, $D(n,t) \sim 1$ unless f is positive. One can easily see that for $\frac{n}{t} > N^* \equiv \frac{\sqrt{a}\sqrt{a+1}}{\ln(\sqrt{a}+\sqrt{a+1})}$, we have $f < 0$, while if $\frac{n}{t} < N^*$, we have $f > 0$. So for $\frac{n}{t} > N^*$, $b_n \sim \frac{1}{2}$, i.e. the effect of the shock is not yet felt. On the other hand for $\frac{n}{t} < N^*$ the analysis of $D(n,t)$ is non-trivial. From now on we restrict ourselves to the case $N \equiv \frac{n}{t} < N^*$. The above discussion shows in particular that the speed of the shock front is

$$N^* = \frac{\sqrt{a}\,\sqrt{a+1}}{\ln(\sqrt{a}+\sqrt{a+1})} \qquad (I.3.12)$$

Each I_k is a Laplace integral which can be evaluated asymptotically by a Laplace method. In the spirit of [LL] and [VD0] we approximate each ψ_k by a non-negative $\psi \in L^1[-a-1, a-1]$, in the limit as $t \to \infty$.

Theorem I.3.1. As $t \to \infty$,

$$I_k = exp\left(\frac{t^2}{\pi} \max_{\psi \in A_k} Q(\psi) + F_k\right) \tag{I.3.13}$$

where

$$Q(\psi) = (f, \psi) + (L\psi, \psi) \tag{I.3.14}$$

and A_k is the set of all $L^1[-a-1, a-1]$ non-negative functions for which

$$\int_{a-1}^{a-1} \psi(\mu)\, d\mu = \frac{k\pi}{t}, \quad k \in \mathbf{Z}. \tag{I.3.15}$$

The 'error' F_k is a function of n, t such that, for any given $\delta > 0$, when $N > \delta$,

$$F_k(n,t) = F(n,t) + (\alpha(N))\, k \tag{I.3.16}$$

where α is a continuous function of $N = \dfrac{n}{t}$ and F is such that:

$$F(n-1,t) - F(n,t) \quad approaches \ a \ constant \ at \ large \ times. \tag{I.3.16a}$$

Furthermore, it is easy to see that setting $\alpha(N) = 0$ will only change our final answer upto a phase difference (cf. [VDO] section 1). From now on, we assume that indeed $\alpha(N) = 0$.

For a justification of Theorem I.3.1, see [VD0].

Thus the problem of evaluating b_n asymptotically is reduced to solving the following maximization problem:

$$\text{maximize } Q(\psi) \quad \text{for } \psi \in A_k. \tag{I.3.17}$$

Writing

$$\psi = \psi^* + \frac{1}{t}\bar{\psi} + \frac{1}{t^2}\bar{\bar{\psi}} + \cdots$$

with $\psi^*, \bar{\psi}, \bar{\bar{\psi}}, \cdots$ independent of t, we obtain

$$Q(\psi) = (f, \psi^*) + (L\psi^*, \psi^*) + \frac{1}{t}(f + 2L\psi^*, \bar{\psi})$$

$$+ \frac{1}{t^2}((f + 2L\psi^*, \bar{\bar{\psi}}) + (L\bar{\psi}, \bar{\psi}))$$

$$+ \frac{1}{t^3}((f + 2L\psi^*, \bar{\bar{\bar{\psi}}}) + (L\bar{\bar{\psi}}, \bar{\bar{\psi}})) + \cdots$$

The procedure is as follows:

(i) *Leading Order Maximization*

We maximize $(f, \psi^*) + (L\psi^*, \psi^*)$ with respect to the condition $\psi^* \geq 0$. Note that the condition (I.3.14) is a higher order condition, so it does not affect the leading order maximization.

The variational condition for this problem is

$$f(\lambda) + 2L\psi^*(\lambda) = 0 \qquad \text{if} \quad \psi^* > 0 \tag{I.3.18a}$$

$$f(\lambda) + 2L\psi^*(\lambda) \le 0 \qquad \text{if} \quad \psi^* = 0 \tag{I.3.18b}$$

Calculations below (see section II.1) show that the inequality in (I.3.18b) is strict:

$$f(\lambda) + 2L\psi^*(\lambda) < 0 \qquad \text{if} \quad \psi^* = 0 \tag{I.3.18c}$$

(ii) *First Order Maximization*

We maximize $(f + 2L\psi^*, \overline{\psi})$ with respect to $\overline{\psi}$, where ψ^* is the solution of the previous problem, subject to

$$\overline{\psi}(\lambda) \ge 0 \quad \text{if} \quad \psi^* = 0$$

and

$$\int_{a-1}^{a-1} (\psi^* + \frac{1}{t}\overline{\psi})d\mu = k\frac{\pi}{t} \; , \; k \in \mathbf{Z} \tag{I.3.19}$$

Since $(f + 2L\psi^*, \overline{\psi}) \le 0$, we see by (I.3.18c) that its maximum is zero and is obtained if

$$\overline{\psi} = 0 \quad \text{when} \quad \psi^* = 0 \tag{I.3.20}$$

(iii) *Second Order Maximization*

We still need to find $\overline{\psi}$, if $\psi^* > 0$. By (I.3.20), $\overline{\overline{\psi}} \ge 0$ if $\psi^* = 0$. The second order maximization now implies as above that $\overline{\overline{\psi}} = 0$ when $\psi^* = 0$. In particular $(f + 2L\psi^*, \overline{\overline{\psi}}) = 0$. To complete the second order maximization we need to maximize $(L\overline{\psi}, \overline{\psi})$ under the constraints

$$supp\,\overline{\psi} \subseteq supp\psi^*$$

and

$$\int_{a-1}^{a-1} (\psi^* + \frac{1}{t}\overline{\psi}) \, d\mu = \frac{k\pi}{t} \; , \; k \in \mathbf{Z} \tag{I.3.21}$$

(iv) *Higher Order Maximization*

As before, $\psi^* = 0$ implies $\dot{\overline{\psi}} = 0$, and $(f + 2L\psi^*, \overline{\overline{\psi}}) = 0$. Maximizing $(L\overline{\overline{\psi}}, \overline{\overline{\psi}})$ we find $\overline{\overline{\psi}} = 0$. Similarly all higher order terms are zero. We combine all the above to obtain

$$\max_{\psi \in A_k} Q(\psi) = \frac{1}{2}(f, \psi^*) + \frac{1}{t^2}(L\overline{\psi}, \overline{\psi}) \tag{I.3.22}$$

So, writing $\overline{\psi}_k = \overline{\psi}$,

$$I_k = exp\,(\frac{t^2}{2\pi}(f, \psi^*) + \frac{1}{\pi}(L\overline{\psi}_k, \overline{\psi}_k) + F_k) \tag{I.3.23}$$

263

For the definition and the properties of F_k, see equations (I.3.16), and (I.3.16a) above.

II. The Case $a = 1$

1. Solution of the Variational Problems

We begin with the solution of the leading order maximization problem. We only need to produce one solution ψ^* of the variational condition (I.3.18). Then this will solve the leading order maximization problem *uniquely*, since L is negative definite and hence Q is strictly convex. Note that for $a = 1$, $[-a-1, a-1] = [-2, 0]$.

Theorem II.1.1 The maximizer ψ^* of the functional $Q(\psi) = (f, \psi) + (L\psi, \psi)$ such that $\psi^* \geq 0$ and $\psi^* \in L^1[-2, 0]$ is given as follows. In the region $N \equiv \frac{n}{t} \in [0, N^*]$ we have

$$\psi^*(\lambda) = -i \frac{P(\lambda)}{R(\lambda)} \qquad \lambda \in [-2, \gamma]$$

$$\psi^*(\lambda) = 0 \qquad \lambda \notin [-2, \gamma] \tag{II.1.1}$$

where

$$R(\lambda) = ((\lambda + 2)(\lambda - \gamma) \lambda (\lambda - 2))^{1/2} \tag{II.1.2}$$

with $R(\lambda)$ chosen to be positive when $\lambda \to \infty$, and

$$P(\lambda) = (\lambda - \gamma)(\lambda + N + \gamma/2). \tag{II.1.3}$$

Here γ is uniquely defined by

$$\int_\gamma^0 \frac{P(\lambda)}{R(\lambda)} \, d\lambda = 0. \tag{II.1.4}$$

In particular, as $N \to 0, \gamma \to 0$ also.

For the proof of Theorem II.1.1 see [Ka].

The theorem above shows how the support of ψ^* varies with N: for $N > N^*$, *supp* $\psi^* = \varnothing$. In this case (see discussion in section I.3) the perturbation of the b_n's is exponentially small and the effect of the shock is not yet felt. For $N < N^*$, *supp* $\psi^* = [-2, \gamma(N)]$, while as $N \to 0$, *supp* ψ^* tends to cover all of $[-2, 0]$. As we shall see later the motion of the lattice will be identified with an algebro-geometric solution corresponding to the continuous spectrum $[-2, \gamma(N)] \cup [0, 2]$. For a fixed N this would be a time-periodic oscillatory motion. As N gets small, that is as the gap $[\gamma(N), 0]$ closes, we see decaying oscillations instead.

It is now easy to calculate (f, ψ^*). We get

$$(f, \psi^*) = AN^2 + BN + C, \tag{II.1.5}$$

where

$$A = -i\int_{-2}^{\gamma} \frac{(\lambda - \gamma)^{\frac{1}{2}} \ln z^2}{((\lambda + 2)\lambda(\lambda - 2))^{\frac{1}{2}}} \, d\lambda \qquad (\text{II.1.6})$$

$$B = -i\int_{-2}^{\gamma} \frac{(\lambda - \gamma)^{\frac{1}{2}} (z - 1/z)}{((\lambda + 2)\lambda(\lambda - 2))^{\frac{1}{2}}} \, d\lambda - i\int_{-2}^{\gamma} \frac{(\lambda - \gamma)^{\frac{1}{2}} \ln z^2 (\lambda + \gamma/2)}{((\lambda + 2)\lambda(\lambda - 2))^{1/2}} \, d\lambda \qquad (\text{II.1.7})$$

$$C = -i\int_{-2}^{\gamma} \frac{(\lambda - \gamma)^{\frac{1}{2}} (z - \frac{1}{z}) (\lambda + \gamma/2)}{((\lambda + 2)\lambda(\lambda - 2))^{\frac{1}{2}}} \, d\lambda \qquad (\text{II.1.8})$$

Each of the quantities A, B, C is a function of γ and hence of N, and can be explicitly expressed in terms of elliptic integrals and elliptic functions. If N is small, we can say even more. From (II.1.4) we know that as $\gamma \to 0$, $N \to 0$. An easy computation shows that

$$\gamma = -\frac{4}{3} N + 0(N^2), \qquad (\text{II.1.9})$$

and expressions for A, B, C are

$$A = 2\ln 2\pi + \frac{\pi^2}{3}N + 0(N^2) \qquad (\text{II.1.10})$$

$$B = -4\pi + \frac{2\pi}{3} N - \frac{\pi^2}{3} N^2 + 0 (N^3) \qquad (\text{II.1.11})$$

$$C = -\frac{2\pi}{3} N^2 + \frac{3\pi}{2}. \qquad (\text{II.1.12})$$

Hence, by (II.1.5), as $N \to 0$,

$$\frac{1}{\pi}(f, \psi^*) = \frac{3}{2} - 4N + 2\ln 2N^2 + 0(N^4). \qquad (\text{II.1.13})$$

Next, we obtain the solutions $\overline{\psi}_k$ of the second order maximization problem. We need to maximize $(L\overline{\psi}_k, \overline{\psi}_k)$ under condition (I.3.15).

Theorem II.1.2: The maximizer $\overline{\psi}_k$ of the functional $(L\psi, \psi)$ such that

$$supp \, \overline{\psi}_k \subseteq supp \, \psi^* ,$$

$$\overline{\psi}_k \in L^1 ,$$

$$\int_{-2}^{0} (\psi^* + \frac{1}{l}\overline{\psi}_k) dy = \frac{k\pi}{l}$$

is given as follows:

$$\overline{\psi}_k(\lambda) = \frac{-iE(k)}{R(\lambda)} \qquad \text{for } \lambda \in supp \, \psi^*$$

$$= 0 \qquad \text{for } \lambda \notin supp \, \psi^* \qquad (\text{II.1.14})$$

where $R(\lambda)$ is defined by (II.1.2) and $E(k)$ is determined by (I.3.15):

265

$$\frac{iE(k)}{\pi} \int_{-2}^{\gamma} \frac{d\lambda}{R(\lambda)} = \sigma t - k \qquad (II.1.15)$$

where

$$\sigma = \frac{1}{\pi} \int_{-2}^{\gamma} \psi^* (\lambda) \, d\lambda. \qquad (II.1.16)$$

Again we omit the proof, which can be found in [Ka].

It is now very easy to compute $(L\overline{\psi}_k, \overline{\psi}_k)$. We have

$$L\overline{\psi}_k(\lambda) = E_k \int_{\gamma}^{0} \frac{d\mu}{R(\mu)} \qquad \lambda \in [-2, \gamma] \qquad (II.1.17)$$

$$= E_k \int_{\lambda}^{0} \frac{d\mu}{R(\mu)} \qquad \lambda \notin [-2, \gamma]$$

From (II.1.14), (II.1.15) , (II.1.17) we finally get

$$(L\overline{\psi}_k, \overline{\psi}_k) = i\tau\pi^2(\sigma t - k)^2 \qquad (II.1.18)$$

where

$$\tau = \frac{\int_{\gamma}^{0} \dfrac{d\lambda}{R(\lambda)}}{\int_{-2}^{\gamma} \dfrac{d\lambda}{R(\lambda)}}. \qquad (II.1.19)$$

Note that $i\tau < 0$, so that $(L\overline{\psi}_k, \overline{\psi}_k) < 0$. We then substitute these formulae back into (I.3.22) to get a formula for I_k.

$$I_k = exp(\frac{t^2}{2\pi}(f, \psi^*) + \frac{1}{\pi}(L\overline{\psi}_k, \overline{\psi}_k) + F_k)$$

$$= exp\left\{ \frac{t^2}{2}(\frac{3}{2} - 4N + 2\ln 2 \, N^2 + 0(N^4)) + F_k \right\}$$

$$\cdot exp\left\{ i\pi\tau(N)(\sigma(N)\, t - k)^2 \right\}, \qquad (II.1.20)$$

where $\tau(N)$ is given by (II.1.19) and $\sigma(N)$ is given by (II.1.16).

By the discussion following the statement of Theorem I.3.1, we see that, as long as we restrict ourselves in the region $\delta < N < N^*$, then F_k can be taken as independent of k (such an assumption will only alter the phase ε_0 in the formulae appearing in the statement of Theorem II.2.1.). Hence, the factor $exp(F_k)$ will drop off later when we consider the ratio $\dfrac{D(n-1,t)}{D(n,t)}$. Thus, we can ignore F_k altogether.

Recall that $D(n,t) = 1 + \sum_{k=1}^{k=\infty} I_k$. We may write

$$D(n,t) \sim \sum_{-\infty}^{\infty} I_k \qquad (II.1.21)$$

266

where I_k is defined by (II.1.20). The resulting error will be exponentially small since the only terms contributing to (II.1.35) are those for which $k \sim \sigma(N)t$, in which case k is certainly positive, and O(t). From (II.1.20) and (II.1.21) we obtain, for small N,

$$D(n,t) = exp\left\{ \frac{t^2}{2} \left(\frac{3}{2} - 4N + 2\ln 2 N^2 + O(N^4) \right) \right\} \cdot$$

$$\sum_{k=-\infty}^{\infty} exp\left\{ i\pi\tau(N) \left(\sigma(N)t - k \right)^2 \right\} . \tag{II.1.22}$$

In general, for $N < N^*$ not necessarily small, we of course have similar expressions

$$D(n,t) = exp\left\{ \frac{t^2}{2} (A(N)N^2 + B(N)N + C(N)) \right\} \cdot$$

$$\sum_{k=-\infty}^{\infty} exp\left\{ i\pi\tau(N) \left(\sigma(N)t - k \right)^2 \right\} \tag{II.1.23}$$

where $A(N), B(N), C(N)$ are still defined by (II.1.6),(II.1.7),(II.1.8), and can be written in terms of elliptic integrals.

2. Identification with an Algebro-geometric Solution of Genus 1

Let $\theta(z \mid \tau) = \sum_{k=-\infty}^{\infty} exp(i\pi\tau k^2) exp(2\pi ikz)$ be the standard theta function. Then

$$\sum_{k=-\infty}^{\infty} exp(i\pi\tau(\sigma t - k)^2) = exp(i\pi\tau\sigma^2 t^2) \theta(\sigma t \mid \tau)$$

Equation (II.1.23) now becomes

$$D(n,t) = exp\left\{ \frac{t^2}{2} (A(N)N^2 + B(N)N + C(N)) \right\} \cdot$$

$$exp(i\pi\tau(N) (\sigma(N))^2 t^2)\theta(\sigma(N)\tau(N)t \mid \tau(N)). \tag{II.2.1}$$

When N is small,

$$D(n,t) = exp\left\{ \frac{t^2}{2} \left(\frac{3}{2} - 4N + 2\ln 2 N^2 + O(N^4) \right) \right\} \cdot$$

$$exp(i\pi\tau(N) (\sigma(N))^2 t^2)\theta(\sigma(N)\tau(N)t \mid \tau(N)). \tag{II.2.2}$$

On the other hand, Jacobi's transformation gives

$$exp(i\pi\tau\sigma^2 t^2) \theta(\sigma t \mid \tau) = \sqrt{\frac{i}{\tau}} \theta(\sigma t \mid -\frac{1}{\tau}).$$

Hence, for N small,

$$D(n,t) = exp\left\{ \frac{t^2}{2} \left(\frac{3}{2} - 4N + 2ln\,2\,N^2 + 0(N^4) \right) \right\} \cdot$$

$$\sqrt{\frac{i}{\tau(N)}}\ \theta(\sigma(N)t\,|-\frac{1}{\tau(N)}). \tag{II.2.3}$$

We obtain an additional expression for $\sigma(N)$ by use of the Riemann bilinear relations.

Proposition II.2.1.

$$\sigma(N) = \left[\int_\gamma^0 \frac{d\lambda}{R(\lambda)} \right]^{-1} \left[1 - N\int_2^\infty \frac{d\lambda}{R(\lambda)} \right]. \tag{II.2.4}$$

For the proof, see [Ka].

When N is small, the asymptotic evaluation of (II.2.4) shows that

$$\sigma = \frac{2}{\pi} + 0(N), \tag{II.2.5}$$

and similarly, from (II.1.19)

$$\tau = i\frac{\pi}{2lnN} + 0(N). \tag{II.2.6}$$

From (II.2.4) we obtain

$$\sigma(N)t = \frac{1}{\int_\gamma^0 \frac{d\lambda}{R(\lambda)}}t - \frac{\int_2^\infty \frac{d\lambda}{R(\lambda)}}{\int_\gamma^0 \frac{d\lambda}{R(\lambda)}}n. \tag{II.2.7}$$

Each of the integrals on the RHS depends on γ and hence on N. Consider

$$\Theta(N,n,t) \equiv \theta\left[\frac{1}{\int_\gamma^0 \frac{d\lambda}{R(\lambda)}}t - \frac{\int_2^\infty \frac{d\lambda}{R(\lambda)}}{\int_\gamma^0 \frac{d\lambda}{R(\lambda)}}n\ \middle|-\frac{1}{\tau(N)} \right]. \tag{II.2.8}$$

Detailed computations show that

$$\Theta(N - \frac{1}{t},n - 1,t) = \Theta(N,n - 1,t) + 0(N) \tag{II.2.9}$$

uniformly in n,t. Thus, using (I.3.2), (II.2.3), (II.2.7) and (II.2.8), we easily compute for N small:

$$x_n = (1 - 2\,n)\,ln\,2 + ln(\frac{\Theta(N,n - 1,t)}{\Theta(N,n,t)}) + 0\,(N^2), \tag{II.2.10}$$

while using (I.1.3) and (II.2.10) we obtain

$$b_n^2 = \frac{\Theta(N,n+1,t)\ \Theta(N,n-1,t)}{(\Theta(N,n,t))^2} + 0(N^2). \tag{II.2.11}$$

We can now identify the solution of our problem with a modulated algebro-geometric solution.

Proposition II.2.2. Let $X(N,n,t) = (1 - 2n) \ln 2 + \ln(\dfrac{\Theta(N,n-1,t)}{\Theta(N,n,t)})$. Then $X(N,n,t)$ considered as a function of n,t depending on the parameter N, solves the Toda equations (I.1.1) (together with $Y(N,n,t) = \dot{X}(N,n,t)$). Similarly, let $B(N,n,t) = \dfrac{\Theta(N,n+1,t)\,\Theta(N,n-1,t)}{(\Theta(N,n,t))^2}$ and $A(N,n,t) = -\dfrac{1}{2}\dot{X}(N,n,t)$. Then A and B solve the Toda equations (I.1.4a,b). In fact they form a one-gap algebro-geometric solution coming from a Riemann surface of genus 1.

Proof: See [Ka].

Theorem II.2.1. Let δ be any small positive number. Then, in the region $\delta < N$, for N small, we have as $t \rightarrow \infty$,

$$x_n(t) = -2\ln 2\, n + \frac{1}{3}\,\frac{n}{t}\,(1 - \cos(4t - \pi n + \varepsilon_0)) + O(N^2), \qquad (\mathrm{II.2.12})$$

$$b_n(t) = 1 + \frac{1}{3}\,\frac{n}{t}\,\cos(4t - \pi n + \varepsilon_0) + O(N^2), \qquad (\mathrm{II.2.13})$$

$$a_n(t) = \frac{2}{3}\,\frac{n}{t}\,\sin(4t - \pi n + \varepsilon_0) + O(N^2), \qquad (\mathrm{II.2.14})$$

where ε_0 is some constant.

Proof: Follows easily from formulae (II.2.10) and (II.2.11) (see [Ka]) .

Using arguments similar to the above we can extract $a_n(t)$ and $b_n(t)$ for any $N<N^*, N>\delta$, as $t \rightarrow \infty$. For example,

$$a_n(t) = \chi(\frac{n}{t})\, sn\, (\omega(\frac{n}{t})t + \varepsilon(\frac{n}{t})) \qquad (\mathrm{II.2.15})$$

where $\chi, \omega, \varepsilon$, and the modulus of sn are given by expressions involving elliptic integrals.

Numerical computations suggest that for any fixed n, Theorem II.2.1 in fact holds for any large t, provided n is not too small. Equations (II.2.12), (II.2.13) and (II.2.14) then express the behavior of a fixed (the nth) particle for large times. On the other hand, using the above method we can obtain the leading asymptotics of the lattice even for $N<\delta$ (this is obvious, since the quantum condition is not necessary for the leading order behavior).

III. The Case $0 < a < 1$

In the subcritical case where $0 < a < 1$, we have $spec(L(t)) = [-a - 1, a + 1]$ where in fact the part $[a - 1, -a + 1]$ has multiplicity 2 with the rest of the spectrum having multiplicity 1.

We still have to solve the maximization problem described in section I.3 and thus a corresponding variational problem.

Theorem II.1.1. still holds with a few changes as follows:

$$\psi^* \in L^1[-a-1, a-1], \tag{III.1}$$

and

$$R(\lambda) = ((\lambda + a + 1)(\lambda - \gamma)(\lambda - a + 1)(\lambda - a - 1))^{1/2}. \tag{III.2}$$

Here γ is uniquely defined by

$$\int_{\gamma}^{a-1} \frac{P(\lambda)}{R(\lambda)} \, d\lambda = 0. \tag{III.3}$$

The proof goes through as in section II.1. The difference is now the following: while in the case where $a = 1$, we see that as $\gamma \to 0$ (i.e. as the gap $[\gamma, 0]$ closes) we have $N \to 0$, in the case where $a < 1$ the gap closes at a particular non-zero value for N, say N_*. We have

$$N = -\frac{\displaystyle\int_{\gamma}^{a-1} \frac{(\lambda - \gamma)^{1/2}(\lambda + \gamma/2)}{((\lambda + a + 1)(\lambda - a + 1)(\lambda - a - 1))^{1/2}} \, d\lambda}{\displaystyle\int_{\gamma}^{a-1} \frac{(\lambda - \gamma)^{1/2}}{((\lambda + a + 1)(\lambda - a + 1)(\lambda - a - 1))^{1/2}} \, d\lambda} \tag{III.4}$$

and as $\gamma \to 0$,

$$N \to N_* \equiv 1 - a. \tag{III.5}$$

This means that the discussion in Chapter II is valid only for $1 - a < N < N^*$. We can thus obtain an asymptotic expression for the solution analogous to (II.2.15) with appropriate expressions for $\chi, \omega, \varepsilon$, in that region. For $N < 1 - a$ the above analysis fails. The quantum condition is no more relevant and we cannot capture the oscillations that (as we know from numerical experiments) still occur. However, we are still able to obtain the leading order behavior of the lattice.

<u>Theorem III.1.</u> $a_n \to 0$ and $b_n \to \frac{1}{2}(a+1)$ as $t \to \infty$.

<u>Proof</u> See [Ka].

Concluding remarks

An alternative approach to the Toda shock problem is undertaken in one of the papers of this volume by A. Bloch and Y. Kodama, based on the analysis of the Whitham equations for the Toda lattice. In particular, they are able to specify the different regions in which the behavior of the solution is different, they calculate the shock front velocity, and they derive independently formulas (1.3.12) and (III.5). It seems though that a complete and rigorous proof of such results (especially the oscillations in the region $N < 1-a$, for $a < 1$) needs a much more powerful method. In fact, it seems that the method of Deift and Zhou (also described in a paper in this volume) can provide such refinements.

Acknowledgements

The author wishes to thank Percy A. Deift and Stephanos Venakides for helpful discussions. This work was partially supported by a Fellowship from the Onassis Foundation of Greece.

References

[DT] E. Date, S. Tanaka, Periodic Multi-Soliton Solutions of KdV Equation and Toda Lattice, Progr. Theor. Physics Suppl. 53, p.107 (1976)

[Dy] F. Dyson, Old and New Approaches to the Inverse Scattering Problem, Essays in Honor of Valentine Bargmann, ed. E. H. Lieb, B. Simon and A. S. Wightman, Princeton, Studies in Math. Physics (1976)

[F] H. Flaschka, On the Toda Lattice, I, Physical Review 9, 1974, 1924-5: II, Progress of Theoretical Physics, v. 51 n.3 p.703 (1974)

[HoFlMcL] B. L. Holian, H. Flaschka, D. W. McLaughlin, Shock Waves in the Toda Lattice: Analysis Phys. Rev. A24 (1981) pp. 2595-2623

[HoS] B. L. Holian, G. K. Straub, Physical Review B 18, 1593 (1978)

[Ka] S. Kamvissis, On the Long Time Behavior of the Doubly Infinite Toda Lattice under Shock Initial Data, Dissertation, Courant Institute 1991

[LL] P. D. Lax, C. D. Levermore, The Small Dispersion Limit of the KdV Equation, I , Communications in Pure and Applied Mathematics, v. 36, pp. 253-290

[O] R. Oba, Doubly Infinite Toda Lattice with Antisymmetric Asymptotics, Dissertation, Courant Institute 1988

[T] M. Toda, Theory of Nonlinear Lattices, Springer 1980

[VDO] S. Venakides, P. Deift, R. Oba, The Toda Shock Problem, to appear in Communications in Pure and Applied Mathematics (December 1991)

[V] S. Venakides, Higher Order Lax-Levermore Theory, I, Communications in Pure and Applied Mathematics, v. 43 , pp. 335-362

GEOMETRIC PHASES AND MONODROMY AT SINGULARITIES

Mark S. Alber[1] and Jerrold E. Marsden[2]

[1]Department of Mathematics
University of Notre Dame
Notre Dame, IN 46556, USA
[2]Department of Mathematics
University of California
Berkeley, CA 94720, USA

1 Introduction

In [14, 25, 22, 28, 40] and [3, 4] methods of the complex analysis and algebraic geometry were applied for investigating the geometry of the phase spaces of the nonlinear systems.

In particular, some new (complex) Hamiltonian soliton and homoclinic structures, phase functions and geometric phases for soliton equations were introduced. In particular, phase phenomena caused by the presence of monodromy at singularities in the space of parameters was studied [3].

Phase functions in the quasi-periodic case have important applications in the theory of evolution equations (where they appear in connection with the Whitham equations). It is natural to investigate them in the singular (soliton, umbilic soliton and homoclinic) cases. In this paper we investigate the umbilic and homoclinic solutions in this context considering them as singular points in the moduli spaces of Jacobi varieties.

Our goal is to introduce geometric models for the new soliton and homoclinic so-

Singular Limits of Dispersive Waves, Edited by
N.M. Ercolani et al., Plenum Press, New York, 1994

lutions of the nonlinear PDE's and then to apply methods of the Riemannian sufraces and asymptotic reduction for their investigation.

In particular, we establish a link between classical problem of umbilic geodesics on quadrics from Riemannian geometry and new soliton-like and homoclinic solutions of the nonlinear equations. Umbilic solutions differ from the usual solitons in that their spectral polynomial (the basic polynomial associated with the Riemann surface) has only positive roots, while they are all negative in the usual case. These umbilic solitons provide an n-dimensional geometric model for homoclinic orbits that approach low dimensional tori instead of homoclinic points as $t \to \pm\infty$.

In the process of investigating the new angle-representations for the homoclinic orbits and the umbilic soliton-like solutions of the nonlinear equations, we deal in a natural way with the phenomena of geometric phases. Recall that in [11], Berry considered a geometric phase factor $\exp(i\gamma)$ for systems that are slowly (adiabatically) transported along closed curves in a space of parameters. In [41] a class of connections was constructed to obtain expressions for the Hannay-Berry phases [13] (these are geometric angle shifts in the classical case) for some integrable problems in terms of the nontrivial holonomy of these connections. In [41], Montgomery gave an example of a phase associated with the presence of singularities in the case of a flat connection. Symmetry and reduction were used to obtain a generalization of geometric phases to the non-integrable case in the form of the holonomy of the Cartan-Hannay-Berry connection [37].

All of the papers mentioned above deal with problems defined on the compact invariant varieties in the phase space. In [3], we introduced angle-representations on associated noncompact Jacobi varieties and used asymptotic reduction to obtain phase functions on the corresponding topologically nontrivial level sets in the phase space. This yields soliton symplectic structure and geometric phases and provides a setting for investigating the effect of slow evolution of the solutions of the nonlinear equations in terms of geometric asymptotics (*i.e.*, semiclassical modes).

In the context of phases one usually puts a given system in action-angle form and considers a shift in the angle variables $\theta = (\theta_1, ..., \theta_n)$ after transporting the system along a closed curve in the space of parameters. This change consists of two parts,

$$\triangle\theta = \triangle_D\theta + \triangle_G\theta$$

called the dynamic and geometric phases. The dynamic phase is due to the dynamical evolution of the system and it is proportional to the period of time (T) during which system is transported along the closed curve. To eliminate the dynamical phase and to retain only the geometric part, the method of averaging is usually used. In the soliton case, the period is infinite and so we use the method of "asymptotic reduction" and the complex phase function instead of the averaging approach.

Our method uses the fact that asymptotic reduction leads to the complex splitting of the spectrum of the soliton problem. We show that phases are obtained in terms of the monodromy at singularities of the phase function, which does not depend on the initial conditions. A connection between the θ-function and τ-function can also be used to find a link between soliton geometric phases and geometric phases for quasi-periodic solutions (see [3] and [6]).

The limiting process and τ-functions on the Jacobi tori were investigated in detail by McKean [39] for the KdV equation. Complex geometry related to the θ-functions was previously studied in KdV case by McKean and Ercolani [39, 24]. The modulational Poisson structure [25] for the sine-Gordon system was derived in terms of conformal ingredients such as differentials on Riemann surfaces and θ-functions and a possible link with the Hamiltonian theory was investigated. On the other hand a connection was established (see [46]) between modulation theory [29] and the small dispersion limit [36]. Moreover, the θ-function of the periodic theory was obtained [47] in the form of a singular limit of the Dyson determinant.

This paper is organized as follows. In §2 we introduce the notion of umbilic solitons and in §3 we recall some of the general methods used in our approach. In §4 we apply these methods to the case of umbilic solitons and solitons with a quasi-periodic background. In §5, we show that the Hamiltonian flow associated with the homoclinic orbits introduced by Devaney [18] for the C. Neumann problem [42] coincides with the soliton x-flow of the KdV equation. This result clarifies a connection between solitons and homoclinic orbits and leads to the introduction of homoclinic geometric phases.

Our soliton Hamiltonians and symplectic structure lead to the introduction of presoliton geometric asymptotics and using the results described above, we obtain in §6 a link with geometric phases in the quantum case. Some links with the geometric quantization of Hamiltonians which are not quadratic in P are pointed out as well.

2   Geometry of the Soliton Solutions

In this section, we construct a class of soliton-like solutions of the KdV and defocusing (d)NLS equations using a singular family of umbilic geodesics on n-dimensional quadrics.

2.1   Umbilic Geodesics

It is known [38, 46, 31, 8] that there are finite dimensional invariant tori in the phase space of completely integrable nonlinear problems and that solutions lying on these tori (*i.e.*, on a hyperelliptic Riemann surface), which are called quasi-periodic solutions, can be described using a pair of commuting Hamiltonian systems written in configuration variables $(\mu_1, ..., \mu_n)$ and momentum variables $(P_1, ..., P_n)$. (For details about geometry of the μ-representations in nonsingular case see [22, 24]).

To obtain the case of solitons of the KdV equation, one shrinks pairs of roots of the basic polynomial of the Riemannian surface [39, 43, 1]. In this paper we generalize this procedure to the problem of geodesics and make use of the link between the problem of geodesics [9] and the KdV equation to obtain Hamiltonians and angle-representations for the umbilic solitons.

We start with the definition of the umbilic points on the 2-dimensional surface.

Definition 2.1 *Let K_1 and K_2 be the largest and smallest principal curvatures of the surface*

$$M : W \to R^3.$$

A point $p_0 \in W$ is called an umbilic if $K_1(p_0) = K_2(p_0)$.

It can be shown that in case of 2-dimensional ellipsoid [34] there are exactly 4 such points. At the same time there is a special family of umbilic geodesics on the ellipsoid going through pairs of umbilic points. They correspond to a special choice of the first integral of the quasi-periodic Hamiltonian flow. Namely, first integral should coinside with the length of the medium semiaxes of the ellipsoid. Investigation of the umbilic geodesics on the 2-dimensional quadrics is one of the classical problems of the Riemannian geometry. In what follows we obtain umbilic Hamiltonians and construct a complete set of the umbilic angle-variables. Then we generalize the notion of the umbilic geodesics to the n-dimensional case. This is important since there is a direct link between generalised Jacobi problem of geodesics on n-dimentional quadrics and many nonlinear integrable problems including KdV equation.

2.2 Umbilic Solitons

Consider a general family of quasi-periodic geodesics on the n-dimensional ellipsoid

$$\sum_{j=1}^{n+1} \frac{X_j^2}{l_j} = 1, \quad 0 < l_{n+1} < ... < l_1,$$

which is described (see [8]) in terms of the root-variables $\{\mu_j\}$ as a solution of the system of equations

$$\frac{\partial \mu_j}{\partial x} = \frac{L_0}{\prod_{i \neq j}(\mu_j - \mu_i)} \sqrt{\frac{\prod_{k=1,k \neq j_0}^{n}(\mu_j - m_k)\prod_{r=1}^{n+1}(\mu_j - l_r)}{(-\mu_j)}}, \quad j = 1, ..., n. \tag{2.1}$$

Here m_j and L_0, where

$$l_{j+1} < m_j < l_j, \quad j \neq j_0, \quad 1 \leq j \leq n,$$

are constants along the solutions of the system (2.1).

Instead of using curvature in the definition of umbilics, we consider μ-representation, solution of the system of differential equations (2.1).

To obtain different families of umbilic geodesics, we consider the limiting process

$$m_k \rightarrow l_{k+1} = b_k, \quad k = 1, ..., (j_0 - 1), \tag{2.2}$$

$$m_k \rightarrow l_k = b_k, \quad k = (j_0 + 1), ..., n, \quad l_{n+1} = b_{n+1}, \tag{2.3}$$

which results in the systems of the equations

$$\mu_j' = \frac{L_0 \sqrt{(\mu_j - l_1)(\mu_j - l_{n+1})} \prod_{r=1,r \neq j_0}^{n}(\mu_j - b_r)}{\sqrt{-\mu_j} \prod_{i \neq j}(\mu_j - \mu_i)}, \quad j = 1, ..., n. \tag{2.4}$$

corresponding to different choices of j_0, where $(1 \leq j_0 \leq n)$. This system yields the following umbilic angle-representations

$$\left. \begin{array}{l} \displaystyle\sum_{j=1}^{n} \frac{\sqrt{-\mu_j}\, \mu_j'}{(\mu_j - b_k)\sqrt{(\mu_j - l_1)(\mu_j - l_{n+1})}} = \frac{L_0 \prod_{r=1,r \neq j_0, r \neq k}^{n}(\mu_j - b_r)}{\prod_{i \neq j}(\mu_j - \mu_i)}, \quad k \neq j_0 \\[6mm] \displaystyle\sum_{j=1}^{n} \frac{\sqrt{-\mu_j}\, \mu_j'}{L_0\sqrt{(\mu_j - l_1)(\mu_j - l_{n+1})}} = \frac{\prod_{r=1,r \neq j_0}^{n}(\mu_j - b_r)}{\prod_{i \neq j}(\mu_j - \mu_i)}, \quad k = j_0. \end{array} \right\} \tag{2.5}$$

The right hand sides of these expressions are constants

$$\theta_k' = 0, \quad k = 1, ..., n, \quad k \neq j_0; \quad \theta_{j_0}' = 1, \quad k = j_0. \tag{2.6}$$

which, after integration, results in the angle-representation

$$\left.\begin{aligned}
\theta_k &= \sum_{j=1}^{n} \int_{\mu_j^0}^{\mu_j} \frac{\sqrt{-\mu_j}\, d\mu_j}{(\mu_j - b_k)\sqrt{(\mu_j - l_1)(\mu_j - l_{n+1})}} = \theta_k^0, \quad k = 1, ..., n; \quad k \neq j_0. \\[2mm]
\theta_k &= \sum_{j=1}^{n} \int_{\mu_j^0}^{\mu_j} \frac{\sqrt{-\mu_j}\, d\mu_j}{L_0 \sqrt{(\mu_j - l_1)(\mu_j - l_{n+1})}} = x + \theta_k^0, \quad k = j_0.
\end{aligned}\right\} \tag{2.7}$$

Therefore, we obtain the following definition.

Definition 2.2 *Umbilic geodesics in n-dimensional case are defined by the following choice of the first integrals: $m_k = b_k$. Umbilic points on the n-dimensional quadrics can be described in terms of the μ-representations as follows: $\mu_k = b_k$, $k = 1,..,n$.*

Theorem 2.3 *The system of equations (2.5) of umbilic geodesics on quadrics is a Hamiltonian system with the Hamiltonian*

$$H = \sum_{j=1}^{n} \frac{(e^{M(\mu_j)P_j} - L_0 \prod_{k=1, k \neq j_0}^{n}(\mu_j - b_k))}{\prod_{r \neq j}(\mu_j - \mu_r))} \tag{2.8}$$

Here

$$M(\mu) = \sqrt{\frac{(\mu - l_1)(\mu - l_{n+1})}{-\mu}}. \tag{2.9}$$

The system (2.8) has a complete set of first integrals

$$P_j = \sum_{k=1, k \neq j_0}^{n} \left(\frac{\log(\mu_j - b_k)}{M(\mu_j)} + \frac{\log L_0}{M(\mu_j)} \right) \tag{2.10}$$

and angle-representation (2.7) that linearize the corresponding Hamiltonian x-flow.

Proof Substituting the expressions (2.10) for the integrals into the Hamiltonian system (2.8) we obtain first part of the proof. Then we consider the action-function

$$S = \sum_{j=1}^{n} \int_{\mu_j^0}^{\mu_j} P_j d\mu_j,$$

which generates a Lagrangian submanifold of the phase space $\mathbf{C}^{2n}$, and consider the following system of variables

$$\left.\begin{aligned}
I_k &= b_k, \quad k = 1, ..., n; \quad k \neq j_0; \quad I_{j_0} = L_0 \\[3mm]
\theta_k &= -\frac{\partial S}{\partial I_k}, \quad k = 1, ..., n.
\end{aligned}\right\} \tag{2.11}$$

Even though there are no invariant tori in the phase space, the hamiltonian flow, can be linearized, as we see from (2.7).

Corollary 2.4 *Using the isomorphism [9] between the x-flow of the problem of geodesics and the x-flow of the KdV equation, we obtain the corresponding umbilic KdV Hamiltonian. Constructing the t-flow on the same invariant variety (2.10), we obtain a new class of soliton-like solutions of the KdV equation that correspond to umbilic geodesics.*

The angle-representation (2.7) has logarithmic singularities. In what follows, we will show that it is similar to the soliton (d)NLS (defocusing nonlinear Schrödinger) representations and reperesentations for the homoclinic orbits of the C. Neumann problem and it can be analyzed using asymptotic reduction (to be described below).

Corollary 2.5 *The system of differential equations (2.4) has a particular solution corresponding to the case when all of the root-variables μ_j but one are constants:*

$$\left.\begin{array}{l} \mu_j' = \dfrac{L_0(\mu_j - l_1)(\mu_j - l_{n+1})}{\sqrt{-\mu_j}} \quad j = j_0, \\[4mm] \mu_j = b_j, \; j = 1, ..., n; \quad j \neq j_0. \end{array}\right\} \tag{2.12}$$

This means that the family of umbilic geodesics can asymptotically approach a one dimensional torus (one of the central ellipses on the n-dimensional ellipsoid). It can be veiwed as a closed geodesic obtained after shrinking the caustics of the family of quasi-periodic geodesics.

3 Soliton Geometric Phases

Below we recall from [3] some Hamiltonians and angle-representations for a class of soliton equations. We use a general approach to deduce from the angle-representation, the corresponding phase functions on associated noncompact Jacobi varieties and we introduce and investigate geometric phases in the complex case.

Now we show that the angle-representation gives the limiting behavior of the n-soliton solution and a system of soliton geometric phases. We call the procedure described below the "soliton analysis" and demonstrate it using the (d)NLS equation as an example.

3.1 The Defocusing Nonlinear Schrödinger Equation

Using a method similar to the approach of the previous section, [3] shows that soliton solutions of the defocusing nonlinear Schrödinger (d)NLS equation

$$i\dot{Q} + \frac{1}{2}Q'' - \bar{Q}Q^2 = 0 \tag{3.1}$$

and KdV equation

$$U_t + 6UU_x + U_{xxx} = 0 \tag{3.2}$$

can be described by the system with the spatial Hamiltonian

$$H_s^s = \frac{\sum_{j=1}^n (e^{2M(\mu_j)P_j} - (\bar{C}(\mu_j))^{1/2}/M(\mu_j))}{\prod_{r\neq j}(\mu_j - \mu_r)} \tag{3.3}$$

and with the dynamic Hamiltonian

$$H_s^d = \frac{\sum_{j=1}^n (-\sum_{l\neq j}\mu_l - \sum_{k=1}^n a_k)(e^{2M(\mu_j)P_j} - (\bar{C}(\mu_j))^{1/2}/M(\mu_j))}{\prod_{r\neq j}(\mu_j - \mu_r)} \tag{3.4}$$

and with first integrals

$$P_j = \frac{\sum_{k=1}^n ln(\mu_j - a_k)}{2M(\mu_j)} \quad j = 1, ..., n. \tag{3.5}$$

Here

$$
\left.
\begin{aligned}
(d)NLS: \quad & M(\mu_j) = \sqrt{-(\mu_j - b_1)(\mu_j - b_2)} \\
\\
KdV: \quad & M(\mu_j) = \sqrt{-\mu_j}
\end{aligned}
\right\} \tag{3.6}
$$

and $\bar{C}(E) = (M(E)\prod_{k=1}^n (E - a_k))^2$ is a polynomial with constant coefficients. The expression (3.3) can be considered as a constraint for (3.4).

The first integrals (3.5) define a Lagrangian submanifold of the phase space $\mathbf{C}^{2n}$ that has the form of the symmetric product

$$\Gamma : (\Re \times \ldots \times \Re)/\sigma_n) \tag{3.7}$$

of n copies of the Riemann surface

$$\Re : \quad P = \frac{\sum_{k=1}^n \log(\mu - a_k)}{2M(\mu)}. \tag{3.8}$$

Notice that if we took a formal limit of the Hamiltonian structure, we would obtain a phase space with singularities (*i.e.*, a pinched torus). The Hamiltonian structure obtained here is presumably a regularization of the pinched torus in an appropriate sense, but we shall not pursue that aspect here.

Introduce the following conjugate variables I_k and θ_k

$$\bar{I}_k = a_k, \quad \bar{\theta}_k = -\frac{\partial S}{\partial \bar{I}_k} = -\frac{\partial S}{\partial a_k} = \sum_{j=1}^{n} \frac{1}{2} \int_{\mu_j^0}^{\mu_j} \frac{d\mu_j}{(M(\mu_j)(\mu_j - a_k))}. \tag{3.9}$$

We call a complete set of variables $\bar{\theta}_j$, $j = 1, ..., n$, an **angle-representation** of the multi-soliton solution on the associated n-dimensional complex Lagrangian submanifolds. It describes a map of Abel-Jacobi type determining a noncompact Jacobi variety.

Theorem 3.1 *In terms of the variables $(\bar{I}_k, \bar{\theta}_k)$, the soliton Hamiltonian flows are linearized*

$$\bar{\theta}_k = x + v_k t + \varphi_k, \quad \bar{I}_k = a_k, \quad v_k = a_k \quad k = 1, ..., n \tag{3.10}$$

on the noncompact Jacobi variety introduced above.

For detailed Proof see [3].

3.2 The Method of Asymptotic Reduction

Now we show that this angle-representation allows one to determine the limiting behavior of the n-soliton solution and to describe a system of soliton geometric phases.

We consider the expression for the n-soliton angle-variables

$$\bar{\theta}_r = \sum_{j=1}^{n} \frac{1}{2} \int_{\mu_j^0}^{\mu_j} \frac{d\mu_j}{(M(\mu_j)(\mu_j - a_r))} = x + v_r t, \quad r = 1, ..., n \tag{3.11}$$

for a particular choice of initial values of the root-variables

$$\mu_j(0, 0) = \mu_j^0 = \frac{a_{j-1} + a_j}{2}. \tag{3.12}$$

Note that in the general case, the base points μ_j^0 of the angle map (3.9) are different from the initial points $\mu_j(0, 0)$.

The term $v_r t$ generates a dynamical soliton phase

$$\Delta_D \bar{\theta}_r = v_r T.$$

In the context of Hannay-Berry phases, T is a period of time during which system is transported along a closed curve in the space of parameters and $\Delta_D \bar{\theta}_r$ is a shift of the angle variables due to the dynamics. Averaging the angle variables is usually used to eliminate this term and to calculate any additional (geometric) phase. Our case is special since we are dealing with an infinite period T. Instead of averaging, we will use the method of asymptotic reduction.

Theorem 3.2 *As $(t \to \infty)$ (or $(t \to -\infty)$) the system of angle variables of the n-soliton solution (3.9) splits into n 1-soliton angle variables. We shall call this process* **asymptotic reduction**.

Proof In what follows we will investigate (3.11) as $(t \to \infty)$ (or $(t \to -\infty)$).

Now we define a direction in the (x, t) plane by fixing

$$x + v_k t = d_k = \text{const} \tag{3.13}$$

for some $r = k$. Then we transform the k-th equation from (3.11) adding and subtracting integrals along different intervals on the real axis to obtain a complete integral from 0 to μ_k

$$
\left.
\begin{aligned}
&\frac{1}{2} \int_0^{\mu_k} \frac{d\mu}{(M(\mu)(\mu - a_k))} = d_k + \frac{1}{2} \sum_{j<k} \int_{\mu_j}^{\mu_{j+1}^o} \frac{d\mu_j}{(M(\mu_j)(\mu_j - a_k))} \\
&+ \frac{1}{2} \int_0^{\mu_1^o} \frac{d\mu_1}{(M(\mu_1)(\mu_1 - a_k))} - \frac{1}{2} \sum_{j>k} \int_{\mu_j^o}^{\mu_j} \frac{d\mu_j}{(M(\mu_j)(\mu_j - a_k))} + \varphi_k^o.
\end{aligned}
\right\} \tag{3.14}
$$

Here φ_k^o is a term chosen to be consistent with the initial data (3.12). Notice that in the real case (when all μ_j and P_j are real), every μ_j is varying once along the cycle l_j over the basic cut $[a_j, a_{j-1}]$ on the Riemann surface (3.5) with the P_j's treated as constants.

The rest of the (3.11) can be described for a fixed d_k as follows

$$\sum_{j=1}^n \frac{1}{2} \int_{\mu_j^o}^{\mu_j} \frac{d\mu_j}{(M(\mu_j)(\mu_j - a_r))} = x + v_r t = d_k + (v_r - v_k)t, \quad r \neq k, \quad r = 1, .., n. \tag{3.15}$$

Here $(v_r - v_k) > 0$, $r < k$, and $(v_r - v_k) < 0$, $r > k$.

System (3.15) yields the following limits

$$
\left.
\begin{aligned}
&1. \quad t \to \infty : \mu_r \to a_{r-1} \; for \; r > k \quad and \quad \mu_r \to a_r \; for \; r < k \\
&2. \quad t \to -\infty : \mu_r \to a_r \; for \; r > k \quad and \quad \mu_r \to a_{r-1} \; for \; r < k.
\end{aligned}
\right\} \tag{3.16}
$$

This transforms (3.14) into the expression for the 1-soliton angle variable

$$\bar{\Theta}_k = \frac{1}{2} \int_0^{\mu_k} \frac{d\mu}{(M(\mu)(\mu - a_k))} = x + v_k t + \Phi_k, \quad k = 1, ..., n, \tag{3.17}$$

where

$$\Phi_k = \varphi_k + \varphi_k^o. \, \square \tag{3.18}$$

3.3   Geometric Phases

Now we recall a link between the angle-map obtained above, the Abel-Jacobi map, and geometric phases. There are three approaches: complex splitting of the spectrum, averaging and the complex phase function, and the τ-function approach. Here we follow the first method.

Definition 3.3 *We define* **soliton geometric phases** *as follows:*

$$\Delta\varphi = \oint_C d_a(\varphi). \tag{3.19}$$

Here d_a and C denote the differential and a closed curve in the space of parameters (a) respectively.

The integral in (3.19) depends on the choice of a connection in the space of parameters. For example, in the case of a flat connection, we get the following result.

Theorem 3.4 *Some of the soliton geometric phases coincide with the phase shift of the k-th soliton of the multi-soliton solution and can be described by the following singular integral*

$$\Delta\varphi_k = \frac{1}{2} \oint_{L_k} \frac{d\mu}{(M(\mu)(\mu - a_k))}, \quad k = 1, ..., n. \tag{3.20}$$

Here L_k is a cycle over the cut $[0, a_k]$ on the Riemann surface.

Lemma 3.5 *Asymptotic reduction results, in particular, in the splitting of every element of the discrete spectrum a_k into a pair of pure imaginary points $(-i\alpha_k, i\alpha_k)$. The phase function φ_k is defined on the covering space of the generalized Jacobian.*

For the proof, see [3] where a new class of classical geometric phases for the systems defined on associated noncompact Jacobi varieties was obtained.

The general approach was also demonstrated [3] in complex case using the angle representations for the n-soliton solutions of the focusing NLS equation and for the the breather and kink-kink solutions of the sine-Gordon equation.

4   Asymtotic Reduction for the Umbilic Solitons

From the results of §2 and §3, we see that the angle-representation for umbilic solitons is similar to the case of "dark-hole" solitons of the (d)NLS equation. Asymtotic reduction for the umbilic solitons can be described as follows.

Note that the angle-representation for the t-flow has the form

$$\left.\begin{array}{l}
\theta_k = \sum_{j=1}^{n} \int_{\mu_j^0}^{\mu_j} \dfrac{\sqrt{\mu_j}\, d\mu_j}{(\mu_j - b_k)\sqrt{(l_1 - \mu_j)(\mu_j - l_{n+1})}} = \theta_k^0 + x + v_k t, \quad k = 1, \ldots, n; \ k \neq j_0 \\[1.2em]
\theta_{j_0} = \sum_{j=1}^{n} \int_{\mu_j^0}^{\mu_j} \dfrac{\sqrt{\mu_j}\, d\mu_j}{L_0\sqrt{(l_1 - \mu_j)(\mu_j - l_{n+1})}} = \theta_{j_0}^0 + x + v_{j_0} t.
\end{array}\right\}$$

(4.1)

Now let us fix the direction d_{j_0}:

$$d_{j_0} = x + v_{j_0} t = const$$

(4.2)

and consider what happens with other elements of the angle-representation as $t \to \infty$

$$\theta_k = d_{j_0} + (v_k - v_{j_0})t.$$

(4.3)

An analysis of this process results in the following limiting values of μ_k :

$$\mu_k \to b_k, \quad k > j_0; \quad \mu_k \to b_{k+1}, \quad k < j_0.$$

(4.4)

Substituting these limiting values in the expression (4.1), one obtains

$$\theta_{j_0} = \sum_{k > j_0} \int_{\mu_k^0}^{b_k} \dfrac{\sqrt{\mu_k}\, d\mu_k}{L_0\sqrt{(l_1 - \mu_j)(\mu_j - l_{n+1})}} + \sum_{k < j_0} \int_{\mu_k^0}^{b_{k+1}} \dfrac{\sqrt{\mu_k}\, d\mu_k}{L_0\sqrt{(l_1 - \mu_j)(\mu_j - l_{n+1})}}$$

$$+ \int_{\mu_{j_0}^0}^{\mu_{j_0}} \dfrac{\sqrt{\mu_{j_0}}\, d\mu_{j_0}}{L_0\sqrt{(l_1 - \mu_{j_0})(\mu_{j_0} - l_{n+1})}} = d_{j_0},$$

(4.5)

meaning that the initial system (4.1) is reduced asymptotically to the one dimensional periodic solution

$$\Theta_{j_0} = \int_{\mu_{j_0}^0}^{\mu_{j_0}} \dfrac{\sqrt{\mu_{j_0}}\, d\mu_{j_0}}{L_0\sqrt{(l_1 - \mu_{j_0})(\mu_{j_0} - l_{n+1})}} = \Theta_{j_0}^0 + x + v_{j_0} t.$$

(4.6)

Now we apply our method for investigating so-called "solitons with a quasi-periodic background". The corresponding angle-representation consists of two different parts and it describes an interaction between solitons and quasi-periodic solutions.

The general limiting process leading to the soliton solutions of this type can be described as follows. Let

$$C(\mu) = - \prod_{k=1}^{2n+1} (\mu - m_k)$$

denote the basic polynomial of the Riemann surface for the quasi-periodic solutions of the KdV equation. Let us consider the limiting proccess:

$$[m_{2k_r}, m_{2k_r+1}] \to b_{k_r}.$$

(4.7)

Here

$$k_r \in (k_1, ..., k_d), \quad 1 \le k_r \le n, \quad 1 \le d \le (n-1).$$

This yields the corresponding angle-representation

$$
\left.
\begin{aligned}
\theta_r &= \sum_{j=1}^{n} \int_{\mu_j^0}^{\mu_j} \frac{\mu_j d\mu_j}{(\mu_j - b_{k_r})\sqrt{-\prod_{m_s \in M}(\mu_j - m_s)}} = \theta_r^0 + x + v_r t, \quad r = 1, ..., d \\[4mm]
\theta_r &= \sum_{j=1}^{n} \int_{\mu_j^0}^{\mu_j} \frac{\mu_j^{n-r} \, d\mu_j}{\sqrt{-\prod_{m_s \in M}(\mu_j - m_s)}} = \theta_r^0 + x + v_r t, \quad r = (d+1), ..., n
\end{aligned}
\right\}
\tag{4.8}
$$

where

$$M = (m_s, \quad 1 \le s \le (2n+1); \ s \ne (2k_r), \ s \ne (2k_r + 1)), \ r = 1, ..., d. \tag{4.9}$$

Asymptotic reduction of the angle-representation leads to the description of an interaction between solitons and quasi-periodic solutions.

Lastly, we consider a different representation for umbilic solitons. We start with the system of differential equations for the quasi-periodic KdV t-flow and use the following change of variables [9]

$$\psi_j = \frac{1}{\mu_j}, \quad l_k = \frac{1}{m_k},$$

together with the limiting process analogous to (2.2),(2.3).

This leads to the introduction of the angle representation

$$
\left.
\begin{aligned}
\theta_k &= \sum_{j=1}^{n} \int_{\psi_j^0}^{\psi_j} \frac{\sqrt{\psi_j} \, d\psi_j}{(\psi_j - b_k)\sqrt{(l_1 - \psi_j)(\psi_j - l_{n+1})}} = \theta_k^0 + \alpha_k x + v_k t, \ k = 1, ..., n; \ k \ne j_0 \\[4mm]
\theta_{j_0} &= \sum_{j=1}^{n} \int_{\psi_j^0}^{\psi_j} \frac{\sqrt{\psi_j} \, d\mu_j}{L_0 \psi_j \sqrt{(l_1 - \psi_j)(\psi_j - l_{n+1})}} = \theta_{j_0}^0 + \alpha_{j_0} x + v_{j_0} t.
\end{aligned}
\right\}
\tag{4.10}
$$

and preserves the natural parameters x and t of the KdV problem. Asymptotic reduction of the (4.10) leads to a 1-dimensional periodic solution.

5 Homoclinic Hamiltonian Systems

Here we show that Hamiltonian flows of the homoclinic orbits described by Devaney [18] for the C. Neumann problem coincide with the soliton Hamiltonian x-flows for the KdV equation. This is important since it provides an understanding of the link between soliton solutions and homoclinic orbits for associated completely integrable systems. In

285

particular, this enables one to treat the homoclinic case in the same manner as the soliton case and to introduce homoclinic Hamiltonians, angle-repersentations, and geometric phases. In particular, using the Knörrer-Moser isomorphism [42] between the Jacobi problem of geodesics and the C. Neumann problem, we introduce new orbits corresponding to the umbilic geodesics and give a geometric interpretation of the homoclinic orbits of Devaney.

5.1 Homoclinics and Solitons

First of all, we note that a system of first integrals inrtoduced by Devaney and Uhlenbeck [18] for the C. Neumann problem, namely

$$\Phi_j(y, \dot{y}) : TS^n \to \mathbf{R}, \tag{5.1}$$

$$\Phi_j(y, \dot{y}) = y_j^2 + \frac{1}{2} \sum_{k \neq j} \frac{(\dot{y}_j y_k - \dot{y}_k y_j)^2}{l_k^2 - l_j^2}, \quad j = 1, ..., n \tag{5.2}$$

play central role in Devaney's description of the transversal homoclinic orbits. Here $l_0 < l_1 < ... < l_n$ and $l_0 = 0$. Namely, he proved the following Lemma.

Lemma 5.1 *The first integrals are identically zero along the orbits; i.e.,*

$$\Phi_j(y, \dot{y}) = 0, \quad j = 1, ..., n. \tag{5.3}$$

We use this fact together with the algebraic-geometric description of the quasi-periodic solutions [9] of the C. Neumann problem to obtain the following result.

Theorem 5.2 *The Hamiltonian flow of the homoclinic orbits of the C. Neumann problem coincide with the soliton Hamiltonian x-flow of the KdV equation. Both flows are defined on a noncompact Jacobi variety.*

Proof In [9], quasi-periodic solutions of the C. Neumann problem were discribed using root-variables μ_j, *i.e.,* solutions of the system

$$\frac{\partial \mu_j}{\partial x} = 2 \frac{\sqrt{-\prod_{r=1}^{n}(\mu_j - m_r) \prod_{k=1}^{n+1}(\mu_j - a_k)}}{\prod_{i \neq j}(\mu_j - \mu_i)}, \quad j = 1, ..., n \tag{5.4}$$

and action-angle variables defined on the Jacobi variety of the symmetric product

$$\Gamma : ((\Re \times ... \times \Re)/\sigma_n) \tag{5.5}$$

of n copies of the Riemannian surface

$$\Re : \quad P^2 = -\prod_{k=1}^{n+1}(\mu - a_k) \prod_{r=1}^{n}(\mu - m_r), \tag{5.6}$$

Γ being a Lagrangian submanifold of the phase space $\mathbf{C}^{2n}$.

In this setting, the first integrals of the problem can be represented in the form

$$\Phi_j = F_j = \frac{\prod_{r=1}^n (a_j - m_r)}{\prod_{k \neq j}(a_j - a_k)}. \tag{5.7}$$

Here,

$$a_j = -2l_j^2, \quad j = 0, ..., n.$$

Condition (5.3) and formula (5.7) yield the following choice of first integrals m_j

$$m_j = a_j, \ j = 1, ..., n; \quad a_0 = 0, \tag{5.8}$$

meaning that all roots of the basic polynomial of the Riemannian surface (5.6) are double negative roots, except $a_0 = 0$.

Therefore, the system (5.4) corresponding to the case of a singular spectrum (5.8) coincides precisely with the system of equations describing soliton solutions of the KdV equation. It leads to the introduction of the exponential homoclinic Hamiltonians (3.3), (3.4) with logarithmic first integrals and angle-representation

$$\theta_r = \sum_{j=1}^n \frac{1}{2} \int_{\mu_j^0}^{\mu_j} \frac{d\mu_j}{\sqrt{-\mu_j(\mu_j - a_r))}} = x + \theta_r^0, \quad r = 1, ..., n \tag{5.9}$$

defined on the noncompact Jacobi variety of the symmetric product of n copies of the Riemannian surface

$$P = \frac{\log(\mu - a_r)}{2\sqrt{-\mu}}. \tag{5.10}$$

□

Corollary 5.3 *As* $(x \to \infty)$ *(or* $x \to -\infty$*), the spectrum of the homoclinic orbit splits into complex pairs*

$$a_j \to (i\alpha_j, -i\alpha_j), \quad a_j = -\alpha_j^2, \quad j = 1, ..., n$$

and an analysis of the angle representation yields an introduction of the following homoclinic points (p), corresponding to the following values of μ_j:

$$\mu_j = a_j, \quad j = 1, ..., n \tag{5.11}$$

with corresponding stable W^s *(and unstable* W^u*) manifolds, which consist of the orbits on the two different sheets of the double covering of the Riemannian surface (5.10) defined by the following change of variables*

$$\xi_j^2 = -\mu_j, \quad j = 1, ..., n. \tag{5.12}$$

Orbits on the stable and ustable manifolds are forward (and backward) asymptotic to the points from the set (p)

$$\xi_j \to \alpha_j, \quad j = 1, ..., n \qquad (5.13)$$

and

$$\xi_j \to -\alpha_j, \quad j = 1, ..., n \qquad (5.14)$$

respectively.

The splitting of the spectrum follows from evaluating the basic integral from (5.9)

$$\frac{1}{2\alpha_j} \log \left(\frac{\xi_j - \alpha_j}{\xi_j + \alpha_j} \right) \qquad (5.15)$$

and analysing the angle-representation (5.9) as $x \to \infty$ (or $x \to -\infty$).

Lastly, we define an additional Hamiltonian t-flow on the invariant variety of the homoclinic orbit and obtain exactly the n-soliton solution of the KdV problem.

Remark 5.4 *Geometric phases are defined in the same way as in soliton case .*

For details see [4].

5.2 Geometry of the Homoclinic Orbits

The isomorphism between classes of solutions of Jacobi problem and C. Neumann problem can be used in the opposite direction. Namely, one can consider a special family of umbilic geodesics on the n-dimensional complex quadrics corresponding to the homoclinic orbits.

In what follows, we use the isomorphism between classes of solutions in the form established in [9]. Let us put in (5.6) one of the parameters $m_r = 0$ and use the following change of variables

$$z_j^2 = \frac{\frac{x_j}{d_j}}{\sum_{k=1}^n \frac{x_j^2}{d_j}}, \quad a_j = \frac{1}{d_j}, \quad \mu_j = \frac{1}{\eta_j}, \quad j = 1, ..., n, \qquad (5.16)$$

which relates parameters and variables of the two systems and enables one to avoid Abeleian integrals of the second kind in the angle-representation. Here z_j and η_j denote variables of the problem of geodesics and d_j are semiaxes of the corresponding quadrics.

In our case all $a_j < 0$. Therefore geodesics which correspond to the homoclinic orbits form a singular family of geodesics on an n-dimensional complex quadrics.

Remark 5.5 *In [4] we also investigated umbilic geodesics on the n-dimensional hyperbolloids. This yields geometric model for a new homoclinic orbit.*

6 Soliton Geometric Asymptotics and Geometric Phases

Another application of the machinery developed here is to geometric asymptotics. In particular, the quantum and classical phases are linked.

Recently, complex geometric asymptotics have been studied [20, 48, 33, 5, 6] in connection with algebraic-geometric methods for nonlinear problems. In particular, quantization involving Maslov method of canonical operator was used [20] for the semi-classical approximation of the finite-gap solutions of the Toda lattice and some other problems. This method which is based on the construction of the Lagrangian manifolds with complex jerms was then developed [48] to obtain asymptotics of the eigenfunctions of the quantum periodic Toda chain over a solvable Lie algebras. (For details about general method of geometric asymptotics see [32]).

In [5, 6] a method is suggested for constructing local semiclassical solutions (modes) in the form of functions of several complex variables (geometric asymptotics) on the Jacobian varieties of compact multisheeted Riemannian surfaces. And quantum conditions are defined as conditions of finiteness on the number of sheets in the Riemannian surface.

In what follows we study quantization of Hamiltonians quadratic in P as the quantizing procidure in the exponential case (see soliton Hamiltonians above) is not clear. However, the work here suggests that this can nevertheless be done on the level of geometric asymptotics and phases. We demonstrate our approach using example of the KdV equation.

To obtain soliton geometric asymptotics, we first consider the quantum problem corresponding to the quasi-periodic Hamiltonians [8, 5, 7] for the KdV equation

$$H^s = -\sum_{j=1}^{n} \frac{(P_j^2 + C_{2n+1}(\mu_j))}{\prod_{r \neq j}(\mu_j - \mu_r)} \tag{6.1}$$

and

$$H^d = -\sum_{j=1}^{n} \frac{2(-\sum_{l \neq j} \mu_l - \sum_{k=1}^{2n+1} m_k)(P_j^2 + C_{2n+1}(\mu_j))}{\prod_{r \neq j}(\mu_j - \mu_r)}. \tag{6.2}$$

These are quadratic Hamiltonians

$$H = \frac{1}{2} \sum_{j=1}^{n} g^{jj} P_j^2 + V(\mu_1, ..., \mu_n) \tag{6.3}$$

defined on the $T^*((\Re \times \ldots \times \Re)/\sigma_n)$. (Here $\Re$ is the Riemann surface: $P^2 = -C_{2n+1}(\mu)$).

In accordance with [5] we use the functions g^{jj}

$$g^{jj} = -\frac{1}{\prod_{r \neq j}(\mu_j - \mu_r)} \tag{6.4}$$

and

$$g^{jj} = -\frac{\left(-\sum_{l \neq j} \mu_l - \sum_{k=1}^{2n+1} m_k\right)}{\prod_{r \neq j}(\mu_j - \mu_r)}. \tag{6.5}$$

as components of the Riemannian metric and construct an operator of Laplace-Beltrami type and then the stationary Schrödinger equation

$$\nabla^j \nabla_j U + w^2(E - V)U = 0, \tag{6.6}$$

defined on the n-dimensional Riemannian manifold. (Here ∇^j and ∇_j are covariant and contravariant derivatives defined by the tensor g^{jj} and w and E are parameters).

Theorem 6.1 *The quantum equation (6.6) corresponding to the Hamiltonian (6.1) (or (6.2)) can be reduced to a differential system of n equations*

$$-\frac{1}{w^2 U_j}\left(\frac{\partial^2 U_j}{\partial \mu_j^2} + w^2 C_{2n+1}(\mu_j)U_j\right) = R_j(\mu_j) \quad j = 1, ..., n. \tag{6.7}$$

connected with each other by means of the symmetry condition

$$\sum_{j=1}^{n} \frac{R_j(\mu_j)G_j}{\prod_{r \neq j}(\mu_j - \mu_r)} = -E. \tag{6.8}$$

Here $G_j = 1$ and $G_j = (-\sum_{l \neq j} \mu_l - \sum_{k=1}^{2n+1} m_k)$ in cases (6.4) and (6.5) respectively.

Here $R_j(\mu)$ are the same polynomial

$$R_j(\mu) = R(\mu) = w_1 \mu^{n-1} + w_2 \mu^{n-2} + ... + w_n, \quad w_1 = -E \tag{6.9}$$

with constant coefficients.

Now we establish a link between equation (6.6) and the initial nonlinear problem by means of geometric asymptotics. Let

$$U = \sum_r A_r(\mu_1, ..., \mu_n) \exp[iwS_r(\mu_1, ..., \mu_n)]$$

$$= \sum_r \prod_{j=1}^{n} U_{rj}(\mu_j) = \sum_r \prod_{j=1}^{n}(A_j(\mu_j)\exp[iwS_{rj}(\mu_j)]) \tag{6.10}$$

which is a function of several complex variables. Substituting (6.10) in (6.7), equating coefficients for w and w^2 and integrating, we obtain the following geometric asymptotics

$$U = \sum_r \frac{A_o}{\prod_{j=1}^{n}(C_{2n+1}(\mu_j))^{\frac{1}{4}}} \exp[iw \sum_{j=1}^{n} \int_{\mu_j^o}^{\mu_j} \sqrt{C_{2n+1}(\mu_j)} \, d\mu_j]. \tag{6.11}$$

Let us consider the special class of quasi-periodic solutions of the initial KdV equation defined by the following choice of basic polynomial

$$C_{2n+1}(\mu) = (-\mu) \prod_{k=1}^{n} (\mu - a_k)^2 + R(\mu). \qquad (6.12)$$

which depends on exactly 2n parameters (a_r, w_k).

Definition 6.2 *We call solutions from this class* **presoliton solutions**.

The main result of this section can be represented in the of a theorem.

Theorem 6.3 *n-soliton solutions correspond to the case when all coefficients w_k of the polynomial $R(\mu)$ from (6.12) are equal to zero.*

For the Proof see [3].

Corollary 6.4 *If $R(\mu) = 0$, then formulae (6.11) describes soliton geometric asymptotics with*

$$U = \sum_l \frac{A_o}{\prod_{j=1}^{n} (\sqrt{-\mu_j} \prod_{r=1}^{n} (\mu_j - a_r))^{\frac{1}{2}}} \exp\left[iw \sum_{j=1}^{n} \int_{\mu_j^o}^{\mu_j} \sqrt{-\mu_j} \prod_{r=1}^{n} (\mu_j - a_r) \ d\mu_j\right]. \qquad (6.13)$$

Investigating the dependence of the soliton geometric asymptotics (6.13) on the slowly changing parameters a_l one can also obtain geometric phases in the quantum case [3]. Similar results are obtained in case of (NLS) and (SG) hierarchies of equations.

In conclusion, note that geometric asymptotics constructed in this section provide a setting for investigating semiclassical theory of the modulational equations.

References

[1] M.J. Ablowitz and H. Segur, Solitons and the Inverse Scattering Transform, SIAM, Philadelphia 1981).

[2] M. Adler and P. Van Moerbeke, Completely integrable systems, Kac-Moody Lie algebras and curves, *Adv.in Math.* 38(3): 267(1980).

[3] M.S. Alber and J.E.Marsden, On Geometric Phases for Soliton Equations, *Commun. Math. Phys.* (to appear).

[4] M.S. Alber and J.E.Marsden, Umbilic Solitons and Homoclinic Geometric Phases, (in print).

[5] M.S. Alber, On integrable systems and semiclassical solutions of the stationary Schrödinger equations, *Inverse Problems* 5: 131 (1989).

[6] M.S. Alber, Complex Geometric Asymptotics, Geometric Phases and Nonlinear Integrable Systems, *in* , Studies in Math.Phys. 3, North-Holland, Elsevier Science Publishers B.V., Amsterdam (1992).

[7] M.S. Alber, Hyperbolic Geometric Asymptotics, *Asymptotic Analysis* 5, 2: 161 (1991).

[8] M.S. Alber and S.J. Alber, Hamiltonian formalism for finite-zone solutions of integrable equations, *C.R. Acad. Sc. Paris* 301: 777 (1985).

[9] M.S. Alber and S.J. Alber, Hamiltonian formalism for nonlinear Schrödinger equations and sine-Gordon equations, *J.London Math.Soc.* (2) 36: 176 (1987).

[10] V.I. Arnold, A remark on the branching of hyperelliptic integrals as functions of the parameters, *Func. Anal. Appl.* 2: 187 (1968).

[11] M.V. Berry, Quantal phase factors accompanying adiabatic changes, *Proc. R. Soc. Lond.A* 392: 45 (1984).

[12] M.V. Berry, Classical adiabatic angles and quantal adiabatic phase, *J. Phys.. A: Math. Gen.* 18: 15 (1985).

[13] M.V. Berry and J.H. Hannay, Classical non-adiabatic angles, *J.Phys.A: Math. Gen.* 21: L325 (1988).

[14] M.V. Berry, Quantum Adiabatic Anholonomy, *in*, Lectures given at the Ferrara School of Theoretical Physics on "Anomalies, defects, phases ...", June 1989, (to be published by Bibliopolis), Naples.

[15] B.Birnir, Singularities of the complex Korteweg-de Vries flows, *Comm. Pure Appl. Math.* 39: 283 (1986).

[16] A.R. Bishop, D.W. McLaughlin and M. Solerno, Global coordinates for the breather-kink (antikink) sine-Gordon phase space: An explicit separatrix as a possible source of chaos, *Phys. Rev. A*, 40, 11: 6463 (1989).

[17] P. Deift, L.C. Li and C. Tomei, Matrix factorizations and integrable systems, *Comm. Pure Appl. Math.* 443 (1989).

[18] R. Devaney, Transversal homoclinic orbits in an integrable system, *Amer. J. Math.*, 100: 631 (1978).

[19] S.U.Dobrohotov and V.P.Maslov, Multiphase asymptotics of nonlinear partial differential equations with a small parameter, *in*, Math. Phys. Reviews, Over. Pub. Ass., Amsterdam, (1982).

[20] S.U.Dobrohotov and V.P.Maslov, Finite-zone, almost-periodic solutions in WKB approximation, *J.Soviet Math.*,16, 6: 1433 (1981).

[21] H.J. Duistermaat, On global Action-Angle Coordinates, *Comm. Pure Appl. Math.* 23: 687 (1980).

[22] N. Ercolani, Generalized Theta functions and homoclinic varieties, *Proc. Symp. Pure Appl. Math.*, 49: 87 (1989).

[23] N. Ercolani and H.P. McKean, Geometry of KdV(4): Abel sums, Jacobi variety, and theta function in the scattering case, *Invent. Math.* 99: 483 (1990).

[24] N. Ercolani and H. Flaschka, The geometry of the Hill equation and of the Neumann system, *Phil. Trans. R. Lond. A* 315:405 (1985).

[25] N. Ercolani, M. Forest, D.W. McLaughlin, and R. Montgomery, Hamiltonian structure for the modulation equations of a sine-Gordon wavetrain, *Duke Math. Journal* 55, 4:949 (1987).

[26] N. Ercolani, M. Forest and D.W. McLaughlin, Notes on Melnikov integrals for models of the periodic driven pendulum chain, (preprint) (1989).

[27] N. Ercolani and M. Forest, The Geometry of Real Sine-Gordon Wavetrains, Comm. Math. Phys. 99(1985)1-49.

[28] N. Ercolani and D.W. McLaughlin, Toward a topological classification of integrable PDE's, *in* , Proc. of a Workshop on The Geometry of Hamiltonian Systems, T.Ratiu, ed., MSRI Publications 22, Springer-Verlag, (1991).

[29] H. Flaschka, D.W. McLaughlin and M.G. Forest, Multiphase averaging and the

inverse spectral solution of the Korteweg-de Vries equation, *Comm. Pure Appl. Math.* 33: 739 (1980).

[30] H. Flaschka, A.C. Newell and T. Ratiu, Kac-Moody Lie algebras and soliton equations II, III, *Physica D* 9: 300 (1983).

[31] M.G. Forest and D.W. McLaughlin, Modulation of Sinh-Gordon and Sine-Gordon Wavetrains, *Studies in Appl. Math.* 68: 11 (1983).

[32] V. Guillemin and S. Sternberg, Geometric Asymptotics, Math.Sutveys 14, AMS, Providence, Rhode Island, (1977).

[33] V. Guillemin and S. Sternberg, The Gelfand-Cetlin system and the quantization of the complex Flag Manifolds, *J. Funct. Anal.* 52 (1): 106 (1983).

[34] W.Klingenberg, *Riemannian Geometry*, Berlin: New York: de Gruyter (1982).

[35] H. Knörrer, Singular fibres of the momentum mapping for integrable Hamiltonian systems, *J. Reine u. Ang. Math.* 355: 67 (1984).

[36] P.D. Lax and C.D. Levermore, The small Dispersion Limit of the Korteweg - de Vries Equation, I, II, III, *Comm. Pure Appl. Math.* 36, 2: 253, 571, 809 (1983).

[37] J.E. Marsden, R. Montgomery and T. Ratiu, Cartan-Hannay-Berry phases and symmetry, Contemporary Mathematics 97: 279 (1989); see also Mem. AMS Vol. 436 (1990).

[38] H.P. McKean, Integrable Systems and Algebraic Curves, Lecture Notes in Mathematics, Springer-Verlag, Berlin, (1979).

[39] H.P. McKean, Theta functions, solitons, and singular curves, *in*, PDE and Geometry, Proc. of Park City Conference, C.I.Byrnes, ed., (1977).

[40] D.W. McLaughlin and E.A. Overman, *Surveys in Appl.Math.* 1 (1992) (to appear).

[41] R. Montgomery, The connection whose holonomy is the classical adiabatic angles of Hannay and Berry and its generalization to the non-integrable case, *Comm. Math. Phys.* 120: 269 (1988).

[42] J. Moser, Integrable Hamiltonian Systems and Spectral Theory, Academia Nazionale dei Lincei, Pisa, (1981).

[43] D. Mumford, Tata Lectures on Theta I and II, Progress in Math.28 and 43, Birkhauser, Boston (1983).

[44] E. Previato, Hyperelliptic quasi-periodic and soliton solutions of the nonlinear Schrödinger equation, *Duke Math.Journal* 52: 2 (1985).

[45] S.N.M. Ruijsenaars, Action-angle maps and scattering theory for some finite-dimensional integrable systems I.The pure soliton case, *Comm. Math. Phys.* 115: 127 (1988).

[46] S. Venakides, The generation of modulated wavetrains in the solution of the Korteweg-de Vries equation, *Comm.Pure and Appl.Math.* 38: 883 (1985).

[47] S. Venakides and T. Zhang, Periodic limit of inverse scattering, Duke University, preprint (1991).

[48] Yu.M.Vorob'ev and S.U.Dobrohotov, Quasiclassical quantisation of the periodic Toda chain from the point of view of Lie algebras, *Theor.and Math.Phys.* 54, 3:312 (1983).

[49] A. Weinstein, Connections of Berry and Hannay type for moving Lagrangian sub-manifolds, *Adv. in Math.* 82, 2: 133 (1990).

NONLINEAR WAVES AND THE 1:1:2 RESONANCE

Walter Craig[1] and C. Eugene Wayne[2]

[1]Mathematics Dept., Brown University, Providence, RI 02912
[2]Mathematics Dept., Pennsylvania State University,
University Park, PA 16802

1. Introduction

In this paper we discuss the existence of periodic solutions to nonlinear wave equations
in the form

$$\partial_t^2 u = \partial_x^2 u - g(x,u) \ , \qquad 0 \le x \le \pi \ , \tag{1.1}$$

imposing either periodic or Dirichlet boundary conditions at the endpoints $x = 0$, π.
The results are perturbative, for solutions near the equilibrium $u \equiv 0$. The nonlinearity is given by

$$g(x,u) = g_1(x)u + g_2(x)u^2 + g_3(x)u^3 + \cdots \tag{1.2}$$

and will be asked to be analytic in both variables. This equation can be viewed as a
dynamical system with infinitely many degrees of freedom, and we are interested in
developing techniques with which some of the basic structures of its phase space can
be studied. We will concentrate on the construction of families of periodic solutions to
(1.1), although the evidence is that the construction of quasiperiodic solutions will be
similar. Due to its infinite dimensional character even the periodic solutions exhibit
the small divisor problem, and a version of the Nash-Moser method is employed to
overcome the accompanying difficulties of convergence. The larger paper (Craig &
Wayne [CW]) gives a complete account of this construction in the generic case, in
which exact linear resonances are avoided, and a 'twist condition', or condition of
genuine nonlinearity holds.

Singular Limits of Dispersive Waves, Edited by
N.M. Ercolani et al., Plenum Press, New York, 1994

In this shorter note we describe several extensions of these results to cases which are linearly resonant, in particular the $1:2$ resonance and the $1:1:2$ resonance of the title. While we present here only several explicit cases, one can see that issues emerge that are parallel to problems with finitely many degrees of freedom, which have been addressed in articles of A. Weinstein [We] and J. Moser [M]. So far we are only able to handle cases with finitely many linear frequencies in resonance; the case of infinite multiplicity of linear resonance is of great interest, and is still open. There is recent work by other authors on the subject of perturbation theory for systems with infinitely many degrees of freedom, in particular S. Kuksin [K], J. Pöschel [P], and C. Albanese, J. Fröhlich & T. Spencer [AFS].

Equation (1.1) is well known to be a Hamiltonian system, with Hamiltonian

$$\mathcal{H}(u, p) = \int_0^\pi \tfrac{1}{2} p^2 + \tfrac{1}{2}(\partial_x u)^2 + G(x, u) \, dx \quad ,$$
$$\partial_u G(x, u) = g(x, u) \quad .$$

(1.3)

Let $z = (u, p)^T$. Then (1.1) is equivalent to

$$\dot{z} = J \operatorname{grad} \mathcal{H} \quad ,$$

(1.4)

with

$$J = \begin{pmatrix} 0 & 1 \\ -1 & 0 \end{pmatrix}$$

(1.5)

the matrix describing the usual symplectic form. This equation, and its Hamiltonian formulation, is a model problem for many other nonlinear equations of physical interest. A short list of the most well known ones includes the nonlinear Schrödinger equation

$$i \partial_t \psi = -\tfrac{1}{2} \Delta \psi + Q(x, \psi) \quad .$$

(1.6)

The Hamiltonian is

$$\mathcal{H}_{NLS}(\psi) = \int \frac{1}{2} |\nabla \psi|^2 + G(x, \psi) \, dx \quad ,$$

(1.7)

where $Q(x, \psi) = \partial_{\overline{\psi}} G(x, \psi)$, and the symplectic structure is given by multiplication by i. The generalized Korteweg deVries equation (KdV) has the Hamiltonian

$$\mathcal{H}_{KdV}(q) = \int \frac{1}{12} (\partial_x q)^2 + G(x, q) \, dx \quad , \quad \partial_q G(x, q) = g(x, q) \quad ,$$

(1.8)

and the symplectic structure given by the operator $-\partial_x$. This gives rise to the partial differential equation

$$\partial_t q = \frac{1}{6} \partial_x^3 q - \partial_x g(x, u) \quad .$$

(1.9)

A third example is the water wave problem, which is a Hamiltonian free surface problem in fluid dynamics, with Hamiltonian due to Zakharov

$$\mathcal{H}_{WW}(\eta, \xi) = \tfrac{1}{2} \int \xi G(\eta) \xi + g\eta^2 \, dx \quad . \tag{1.10}$$

In this expression η represents the free surface elevation, ξ the boundary values of the velocity potential function, and the operator $G(\eta)$ is the Dirichlet-Neumann operator for the fluid domain. The water wave equations take the form (1.4) with the usual symplectic operator J of (1.5).

We are hopeful that the techniques that have been developed for the nonlinear wave equation will be able to be extended to many of the above physical problems, a program of research that we are currently pursuing.

2. The harmonic oscillator

To motivate the question and the statement of results, it is useful to solve equation (1.1), linearized about the trivial solution $u \equiv 0$.

$$\partial_t^2 v = \partial_x^2 v - g_1(x)v \quad . \tag{2.1}$$

This is simply done by separation of variables. Let $\{(\psi_j(x), \omega_j^2)\}_{j=1}^\infty$ be eigenfunction–eigenvalue pairs for the Sturm Liouville operator

$$-\frac{d^2}{dx^2} + g_1(x) \tag{2.2}$$

with the appropriate boundary conditions. Then (as long as ω_j is real) a periodic solution of (2.1) is given by

$$v_j(x, t) = r \cos(\omega_j t + \xi) \psi_j(x) \tag{2.3}$$

which is parametrized by amplitude r (action r^2) and angle ξ. A more general solution is given by

$$v(x, t) = \sum_{j:\, \omega_j \in \mathbf{R}} r_j \cos(\omega_j t + \xi_j) \psi_j(x)$$
$$\sum_{j=1}^\infty r_j^2 < \infty \quad , \tag{2.4}$$

but this is typically quasiperiodic or almost periodic in time, unless a full set of resonance conditions is satisfied among the frequencies ω_ℓ such that the amplitudes r_ℓ are nonzero. From this expression we easily deduce the full structure of phase space, particularly that if $\{c_j\}_{j=1}^\infty \in \ell^1$ then the torus $\{r_j^2 = c_j : \omega_j \in \mathbf{R}\}$ is a family of L^2 functions invariant under the flow of (2.1); solutions with temporal frequencies $\{\omega_j\}$. This motivates the natural question whether some of these periodic, quasiperiodic or almost periodic solutions persist for the nonlinear equation.

The first issue that arises in attempting to answer this question is whether the linear frequencies are in resonance. Define a **resonant group** to be the set

$$N \equiv \{(j_\ell, \pm k_\ell) \in \mathbf{Z}^2; \omega_{j_1}/k_1 = \omega_{j_2}/k_2 = \ldots = \omega_{j_M}/k_M \ldots\} \qquad (2.5)$$

for integers $k_1, k_2 \ldots k_M \ldots$. Ordering $\omega_{j_1} \le \omega_{j_2} \le \ldots \le \omega_{j_M}$ we only need consider the case $k_1 = 1$, and we call $\omega = \omega_{j_1} = \ldots = (\omega_{j_M}/k_M)$ the **fundamental frequency**. This resonant group may be finite or infinite; we will assume for the results of this talk that it is finite. Note that the other ratios $\{(\omega_j/k); (j, \pm k) \notin N\}$ typically form a dense set in $\mathbf{R}$. The periodic solutions of (2.1) with fundamental frequency ω are given by

$$v(x,t) = \sum_{\ell:(j_\ell, k_\ell) \in N} r_\ell \cos(\omega_{j_\ell} t + \xi_\ell)\psi_{j_\ell}(x) \quad ; \qquad (2.6)$$

they form a $2M$ dimensional family parametrized by actions $\{r_\ell^2\}_{\ell=1}^M$ and angles $\{\xi_\ell\}_{\ell=1}^M$, a foliation by M-dimensional invariant tori. For the nonlinear problem most of this family will not be preserved, but we will see that certain tori will persist in a neighborhood of the origin.

3. Nonlinear periodic solution

Our assumptions on the nonlinearity $g(x,u)$ are that it is: (1) periodic in x with period π, (2) analytic in a strip about the real-x axis, and in a neighborhood of the origin in u, and (3) in the case of Dirichlet boundary conditions we assume that it is odd in the (x,u) plane. The first simple result to be stated is one of the conclusions of [CW], section 6.

Proposition 3.1. *For a generic set of nonlinearities all resonant groups have multiplicity $M = 1$.*

The above reference [CW] describes the proof of the following theorem in the case $M = 1$.

Theorem 3.2. *For a generic set of nonlinearities g there exists a constant $r_0 > 0$, a Cantor set $\mathcal{C} \subseteq [0, r_0)$, and a C^∞ family*

$$(u(x, t; r), \Omega(r)) \tag{3.1}$$

such that

(i) *For $r \in \mathcal{C}$, u is a periodic solution of (1.1) with frequency $\Omega(r)$.*

(ii) *There is a choice of angle $\xi = \xi(r)$ such that*

$$|u(x, t; r) - r\cos(\Omega(r)t + \xi)\psi_j(x)| < Cr^2$$
$$|\omega_j - \Omega(r)| < Cr^2 \ . \tag{3.2}$$

(iii) *The function u is analytic in (x, t).*

Furthermore, the admissibility of $g(x, u)$ depends only upon its $3 - jet$, that is upon $g_1(x), g_2(x), g_3(x)$.

This is the analog of the Lyapounov center theorem for the nonlinear wave equation (1.1). This talk will focus now upon cases in which the resonant group has size $M > 1$. Let L_0 be a constant which is sufficiently large. Our hypotheses, in addition to (1),(2),(3) are that

(H1) The resonant group N is finite;

$$\omega = \omega_{j_1} = \frac{\omega_{j_2}}{k_2} = \cdots = \frac{\omega_{j_M}}{k_M} \ . \tag{3.3}$$

(H2) Other linear frequencies are sufficiently nonresonant with ω;

$$|\omega - \frac{\omega_j}{k}| > d_0 \tag{3.4}$$

for all $(j, k) \notin N$, with $|j| + |k| \le L_0$, and $d_0 > L_0^{-\frac{1}{2}}$.

(H3) A condition of genuine nonlinearity, which we will explain below.

Whenever these conditions hold, we have the following theorem.

Theorem 3.3. *There exist $s_0 > 0$, Cantor sets $\mathcal{C}_\ell \subseteq [0, s_0)$ for $\ell = 1, \ldots \overline{M}$ and C^∞ families*

$$(u_\ell(x, t; s), \Omega_\ell(s)) \ , \quad \ell = 1, \ldots, \overline{M} \tag{3.5}$$

such that

(i) *For $s \in \mathcal{C}_\ell$, u_ℓ is a periodic solution of (1.1) with frequency $\Omega_\ell(s)$.*

(ii) *The functions $u_\ell(x, t; s)$ are analytic in (x, t).*

(iii) There exist directions $(\vec{r}_\ell, \vec{\xi}_\ell) \in \mathbf{R}^M \times \mathbf{T}^M$ such that

$$\frac{1}{2}\sum_{j=1}^{M} k_j^2 (r_\ell^j)^2 = 1 \quad ,$$

$$|u_\ell(x,t;s) - s\sum_{j=1}^{M} r_\ell^j \cos(k_j \Omega_\ell(s)t + \xi_\ell^j)\psi_j(x)| < c|s|^2 \quad , \tag{3.6}$$

$$|\Omega_\ell(s) - \omega| < C|s| \quad .$$

Additionally, the number of families of solutions is such that $\overline{M} \geq M$.

This result only addresses periodic solutions to (1.1). Other authors have worked on both periodic and quasiperiodic motion of infinite dimensional Hamiltonian systems, principally by extending KAM results for lower (finite) dimensional invariant tori [E][P]. The latter process has as starting point a partial Birkhoff normal form, which has no resonant cubic terms and only the generic quartic resonant terms. Thus in resonant cases Theorem 3.3 will not be available by their methods. The extension of Theorem 3.3 to quasiperiodic motion will also be of interest in finite degrees of freedom, as it will provide an alternate proof of the existence of invariant tori, which will extend to the situation in which the Birkhoff normal form has resonant cubic terms and nongeneric resonant quartic terms.

4. Proof of Theorem 3.3

A brief description of the proof of Theorem 3.3 is given below. A more complete description, as well as a more general approach to the resonant case will appear in a subsequent publication [CW2]. There are three basic steps in the method of solution. First the problem is transformed to a nonlinear problem on a lattice. Then the resonant group is isolated by a Lyapounov-Schmidt decomposition, and a solution of the infinite dimensional part of the problem is given using a Nash-Moser iteration. This is the step which inherits the small divisor problem and thus needs detailed attention; full details appear in [CW]. Finally, the resonant part of the problem is studied, using a modification of the principle of least action.

4.1. Reduction to the lattice

A periodic solution of (1.1) can be viewed as an embedded circle in the phase space $(H^1(0,\pi) \times L^2(0,\pi))$ (whose elements consist of functions (u,p)), which is invariant under the nonlinear evolution. An embedded circle is described by

$$S(x,\xi) = \sum_{j=1}^{\infty} s_j(\xi)\psi_j(x) , \qquad s_j(\xi + 2\pi) = s_j(\xi) . \tag{4.1}$$

If it is invariant, and traversed with frequency Ω, then S satisfies

$$\Omega^2 \partial_\xi^2 S - \partial_x^2 S + g(x, S) = 0 . \tag{4.2}$$

Expand each $s_j(\xi)$ in Fourier series,

$$S(x,\xi) = \sum_{j,k} s(j,k)e^{ik\xi}\psi_j(x) , \tag{4.3}$$

where the sequence $\{s(j,k)\} \in \ell^2(\mathbf{Z}^2)$ (for notational convenience we set $s(j,k) = 0$ whenever $j \leq 0$). The equation (4.2) is equivalent to the following **mode interaction equation**

$$V(\Omega)s + W(s) = 0 . \tag{4.4}$$

The operator $V(\Omega)s(j,k) = (\omega_j^2 - \Omega^2 k^2)s(j,k)$ is diagonal in the basis we have given, and the term $W(s)$ accounts for the nonlinear interactions in (4.2), expressed in this basis. Our central theorem will actually be a result for such nonlinear lattice systems, in scales of Hilbert spaces which control exponential decaying sequences on the lattice. Under hypotheses (1),(2), and (3), if $S(x,\xi)$ is analytic then the sequence $\{s(j,k)\}$ decays exponentially in its indices (j,k), indeed

$$\|s\|_\sigma^2 \equiv \sum_{j,k} |s(j,k)|^2 e^{2\sigma(|j|+|k|)} < \infty \tag{4.5}$$

for some choice of exponent σ. We consider nonlinear lattice problems in which the nonlinear term has certain properties of analyticity with respect to this norm. Define the scale of Hilbert spaces $\mathcal{H}_\sigma \equiv \{s \in \ell^2(\mathbf{Z}^2); \|s\|_\sigma < \infty\} \subseteq \ell^2(\mathbf{Z}^2)$. We require that $W \in C^\omega(\mathcal{H}_\sigma; \mathcal{H}_{\sigma-\gamma})$ for all $0 < \gamma \leq \sigma$, and furthermore that W satisfies the estimates

$$\|W(s)\|_{\sigma-\gamma} \leq \frac{C_W}{\gamma^3}\|s\|_\sigma^2 ,$$

$$\|D_s W(s)v\|_{\sigma-\gamma} \leq \frac{C_W}{\gamma^3}\|s\|_\sigma\|v\|_{\sigma-\gamma} , \tag{4.6}$$

$$\|D_s^2 W(s)[v,w]\|_{\sigma-\gamma} \leq \frac{C_W}{\gamma^3}\|v\|_\sigma\|w\|_{\sigma-\gamma} .$$

This is the case for the nonlinear wave equation (1.1), for all $\sigma < \overline{\sigma}$, where $\overline{\sigma}$ is related to the width of the complex strip in which the nonlinearity $g(x,u)$ is analytic.

4.2. Lyapounov-Schmidt decomposition

For $\Omega = \omega$ our fundamental frequency, the operator $V(\Omega)$ has a null space $\ell^2(N)$, where N is the resonant group. Define the orthogonal projections Q onto $\ell^2(N)$, and $P = (1 - Q)$. Equation (4.4) is equivalent to

$$
\begin{aligned}
P\big(V(\Omega)s + W(s)\big) &= 0 \\
Q\big(V(\Omega)s + W(s)\big) &= 0 \; .
\end{aligned}
\tag{4.7}
$$

Correspondingly, we decompose $s = \varphi + u$, with $\varphi \in \ell^2(N)$, and seek a solution $u = u(\varphi, \Omega)$ of the first equation of (4.7), parametrized by (φ, Ω).

Theorem 4.1. *There exists* $s_0 > 0$, *a neighborhood* $\mathcal{N}_0 = \{|\varphi| < s_0, |\Omega - \omega| < s_0^2\}$, *a function in* $\ell^2(\mathbf{Z}^2 \backslash N)$

$$
u(\varphi, \Omega) \in C^\infty(\mathcal{N}_0)
\tag{4.8}
$$

and a Cantor set $\mathcal{N} \subseteq \mathcal{N}_0$ *such that for* $(\varphi, \Omega) \in \mathcal{N}$, *the function* $u(\varphi, \Omega)$ *is a solution of the first equation of (4.7). Furthermore,*

$$
\|u\|_{\bar\sigma/2} \leq C|\varphi|^2 \; .
\tag{4.9}
$$

In fact the resonant group N need not be finite in order to prove Theorem 4.1. It will suffice if a finite number of lattice sites of N are well separated from other resonances and near resonances. That is, it is sufficient that $N_0 \equiv N \cap \{(j, k); |j| + |k| \leq L_0\}$ have its lattice sites satisfy the hypothesis (H2), with $(j_\ell, k_\ell) \in N_0$ satisfying $|j_\ell| + |k_\ell| \leq o(L_0)$. Theorem 3.3 will also hold in this situation, however it may be that the point $(\varphi, \Omega) = (0, \omega)$ is not in the Cantor set $\mathcal{N}$, so that the periodic solutions of the families $\mathcal{C}_\ell$ will not accumulate at $\varphi = 0$.

Proof. The first equation of (4.7) is the one that inherits the small divisor problem of (1.1). The proof must account for that, using a Nash-Moser iteration method with its supergeometric rate of convergence. The resulting linearized problems are solved in the spaces $\mathcal{H}_\sigma$, using Fröhlich-Spencer type estimates for the lattice Green's functions. The construction of the Green's functions must avoid resonances, thus portions of the parameter neighborhood $\mathcal{N}_0$ must be inductively excised, ultimately resulting in the Cantor set $\mathcal{N} \subseteq \mathcal{N}_0$. Complete details of this delicate process are given in reference [CW]. $\qquad\square$

This Lyapounov-Schmidt decomposition reduces the full problem to that of finding solutions of the second equation of (4.7), for parameter values which also lie in the set $\mathcal{N}$ of solutions to the first equation of (4.7). The induction and the excision process give rise to $\mathcal{N}$ which is a Cantor set, yet its relative measure is large and we have control of its geometry. This information is used to insure that the solution set of the second bifurcation equation will intersect $\mathcal{N}$.

5. The principle of least action

In order to describe solutions of the second bifurcation equation in (4.7), we find it convenient to appeal to some formalism of Hamiltonian mechanics. The **action** for periodic solutions of the nonlinear wave equation (1.1) is the functional

$$A(u) = \int_0^{2\pi/\Omega} \int_0^\pi \tfrac{1}{2}(\partial_t u)^2 - \tfrac{1}{2}(\partial_x u)^2 - G(x, u) \ dx dt \ , \tag{5.1}$$

for functions which satisfy the appropriate boundary conditions, and are $2\pi/\Omega$ periodic in time. This can be obtained, for example, from the Legendre transform of the Hamiltonian $\mathcal{H}$. The principle of least action states that critical points of the action,

$$\delta A(u) = 0 \tag{5.2}$$

are solutions to (1.1). If we normalize $\xi = \Omega t$, and set

$$K(u) = \int_0^{2\pi} \int_0^\pi \tfrac{1}{2}(\partial_\xi u)^2 \ dx d\xi \tag{5.3}$$

the average kinetic energy, and

$$\Phi(u) = \int_0^{2\pi} \int_0^\pi \tfrac{1}{2}(\partial_x u)^2 + G(x, u) \ dx d\xi \tag{5.4}$$

the average potential energy, then there is an equivalent principle which is preferable to use. Let $\mathcal{M} = \{u; K(u) = s^2\}$ be the set of functions with fixed average kinetic energy. Critical points of Φ over the set $\mathcal{M}$ give

$$\delta \Phi(u) = \Omega^2 \delta K(u) \ , \tag{5.5}$$

with Ω^2 the Lagrange multiplier. Formally such critical points, if the Lagrange multiplier is positive, give rise to solutions of (1.1), with frequency Ω. Incidentally, this equivalent principle is an alternative to the least action principle whenever the kinetic energy in the Hamiltonian is given by a nondegenerate quadratic form.

Unfortunately this direct method for the solution of the nonlinear wave equation seems quite intractable, mainly for analytic reasons. We will use a modification of it however to solve the second bifurcation equation. Define a restricted principle of least action for $(\varphi, \Omega) \in \mathcal{N}_0$, using the function $u(\varphi, \Omega)$ of Theorem 4.1;

$$
\begin{aligned}
\Phi(\varphi, \Omega) &= \Phi(\varphi + u(\varphi, \Omega)) \\
K(\varphi, \Omega) &= K(\varphi + u(\varphi, \Omega)) \ .
\end{aligned}
\tag{5.6}
$$

The set $\mathcal{M}_s \equiv \{s; K(\varphi, \Omega) = s^2\}$ for fixed Ω is topologically a sphere in $\mathbf{R}^{2M}$. Stemming from the fact that (1.1) is autonomous, there is an S^1 action on $\ell^2(N)$ under which both K and Φ are invariant. Explicitely, if we write $\varphi = \sum_N p_\ell \delta(j_\ell, k_\ell) + \overline{p}_\ell \delta(j_\ell, -k_\ell)$ then for $\xi \in [0, 2\pi)$ the S^1 action is given by $T_\xi \varphi = \sum_N e^{i\xi k_\ell} p_\ell \delta(j_\ell, k_\ell) + e^{-i\xi k_\ell} \overline{p}_\ell \delta(j_\ell, -k_\ell)$. This group action represents translation in time in the original problem.

The variational problem (5.5), using the restricted functionals of (5.6), gives rise to the same situation as that which has been addressed in the papers of A. Weinstein [We] and J. Moser [M]. From a theorem of Krasnoselskii one concludes that for each level set $\mathcal{M}_s$ there are at least M many critical points of Φ. For s small it can be checked that the Lagrange multiplier of each of these is positive. This gives rise to the count $\overline{M} \geq M$ appearing in Theorem 3.3. This does not complete the story, for in the infinite dimensional situation we need additionally to show that some of the critical points are in $\mathcal{N}$, and thus the first bifurcation equation (4.7) is also solved. This will be assured if a condition of genuine nonlinearity is satisfied. We will leave a discussion of this condition in the general resonant case to a further publication, and confine ourselves here to several particular resonant cases, in which the critical points of (5.5) can be explicitly analysed.

6. The 1:2 and 1:1:2 resonances

Using the solution $u(j, k; \varphi, \Omega)$ of equation (4.7) the embedded circle (4.3) is given by the sequence

$$
\begin{aligned}
s(j_\ell, k_\ell) &= p_\ell, \quad (j_\ell, k_\ell) \in N \ , \\
s(j, k) &= u(j, k; \mathbf{p}, \Omega), \quad (j, k) \notin N \ .
\end{aligned}
\tag{6.1}
$$

In terms of this sequence s, the average kinetic and potential energies are

$$
\begin{aligned}
K &= \tfrac{1}{2} \sum_{j,k} k^2 |s(j,k)|^2 \\
\Phi &= \tfrac{1}{2} \sum_{j,k} \omega_j^2 |s(j,k)|^2 + \sum_{m \geq 3} \left(\sum_{|\beta| = m} C_\beta s(\beta_1) s(\beta_2) \cdots s(\beta_m) \right) \ ,
\end{aligned}
\tag{6.2}
$$

where $\beta = (\beta_1, \ldots, \beta_m)$ are multiindices. The average potential energy Φ is given as a sum of terms $\Phi_m(s) \equiv \sum_{|\beta|=m} C_\beta \prod_{\ell=1}^m s(\beta_\ell)$, which are homogeneous of degree m in s. The interaction coefficients C_β, $|\beta| = m$ depend upon $g_m(x)$ in a way determined from expression (5.4). One sees immediately the following simple result.

Proposition 6.1. *The interaction coefficient C_β is nonzero only if*

$$k_1 + \cdots + k_m = 0 \; . \tag{6.3}$$

When all $\beta_\ell = (j_\ell, k_\ell) \in N$ this implies that there is the relationship $\omega_{j_1} \pm \omega_{j_2} \pm \ldots \pm \omega_{j_m} = 0$.

Examples (i) The $1 : 2$ resonance case is where

$$\omega_{j_1} = \frac{\omega_{j_2}}{2} \; . \tag{6.4}$$

Let $\beta_{\pm 1} = (\omega_{j_1}, \pm k_1)$, $\beta_{\pm 2} = (\omega_{j_2}, \pm k_2)$, then the only nonzero interaction coefficients with $m = 3$ are the term $C_{\beta_1, \beta_1, \beta_{-2}}$ and its complex conjugate $C_{\beta_{-1}, \beta_{-1}, \beta_2}$.

(ii) The $1 : 1 : 2$ resonance is the case in which there is the relationship

$$\omega_{j_1} = \omega_{j_2} = \frac{\omega_{j_3}}{2} \; . \tag{6.5}$$

The possible nonzero interaction coefficients are $C_{\beta_1, \beta_1, \beta_{-3}}$, $C_{\beta_1, \beta_2, \beta_{-3}}$, and $C_{\beta_2, \beta_2, \beta_{-3}}$ and their complex conjugates.

Proposition 6.2. *The Dirichlet problem for equation (1.1) is never in $1 : 1 : 2$ resonance.*

This is clear, since Sturm-Liouville problems posed on $[0, \pi]$ have only simple eigenvalues. There are many other resonances for $m = 3$ that do occur however, including the possibilities $(1 : 2 : 3)$, $(1 : 2 : 4)$, $(1 : q : q + 1)$ and $(1 : q : 2q)$.

We remark that while the coefficients C_β are generically zero for $|\beta|$ odd, when $|\beta| = 2n$ the generic situation is that there are nonzero resonant terms present. These take the form C_α, with $\alpha = ((j_1, k_1), (j_1, -k_1), \ldots (j_n, k_n), (j_n, -k_n))$, and the resonant monomials are in the form $C_\alpha |s(\alpha_1)|^2 \cdots |s(\alpha_n)|^2$. This is as is expected from the Birkhoff normal form.

To understand the structure of the set of periodic solutions of frequency near ω, one studies the behavior of the critical points of the principle (5.5) in both parameters φ and Ω, for the restricted functionals (6.2), with s near zero. Since there is the estimate (4.9), in the two resonant cases $1 : 2$ and $1 : 1 : 2$ this can be deduced from K and $\Phi_2 + \Phi_3$.

6.1. The 1:2 case

The parameter is given by $\sum_{\ell=1}^{2} p_{\ell} \delta(j_{\ell}, k_{\ell}) + \overline{p}_{\ell} \delta(j_{\ell}, -k_{\ell})$, and the functionals are accordingly

$$
\begin{aligned}
K(\varphi) &= \tfrac{1}{2}(|p_1|^2 + 4|p_2|^2) + O(|p|^4) \\
\Phi_2 + \Phi_3 &= \frac{\omega^2}{2}(|p_1|^2 + 4|p_2|^2) + \mathrm{re}\,(C_{(\beta_1,\beta_1,\beta_{-2})} p_1^2 \overline{p}_2) + O(|p|^4) \ .
\end{aligned}
\tag{6.6}
$$

For s near zero the branches of families of critical points are obtained which are approximated by solutions of the finite dimensional problem $\nabla_p(\Phi_2 + \Phi_3)(p) = \Omega^2 \nabla K(p)$, in which $u(j,k)$, $(j,k) \notin N$ are omitted. Solution by direct means shows that there are three families of solutions, thus $\overline{M} = 3 > M = 2$. Since K and all Φ_m are invariant under the circle action T_ξ, these families of solutions are as well. There are two families of 'slow' solutions, with frequency close to ω (in fact one frequency is slightly slower, and one slightly faster). In coordinates, the first order approximation to the solutions is:

$$
\begin{aligned}
p_1 &= s\sqrt{\frac{4}{3}} e^{i\xi} \\
p_2 &= \pm\, s\frac{1}{\sqrt{6}} \frac{C_{(\beta_1,\beta_1,\beta_{-2})}}{|C_{(\beta_1,\beta_1,\beta_{-2})}|} e^{2i\xi} \\
\Omega^2 &= \omega^2 \pm s\sqrt{\frac{2}{3}} |C_{(\beta_1,\beta_1,\beta_{-2})}| \ .
\end{aligned}
\tag{6.7}
$$

Features of note are that the frequency of the solution changes linearly with amplitude, and that the angle of the plane of the approximate solution from the p_1-plane is $\arctan(\sqrt{1/8})$, which is a universal quantity, independent of the interaction coefficient $C_{(\beta_1,\beta_1,\beta_{-2})}$ and only depending upon the presence of the resonance.

The third family of solutions has frequency close to 2ω, and is given to highest order by

$$
\begin{aligned}
p_1 &= 0 \\
p_2 &= s\frac{1}{\sqrt{2}} e^{2i\xi} \\
\Omega^2 &= (2\omega)^2 + \kappa s^2 \ .
\end{aligned}
\tag{6.8}
$$

The frequency varies only quadratically with the amplitude of the solution, as in the nonresonant cases. The expression for the curvature κ is given in [CW], we give it here for completeness. Let $\beta_2 = (j_2, k_2)$, $\beta_{-2} = (j_2, -k_2)$, $\beta = (\beta_2, \beta_2, \beta_{-2}, \beta_{-2})$, and define $C_{2,2} = C_\beta$. Furthermore let $u((j,k); p, \Omega) = p_2^{k/2} u_0(j,k) + o(|p|^2)$ for k even, and define $a_0 = \sum_j C_{(\beta_2, \beta_{-2}, (j,0))} u_0(j,0)$, $a_4 = \sum_j C_{(\beta_2, \beta_{-2}, (j,-4))} u_0(j,-4)$. Then

$$
\kappa = \tfrac{1}{2}(C_{2,2} + \mathrm{re}\,(a_0 + a_4)) \ .
\tag{6.9}
$$

Fix the exponent $1/2 < \eta < 1$. To insure that a solution family of (5.5) intersects the solution set $\mathcal{N}$ of the first bifurcation equation (4.7), the condition of genuine nonlinearity (H3) is that

$$|C_{(\beta_1,\beta_1,\beta_{-2})}| > L_0^{-\eta/2} \tag{6.10}$$

for the two 'slow' families of solutions, and that

$$|\kappa| > L_0^{-\eta} \tag{6.11}$$

for the third family. In fact due to the nonresonance condition (H2), the third family has 2ω as fundamental frequency, and (ω_{j_2}, k_2) satisfies the hypotheses for $M = 1$, which is a case covered in the previous paper [CW].

6.2. The 1:1:2 case

The subspace $\ell^2(N) \simeq \mathbf{C}^3$ is parametrized by $\varphi = \sum_{\ell=1}^{3} p_\ell \delta(j_\ell, k_\ell) + \overline{p}_\ell \delta(j_\ell, -k_\ell)$, and the relevant functionals are

$$
\begin{aligned}
K(\varphi) =& \tfrac{1}{2}(|p_1|^2 + |p_2|^2 + 4|p_3|^2) + O(|p|^4) \\
\Phi(\varphi) =& \frac{\omega^2}{2}(|p_1|^2 + |p_2|^2 + 4|p_3|^2) \\
& + \mathrm{re}\left((C_{(\beta_1,\beta_1,\beta_{-3})} p_1^2 + C_{(\beta_1,\beta_2,\beta_{-3})} p_1 p_2 + C_{(\beta_2,\beta_2,\beta_{-3})} p_2^2) \overline{p}_3 \right) + O(|p|^4) \ .
\end{aligned}
\tag{6.12}
$$

By a rotation of the (p_1, p_2)-plane we may assume that $C_{(\beta_1,\beta_2,\beta_{-3})} = 0$. As in the $1 : 2$ case, we may approximate solutions by the above functionals without the $O(|p|^4)$ terms. In case $|C_{(\beta_1,\beta_1,\beta_{-3})}| \neq |C_{(\beta_2,\beta_2,\beta_{-3})}|$ there are five families of approximate solutions, given in the following list:

$$
\begin{aligned}
p_1 =& s\sqrt{\frac{4}{3}} e^{i\xi} \\
p_3 =& \pm s\frac{1}{\sqrt{6}} \frac{C_{(\beta_1,\beta_1,\beta_{-3})}}{|C_{(\beta_1,\beta_1,\beta_{-3})}|} e^{2i\xi} \\
\Omega^2 =& \omega^2 \pm s\sqrt{\frac{2}{3}}|C_{(\beta_1,\beta_1,\beta_{-3})}| \ ,
\end{aligned}
\tag{6.13}
$$

and

$$
\begin{aligned}
p_1 =& s\sqrt{\frac{4}{3}} e^{i\xi} \\
p_3 =& \pm s\frac{1}{\sqrt{6}} \frac{C_{(\beta_2,\beta_2,\beta_{-3})}}{|C_{(\beta_2,\beta_2,\beta_{-3})}|} e^{2i\xi} \\
\Omega^2 =& \omega^2 \pm s\sqrt{\frac{2}{3}}|C_{(\beta_2,\beta_2,\beta_{-3})}| \ .
\end{aligned}
\tag{6.14}
$$

There is a fifth faster solution family, with frequency close to 2ω.

$$
\begin{aligned}
p_1 &= 0 \\
p_2 &= 0 \\
p_3 &= \frac{s}{\sqrt{2}} e^{2i\xi} \\
\Omega^2 &= 4\omega^2 + s^2\kappa \quad ,
\end{aligned}
\tag{6.15}
$$

with κ given by an expression analogous to (6.9). Each family is foliated by circles, due to the invariance of the problem under the action of T_ξ. The number of families of solutions is $\overline{M} = 5 > M = 3$.

In case $|C_{(\beta_1,\beta_1,\beta_{-3})}| = |C_{(\beta_2,\beta_2,\beta_{-3})}|$ (always in coordinates in which $C_{(\beta_1,\beta_2,\beta_{-3})}$ $= 0$), the above approximate problem has a full two dimensional torus of critical points for each value s^2 of the average kinetic energy, and the relationship to the critical points of the full problem is determined only after including higher order terms in the variational problem. There will be at least $\overline{M} = 3$ families of solutions in this case. However we expect in the generic case that we will have $\overline{M} = 5$. Additionally, depending on the nature of the higher order terms, there can be arbitrarily large numbers of families of periodic solutions. This is similar to the situation that we will see in the statements for the nonresonant cases below.

The condition of genuine nonlinearity for the four 'slow' solutions is that the discriminant

$$
\Delta = \left(C_{(\beta_1,\beta_2,\beta_{-3})} \right)^2 - 4 C_{(\beta_1,\beta_1,\beta_{-3})} C_{(\beta_2,\beta_2,\beta_{-3})}
\tag{6.16}
$$

satisfy $\Delta > L_0^{-\eta}$. For the fifth family the condition is the same as in [CW], that is $\kappa > L_0^{-\eta}$.

In the above examples we have seen that the presence of resonance has not eliminated the periodic solutions, in fact the contrary has occurred. The number of families of periodic solutions has increased in the above two cases, something that we believe is a general phenomenon in both finite and infinite degrees of freedom. Additionally, many of the families are such that the frequency varies linearly with amplitude, as opposed to quadratic or higher order behavior in nonresonant cases. As the frequency must 'pass through' resonant gaps in the Cantor set $\mathcal{N}$ in order solve the full problem (4.7), this increases the measure of the set of surviving periodic solutions, as compared to the nonresonant case.

6.3. Nonresonant cases

In the last section of this paper we discuss a more general class of problems, where the resonant group has $M > 1$, but there are no exceptional cubic or quartic resonance relations. The analysis procedes as above, reducing the problem through Theorem 4.1 to the second bifurcation equation (4.7). This is equivalent to finding critical points of the restricted principle of least action. As these critical points must lie within the Cantor set $\mathcal{N}$, the effort of the existence theorem is directed at identifying the condition of genuine nonlinearity that will guarantee this, at least for a Cantor set of parameter values of large relative measure.

For simplicity, we address the class of problems such that

$$g_2(x) = 0 \ , \tag{6.17}$$

so that in the average potential energy in (6.2) any cubic coefficients vanish, $C_\beta = 0$, $|\beta| = 3$. We will also assume that there are only generic fourth order resonance relations. That is, among $(j_\ell, k_\ell) \in N$, there is no relation

$$k_1 + k_2 + k_3 + k_4 = 0 \tag{6.18}$$

unless $k_1 = -k_2$, $k_3 = -k_4$, or some other permutation. A further consequence of (6.17) is the estimate of the solution u of the first bifurcation equation (4.7);

$$\|u\|_{\overline{\sigma}} \leq C|\varphi|^3 \ . \tag{6.19}$$

The restricted variational problem becomes

$$K(p) = \tfrac{1}{2} \sum_{\ell=1}^{M} k_\ell^2 |p_\ell|^2 + O(|p|^6)$$

$$\Phi(p) = \frac{\omega^2}{2} \sum_{\ell=1}^{M} k_\ell^2 |p_\ell|^2 \tag{6.20}$$

$$+ \sum_{\ell,m=1}^{M} C_{\ell,m} |p_\ell|^2 |p_m|^2 + O(|p|^6)$$

with matrix elements $C_{\ell,m} = C_\beta$, where $\beta_{\pm\ell} = (j_\ell, \pm k_\ell) \in N$ and $\beta = (\beta_\ell, \beta_{-\ell}, \beta_m, \beta_{-m})$.

We will estimate the position of any critical points of equation (5.5) using the truncated functionals with the $O|p|^6$ terms dropped.

$$\partial_{\overline{p}} K = k_\ell^2 p_\ell$$

$$\partial_{\overline{p}} \Phi = \omega^2 k_\ell^2 p_\ell \tag{6.21}$$

$$+ \sum_m C_{\ell,m} |p_m|^2 p_\ell$$

Thus solutions of $\delta\Phi = \Omega^2 \delta K$ satisfy

$$(\Omega^2 - \omega^2)k_\ell^2 = \sum_m C_{\ell,m}|p_m|^2 \quad . \tag{6.22}$$

The natural condition to require is that the vector (k_ℓ^2) be in the range of the matrix $(C_{\ell,m})$. Thus there is a left inverse C^{-1} defined on $\vec{k}$ and we have the solution

$$|p_m|^2 = (\Omega^2 - \omega^2)\sum_\ell C_{\ell,m}^{-1}k_\ell^2. \tag{6.23}$$

Additionally the constraint that $K(p) = s^2$ implies that

$$(\Omega^2 - \omega^2) = 2s^2\Big(\sum_{\ell,m} C_{\ell,m}^{-1}k_\ell^2 k_m^2\Big)^{-1} \quad . \tag{6.24}$$

The condition of genuine nonlinearity is that the coefficient of s^2 of the right hand side of (6.24) be of magnitude greater than $L_0^{-\eta}$.

The critical points (6.23) have given amplitude but the angle is not specified. That is, there is a full M dimensional torus of critical points for the truncated problem. The actual critical set for the full problem (consisting of circles invariant under T_ξ) will have their position determined by the higher order terms in the restricted average potential energy. These can be obtained as a variational problem over the $(M-1)$-torus. There will be at least M of them, as the Lyusternik-Schnirlman category of T^{M-1} is M. Generically however, 2^{M-1} critical circles appear, due to the Morse inequalities, and there is the possibility of many more, depending again on the details of the higher order terms.

Acknowledgements: The authors would like to thank the Département de Physique Thèorique, Université de Genève and the Mathematical Institute of Oxford University for their hospitality. Research for this paper was supported in part by the National Science Foundation under grants # DMS-8920624 and # DMS-9002059. Additionally the first author is partially supported by a Sloan Foundation Fellowship.

References

[AFS] Albanese, C. and Fröhlich, J. and Albanese, C. Fröhlich, J. and Spencer, T.: Periodic solutions of some infinite-dimensional hamiltonian systems associated with non-linear partial difference equations: Parts I and II. Commun. Math. Phys. **116**, 475-502 **119**, 677-699 (1988).

[CW] Craig, W. and Wayne, C.E.: Newton's method and periodic solutions of nonlinear wave equations. Commun. Pure Applied Math. **XLVI**, 1409-1501 (1993).

[CW2] Craig, W. and Wayne, C.E.: preprint in peparation. (1993).

[CW3] Craig, W. and Wayne, C.E.: Nonlinear waves and the KAM theorem: nonlinear degeneracies. Large scale structures in nonlinear physics; Springer Lecture Notes in Physics, eds. J.-D. Fournier and P.-L. Sulem **392**, 37-49 (1991).

[E] Elliasson, H.: Perturbations of stable invariant tori.. Annali Sc. Norm. Sup.-Pisa **IV Ser. 15**, 115-147 (1988).

[K] Kuksin, S.: Perturbation of quasiperiodic solutions of infinite-dimensional linear systems with an imaginary spectrum. Preprint of Max-Plank-Institut, Bonn **21**, 192-205 (1987).
Perturbation theroy for quasiperiodic solutions of infinite-dimensional hamiltonian systems; Parts I-III. (1990).

[M] Moser, J.: Periodic orbits near an equilibrium and a theorem by Alan Weinstein. Commun. Pure Applied Math. **XXIX**, 727-747 (1976).

[P] Pöschel, J.: Small divisors with spatial structure in infinite dimensional hamiltonian systems. Commun. Math. Physics **127**, 351-393 (1990).

[W] Wayne, C.E.: Periodic and quasi-periodic solutions of nonlinear wave equations via KAM theory. Commun. Math. Phys. **127**, 479-528 (1990).

[We] Weinstein, A.: Normal modes for non-linear hamiltonian systems. Inventiones Math. **20**, 47-57 (1973).

[Ze] Zehnder, E.: Generalized implicit function theorems with applications to some small divisor problems I. Commun. Pure and Applied Math. **XXVIII**, 91-140 (1975).

DEFECTS OF ONE - DIMENSIONAL VORTEX LATTICES

A.I. Chernykh[1], I.R. Gabitov[2] and E.A. Kuznetsov[1]

[1]Institute of Automation and Electrometry
Sib. Branch Russian Academy of Sciences
630090, Novosibirsk, Russia

[2]L.D. Landau Institute for Theoretical Physics
Russian Academy of Sciences
Kosygina St. 2, 117334, Moscow, Russia

Abstract

In the framework of the equation

$$\psi_t = \psi_{xx} + \psi - |\psi|^2\psi \,,$$

the dynamics of one-dimensional lattices of Taylor vortices in Couette flow and of rolls in weak supercritical convection is studied. It is shown that the propagation of the defects as transition areas between stable (according to Eckhaus) and unstable lattices depends significantly on the topological properties of the field $\psi(x)$, i.e. the degree of mapping $R^1 \to S^1$. The velocity of such defects has been determined. It has been clarified that the defects between stable lattices spread diffusively due to the conservation of the topological invariant.

1. INTRODUCTION

This article studies the dynamics of defects described by the equation

$$\psi_t = \psi_{xx} + (1 - |\psi|^2)\psi \,. \tag{1}$$

This equation arises when we study the weak modulation of a $1D$ vortex lattice over x in the vicinity of the instability threshold in systems whose laminar state had the translational symmetry along some axis. Such systems comprise a flow between two coaxial cylinders with a fixed external and a rotating internal, Couette flow, losing their stability at a certain critical Reynolds number Re, which results in the formation of Taylor vortices, and a flow emerging beyond the threshold of convective instability

Singular Limits of Dispersive Waves, Edited by
N.M. Ercolani et al., Plenum Press, New York, 1994

of the liquid between two horizontal rigid planes. In these cases the laminar state, i.e., the Couette flow or the stationary heat transfer due to the heat conductivity from lower plane to the upper, is invariant with respect to spatial shift. Therefore, in the vicinity of the threshold the instability growth rate γ_k, dependent on the wave number k, will have, at a certain $k = k_0$, a maximum whose value for a weak supercriticality $\epsilon = (Re - Re_c)/Re_c$ is proportional to ϵ. In the vicinity of the maximum in this case the growth rate can be approximated by the quadratic dependence

$$\gamma_k = \gamma_0 - \alpha(k - k_0)^2, \qquad \gamma_0 \sim \epsilon.$$

This formula shows that at the linear stage a whole range of perturbations with small width $\Delta k \sim \sqrt{\gamma_0} \ll k_0$ is excited. Therefore, to find out the structure of the nonlinear term leading to the saturation of the instability, it is sufficient to average the original equations of motion over "fast" spatial oscillations. This averaging, after some simple rescaling, leads to Eq. (1) for the amplitude. (This is how this equation was derived by Newell and Whitehead [1] for the weak supercritical convection.) Subsequently, the same equation was obtained for the description of a modulation of a $1D$ chain of Taylor vortices near the instability threshold for the Couette flow [2].

It is well known that Eq. (1) can be represented in the variational form as

$$\partial \psi / \partial t = -\delta F / \delta \psi^*, \tag{2}$$

where $F = \int \mathcal{F} dx$ has the meaning of the free energy, with density

$$\mathcal{F} = -|\psi|^2 + |\psi|^4/2 + |\psi_x|^2.$$

The stationary stable state corresponds to the free energy minimum. From (2) it follows that

$$\partial F / \partial t = -2 \int |\delta F / \delta \psi|^2 dx < 0. \tag{3}$$

Hence, it is clear that F is a Lyapunov functional, and the state corresponding to the global minimum will, according to the Lyapunov theorem, be stable. To find the stationary point of the functional F it is essential to know the boundary conditions. If one studies (1) on the whole axis ($-\infty < x < +\infty$), the expression for free energy for distributions nonvanishing at infinity since, only such distributions make physical sense, will linearly diverge with an increasing size. Therefore, in this case in order to define the minimum of F it is sufficient to compare the free energy densities $\mathcal{F}$. If we consider the simplest stationary solutions of (1) in the form

$$\psi = (1 - k^2)^{1/2} e^{ikx}, \qquad 0 < k^2 < 1, \tag{4}$$

the free energy density

$$\mathcal{F} = -(1 - k^2)^2/2,$$

will evidently be minimal at $k = 0$. In this state, from a class of functions nonincreasing at infinity, F will have a global minimum and, therefore, it is stable with respect to not only small but also finite perturbations. Nevertheless, apart from the solution $\psi = 1$ there exist other stable solutions realizing local minima of F. The study of the stability problem with respect to small perturbations leads to the so-called Eckhaus criterion [3]

$$k^2 \leq 1/3, \tag{5}$$

of the stability region of solutions (4). It is important to emphasize that the analysis of the stability of arbitrary stationary solutions of (1) showed that there are no other stable solutions except those described above [4].

In the case when the region is finite, which is actually realized in experiments, the question of defining stable solutions remains open. This problem is of particular importance when we perform numerical simulations of (1) i.e., we must necessarily solve the boundary problem.

Henceforth, we shall confine ourselves to the discussion of three versions of the boundary conditions:

1) zero, when $\psi(0) = \psi(l) = 0$;

2) periodic;

3) $\psi(0) = a_0$, $\psi(l) = a_1$, where a_0 and a_1 are time-independent constants.

From the point of view of applications, the zero boundary conditions are typical for hydrodynamics. For convection this means that the edge vortex near the wall has a zero amplitude. For the Couette flow, however, the boundary conditions for ψ should be equal to a certain constant determined by the frequency of the rotation of the external cylinder than equal to zero. Near the edges of the interval for the first and third boundary conditions, the stationary solution realizing a minimum of F will significantly differ from $\psi = 1$. If the boundary conditions are periodic, then, obviously, the minimum of F is realized at $\psi = 1$. In this case among the solutions of (4) only those hold up for which $k_n^2 = (2\pi n/l)^2$ with integer n. At $k_n^2 > 1/3$ these solutions will be unstable.

In this paper we study the dynamics of defects of vortex lattices which represent transition domains between i) stable distributions of (4) and ii) unstable solutions of (4). We investigate the development of both the nonlinear stage of instability of solutions (4) and the influence of the boundary conditions on it. We show that the dynamics of such defects essentially depends on topological characteristics of the field $\psi(x)$. In the case when a defect is a region of transition between stable states, this defect expands diffusively. In contrast to such defects, a defect between a stable ($\psi = 1$) and unstable states propagates with a certain velocity oscillating in time. The mean velocity and the frequency of oscillations are analytically determined and the comparison is made with the data of numerical experiments. It is important to emphasize that the problem of the propagation by its formulation represents a generalization of the problem posed and solved by Kolmogorov, Petrovsky and Piskunov [5] for an equation of the form (1) for a real field $\psi(x)$. The method which we have used to solve this problem is borrowed by us from the work by Kamensky and Manakov [6]. It should be noted that later and independently this method was recovered in the paper[7].

2. STATIONARY STATES AND LINEAR STABILITY

Let us study the problem of stationary states and their stability.

First, let us give a brief solution of the problem of the stability of stationary states (4) assuming the interval to be infinite. Represent ψ as

$$\psi = (\tilde{\psi}_0 + \chi' + i\chi'')e^{ikx}, \tag{6}$$

where $\tilde{\psi}_0 = (1 - k^2)^{1/2}$, χ' and χ'' are the real and imaginary part of the perturbation and $|\chi'|$, $|\chi''| \ll |\tilde{\psi}_0|$. Then in the linear approximation we have

$$\begin{aligned} \chi'_t &= \chi'_{xx} - 2\tilde{\psi}_0^2\chi' - 2k\chi''_x, \\ \chi''_t &= \chi''_{xx} + 2k\chi'_x. \end{aligned} \tag{7}$$

Assuming that χ', $\chi'' \sim e^{\Gamma t + i\sigma x}$, for the growth rate Γ we get

$$\Gamma_{1,2} = -\sigma^2 - \tilde{\psi}_0^2 \pm \left(\tilde{\psi}_0^4 + 4k^2\sigma^2\right)^{1/2}. \tag{8}$$

Hence, in particular, it follows that instability ($\Gamma > 0$) takes place [3] at

$$k^2 > 1/3, \tag{9}$$

and that stability occurs in the opposite case. In the case of a finite interval L and periodic boundary conditions the criterion (9) remains if we put $k = 2\pi n/L$, where n is an integer.

Now consider the case of zero boundary conditions, assuming that the length of the interval is $L \gg 1$. Then far from the boundary the minimum of F will be realized by a function close to $\psi = 1$ with an exponential accuracy as will be shown below. Thus, one needs to find a stationary solution which tends to $\psi = 1$ far from $x = 0$, and turns into zero at $x = 0$. It is easy to see that this solution is

$$\psi_0(x) = \tanh(x/\sqrt{2}). \tag{10}$$

Now let us demonstrate that this solution is stable in the class of functions with $\psi(0) = 0$ and $\psi \to 1$ as $x \to \infty$. In the linear approximation, $\psi = \psi_0 + \xi$, for perturbations $\xi = \xi' + i\xi'' \sim e^{-Et}$ the following spectral problem arises

$$E\xi' = -\xi'_{x'x'}/2 - (1 - 3\psi_0^2)\xi', \tag{11}$$

$$E\xi'' = -\xi''_{x'x'}/2 - (1 - \psi_0^2)\xi'', \tag{12}$$

where $x' = x/\sqrt{2}$. The first equation, Schroedinger equation, has a stable solution in the form of the shift mode $\xi'_0 = \partial\psi_0/\partial x = 1/(\sqrt{2}\cosh^2 x')$, corresponding to the ground state $E = 0$. All other eigenfunctions of (11) have $E > 0$ and, consequently, are stable. The second equation of the system (11)-(12) has one bound state $\xi' = \cosh^{-1} x'$, with $E = -1/2$ and a continuous spectrum with $E > 0$. In the absence of the boundary these perturbations grow as $e^{t/2}$. However, in the presence of the boundary when $\psi(0) = 0$, symmetric solutions do not survive; there are only anti-symmetric solutions with $E > 0$, and perturbations prove to be stable. This confirms the linear stability of this solution.

It is easy to show that this solution realizes the minimum of the functional

$$F = \int_0^\infty |\psi_x|^2 + \frac{1}{2}(|\psi|^2 - 1)^2 \, dx \,, \tag{13}$$

on the class of functions with $\psi(0) = 0$ and $\psi(\infty) = 1$, where at F in comparison with (2) the constant term corresponding to $\psi = 1$ is subtracted.

To prove this statement it is sufficient to examine all stationary points of the functional (13) and select those that obey the necessary boundary conditions. The stationary equation associated with (1) is

$$\psi_{xx} + \psi - |\psi|^2 \psi = 0 \,;$$

it is easy to check that it has only one solution satisfying the necessary boundary conditions, i.e., $\psi_0 = \tanh(x/\sqrt{2})$. Hence, it follows that $\psi_0(x)$ realizes the absolute minimum of F (13) and consequently is stable according to the Lyapunov theorem.

In the case when the interval size l is large ($l \gg 1$) and the boundary conditions are zero, the solution $\psi_0(x)$ will be very close to $\tanh(x/\sqrt{2})$ in the vicinity of $x = 0$ and to $\tanh((l - x)/\sqrt{2})$ near $x = l$. In the middle of the region the solution will approximate $\psi = 1$ with an exponential accuracy. In Fig.1 we see the dependence of $|\psi|$ on x for $l = 25$ of the stationary state which arises as a result of the development of instability of small initial data. Towards the middle of the interval, the difference of ψ from 1 amounts to 10^{-6} whereas on the edges $\psi(x)$ is described with good accuracy by the dependence $\tanh(x/\sqrt{2})$.

Let the initial condition $\psi_0(x)$ represent a defect for the infinite interval. As $x \to \infty$, $\psi_0(x)$ approaches the absolutely stable solution $\psi = 1$, and at the other infinity tends to one of the solutions of (4). For vortices this defect is a region of transition between a chain of vortices having an optimal size corresponding to $k = k_0$ and a system of vortices compressed or stretched in comparison with the optimal size of the vortex. Since the state (4) has a larger value of the free energy density $\mathcal{F}$ than $\psi = 1$ such a defect will propagate in accordance with (3) towards a decrease of F, i.e., in the given case to the right. It is clear that in a finite but sufficiently large system the influence of the boundaries will not effect the defect if its size Δl is small in comparison with l. However, we should stress that over a long period of time the influence of the boundaries will become significant. The most important factor determining the dynamics of the defect is connected with topological restrictions.

3. TOPOLOGICAL CONSTRAINTS

Let us assume the boundary conditions are periodic; then the difference of phases on the boundary of the interval $\Phi = \phi(l) - \phi(0) = \int_0^l \partial\phi/\partial x \, dx$ of the field $\psi(x) = Ae^{i\phi}$ (A is the amplitude and ϕ is the phase) must be equal to $2\pi N$ for some integer N. In other words, this means that the edge of the two-dimensional vector $\mathbf{A} = (\psi', \psi'')$, where ψ' and ψ'' are the real and imaginary parts of ψ respectively, while the "motion" along the axis x from $x = 0$ to $x = l$ describes some helical line, performing N rotations around the x axis. In the case when the amplitude does not vanish anywhere, N coincides with a degree of the mapping $R^1 \rightarrow S^1$. If as a result of the evolution in t the vector $\mathbf{A}$ does not turn into zero at any point, then N is a certain integral of motion. However, if in a certain moment of time t_0 the vector $\mathbf{A}$ does turn into zero in a certain point x_0, then in this case Φ will change by 2π. From the geometrical point of view, this corresponds to the intersection of the curve described by the edge of the vector $\mathbf{A}$ and x-axis. As a result of such an intersection, the number of rotations N will change by one unit. This consideration shows that this effect does not depend on the type of boundary conditions. In any case, such a topological property, if the field $\psi(x)$ exists, its change will occur in the same manner.

Let us now consider what the solutions of (4) are from this geometrical point of view. For these solutions the vector $\mathbf{A}$ describes a helix with a constant step $h = 2\pi/k$. On the other hand, as has already been pointed out, the solutions of (4) have a smaller value of free energy than $\psi = 1$ corresponding in the three-dimensional space (ψ', ψ'', x) to a straight parallel to x-axis. From the point of view of energy, it is preferable for the solutions (4) to transit into the state $\psi = 1$. This transition must be accompanied by a phase jump by the integer N in units of 2π and by vanishing of ψ in certain moments of time t_{0i} at a certain point x_{0i}. This transition is possible if a strong instability exists since the state $\psi = 0$ is unstable. We should remember that for the solution $\psi_0 = \tanh(x/\sqrt{2})$ the point $x = 0$ occurs as a saddle: along one direction (real) there is attraction; along the other (imaginary) we have repulsion. In Figs.2-4 are the results of simulations of Eq. (1) with periodic boundary conditions for the initial data $\psi = (\tilde{\psi}_{0k} + \epsilon\cos x)e^{ikx}$ with $k = 14 * 2\pi/125$ from the unstable region, $\sigma = 10 * 2\pi/125$ and $\epsilon = 0.02$. Fig.2 shows the dependence of $|\psi|$ on x for three moments of time when $|\psi|$ becomes zero. Fig.3 shows the time dependence of $U = min_x|\psi|$. In the moment of time when u touches the x-axis, the phase Φ changes to 2π by a jump. The phase reduction occurs sufficiently quickly, reaching a certain stable N_{st} corresponding to $k_{st} = 2\pi N_{st}/l$. For this run $N_{st} = 2$. After this process we can observe a slower diffusive relaxation tending to the state $\psi = \psi(k_{st})$.

4. DEFECT PROPAGATION

From the aforementioned it becomes clear that the propagation of the defects with some velocity is possible only if one of the states is unstable. If both states

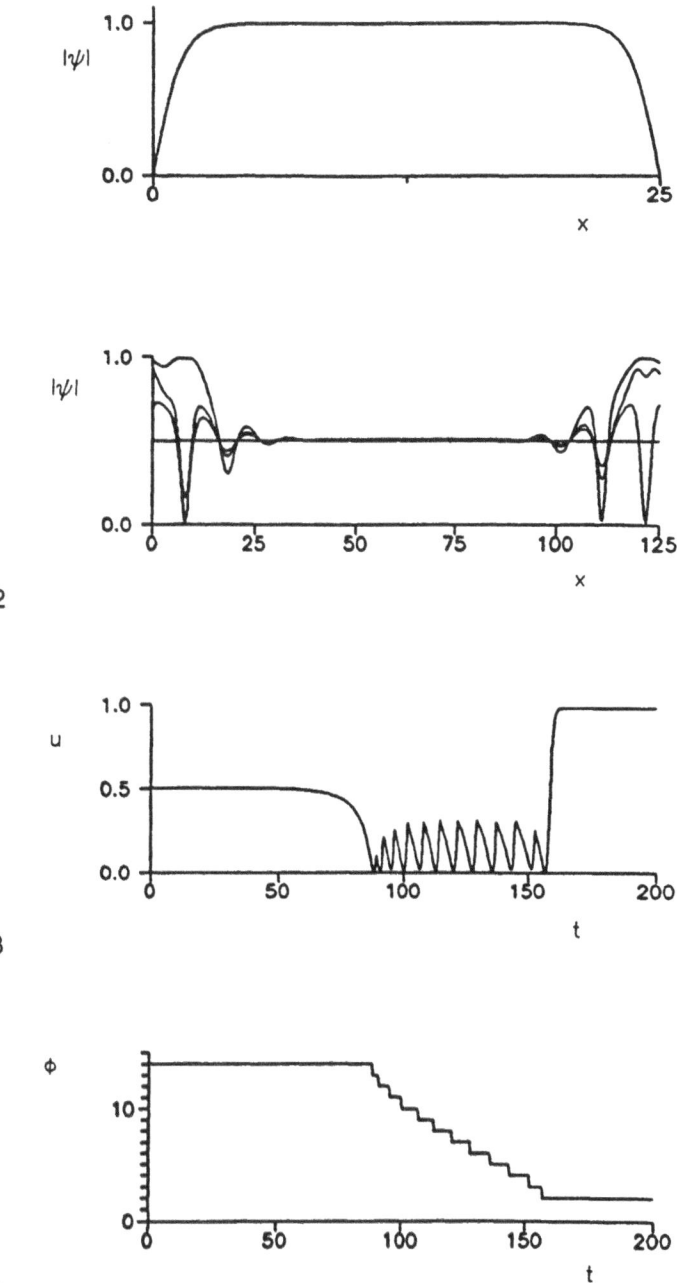

Fig.1-4. Dependencies of $|\psi|$ in successive moments of time, dependencies of $U = min_x|\psi|$ and phase Φ in 2π units as functions of time for the instability development of the solution $\psi_k = (1 - k^2)^{1/2}e^{ikx}$ with $k = 14 * 2\pi/125$. The vanishing of U and the jumps in phase Φ of 2π take place at the same moments of time.

are stable, there is no reason for the defect to move. Its dynamics will be essentially different from the first case.

Let the initial conditions be such that $\psi_0 \to 1$ as $x \to -\infty$ and $\psi_0 \to \psi_k = (1 - k^2)^{1/2} e^{ikx}$ as $x \to \infty$, $|k| > 1/\sqrt{3}$. Then the interface will propagate to the right with a certain velocity. To find the velocity we shall assume that far from the front in the unstable phase the initial conditions will differ only slightly from ψ_k. Slightly means that $\Lambda = \ln(|\psi_k/\delta\psi|) \gg 1$ where $\delta\psi$ is the perturbation. This assumption permits us to consider that the linear stage of the instability (8) lasts long enough and can be described by means of the saddle point method. Knowing the solution before the front it is necessary to match it to the main wave coming onto the asymptotics $\psi \to 1$ as $x \to -\infty$. This problem is analogous by its formulation to the Kolmogorov-Petrovsky-Piskunov problem [5] but has a principal difference. As is shown by the numerical experiment, the defect moves, on average, with a constant velocity V. In Figs.5-9 are the data for version $k = 0.95$. At the motion of the defect in its front, we can observe periodic variations with a frequency which can be estimated as $\omega_D = kV$. In this situation, after each period of oscillations there occurs a phase-slip of 2π, i.e., the spiral uncoils (Figs.5,6). Thus on the front of the wave there are complicated nonlinear oscillations. Apart from them in the value of the velocity we can also observe certain oscillations, but with a smaller frequency. After the main front is gone, the state after the front is different from $\psi = 1$, and there is still a certain residual rotation (Figs.8,9). To define the value of the mean velocity of the defect let us use the method proposed in [6]. For this purpose, consider a solution far from the front in the region of instability, where perturbations are small: $\Lambda = \ln|\psi_0/\delta\psi| \gg 1$. To find the velocity, let us require that in the reference system, moving with the velocity of the defect V, perturbations do not grow exponentially in time. In this instance, we shall select the solutions that will grow exponentially over x while approaching the main front of the defect. It is principally important that to find the velocity of the defect it is not necessary to solve the problem of matching with the main wave; the velocity is obtained from the analysis of only the linear problem.

So, transforming to the reference system moving with the velocity V, from Eqs.(7)-(8) for the perturbations χ', χ'' we get

$$
\begin{pmatrix} \chi' \\ \chi'' \end{pmatrix} = \int_{-\infty}^{\infty} \left[\begin{pmatrix} \Gamma_1 + \sigma^2 \\ 2ik\sigma \end{pmatrix} C_1(\sigma) \exp\left(\Gamma_1(\sigma)t + i\sigma x' + i\sigma Vt\right) \right.
$$
$$
\left. + \begin{pmatrix} \Gamma_2 + \sigma^2 \\ 2ik\sigma \end{pmatrix} C_2(\sigma) \exp\left(\Gamma_2(\sigma)t + i\sigma x' + i\sigma Vt\right) \right] d\sigma . \tag{14}
$$

Here $x' = x - Vt$, and the functions $C_1(\sigma)$ and $C_2(\sigma)$ are defined from the initial conditions.

Since the initial noise is small ($\Lambda \gg 1$), the linear stage of the instability lasts a long time $\Lambda\Gamma_{\max}^{-1}$. Therefore, far from the front we can confine ourselves to the linear

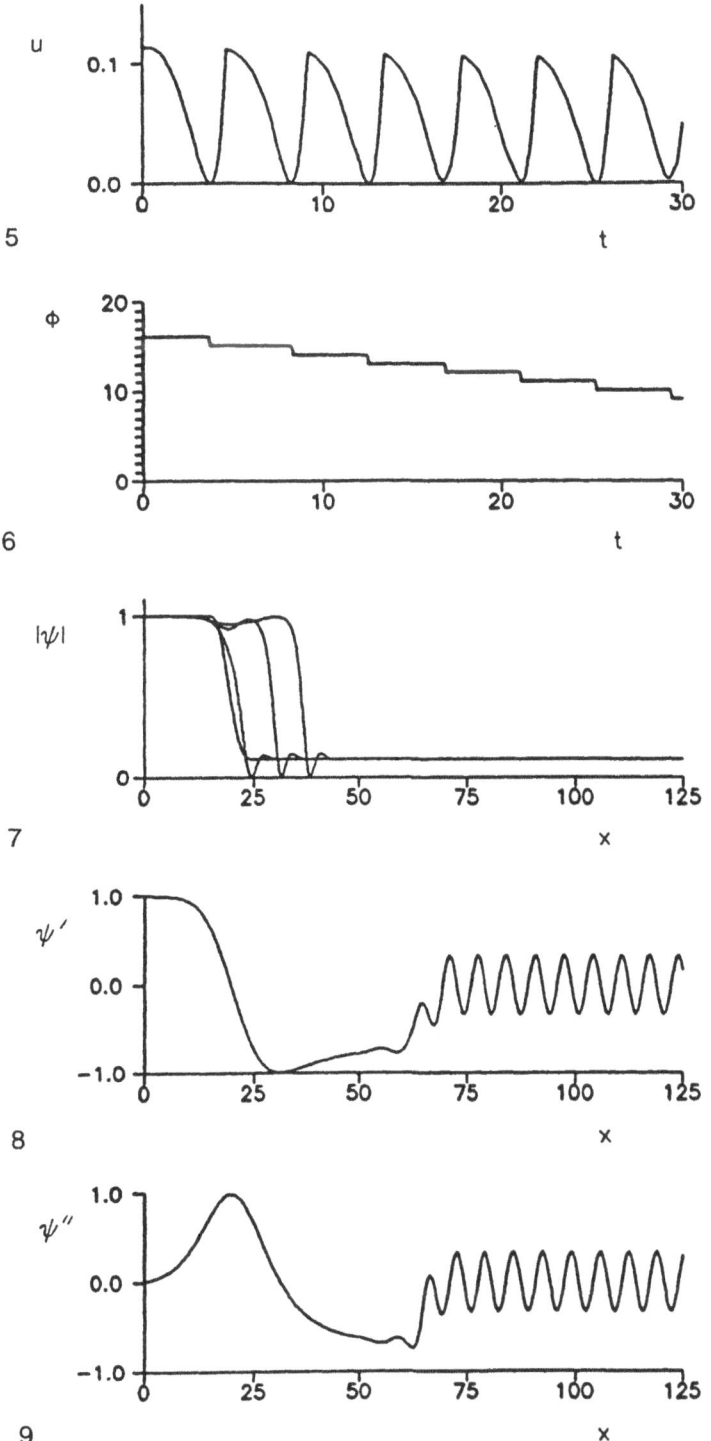

Fig.5-9. Dependencies of U, Φ, $|\psi|$, ψ' and ψ'' for the propagation of the defect with $k = 0.95$. Boundary conditions: $\psi(0) = 1$, $\psi(l) = \psi_k(l)$.

approximation and find the solution in terms of the integral (14) as $t \to \infty$ by using the saddle point method. In this case, it is necessary to make the cut through the points $\sigma = \pm i\psi_0^2/2k$ and to glue the edges of the cut. After this, we can use the saddle point method.

The saddle point is found from the condition

$$\Gamma'(\sigma) + iV = 0 \,. \tag{15}$$

This equation reduces to the equation of the fourth power with respect to σ. Among the roots of this equation we should choose a σ_m for which the value of $Re(\Gamma(\sigma) + i\sigma V)$ could be maximal.

This saddle point gives the maximal contribution to the integral (14). As a result, the perturbation with the accuracy up to the preexponential factor (irrelevant for our further investigation) behaves as

$$\exp\left\{ Re(\Gamma(\sigma_m) + i\sigma_m V)t + i\omega_m x' + ik(x' + Vt) \right\}, \tag{16}$$

$$\omega_m = Im(\Gamma(\sigma_m) + i\sigma_m V) \,.$$

We have included in this expression the exponential factor connected with the change (6).

The absence of the exponential growth over t in (16) determines the value of the velocity of the wave:

$$Re(\Gamma(\sigma_m) + i\sigma_m V) = 0 \,. \tag{17}$$

At a velocity smaller than V, defined from (17), perturbations will grow exponentially and, consequently, the process of the propagation will not be of a quasi-stationary character. Given a large V, perturbations will not have time to develop into a wave. Therefore, the requirement (17) defines the value of the velocity of the wave. It should be noted that $Im\,\sigma_m > 0$. This means that, at the approach to the front of the defect, the solution grows exponentially as a function of x. From (16) it is also evident that perturbations, apart from the Doppler frequency kV, have the frequency ω_m.

Eqs.(15),(17) can be studied analytically in two limits:

$$1 - k^2 = |\psi_k|^2 \ll 1 \,,$$

and

$$k^2 - 1/3 = \epsilon \ll 1 \,.$$

In the first limit $\Gamma(\sigma)$ can be written approximately in the form

$$\Gamma(\sigma) \approx -\sigma^2 - 2k\sigma - \psi_k^2 \,.$$

Substituting this expression in (15), we find the quantity

$$\sigma_m = -iV/2 - k \,.$$

The insertion of σ_m into (17) for the velocity V yields

$$V = 2(1 - |\psi_k|^2).$$

Then the Doppler frequency equals

$$\omega = 2k(1 - |\psi_k|^2).$$

To calculate the frequency ω_m, it is necessary to retain the following corrections over $|\psi_k|^2$ in the expression for $\Gamma(\sigma)$. Simple calculations yield

$$\omega_m = |\psi_k|^4/4.$$

In the other limit the growth rate can be replaced by

$$\Gamma(\sigma) = \frac{9}{2}\epsilon\sigma^2 - \frac{3}{16}\sigma^4.$$

In this case it is convenient in the expression

$$\Gamma(\sigma) + i\sigma V = f(\sigma),$$

to introduce new variables θ and y: $\sigma = (12\epsilon)^{1/2}y$, $V = \theta\epsilon^{3/2}9\sqrt{12}$. As a result,

$$f(\sigma) = 108\epsilon^2 g(y),$$

$$g(y) = y^2/2 - y^4/4 + i\theta y.$$

After simple transformations, Eqs.(15,17) can be solved as

$$y = \left((\sqrt{7}+3)/4\right)^{1/2} + i\left((\sqrt{7}-1)/12\right)^{1/2},$$
$$\theta = (\sqrt{7}+2)(\sqrt{7}-1/3)^{1/2}/3.$$

The results of the numerical calculation of Eqs.(15),(17) are given in Figs.10-12. In Fig.10 the solid line corresponds to the values of the velocity V, calculated from (15), (17). The asterisks mark the values of the velocity measured in numerical experiments. The mean velocity has been defined as the ratio of the distance propagated by a defect per a period to the period of oscillations. This period of oscillations has been measured as the time between two successive phase jumps (see Fig.6). In Fig.11 the solid line stands for the Doppler frequency, calculated by means of Eqs.(15),(17). The asterisks mark the frequency determined by the period of oscillations of the quantity $u = \min_x |\psi(x,t)|$. In both graphs for V and ω_D, there is good agreement between theory and numerical experiments. Finally, Fig.12 shows the dependence of the frequency ω_m on the wave number k. In numerical experiments we have observed a frequency close to ω_m, which corresponds to oscillations of the velocity V with respect to the mean value. These oscillations amount to 5% and have a tendency to decay.

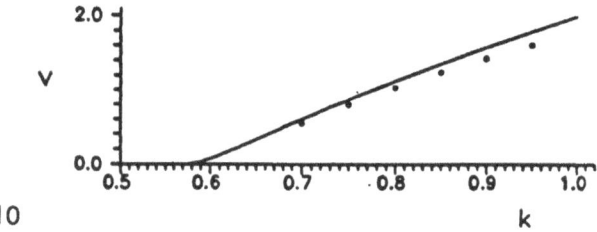

10

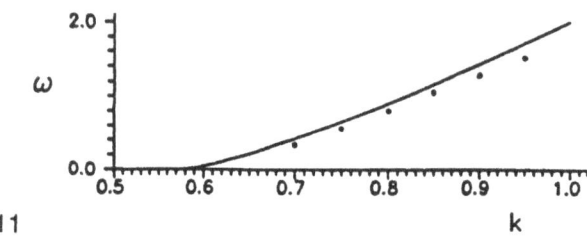

11

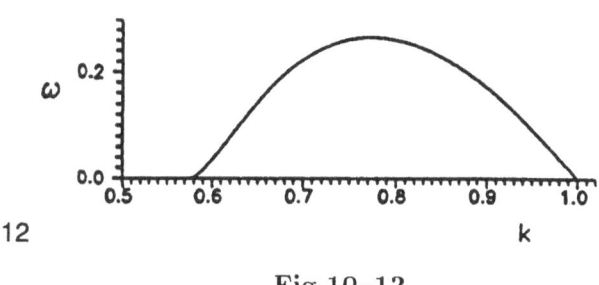

12

Fig.10–12.

Now let us briefly study what effects occur for the defects between two stable vortex lattices. The principal distinction of the defects studied above consists of the absence of a strong instability. Therefore the defect cannot propagate by virtue of the topological constraints. In this case, there occurs a slow diffusive unwinding of the helix. Figs.13-15 give the spatial distributions for $|\psi|$, the real part ψ', and the imaginary part ψ''. On the edges of the interval $\psi(x)$ was constant: $\psi|_{x=0} = 1$, $\psi|_{x=l} = \psi_k(l)$. The transition region was expanding in time, which corresponded to the unwinding of the helix. In the region where ψ was equal to one, there was rotation, which is absolutely clear from Fig.14-15. The real part in this region became smaller than one, and the imaginary part, on the contrary, grew. These results are in full agreement with the conclusion of the paper [8] — such defects expand diffusively in time.

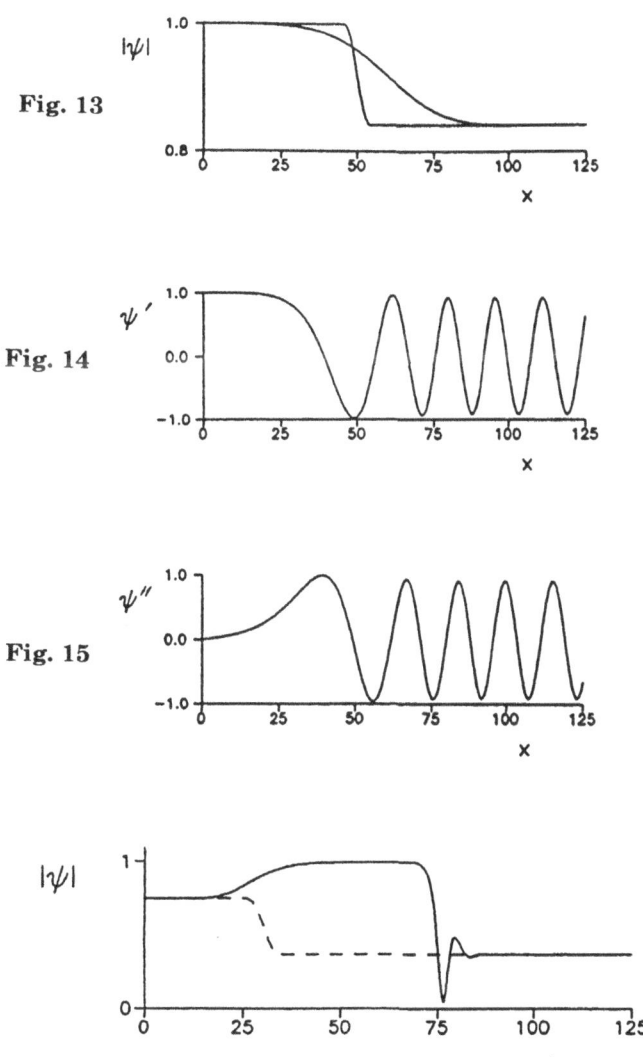

Fig. 13

Fig. 14

Fig. 15

Fig. 16. Distributions of $|\psi|$ for the decay of the initial condition (dotted line) in the form of defect with $k_1 = 0.5$ and $k_2 = 0.8$. Boundary conditions: $\psi(0) = \psi_{k_1}(0)$, $\psi(l) = \psi_{k_2}(l)$.

Let us consider now how the defect with arbitrary values k_1 and k_2 will decay if k_1 lies in the unstable region but if k_2 belongs to the stable one. As numerical experiments showed (see Fig.16), the defect begins to propagate into the unstable region with the parameters defined with the help of formulae (16),(17). The state behind the defect front has the amplitude close to $|\psi| = 1$ with some residual rotation. Between this state and the wave defined by k_1 from the stable region, the defect of the first type forms with a diffusive spreading front. The decay of such initial conditions leads, thus, to the formation of two types of defects. The parameters of the defects can be defined independently.

5. CONCLUSION

We have clarified that the dynamical properties of the defects for vortex lattices depend on whether the topological invariant of the complex field $\psi(x)$, coinciding up to the constant multiplier 2π with the integral phase difference of $\psi(x)$ on the interval edges, conserves or does not conserve. The reason for defect propagation is just connected with nonconservation of the invariant, i.e., its reduction. If the topological invariant conserves then the defect will spread diffusively.

It should be noted also the considered in this paper topological effects are intrinsic for two-dimensional (1+1) models for a complex field. For such cases the transition mechanism from we state with the invariant N_1 to the other with N_2 is common — the phase of this field ψ has to change by 2π with each touching of the x-axis by $|\psi|$. It should be emphasized that such a transition is only possible for the strong instability existence. The nonlinear Schroedinger equation with attraction and repulsion and its different generalizations represent the examples of such systems.

Acknowledgment: I.R. Gabitov would like to thank the Department of Mathematics of the University of Arizona for its hospitality during his visit. Work was supported in part by the Arizona Center for Mathematical Sciences, sponsored by AFOSR Contract FQ 8671-900589 with the University Research Initiative Program at the University of Arizona.

References

1. A.C. Newell and J.A. Whitehead, J. Fluid Mech. **38** (1969), 279.
2. S. Kogelman and R.S. DiPrima, Phys. Fluids **13** (1970), 1.
3. W. Eckhaus, J. Mecanique **2** (1963), 153.
4. Yu.G. Vasilenko, E.A. Kuznetsov, V.S. Lvov et al, Zhurnal Prikladnoi Mekhaniki i Technicheckoi Phisiki **2** (1980), 58 (in Russian).
5. A.N.Kolmogorov, I.G.Petrovsky and N.S.Piskunov, Bull. MGU. Matematika i Mekhanika **1** (1937), 1 (in Russian).
6. V.G.Kamensky and S.V.Manakov, Pis'ma v ZhETF (JETP Letters) **45** (1987), 499 (in Russian).
7. W. van Saarloos, Phys Rev. A **37** (1988), 211.
8. A.A. Nepomnyaschy, *Defekty v Konvektivnykh Strukturakh (Defects in Convective Structures)*, Nauchnye Doklady, Institut Mekhaniki Sploschnykh Sred, Sverdlovsk (1988), (in Russian).

OSCILLATIONS ARISING IN NUMERICAL EXPERIMENTS

C. David Levermore[1] and Jian–Guo Liu[2]

[1] Department of Mathematics, University of Arizona
Tucson, Arizona 85721, USA

[2] Courant Institute of Mathematical Sciences, New York University
251 Mercer St., New York, New York 10012, USA

Abstract

We present some numerical experiments that illustrate the emergence of oscillatory behavior in solutions of a dispersive numerical scheme for the Hopf equation. These oscillations arise at the time that the classical solution of the Hopf equation develops a singularity. Modulation equations are derived for period-two oscillations and are found to have both a hyperbolic and an elliptic region; the period-two oscillatory solutions break down after they enter that region. This kind of phenomenon has not been observed for integrable schemes. In addition, when an envelope of modulated period-two oscillations develops a contact discontinuity, it is shown to be unstable to shifts in the underlying mesh by a fraction of a cell width.

1. INTRODUCTION

In 1943, von Neumann used a central difference scheme to compute compressible fluids flows containing strong shocks and found oscillatory behavior after shock formation. These oscillations arose due to the dispersive nature of the numerical scheme he chose and are characteristic of all such schemes. Continuous analogues of this phenomenon have been studied recently by Lax and Levermore [12], Venakides [17], Flaschka, Forrest and McLaughlin [2], and Jin, Levermore and McLaughlin [9]. There have also been detailed studies of dispersive numerical schemes, for example, those by Holian and Straub [7], Holian, Flaschka and McLaughlin [6], Trefethen [16], Lax [11], Goodman and Lax [3], Herbst and Ablowitz [5], Hou and Lax [8], Venakides, Deift and Oba [18] and Greenberg [4].

In this paper, we examine the large oscillations arising in the numerical approximation of the Hopf (inviscid Burgers) initial-value problem

$$\partial_t u + u\,\partial_x u = 0\,, \qquad u(x,0) = u^{in}(x)\,, \tag{1.1}$$

Singular Limits of Dispersive Waves, Edited by
N.M. Ercolani et al., Plenum Press, New York, 1994

given by the semidiscrete dispersive difference scheme

$$\frac{du_j}{dt} + \tfrac{1}{3}(u_{j-1} + u_j + u_{j+1})\frac{u_{j+1} - u_{j-1}}{2h} = 0, \qquad u_j(0) = u^{in}(x_j), \qquad (1.2)$$

where $x_j = jh$ and $h = x_{j+1} - x_j$ is the spatial grid size. While the origin of this scheme is somewhat clouded, it has a long and distinguished history. For example, it was used by Zabusky and Kruskal [21] in 1965 when they discovered the remarkable interaction properties of soliton solutions of the Korteweg-de Vries equation. The historical appeal of this scheme derives from the fact that it possesses the two semidiscrete local conservation laws

$$\frac{du_j}{dt} + \frac{f_{j+\frac{1}{2}} - f_{j-\frac{1}{2}}}{h} = 0, \qquad f_{j+\frac{1}{2}} = \tfrac{1}{6}\left(u_j^2 + u_j u_{j+1} + u_{j+1}^2\right),$$
$$(1.3a)$$

$$\frac{du_j^2}{dt} + \frac{g_{j+\frac{1}{2}} - g_{j-\frac{1}{2}}}{h} = 0, \qquad g_{j+\frac{1}{2}} = \tfrac{1}{3}\left(u_j^2 u_{j+1} + u_j u_{j+1}^2\right), \qquad (1.3b)$$

hence reflecting the local conservation of u and u^2 by classical solutions of the Hopf equation (1.1).

Taylor expanding the finite difference approximation in (1.2), the truncation error of the scheme is found to be

$$\partial_t u + u\,\partial_x u + \tfrac{1}{18}h^2\left(u\,\partial_{xxx}u + \partial_{xxx}(u^2)\right) = O(h^4). \qquad (1.4)$$

Consequently, its continuum limit ($h \to 0$) has many similarities with the zero dispersion limit ($\varepsilon \to 0$) of the Korteweg-de Vries initial-value problem,

$$\partial_t u + u\,\partial_x u + \varepsilon^2 \partial_{xxx}u = 0, \qquad u(x,0) = u^{in}(x), \qquad (1.5)$$

a problem that has been well understood [12,17].

It is well known that a classical solution of the Hopf initial-value problem (1.1) develops an infinite derivative after a finite time for any initial data with a decreasing part, even for smooth initial data. However, so long as this solution remains classical, a Strang-type convergence theorem [14] states that the solutions of both (1.2) and (1.5) will converge strongly to it. After the singularity formulation however, large oscillations are developed in the approximating solutions of (1.5). While the wavelength of these oscillations is of order $O(\varepsilon)$, their amplitude does not vanish as ε tends to zero [12,17]. We expect that solutions of the scheme (1.2) also develop oscillations with a wavelength of order $O(h)$ with a nonvanishing amplitude. In either case, after the breaktime the solutions can at best be expected to have a weak limit as ε or h tends to zero.

This limiting behavior contrasts sharply with that for solutions of any zero dissipation limit, say as ε tends to zero in the Burgers equation

$$\partial_t u + u\,\partial_x u - \varepsilon^2 \partial_{xx}u = 0. \qquad (1.6)$$

In this case the solutions converge almost everywhere and strongly to a weak solution of (1.1) with shock discontinuities. After the formation of shocks this limiting solution locally dissipates u^2, a residual of the fact that the approximating solutions do so. On the other hand, u^2 is locally conserved by solutions of either (1.2) or (1.5). As a result, the weak limit of solutions of either (1.2) or (1.5), if it exists, can not be this zero-dissipation solution of the Hopf equation (1.1).

So far, much of the analysis of dispersive numerical schemes depends heavily on their integrability [3,5,6,10,11,15,18]. In fact, only two integrable lattices, those of Toda [15] and Kac-van Moerbeke [10], have been carefully studied, and these are essentially identical. The study of near-integrable and nonintegrable numerical schemes is important for our general understanding of dispersive numerical phenomena. We chose scheme (1.2) as prototypical (we don't know whether or not it is integrable, but numerical evidence indicates that it is not) and compared it with an integrable scheme. We find notable differences in the behavior of these two schemes.

The simplest oscillatory behavior is the period-two (or binary) oscillation, but even this exhibits many interesting phenomena [13]. In the spirit of the Whitham averaging method [20], in the next section we use the local conservation laws (1.3) to derive equations that describe the evolution of an envelope of period-two oscillations. These so-called modulation equations are a 2×2 system of conservation laws that is strictly hyperbolic in a region containing the initial data and is elliptic in the remaining region. As soon as the period-two oscillations evolve out of the hyperbolic region at some location, they break-down locally. The breakdown region exhibits chaotic small scale behavior, a phenomenon not shared with the observed behavior of integrable schemes.

2. THE MODULATION EQUATIONS FOR BINARY OSCILLATIONS

Due to its spatial central difference, the scheme (1.2) has as an exact stationary solutions any period-two spatial oscillation. Solutions in this family are determined up to a phase by their mean value v and mean square w as defined by

$$v = \tfrac{1}{2}(u_j + u_{j+1}), \qquad w = \tfrac{1}{2}(u_j^2 + u_{j+1}^2).$$ (2.1)

The oscillatory solution u_j can the be recovered (up to a phase) by

$$u_j = v \pm (-1)^j \sqrt{w - v^2}.$$ (2.2)

In order to describe a solution of (1.2) that is a modulation of this family of period-two oscillations it is therefore natural to introduce the variables

$$v_{j+\frac{1}{2}} = \tfrac{1}{2}(u_j + u_{j+1}), \qquad w_{j+\frac{1}{2}} = \tfrac{1}{2}(u_j^2 + u_{j+1}^2).$$ (2.3)

Both $v_{j+\frac{1}{2}}$ and $w_{j+\frac{1}{2}}$ are conserved densities of the scheme (1.2) and are smoothly varying for modulated period-two oscillations.

The evolution of $v_{j+\frac{1}{2}}$ and $w_{j+\frac{1}{2}}$ can be simply expressed using the local conservation laws (1.3). First notice that the fluxes of (1.3) can be expressed in terms of the variables $v_{j+\frac{1}{2}}$ and $w_{j+\frac{1}{2}}$ as

$$\begin{aligned}
f_{j+\frac{1}{2}} &= \tfrac{1}{3}v_{j+\frac{1}{2}}^2 + \tfrac{1}{6}w_{j+\frac{1}{2}}, \\
g_{j+\frac{1}{2}} &= \tfrac{4}{3}v_{j+\frac{1}{2}}^3 - \tfrac{2}{3}v_{j+\frac{1}{2}}w_{j+\frac{1}{2}}.
\end{aligned}$$ (2.4)

Averaging the local conservation laws (1.3) over adjacent spatial points yields

$$\frac{dv_{j+\frac{1}{2}}}{dt} + \frac{f_{j+\frac{3}{2}} - f_{j-\frac{1}{2}}}{2h} = 0,$$

$$\frac{dw_{j+\frac{1}{2}}}{dt} + \frac{g_{j+\frac{3}{2}} - g_{j-\frac{1}{2}}}{2h} = 0. \qquad (2.5)$$

Equation (2.5) can be viewed as the central difference approximation to the 2×2-system

$$\partial_t v + \partial_x f(v, w) = 0, \qquad f(v, w) \equiv \tfrac{1}{3}v^2 + \tfrac{1}{6}w,$$

$$\partial_t w + \partial_x g(v, w) = 0, \qquad g(u, v) \equiv \tfrac{4}{3}v^3 - \tfrac{2}{3}vw. \qquad (2.6)$$

Since $v_{j+\frac{1}{2}}$ and $w_{j+\frac{1}{2}}$ are smoothly varying for modulated period-two oscillations, formally at least, their continuum limits v and w will satisfy (2.6), the so-called modulation equations for period-two oscillations. Furthermore, v and w are then the weak limits of u_j and u_j^2 respectively.

The Jacobian matrix of flux functions of (2.6) is

$$A = \begin{pmatrix} \partial_v f & \partial_w f \\ \partial_v g & \partial_w g \end{pmatrix} = \tfrac{1}{6} \begin{pmatrix} 4v & 1 \\ 24v^2 - 4w & -4v \end{pmatrix}, \qquad (2.7)$$

and its eigenvalues are

$$\lambda_\pm = \pm \tfrac{1}{3}\sqrt{10v^2 - w}. \qquad (2.8)$$

Clearly, system (2.6) is hyperbolic in the region

$$\left\{ (v, w) \ \middle| \ w \le 10v^2 \right\}, \qquad (2.9)$$

and is elliptic in the region

$$\left\{ (v, w) \ \middle| \ 10v^2 < w \right\}. \qquad (2.10)$$

It can be shown that classical solutions of the modulation equations (2.6) satisfy

$$\partial_t (w - v^2) - v \, \partial_x (w - v^2) - \tfrac{2}{3}(w - v^2)\partial_x v = 0. \qquad (2.11)$$

Hence, the region $v^2 \le w$ is invariant for these solutions, a fact that is consistent with the origins (2.3) of v and w and indeed allows the reconstruction of u (up to phase) through formula (2.2). However, as we demonstrate in the next section, the hyperbolic region is not invariant for these solutions.

In [13], we show that the central difference approximating a general system of hyperbolic conservation laws is L^2 stable. Hence, by a theorem of Strang, the modulation equations (2.6) give a good description of the envelope of period-two oscillations in the dispersive scheme (1.2) provided the solution of (2.6) is Lipschitz continuous and remains within the hyperbolic region (2.9).

3. SOME NUMERICAL EXPERIMENTS

3.1 The Emergence of Period-Two Oscillations

Let's begin with the some numerical experiments of the difference scheme (1.2) approximating the Hopf equation (1.1) with initial data,

$$u^{in}(x) = -\sqrt[3]{x} \, . \tag{3.1}$$

It is known that the zero-dissipation solution of Hopf equation with the above initial data has a stationary shock at $x = 0$ at any positive time. Below we show that period-two oscillations are developed in the numerical solution of (1.2) and fan out in both directions from a jump-discontinuity at $x = 0$.

In Figures 1(a-d), we plot the numerical solutions of scheme (1.2) with initial data (3.1) for a grid size of $h = 0.01$ at times $t = 0.0$, $t = 0.3$, $t = 0.6$ and $t = 0.9$ respectively. The ordinary differential equation (ODE) (1.2) is solved using an Adams-Bashfort method with the Courant-Friedrichs-Lewy (CFL) number equal to 0.2. The numerical boundary condition is implemented by imposing the endvalues at every time step. We also used a four-step Rung-Kutta method to solve the ODE (1.2) and obtained indistinguishable plots.

The evident discontinuity in the envelope of period-two oscillations at the center of Figure 1(b) does not correspond to a shock solution of the modulation equations (2.6), but rather to a contact discontinuity since there is no flux across it. Figures 1(a-d) clearly show that the envelope of period-two oscillations fans out in both direction from this discontinuity as the time increases. The envelope is smooth except at the origin and at two endpoints of the oscillatory region where it connects to the smooth solution with a square root profile.

We plot the weak limit of the numerical solutions for the above experiments results long with the exact zero-dissipation weak solutions of the Hopf equation in Figures 2(a-d). The weak limit is computed by averaging values at adjacent spatial points of the approximate solution. Since that solution is a modulation of period-two oscillations, this local average gives a good approximation to the weak limit of the solutions. We can see in Figures 2(a-d) that the weak limit of the numerical solution of the scheme (1.2) is not equal the entropy solution of Hopf equation (1.1) in the oscillatory region.

Some analysis of the above numerical experiments are given in the following subsections and in a forthcoming paper [13]. We will give more numerical experiments later on to incorporate with the analysis to explain the oscillatory behavior of scheme (1.2). We should also point out that period-two oscillations in Toda lattices with certain Riemann shock initial data were discovered numerically by Holian, Flaschka and McLaughlin [6] and were proved analytically by Venakides, Deift and Oba [18]. Similar phenomenon were also found by Hou and Lax [8] in the von Neumann scheme for gas dynamics with Riemann shock data.

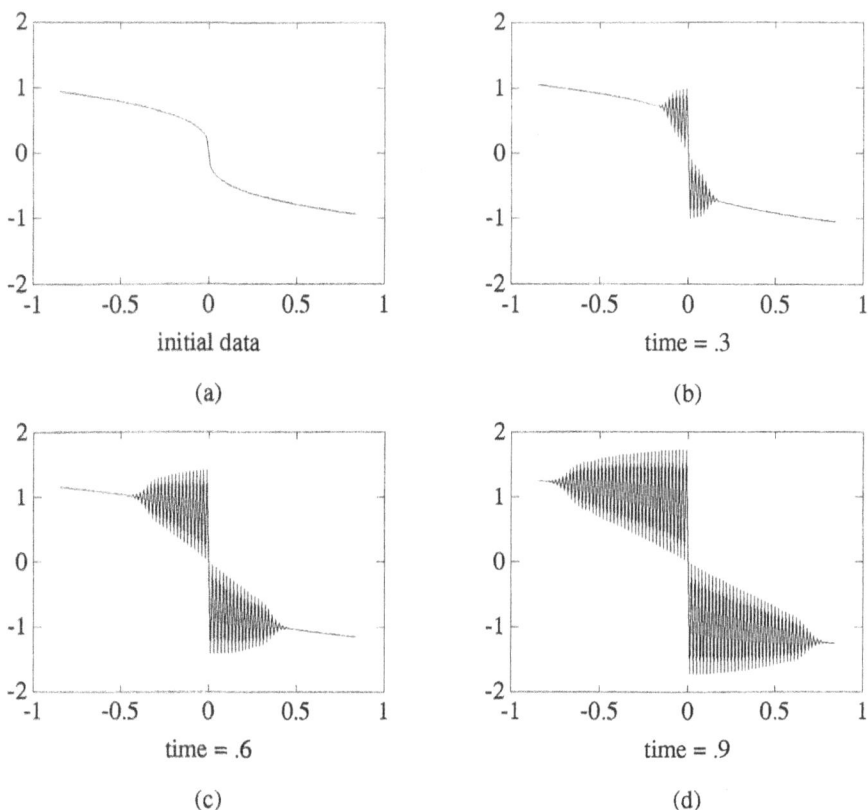

Figure 1. Numerical solutions of the dispersive scheme (1.2) approximating the Hopf equation (1.1) with initial data $u^{in}(x) = -\sqrt[3]{x}$ at times (a) $t = 0$, (b) $t = 0.3$, (c) $t = 0.6$ and (d) $t = 0.9$. The gridsize is $h = 0.01$ and the scheme (1.2) is implemented using an Adams-Bashfort ODE solver with a CFL number 0.2. The numerical solution develops period-two oscillations at the origin and these oscillations self-similarly fan out as time increases.

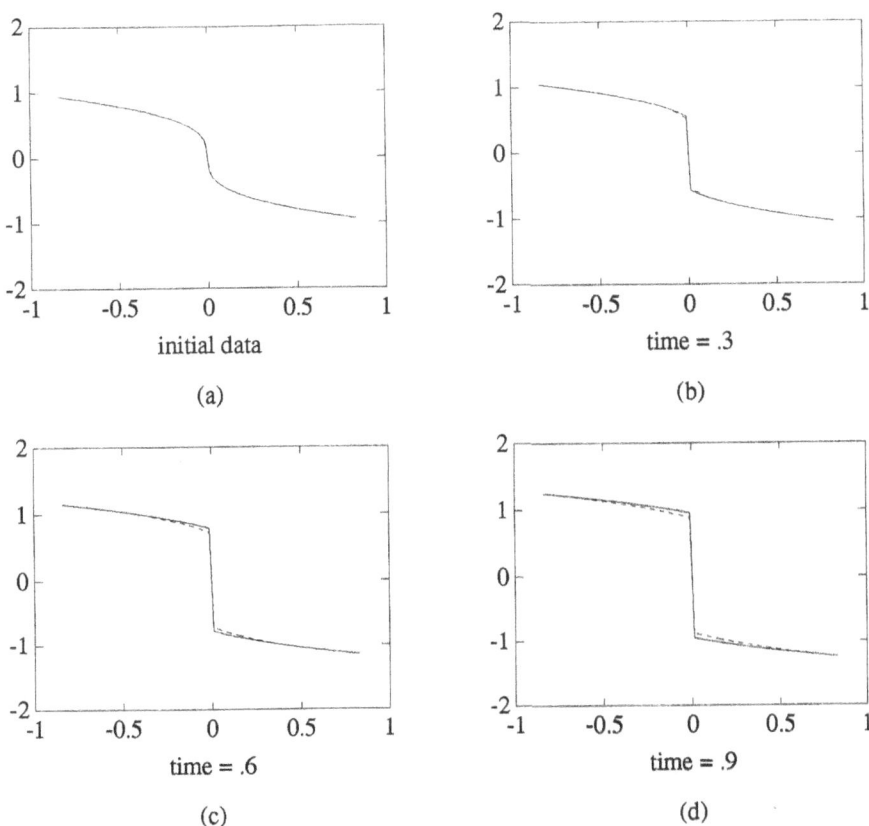

Figure 2. The solid lines are the zero-viscosity solution of the Hopf equation (1.1) at times (a) $t = 0$, (b) $t = 0.3$, (c) $t = 0.6$ and (d) $t = 0.9$ obtained with an ENO scheme. The dashed lines indicate the weak limit of the dispersive scheme (1.2), which is computed by averaging adjacent points of the same approximate solution shown in Figure 1. As time increases, the difference between the weak limit of the numerical solution and the zero-viscosity solution also increases.

3.2 The Loss of Hyperbolicity in Period-Two Oscillations

The following numerical experiment illustrates what happens when a period-two oscillation evolves into the elliptic region (2.10) at some location. We use the dispersive approximation (1.2) to the Hopf equation (1.1) with initial data

$$u^{in}(x) = -0.3\sin(\pi x), \qquad (3.2)$$

Clearly, singularities will form in the solution when $t = 1$ at $x = 0, \pm 1, \pm 2, \cdots$. For a grid size $h = 0.005$, we plot the numerical solution at the times $t = 6$, $t = 11$, and $t = 12$ in Figures 3(a-c) respectively. The ODE (1.2) is solved using the Adams-Bashfort method with the CFL number equal to 0.2.

The discontinuity in the envelope of period-two oscillations at the center of Figure 3(a) is again a contact discontinuity (with no flux across it). The period-two oscillations are generated at the central contact discontinuity and fan out in both directions as time increases. The oscillations are slowly varying in space and time, and their envelope is smooth expect at the central contact discontinuity and at the endpoints of the oscillating region where the envelope has a square root behavior. Figure 3(b) shows the growth of the oscillatory region and the development of two small defects in the envelope at $x \sim 0.3$ where the solution has just moved out of the hyperbolic region (2.9). The resulting dramatic local breakdown of the period-two oscillations is shown in Figure 3(c), where the solution is seen to remain a perfect period-two oscillation throughout much of space. Within the breakdown region the solution becomes rapidly varying in both space and time, behaving in a chaotic way. Notice that this breakdown occurs far from the central contact discontinuity. In order to ensure that this phenomenon is not simply generated by the numerical ODE solver, we halved the time step and obtained indistinguishable results. We believe that Figure 3(c) depicts the dynamical behavior of ODE (1.2). We refer to [13] for a more detailed discussion.

To further illustrate this connection between the local loss of hyperbolicity and the breakdown of the modulated period-two solutions, in Figures 4(a-d) we depict the solution of Figures 3(a-c) in the (v, w)-plane. The initial data in Figure 4(a) lies on the curve $w = v^2$. As the period-two oscillations are developed, Figure 4(b) shows that the solution enters the hyperbolic region (2.9). At $t = 11$, Figure 4(c) shows that the solution has just passed the curve $w = 10v^2$ and entered the elliptic region (2.10). While the breakdown time of the period-two oscillations is seen to coincide with the time the solution enters the elliptic region, notice in Figure 4(d) that some period-two structure persists into the elliptic region. The cause for the persistent structure is not well understood.

The breakdown of period-two oscillations in this way has not observed in integrable schemes. For comparison, we approximate the Hopf equation (1.1) with initial data (3.2) by the dispersive scheme

$$\frac{du_j}{dt} + u_j \frac{u_{j+1} - u_{j-1}}{2h} = 0, \qquad u_j(0) = u^{in}(x_j), \qquad (3.3)$$

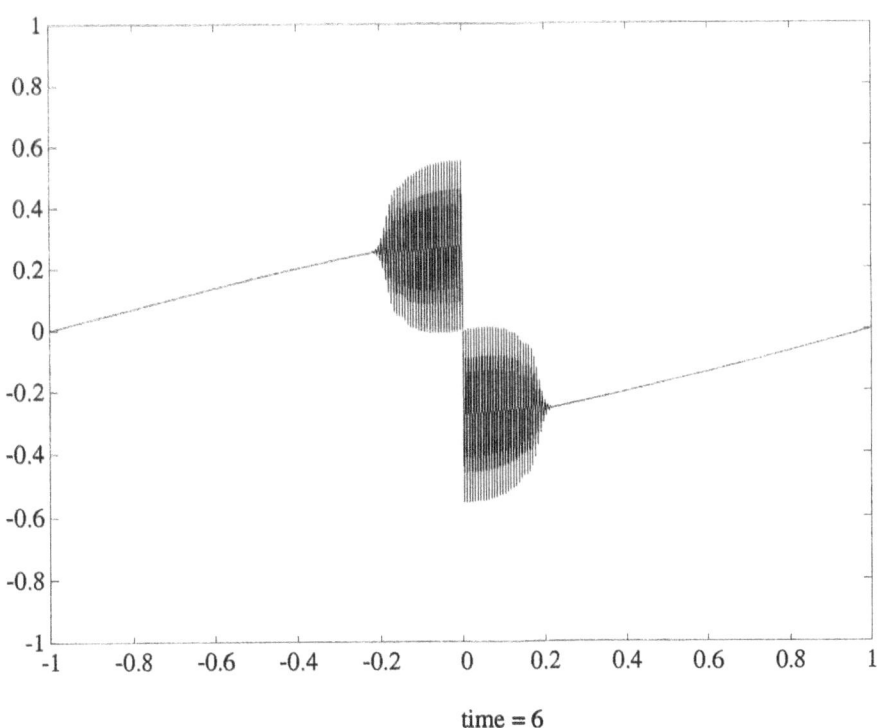

time = 6

Figure 3a. The numerical solutions of the dispersive scheme (1.2) approximating the Hopf equation (1.1) with initial data $u^{in}(x) = -0.3\sin(\pi x)$ at time $t = 6$. The gridsize is $h = 0.005$ and the scheme (1.2) is implemented using an Adams-Bashfort ODE solver with a CFL number 0.2. The solution develops period-two oscillations when the solution of the Hopf equation (1.1) develops a singularity. Their envelope forms a contact discontinuity at the origin and fans out as time increases.

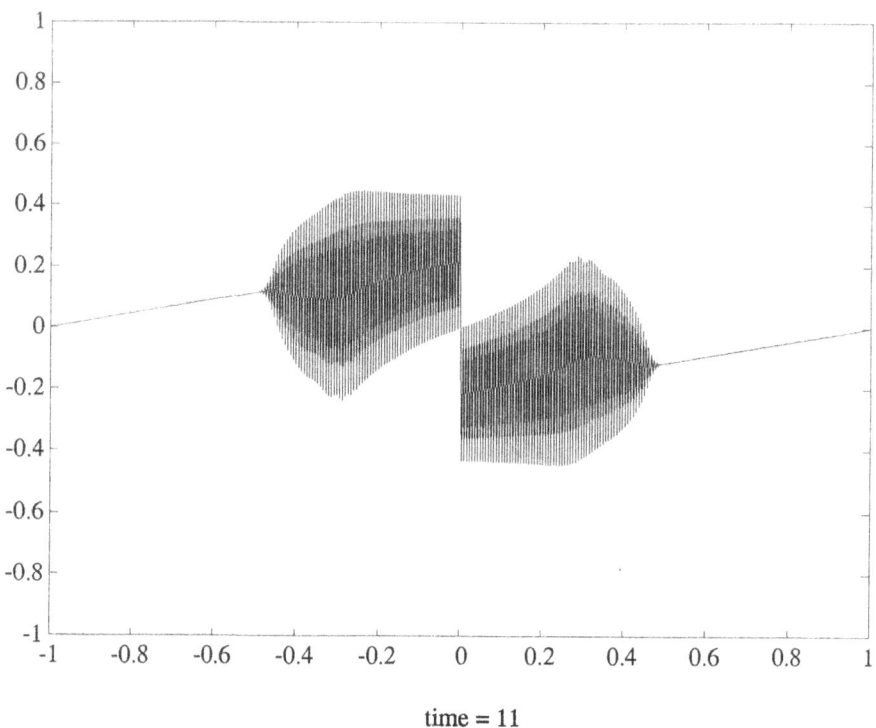

time = 11

Figure 3b. A continuation of the run described in Figure 3(a), now shown at time $t = 11$. The envelope of period-two oscillations has evolved to a point where it has just left the hyperbolic region (2.9). Notice the slight irregularity in the oscillations near $x = \pm 0.3$ where the hyperbolicity condition has been violated.

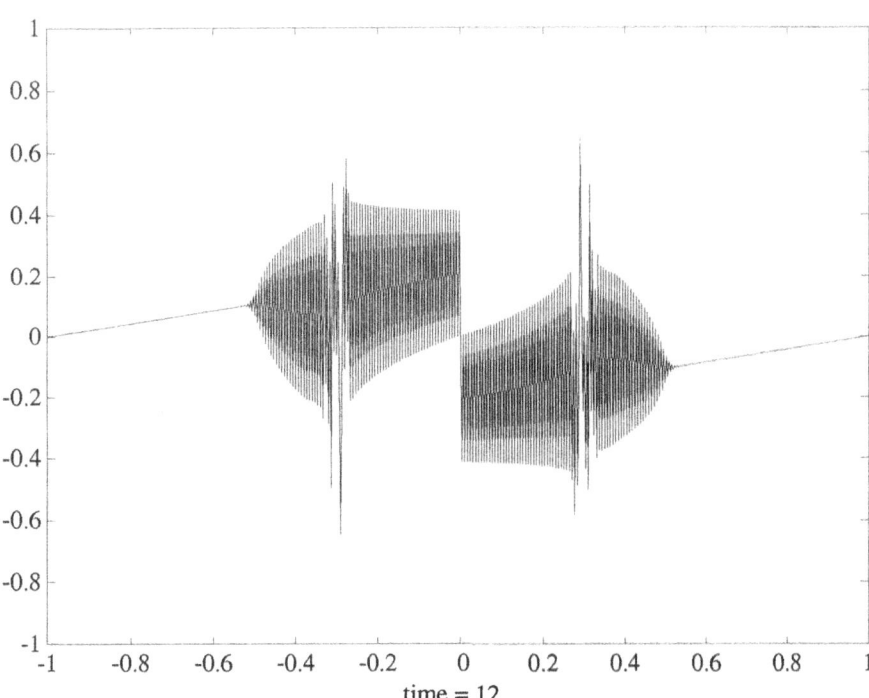

time = 12

Figure 3c. A continuation of the run described in Figure 3(a), now shown at time $t = 12$. The envelope of period-two oscillations has broken down and has developed a spatially and temporally chaotic region. Notice though that a well-define region of period-two oscillations persists.

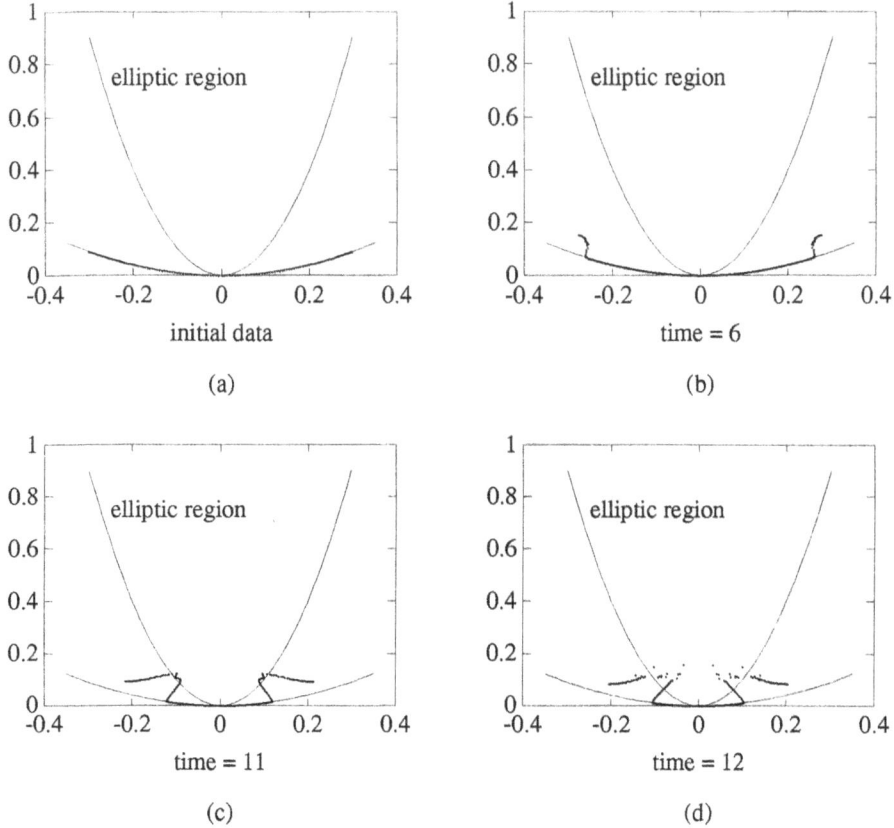

Figure 4. The numerical results of Figures 3(a-c) are depicted in the (v, w)-plane. The values v and w are computed by formula (2.3). The initial data (a) lies on the curve $w = v^2$. The solution at $t = 6$ (b) lies within the hyperbolic region (2.9) as period-two oscillations develop. The solution at $t = 11$ (c) has just crossed the curve $w = 10v^2$ and entered the elliptic region (2.10), triggering the onset of the local breakdown of period-two oscillations. The solution at $t = 12$ (d) is well into the elliptic region and the period-two oscillations have broken down. Near the boundary curve $w = 10v^2$, parts of the solution maintain a period-two structure even in the elliptic region.

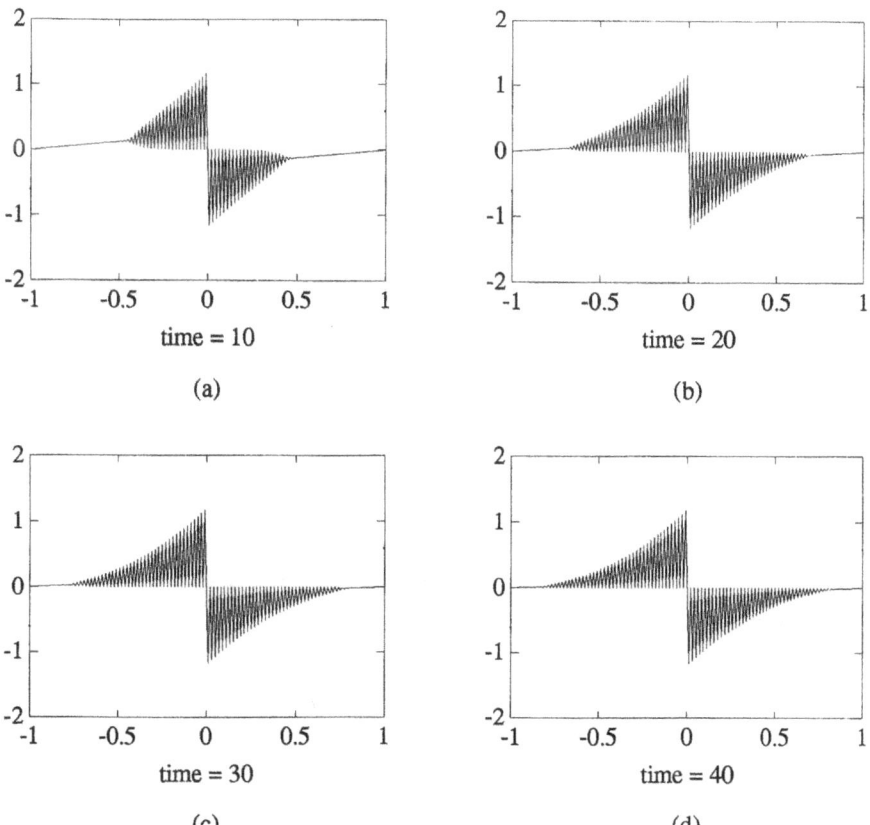

Figure 5. The numerical solution of an integrable scheme (the Kac-Moerbeke lattice) approximating the Hopf equation (1.1) with initial data $u^{in}(x) = -0.3\sin(\pi x)$ at times (a) $t = 10$, (b) $t = 20$, (c) $t = 30$ and (d) $t = 40$. The gridsize is $h = 0.01$ and the scheme (1.2) is implemented using an Adams-Bashfort ODE solver with a CFL number 0.2. The solution develops period-two oscillations when the solution of Hopf equation (1.1) forms a singularity. A contact discontinuity forms in their envelope and they fan out as time increases. A sharp difference with Figures 3(a-c) is that the period-two oscillatory solution never breaks down.

and plot the numerical solutions at times $t = 10$, $t = 20$, $t = 30$ and $t = 40$ in Figures 5(a-d), respectively. The above scheme were shown by Goodman and Lax [3] to be equivalent to the Kac-von Moerbeke lattice [10], provided the initial data has same sign, and hence is integrable. In [13] we derive a modulation equation for period-two oscillations from the local conserved qualities u_j and $\log u_j$ and show that they maintain hyperbolicity. Figures 5(a-d) show that the period-two oscillation never breaks down.

The relation between the stability of numerical schemes and their integrability was also studied by Herbst and Ablowitz [5] through a finite difference approximation to the cubic Schrödinger equation. They found that the standard central difference scheme (which is nonintegrable) could induce numerical chaos in the solution while an integrable scheme (obtained through a minor modification in the approximation of the cubic term) performs well. Of course, some integrable systems have elliptic modulation equations, but we know of no integrable systems with modulation equations that dynamically change type as was shown for equations (2.6).

3.3 A Phase Instability in Period-Two Oscillations

The contact discontinuity in the solution of the modulation equations that arises at the origin in the last two experiments is a delicate point. We now illustrate the instability of the solution at that point to fractional shifts of the grid. The above two numerical experiments are reperformed with the computational grid points $x_j = (j + \alpha)h$, where $0 < \alpha < 1$ and $\alpha \neq 0.5$. The initial data will then no longer have an odd symmetry with respect to the grid. The initial data used in Figures 6(a-d) and Figures 7(a-d) are the same as that used in Figures 1(a-d) and Figures 3(a-c) respectively, except that the grid points have been shifted by choosing $\alpha = \sqrt{2} - 1$.

Figures 6(a-d) and 7(a-d) clearly demonstrate again that the period-two oscillations generated at the origin fan out as time increases. However, the period-two oscillations are destroyed near the origin and the region of breakdown for the period-two oscillations also fans out as time increases. The period-two oscillations always move fastest, and so are found furthest from the origin. The breakdown region of period-two oscillations seen in Figures 3(b-c) also appears in Figures 6(c-d). Moreover, beyond this region one only finds period-two oscillations.

It can be shown [13] that wherever the modulation equations (2.6) are strictly hyperbolic and genuinely nonlinear their classical solutions describe stable period-two oscillations that are insensitive to shifts in the underlying grid. Hence, the differences seen in the corresponding runs above are generated at the contact discontinuity located at the origin and then propagate out from there.

4. CONCLUSIONS

We have studied the oscillatory behavior of a dispersive numerical scheme approximating the Hopf equation. Oscillations are generated in the numerical approximation solution when the classical solution of the Hopf equation developments a singularity.

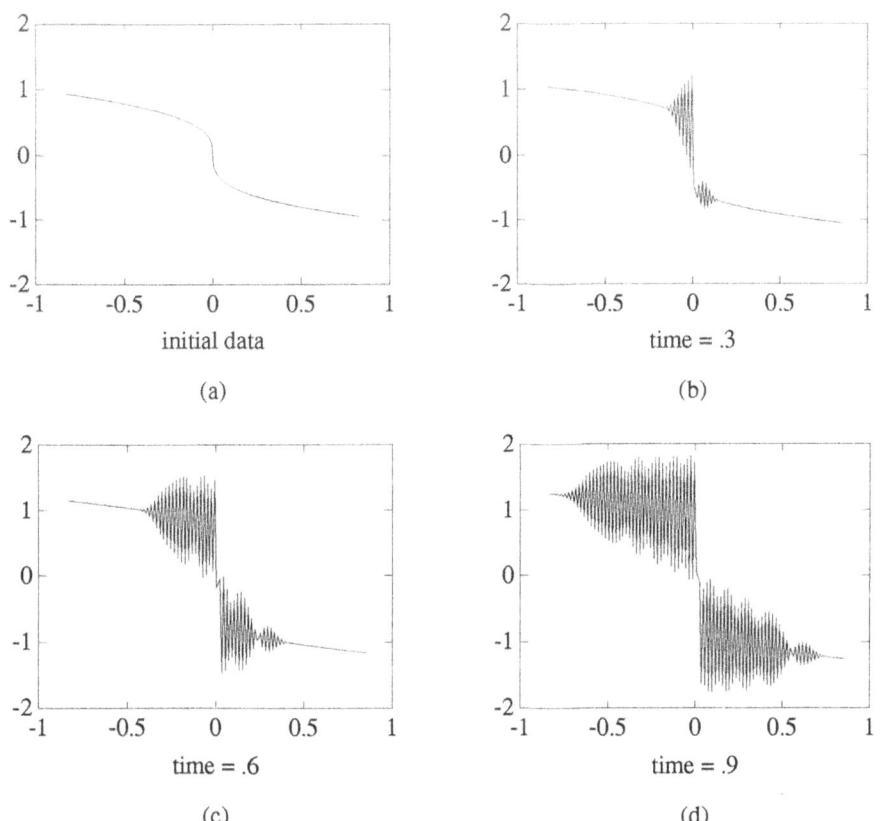

initial data

(a)

time = .3

(b)

time = .6

(c)

time = .9

(d)

Figure 6. The numerical results in this figure are the same as in Figures 1(a-d) except that the computational grid is shifted by $(\sqrt{2}-1)h$. The odd symmetry is thus broken and an instability is formed at the singularity of the solution of Hopf equation. The solution is still period-two oscillations, but there is now a long wavelength structure in the amplitude.

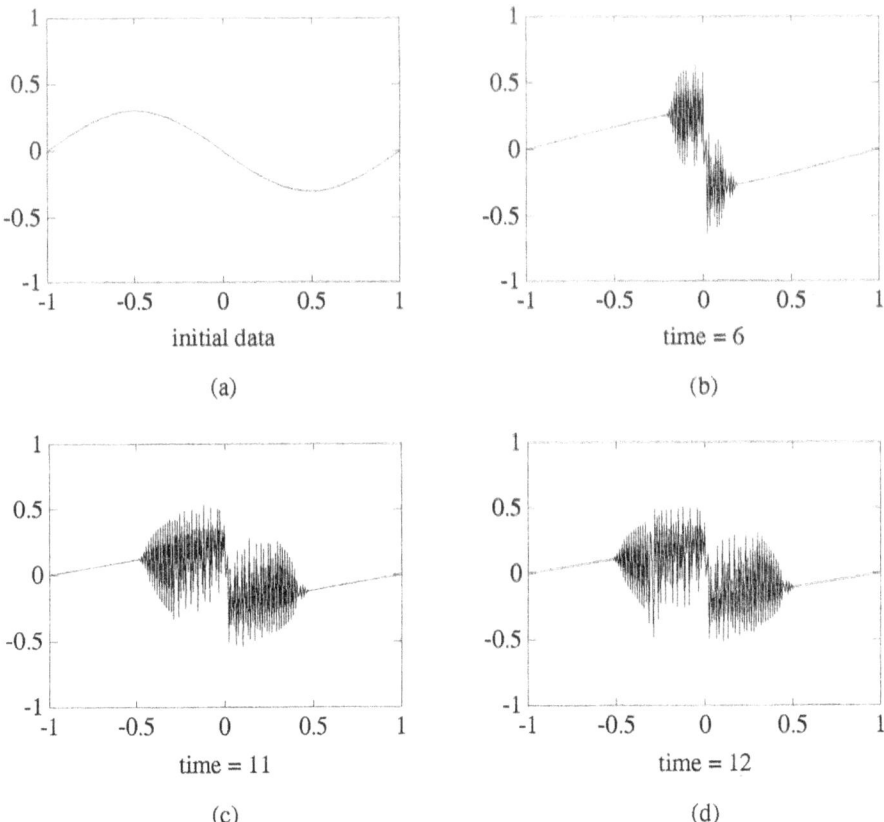

Figure 7. The numerical results in this figure are same as in Figures 3(a-c) except that the computational grid is shifted as $(\sqrt{2}-1)h$. The odd symmetry is thus broken and an instability is formed at the singularity of the solution of Hopf equation. The solution is not a period-two oscillation and there is a no long wavelength structure in the amplitude near the origin.

In some cases, the oscillations are period-two and we derived a modulation equation to describe it. The modulation equation is a 2×2 system of conservation laws that is hyperbolic in a region containing the initial data. The period-two oscillations will breakdown as they evolve out of the hyperbolic region. This phenomenon has not been observed in integrable schemes where the modulation equations have not been seen to dynamically change type. We have also showed the instability of modulated period-two oscillations is related to the loss of strict hyperbolicity or the existence of a contact discontinuity in the solution of the modulation equation describing the envelope.

In our forthcoming paper [13], we examine this oscillatory behavior and its modulation equations in a more detail. We analyze the onset of oscillations and show that the period-two oscillations have the fastest group velocity and always appear at the onset of oscillations. In the phase-transition region the amplitude of the oscillations is shown to satisfy a modified Korteweg-de Vries equation that matches nonoscillatory to oscillatory regions through a solution given by the second Painlevé transcendent. We also discuss the onset of period-three and period-four oscillations and derive their modulation equations.

Acknowledgments: C.D.L. was supported in part by NSF Grant DMS-8505550 while visiting the MSRI at Berkeley and by NSF Grant DMS-8914420. J.-G.L. was supported in part by NSF Grant DMS-8505550 as a postdoctoral fellow at the MSRI at Berkeley and by DOE Grant DE-FG02-88ER25053 as a visiting member of the Courant Institute.

REFERENCES

1. B. Engquist and J.-G. Liu, *Numerical methods for oscillatory solutions to hyperbolic problems*, Preprint (1991).
2. H. Flaschka, M.G. Forest, and D.W. McLaughlin, *Multiphase Averaging and the Inverse Spectral Solutions of the Korteweg-de Vries equation*, Comm. Pure Appl. Math. **33** (1980), 739–784.
3. J. Goodman and P.D. Lax, *On Dispersive Difference Schemes I*, Comm. Pure Appl. Math. **41** (1988), 591–613.
4. J.M. Greenberg, *The Shock Generation Problem for a Discrete Gas with Short Range Repulsive Force*, Comm. Pure Appl. Math. **45** (1992), to appear.
5. B.M. Herbst and M.J. Ablowitz, *Numerically induced chaos in the nonlinear Schrödinger equation*, Phys. Rev. Lett. **62** (1989), 2065–2068.
6. B.L. Holian, H. Flaschka and D.W. McLaughlin, *Shock Waves in the Toda Lattice: Analysis*, Phys. Rev. A. **24** (1981), 2595–2623.
7. B.L. Holian, G.K. Straub, *Molecular Dynamics of Shock Waves in One Dimensional Chains*, Phys. Rev. B. **18** (1978), 1593–1608.
8. T.Y. Hou and P.D. Lax, *Dispersive approximations in fluid dynamics*, Comm. Pure Appl. Math. **44** (1991), 1–40.
9. S. Jin, C.D. Levermore and D.W. McLaughlin, *The Semiclassical Limit for the Defocusing Nonlinear Schrödinger Hierarchy*, Preprint (1991); *The behavior of Solutions of the NLS Equation in the Semiclassical Limit*, this volume.
10. M. Kac and P. von Moerbeke, *On an explicitly soluble system of nonlinear differential equations related to certain Toda lattices*, Adv. in Math. **16** (1975), 160–169.
11. P.D. Lax, *On dispersive difference schemes*, Physica D **18** (1986), 250–254.
12. P.D. Lax and C.D. Levermore, *The Zero Dispersion Limit of the Korteweg-de Vries Equation*, Proc. Nat. Acad. Sci. USA **76(8)** (1979), 3602–3606; *The small dispersion limit of the Korteweg-de Vries equation I, II, III*, Comm. Pure Appl. Math. **36** (1983), 253–290, 571–593, 809–829.

13. C.D. Levermore and J.-G. Liu, *Large oscillations arising in a dispersive numerical scheme*, Preprint (1992).

14. G. Strang, *Accurate partial differential methods II*, Num. Math. **6** (1964), 37–46.

15. M. Toda, *Theory of Nonlinear Lattices*, Springer Series in Solid-State Sciences 20, second enlarged edition, New York, 1988.

16. L. Trefethen, *Instability of difference models for hyperbolic initial boundary value problems*, Comm. Pure Appl. Math. **37** (1984), 329–367.

17. S. Venakides, *The Zero Dispersion Limit of the Korteweg-de Vries Equation with Nontrivial Reflection Coefficient*, Comm. Pure Appl. Math. **38** (1985), 125–155; *The Zero Dispersion Limit of the Korteweg-de Vries Equation with Periodic Initial Data*, AMS Trans. **301** (1987), 189–225.

18. S. Venakides, P. Deift and R. Oba, *The Toda Shock Problem*, Comm. Pure Appl. Math. **44** (1990), 1171-1242.

19. J. von Neumann, *Proposal and analysis of a new numerical method in the treatment of hydrodynamical shock problems*, Collected Works **VI** (1961), Pergamon, New York.

20. G.B. Whitham, *Linear and Nonlinear Waves*, J. Wiley, New York, 1974.

21. N.J. Zabusky and M.D. Kruskal, *Interaction of "solitons" in a collisionless plasma and the recurrence of initial states*, Phys. Rev. Lett. **15** (1965), 240–243.

ON THE HIERARCHY OF THE GENERALIZED KdV EQUATIONS

Carlos E. Kenig

Department of Mathematics
University of Chicago
Chicago, IL 60637, USA

Gustavo Ponce

Department of Mathematics
University of California
Santa Barbara, CA 93106, USA

and

Luis Vega

Facultad de Ciencias
Universidad Autonoma de Madrid
Cantoblanco, Madrid 28049, Spain

ABSTRACT. *We consider a sequence of one-dimensional dispersive equations. These equations contain the KdV hierarchy as well as several higher order models arising in both physics and mathematics. We obtain conditions which guarantee that the corresponding initial value problem is locally and globally well-posed in appropiated function spaces. Our method is quite general and can be used to study other dispersive systems and related problems.*

Singular Limits of Dispersive Waves, Edited by
N.M. Ercolani et al., Plenum Press, New York, 1994

INTRODUCTION

This note is concerned with the initial value problem (IVP)

$$\begin{cases} \partial_t u + \partial_x^{2j+1} u + Q(u, \partial_x u, \ldots, \partial_x^{2j} u) = 0 & x, t \in \mathbb{R} \quad j \in Z^+ \\ u(x, 0) = u_0(x) \end{cases} \tag{1}$$

where $\partial_t = \partial/\partial t$, $\partial_x = \partial/\partial x$, $u(x, t)$ is a real (or complex) valued function and

$$Q : \mathbb{R}^{2j+1} \to \mathbb{R} \quad (\text{or } Q : C^{2j+1} \to C)$$

is a polynomial having no constant or linear terms, i.e.

$$Q(z) = \sum_{|\alpha| \geq k}^{\rho} a_\alpha z^\alpha \quad \text{with } k \geq 2 \tag{2}$$

for $z = (z_1, \ldots, z_{2j+1})$.

The equation in (1) generalizes the KdV hierarchy introduced by Lax[9]. Gardner el al.[2] proved that the eigenvalues of the time independent Schrödinger operator

$$L(q) = \frac{d^2}{dx^2} - q(x)$$

remains unchanged when the potential $q(\cdot) = u(\cdot, t)$ evolves according to the Korteweg-de Vries equation

$$\partial_t u + u \partial_x u + \partial_x^3 u = 0.$$

This remarkable discovery was the starting point of the inverse scattering method. Lax[9] showed that the same principle holds for the sequence known as the KdV hierarchy

$$\partial_t u + [B_j; L(u)] = 0 \tag{3}$$

(with $[A; M] = AM - MA$ the commutator of the operators A and M) where B_j denotes the skew-symmetric operator

$$B_j = \alpha_j \frac{d^{2j+1}}{dx^{2j+1}} + \sum_{\ell=0}^{j-1} \left(b_{j\ell} \frac{d^{2\ell+1}}{dx^{2\ell+1}} + \frac{d^{2\ell+1}}{dx^{2\ell+1}} b_{j\ell} \right).$$

The coefficients $b_{j\ell} = b_{j\ell}(u)$ are chosen such that the differential operator $[B_j; L(u)]$ has order zero. Thus for $j = 1$ and

$$B_1 = \alpha_1 \frac{d^3}{dx^3} + b \left(u \frac{d}{dx} + \frac{d}{dx} u \right)$$

the operator $[B_1; L(u)]$ becomes multiplication by $u \partial_x u + \partial_x^3 u$, and one obtains the KdV equation. After changing variable and rescaling the equation in (3) with $j = 2$ takes the form

$$\partial_t u + 30 u^2 \partial_x u - 20 \partial_x u \partial_x^2 u - 10 u \partial_x^3 u + \partial_x^5 u = 0.$$

Equations of the type described in (1) also arise as higher order models in water waves problems, elastic media with microstructure and in other physical problems (see Ref. 8 and its bibliography).

Our purpose here is to study well-posedness of the IVP (1). In well-posedness we include existence, uniqueness, persistence property (i.e. the solution $u(\cdot)$ describes a continuous curve in the function space X whenever $u_0 \in X$) and continuous dependence of the solution upon the data. Under decay, smoothness and size assumptions on the data together with some algebraic hypothesis on the polynomial $Q(\cdot)$ we obtain results which guarantee that the IVP (1) is locally and globally well-posed in an appropriate function space X.

It is interesting to remark that classical approaches (i.e. energy estimate L^2-theory, abstract semi-group, $L^p - L^q$-theory, etc.) used to study other evolution equations cannot be applied to the IVP (1) except for very particular class of polynomials $Q(\cdot)$.

Our results can be gathered in the following theorem.

Theorem A.

-i) *For any polynomial $Q(\cdot)$ in the class described in (2) there exist $m, s_0 \in Z^+$ and $\delta > 0$ such that for any $s \geq s_0$ and any $u_0 \in H^s(\mathbb{R}) \cap L^2(\mid x \mid^m dx) \equiv X_{s,m}$ with $\parallel u_0 \parallel_{X_{s_0,m}} \leq \delta$ there exist $T = T(Q; \parallel u_0 \parallel_{X_{s_0,m}}) > 0$ (where $T = T(Q; \rho) \to \infty$ as $\rho \to 0$) and a unique solution $u(\cdot)$ of the IVP (1) satisfying*

$$u \in C([-T, T] : X_{s,m}) \tag{4}$$

and

$$\sup_x \int_{-T}^{T} \mid \partial_x^{s+j} u(x,t) \mid^2 dt < \infty. \tag{5}$$

Moreover for any $T' \in (0, T)$ there exists a neighborhood V_{u_0} of u_0 in $X_{s,m}$ such that the map $\tilde{u}_0 \mapsto \tilde{u}(t)$ from V_{u_0} into the class defined by (4)-(5) (with T' instead of T) is Lipschitz.

-ii) *Let $Q(\cdot)$ be a polynomial in the class described in (2) with $k \geq 3$. Then the results in (i) hold with $m = 0$, (i.e. the use of weighted spaces is unnecessary).*

-iii) *Let $Q(\cdot)$ be a polynomial in the class described in (2) which does not depend on z_{2j+1}. Then the results in (i), (ii) hold without any smallness assumption on the data u_0.*

-iv) *The results in (ii) are global in time if one of the following two hypothesis is verified:*

  a) $k \geq 4j + 3$ in (2)

  or

  b) $k \geq 5$ and $Q(z) = Q(z_\ell, \ldots, z_{2j+1})$ with $\ell \geq (2j - 1)/4$.-

To illustrate our result it is convenient to consider the equation

$$\partial_t u + \partial_x^{2j+1} u + u^\ell \partial_x^2 u = 0. \tag{6}$$

For $j = 1$ and $\ell = 1$ we are in case (i) of Theorem A, hence local well-posedness for small data in a weighted Sobolev space is guaranteed. If $\ell \geq 2$ (case (ii)) the weighted assumption is not essential. Thus one obtains local results in a neighborhood of the origin in $H^s(\mathbb{R})$. If $j = 2$ (i.e. stronger dispersion is considered) and $\ell \geq 2$ we have that the IVP for the equation in (6) is locally well-posed in $H^s(\mathbb{R})$ with $s \geq s_0(j, \ell)$ (no smallness assumption, case (iii) of Theorem A). Moreover if $\ell \geq 11$ one has global well-posedness for small data in $H^s(\mathbb{R})$.

Roughly speaking Theorem A gives conditions which guarantee that the nonlinear term is small enough such that the behavior (local or global) of the solution is controlled by the linear part of the equation.

The KdV hierarchy (3) can be written as

$$\partial_t u + \partial_x^{2j+1} u + c_j u \partial_x^{2j-1} u + \tilde{Q}(u, \ldots, \partial_x^{2j-2} u) = 0.$$

Thus $Q(\cdot)$ satisfies the hypothesis of parts (iii) and (i) of Theorem A. Hence our results contain (except for the use of weighted spaces) those obtained by Saut[11] and Schwarz[12] concerning existence and uniqueness for the KdV hierarchy (respectively) . In this case global well-posedness follows by combining local results with the conservation laws satisfied by solutions of these equations. Also Theorem A contains (except for the use of weighted spaces) the results of Ponce[10] related with general fifth order models in hamiltonian and non-hamiltonian form.

We remark that Craig $et\ al.$[1] studied regularity of solutions to the IVP for the fully non-linear equation of KdV type

$$\partial_t u + F(u, \partial_x u, \partial_x^2 u, \partial_x^3 u) = 0,$$

with data decaying at infinity and $F(u, v, w, z)$ such that

$$\frac{\partial F}{\partial z} \geq c > 0 \quad \text{and} \quad \frac{\partial F}{\partial w} \leq 0. \tag{7}$$

Our analysis although based on properties of the associated linear equation can handle nonlinear terms of the form $\pm u^2 \partial_x^2 u$ (see example 6).

Also we observe that part (iv) of Theorem A can be used to treat other cases between those considered in (iv-a) and (iv-b) (i.e. higher nonlinearity and structure of the polynomial $Q(\cdot)$).

The results in Theorem A extend to equations of the form (generalized Benjamin-Ono)

$$\partial_t u + H \partial_x^{2j} u + Q(u, Hu, \ldots, \partial_x^{2j-1} u, H \partial_x^{2j-1} u) = 0$$

or

$$\partial_t u + \partial_x^{2j+1} u + Q(u, Hu, \ldots, \partial_x^{2j} u, H \partial_x^{2j} u) = 0$$

where H denotes the Hilbert transform in the space variable. Also our approach here can be used to study the IVP for dispersive models in higher dimensions, for example that associated to the equation

$$\partial_t u + i(-\Delta)^j u + Q\left((\partial_x^\alpha u)_{|x| \leq 2j-1}, (\partial_x^\alpha \bar{u})_{|\alpha| \leq 2j-1}\right) = 0.$$

The case $j = 1$ was considered in Ref. 5. The general case $j \geq 1$ follows by combining known arguments[5] with those below.

Also we shall remark that all the results commented above apply to the corresponding systems in "diagonal" form.

The method of proof is based on several sharp linear estimates for solutions of the associated linear problem used in the integral equation version of (1) together with the contraction principle. A related approach[6] was used to obtain what we believe are definitive results concerning the well-posedness problem for the generalized KdV equation in classical Sobolev spaces $H^s(\mathbb{R})$.

PROOF OF THEOREM A

It is convenient to consider first the associated linear problem

$$
\begin{cases}
\partial_t v + \partial_x^{2j+1} v = 0 & x, t \in \mathbb{R} \\
v(x, 0) = v_0(x)
\end{cases}
$$

whose solution is given by the group $\{W^j(t)\}_{-\infty}^{\infty}$ i.e. $v(x,t) = W^j(t)v_0(x) = S_t^j * v_0(x)$ with $S_t^j(x)$ defined by the oscillatory integral

$$
S_t^j(x) = c \int\limits_{-\infty}^{\infty} e^{ix\xi} e^{it\xi^{2j+1}} \, d\xi.
$$

Our first tools are sharp versions of the smoothing effect first established by Kato[3] for solutions of the KdV equations. It was established[4] that there exists c_j such that for any $x \in \mathbb{R}$

$$
\left(\int\limits_{-\infty}^{\infty} \left| \frac{\partial^j}{\partial x^j} W^j(t) v_0(x) \right|^2 dt \right)^{1/2} = c_j \parallel v_0 \parallel_2 . \tag{8}
$$

Notice that in the left hand side of (8) the norm used was $L_x^{\infty}(L_t^2)$. The inhomogeneous version of (8) is given by the following inequality[5,6]

$$
\left\| \frac{\partial^{2j}}{\partial x^{2j}} \int\limits_0^t W^j(t - \tau) G(\tau) d\tau \right\|_{L_x^{\infty}(\mathbb{R}:L_t^2(\mathbb{R}))} \tag{9}
$$

$$
= \sup_x \left(\int\limits_{-\infty}^{\infty} \left| \frac{\partial^{2j}}{\partial x^{2j}} \int\limits_0^t W^j(t - \tau) G(\tau) d\tau \right|^2 dt \right)^{1/2}
$$

$$
\leq c \parallel G \parallel_{L_x^1(\mathbb{R}:L_t^2(\mathbb{R}))} .
$$

On the other hand combining (8) with Minkowski's integral inequality and Cauchy-Schwarz inequality one sees that

$$
\left\| \frac{\partial^j}{\partial x^j} \int\limits_0^t W^j(t - \tau) G(\tau) d\tau \right\|_{L_x^{\infty}(\mathbb{R}:L_t^2([-T,T]))} \tag{10}
$$

$$
\leq c \int\limits_T^T \parallel G(\cdot, t) \parallel_2 dt
$$

$$
\leq c \, T^{1/2} \parallel G \parallel_{L_x^2(\mathbb{R}:L_t^2([-T,T]))} .
$$

Hence interpolation between (9)-(10) shows that

$$\left\| \frac{\partial^{j+\ell}}{\partial x^{j+\ell}} \int_0^t W^j(t-\tau) G(\tau) d\tau \right\|_{L_x^\infty(\mathbb{R}:L_t^2([-T,T]))} \tag{11}$$

$$\leq c\, T^\gamma \left\| G \right\|_{L_x^p(\mathbb{R}:L_t^2([-T,T])}$$

for $\ell = 0, 1, \ldots, j$ with $\gamma = \dfrac{j-\ell}{2j}$ and $p = \dfrac{2j}{j+\ell}$.

For further comments and references concerned with the Kato smoothing effect and related results see Ref. 4, section 4.

To complement (8)-(11) one needs estimates for the maximal function $\sup_{[-T,T]} | W^j(t)u_0 |$ in the $L_x^p(\mathbb{R})$-norm. In the proof of part (ii) in Theorem A we shall use that

$$\left\| \sup_{[-T,T]} | W^j(t)v_0 | \right\|_2 \leq c\,(1+T)^\rho \left\| v_0 \right\|_{s,2} \tag{12}$$

for any $\rho > 3/4$ and any $s > (2j+1)/4$ where $\| \cdot \|_{s,2}$ denotes the norm in the classical Sobolev space $H^s(\mathbb{R})$. For the proof of (12) we refer to Refs. 4, 14. To obtain the global results in Theorem A (case (iv) part b) we shall use the following global (in time) estimate due to Kenig et al.[7] (see also Ref. 4, Theorem 2.5)

$$\left(\int_{-\infty}^\infty \sup_{-\infty < t < \infty} | W^j(t)v_0 |^4 (x)\, dx \right)^{1/4} \leq c \left\| D^{1/4} v_0 \right\|_2 \tag{13}$$

where $D^\alpha f = (c\, |\, \xi\, |^\alpha\, \hat{f})^\vee$. In part (i) of Theorem A one needs to estimate in the L^1-norm, the maximal function i.e.

$$\left\| \sup_{[-T,T]} | W^j(t)v_0(x) | \right\|_1 \equiv E_1(v_0).$$

By the theorems of Sobolev and Fubini it is easy to see that

$$E_1(v_0) \leq c\, \frac{1}{T} \int_{-T}^T \int_{-\infty}^\infty | W^j(t)v_0(x) |\, dx\, dt$$

$$+ c \int_{-T}^T \int_{-\infty}^\infty | W^j(t)\partial_x^{2j+1} v_0(x) |\, dx\, dt.$$

Combining the inequality

$$\| g \|_1 \leq c\,(\| g \|_2 + \| xg \|_2),$$

with the identity

$$x W^j(t)v_0 = W^j(t)(xv_0) + (2j+1)t W^j(t)\partial_x^{2j} v_0 \tag{14}$$

we can conclude that

$$\left\| \sup_{[-T,T]} \mid W^j(t)v_0(x) \mid \right\|_1 \leq c(1+T^2) \parallel u_0 \parallel_{4j+1,2} \tag{15}$$
$$+ c(1+T) \parallel u_0 \parallel_{2j+1,2,1} \equiv \||u_0\||_T$$

where

$$\parallel f \parallel_{\ell,p,k} = \sum_{i=0}^{\ell} \parallel x^k \partial_x^i f \parallel_p . \tag{16}$$

The last tool needed in the proof of Theorem A is the following version of the Strichartz[13] type of estimate found in Ref. 4 (Theorem 2.4):

For any $(\theta, \alpha) \in [0,1] \times [0, (2j-1)/2]$

$$\left(\int_{-\infty}^{\infty} \parallel D_x^{\theta\alpha} W^j(t)v_0 \parallel_p^q \ dt \right)^{1/q} \leq c \parallel v_0 \parallel_2 \tag{17}$$

where $(q,p) = (2(2j+1)/\theta(\alpha+1), \ 2/(1-\theta))$. For further comments and references see Ref. 4 (sections 2-3).

With these estimates in hand we can sketch the proof of Theorem A case (i). For the quadratic case (i) we define for each $T > 0$

$$\beta_1^T(w) = \sup_{[0,T]} \parallel w(t) \parallel_{s,2},$$

$$\beta_2^T(w) = \max_{k=1,\ldots,j} \sup_x \left(\int_0^T \mid \partial_x^{k+s} w(x,t) \mid^2 dt \right)^{1/2},$$

$$\beta_3^T(w) = \max_{k=0,\ldots,4j} \int_{-\infty}^{\infty} \sup_{[0,T]} \mid \partial_x^k w(x,t) \mid \ dx$$

and

$$\beta_4^T(w) = \sup_{[0,T]} \parallel w(t) \parallel_{6j+1,2,j},$$

with $s \geq 10j + 1$. Also define

$$\Omega^T(w) = \max \left\{ \beta_j^T(w) / j = 1, \ldots, 4 \right\},$$

$$X_T^a = \left\{ w \in \mathbb{R} \times [0,T] \to \mathbb{R} / \Omega^T(w) \leq a \right\}$$

and for $u_0 \in H^s(\mathbb{R})$ the operator

$$\Phi_{u_0}(w)(t) = W^j(t)u_0 + \int_0^t W^j(t-t')Q(w,\ldots)(t')dt'. \tag{18}$$

To simplify the exposition we shall carry out the details only for the highest and lowest derivatives. Inserting (8), (9) in (18) it follows that

$$\sup_x \left(\int_0^T | \partial_x^{j+s} \Phi(w) |^2 \, dt \right)^{1/2} \tag{19}$$

$$\leq c \parallel u_0 \parallel_{s,2} + \parallel \partial_x^{s-j} Q(w, \ldots, \partial_x^{2j} w) \parallel_{L_x^1(\mathbb{R}: L_t^2([0,T]))}$$

$$\leq c \parallel u_0 \parallel_{s,2} + \parallel \widehat{Q}(w, \ldots, \partial_x^{2j} w) \partial_x^{s+j} w \parallel_{L_x^1(\mathbb{R}: L_t^2([0,T]))}$$

$$+ \text{ lower order terms.}$$

where $\widehat{Q}$ denotes the polynomial coefficient of the term involving the highest derivative obtained after using the Leibniz rule. But

$$\parallel \widehat{Q}(w, \ldots, \partial_x^{2j} w) \partial_x^{s+j} w \parallel_{L_x^1(\mathbb{R}: L_t^2([0,T]))} \tag{20}$$

$$\leq c \max_{k=0,\ldots,2j} \left(\int_{-\infty}^{\infty} \sup_{[0,T]} | \partial_x^k w(x,t) | \, dx \right)$$

$$\sup_x \left(\int_0^T | \partial_x^{s+j} w |^2 \, dt \right)^{1/2}$$

$$\leq c \, \beta_2^T(w) \beta_3^T(w) \left(1 + (\beta_3^T(w))^{\rho-2} \right).$$

Similarly combining (14), (15) it is not hard to see that

$$\int_{-\infty}^{\infty} \sup_{[0,T]} | \Phi(w) | \, dx \leq c \, |||u_0|||_T + c \int_0^T |||Q(w, \ldots, \partial_x^{2j} w)|||_T \, dt \tag{21}$$

$$\leq c \, |||u_0|||_T + c (1 + T^2) T \sup_{[0,T]} \parallel w(t) \parallel_{2j+1,2}$$

$$\left(1 + (\sup_{[0,T]} \parallel w(t) \parallel_{2j+1,2})^{\rho-2} \right) \sup_{[0,T]} \parallel w \parallel_{4j,2,1}$$

$$\leq c \, |||u_0|||_T + c(1 + T^2) T \, \beta_1^T(w) \left(1 + (\beta_1^T(w))^{\rho-2} \right) \beta_4^T(w),$$

(see the notation in (16)). The estimate for the term

$$\sup_{[0,T]} \parallel \partial_x^s \Phi(w) \parallel_2 \tag{22}$$

is similar to that given in (19)-(20). One just needs to combine the dual version of

(8) with the group properties. Finally from (14) we have that

$$\sup_{[0,T]} \| \Phi(w) \|_{0,2,1} \le c \| x u_0 \|_2 \tag{23}$$

$$+ \int_0^T \| x Q(w, \dots, \partial_x^{2j} w) \|_2 \ dt$$

$$\le c \| x u_0 \|_2 + c T \sup_{[0,T]} \| w \|_{2j,2}$$

$$\left(1 + (\sup_{[0,T]} \| w \|_{2j,2})^{\rho-2} \right) \max_{k=0,\dots,2j} \| w \|_{k,2,1}$$

$$\le c \| x u_0 \|_2 + c T \, \beta_1^T(w) \left(1 + (\beta_1^T(w))^{\rho-2} \right) \beta_4^T(w).$$

Hence collecting the results in (19)-(23) and using the above notation it follows that

$$\Omega^T(\Phi(w)) \le c \left(\| u_0 \|_{s_0,2} + \| x^m u_0 \|_2 \right) + c (1 + T^4)(\Omega^T(w))^2 (1 + (\Omega^T(w))^{\rho-2}). \tag{24}$$

A similar argument shows that

$$\Omega^T(\Phi(w) - \Phi(\tilde{w})) \le c (1 + T^4)\Omega^T(w - \tilde{w})(\Omega^T(w) + \Omega^T(\tilde{w})) \tag{25}$$

$$(1 + (\Omega^T(w) + \Omega^T(\tilde{w}))^{\rho-2}).$$

From (24)-(25) it is easy to obtain part (i) of Theorem A.

The proof of part (ii) is easier. In fact, instead of using $\beta_3^T(\cdot), \beta_4^T(\cdot)$ one just needs to consider

$$\beta_5^T(w) = \max_{k=0,\dots,4j} \left(\int_{-\infty}^{\infty} \sup_{[0,T]} | \partial_x^k w(x,t) |^2 \ dx \right)^{1/2}$$

and to apply (12) instead of (15). In part (iii) the hypothesis guarantees that one can rely on the estimate (11) with $\ell < j$. Thus a factor of the form T^γ appears in the second term in the right hand side of (24)-(25). This allows us to show that the map $\Phi : X_T^a \to X_T^a$ is a contraction (for appropriate values of a and T) without any smallness assumption on the data u_0.

Finally, to prove part (iv) of Theorem A one uses the global estimates (8), (9), (13) and (17).

ACKNOWLEDGEMENT

C. E. Kenig and G. Ponce were supported by NSF grants while L. Vega was supported by a DGICYT grant.

REFERENCES

1. W. Craig, T. Kappeler and W. A. Strauss, *Gain of regularity for equations of KdV type, preprint*, to appear in Ann. IHP, Analyse Nonlineaire.

2. C. S. Gardner, J. M. Greene, M. D. Kruskal and R. M. Miura, *A method for solving the Korteweg-de Vries equation*, Phys. Rev. Letters **19** (1967), 1095–1097.

3. T. Kato, *On the Cauchy problem for the (generalized) Korteweg-de Vries equation*, Advances in Mathematics Supplementary Studies, Studies in Applied Math. **8** (1983), 93–128.

4. C. E. Kenig, G. Ponce and L. Vega, *Oscillatory Integrals and Regularity of Dispersive Equations*, Indiana University Math. J. **40** (1991), 33–69.

5. _____, *Small solutions to nonlinear Schrödinger equations*, to appear in Ann. IHP, Analyse Nonlineaire.

6. _____, *Well-posedness and scattering results for the generalized Korteweg-de Vries equation via the contraction principle*, preprint.

7. C. E. Kenig and A. Ruiz, *A strong type (2,2) estimate for the maximal function associated to the Schrödinger equation*, Trans. Amer. Math. Soc. **230** (1983), 239–246.

8. S. Kichenassamy and P. J. Olver, *Existence and Non-existence of solitary waves solutions to higher order model evolution equations*, preprint.

9. P. D. Lax, *Integrals of nonlinear equations of evolution and solitary waves*, Comm. Pure Appl. Math. **21** (1965), 467–490.

10. G. Ponce, *Lax pairs and higher order models for water waves*, to appear in J. Diff. Eqs.

11. J.-C. Saut, *Sur quelques généralisations de l' équations de Korteweg-de Vries, II*, J. Diff. Eqs. **33** (1979), 320–335.

12. M. Schwarz Jr., *The initial value problem for the sequence of generalized Korteweg-de Vries equation*, Advances in Math. **54** (1984), 22–56.

13. R. S. Strichartz, *Restrictions of Fourier transforms to quadratic surfaces and decay of solutions of wave equations*, Duke Math. J. **44** (1977), 705–714.

14. L. Vega, *Doctoral Thesis*, Universidad Autonoma de Madrid, Spain (1987).

A NEW THEORY OF SHOCK DYNAMICS

Phoolan Prasad

Department of Mathematics
Indian Institute of Science
Bangalore, 560012, India

1. INTRODUCTION

Consider a system of conservation laws with given initial or boundary conditions such that the solution contains a single shock surface whose initial position and shape is known. <u>Shock dynamics</u> is a mathematical theory to calculate successive positions and geometry of the shock surface without resort to finding the solution behind the shock. Such a theory based on intuitive arguments, was provided by Whitham (1957, 1959). The only justification of this theory came from the agreement of its results with those obtained from experiments and some purely numerical methods. However, it has only recently been shown by Prasad (1990); and Prasad Ravindran and Sau (1991) that Whitham's shock dynamics cannot be theoretically justified and can give completely wrong results. We present here a new theory of shock dynamics by Ravindran and Prasad (1990) which is based on exact mathematical results on the compatibility conditions on the shock manifold.

2. COMPATIBILITY CONDITIONS ON A SHOCK MANIFOLD IN SPACE-TIME

Jump conditions, which are well known as Rankine-Hugoniot conditions in gas dynamics, across a shock were suggested by Stokes (1848) and derived by Rankine (1870) and Hugoniot (1889). However, the fact that one can derive an infinite system of compatibility conditions on a shock manifold, was first proposed independently only recently by Grinfel'd (1978) and Maslov (1978). These compatibility conditions for gasdynamic equations are very complex and the evaluation of their complete form requires difficult mathematical calculations (Srinivasan and Prasad (1985), Anile and Russo

(1986, 1988), Ravindran and Prasad (1992), Lazarev, Ravindran and Prasad (1992)). It is convenient to express these compatibility conditions in terms of the derivatives of a single variable (say, the density ρ) behind the shock. Let $\Omega : s(\underline{x},t) = 0$ be shock manifold, ρ_0 be the density (not necessarily constant) ahead of the shock $s > 0$ and $\underline{N}$ the unit normal to the shock surface Ω_t at a given time t. We extend the function defined behind the shock (i.e, $s < 0$) and also ρ_0 as C^∞ functions in a neighbourhood $G \subset R^4$ of Ω and define

$$\mu = \frac{\rho - \rho_0}{\rho_0} \in C^\infty \ (G).$$

(2.1)

We further define quantities $v_0, v_1, v_2, v_3, \dots$ on Ω by

$$v_r = \frac{1}{r!} \left(\sum_{i_1, \dots, i_r = 1}^{3} N_{i_1} N_{i_2} \dots N_{i_r} \frac{\partial^r \mu}{\partial x_{i_1} \partial x_{i_2} \dots \partial x_{i_r}} \right) \Bigg|_\Omega .$$

(2.2)

The ith compatibility condition on the shock manifold is of the form

$$\frac{dv_i}{dt} = d_i \ v_{i+1} + f_i, \ i = 0,1,2,\dots$$

(2.3)

where $\frac{d}{dt}$ represents time rate of change along a shock ray defined below, d_i is a coupling coefficient between the ith and the (i+1)th compatibility conditions and f_i is a nonlinear function of the known state ahead of the shock; $v_0, v_1, \dots, v_i$; N_1, N_2, N_3 ; and some interior derivatives of v_0, $v_1, \dots, v_i$ and N_1, N_2, N_3 in the shock surface Ω_t. In the first compatibility condition, the function f_0 is product of the mean curvature of Ω_t and a nonlinear function of the shock amplitude v_0. The coupling coefficient d_i tends to zero as the shock strength v_0 tends to zero. If the fluid velocity ahead of the shock is $\underline{u}_0$, then the shock ray derivative is given by

$$\frac{d}{dt} = \frac{\partial}{\partial t} + (\underline{u}_0 + \underline{N} \ C) \cdot \nabla_x$$

(2.4)

where C is the shock velocity relative to the state ahead and $\nabla_x = (\frac{\partial}{\partial x_1}, \frac{\partial}{\partial x_2}, \frac{\partial}{\partial x_3})$.

(2.3) form an infinite system of equations which is not closed at any stage. The system is also not complete due to the presence of N_1, N_2 and N_3 . To get a complete system of equations, we either take help of the shock manifold partial differential equation introduced by Prasad (1982) or the geometrical and kinematical compatibility conditions (see Thomas (1961), Grinfel'd (1978)). Let us first define an operator

$$L = \nabla_x - \underline{N} < \underline{N} , \nabla_x >$$ (2.5)

Since $< \underline{N}, L > = 0$, the components of L represent tangential derivatives in the shock surface Ω_t. Then the shock ray equations are

$$\frac{d\underline{x}}{dt} = \underline{u}_0 + \underline{N} C$$ (2.6a)

and

$$\frac{d\underline{N}}{dt} = \sum_{\alpha=1}^{3} \underline{N}_\alpha \ L \ u_{\alpha_0} + LC .$$ (2.6b)

In the space of C^∞ functions, the solution of the infinite system (2.3) - (2.6b) is not unique. This was shown by Grinfeld(1978)from the particular case of the Hopf equation $u_t + (\frac{1}{2} u^2)_x = 0$ in one space dimension. Since d_i has a factor v_0, the infinite system becomes a closed system of finite number of equations for problems involving weak shocks (Grinfel'd (1978) and Maslov (1978)). The infinite system has a unique local analytic solution if the initial data is also analytic (Ravindran and Prasad (1990), Prasad (1992)). The problem of continuation of this local solution is open.

3. NEW THEORY OF SHOCK DYNAMICS

We present the theory for the one dimensional model equation

$$u_t + (\frac{1}{2} u^2)_x = 0, \ (x,t) \in R \times R_+$$ (3.1)

and assume that the state $u_r(x,t)$ ahead of the shock curve $\Omega : x-X(t) = 0$ to be known. With analytic initial data behind the initial position X_0 of the shock, the solution behind the shock is also analytic i.e.

$$u(x,t) = \sum_{i=0}^{\infty} v_i(t) \ (x-X(t))^i, \ x < X(t).$$ (3.2)

The shock path is given by

$$\frac{dX(t)}{dt} = \frac{1}{2} (v_0(t) + u_r(x(t),t))$$ (3.3)

which replaces the shock ray equations (2.6). Substituting (3.2) in $u_t + u u_x = 0$, equating coefficients of various powers of $(x - X(t))$ and using (3.3), we get the infinite system of compatibility conditions.

If we set $v_{n+1} = 0$ in the (n+1)th compatibility condition, the first n+1 equations decouple from the rest of the equations and we get a finite system of (n+2) equations. Denoting the solution of this finite system by $\bar{x}(t)$ and $\bar{v}_i(t)$, $i = 0, 1, 2, \ldots, n$; we get

$$\frac{d\bar{x}}{dt} = \frac{1}{2} (\bar{v}_0 + u_r) \tag{3.4}$$

$$\frac{d\bar{v}_0}{dt} = -\frac{1}{2} (\bar{v}_0 - u_r)\bar{v}_1 \tag{3.5}$$

$$\frac{d\bar{v}_i}{dt} = -\frac{i+1}{2} (\bar{v}_0 - u_r) \bar{v}_{i+1} - \frac{i+1}{2} \sum_{j=1}^{i} \bar{v}_j \bar{v}_{i-j+1} \ ,$$

$$i = 1, 2, \ldots, n - 1 \tag{3.6}$$

and

$$\frac{d\bar{v}_n}{dt} = -\frac{n+1}{2} \sum_{j=1}^{n} \bar{v}_j \bar{v}_{n-j+1} \tag{3.7}$$

with initial conditions

$$\bar{x}(0) = x_0 \ , \ \bar{v}_i(0) = v_{i_0}, \ i = 0, 1, 2 \ldots, n \tag{3.8}$$

which can be determined from the original initial condition for (3.1).

The initial value problem (3.4) – (3.8) has unique analytic solution in a neighbourhood of t = 0 and it has been proved by Ravindran and Prasad (1990) that for a fixed value of t, this analytic solution tends to the unique analytic solution of the infinite system as $n \to \infty$. Given a solution of (3.4) – (3.8), we can construct a function

$$\bar{u}(x,t) = \begin{cases} \sum_{i=0}^{n} \bar{v}_i(t) \ (x - \bar{x}(t))^i, & x < \bar{x}(t) \\ u_r(x,t) = \text{known}, & \bar{x}(t) < x. \end{cases} \tag{3.9}$$

The function $\bar{u}(x,t)$ satisfies the conservation law (3.1) approximately, since for $x < \bar{x}(t)$

$$\bar{u}_t + \bar{u} \ \bar{u}_x = (x - x(t))^n h(x,t) \tag{3.10}$$

where h(x,t) is defined in terms of $\bar{v}_i (i = 0,1,\ldots,n)$ and $\bar{x}(t)$. Near the shock, the right hand side of (3.10) tends to zero as $n \to \infty$.

Prasad and Ravindran (1990) have taken two examples to show the validity of the above procedure. Consider two initial data

$$u(x,0) = \Phi_1(x) = \begin{cases} \dfrac{(x+\eta)}{1+\eta} \ , & \eta > 0, \ x \in (-\eta, 1) \\ \dot{0} \ , & x < -\eta \text{ and } 1 < x \end{cases} \tag{3.11}$$

and

$$u(x,t) = \Phi_2(x) = \begin{cases} \propto e^{\beta x} & , \ x < 0 \\ 0 & , \ x > 0 . \end{cases} \qquad (3.12)$$

Initial data (3.11) and (3.12) have a shock at $X_0 = 1$ and $X_0 = 0$ respectively. In (3.12) we choose $\propto = 1$, $\beta = 1$. Rescaling of independent and dependent variables shows that without loss of any generality, we can choose $\propto = 1$ $\beta = 1$. The exact strength v_0 and position X of a shock with an arbitrary initial data $\phi(x)$ for (3.1) can be obtained from $v_0 = \Phi(\xi)$, $X = \xi + v_0 t$, where ξ satisfies

$$t \ \phi^2(\xi) + 2 \int_{X_0}^{\xi} \phi(\mu) \ d\mu = 0., \qquad (3.13)$$

Tables 1 and 2 give the values of v_0 corresponding to initial values ϕ_1 and ϕ_2 at t = 1.0, 5.0 and 10.0. n = k denotes the results obtained from the new theory of shock dynamics i.e. by solving the problem (3.4) – (3.8) with n = k. For n = 1, the error in v_0 is significant, but for n = 2, the error drops rapidly (less than 1% in the Table 2); while for n=3, it is uniformly very small. For n = 5, 8 and 25 the error is less than 0.1%.

TABLE 1 . $\eta = -0.5$

	t = 1.0		t = 5.0		t = 10.0	
	v_0	% error	v_0	% error	v_0	% error
Exact	.47390445	–	.24231081	–	.17572092	–
n = 1	.44721360	-5.6	.21821789	-9.9	.15617376	-11.0
n = 2	.47171239	-.46	.23787367	-1.8	.17140803	- 2.5
n = 3	.47366942	$-.50 \times 10^{-1}$	.24120493	-.46	.17440411	- .75
n = 5	.47390183	$-.56 \times 10^{-3}$	.24221465	$-.40 \times 10^{-1}$	.17554484	- .10
n = 8	.47390560	$.24 \times 10^{-3}$	.24230783	$-.12 \times 10^{-2}$	.17570887	$-.68 \times 10^{-2}$
n =25	.47390561	$.24 \times 10^{-3}$	.24231136	$.24 \times 10^{-3}$	.17572129	$.22 \times 10^{-3}$

TABLE 2. $\Phi = e^x$, $x < 0$; $\Phi = 0$, $x > 0$

	t = 1.0		t = 5.0		t = 10.0	
	v_0	% error	v_0	% error	v_0	% error
Exact	.73205081	–	.46332497	–	.35825757	–
n = 1	.70710678	-3.4	.40824829	-12	.30151134	-16
n = 2	.73372900	.23	.46777169	.96	.36157950	.93
n = 3	.73200502	$-.63 \times 10^{-2}$	.46355666	$.50 \times 10^{-1}$	.35872978	.13
n = 5	.73205096	$.26 \times 10^{-4}$	.46331988	$-.11 \times 10^{-2}$	.35825020	$-.21 \times 10^{-2}$
n = 8	.73205081	0	.46332497	0	.35825765	$.22 \times 10^{-4}$
n =25	.73205081	0	.46332496	0	.35825757	0

Thus, the new theory of shock dynamics proposed here, appears to be a promising method for finding successive positions of a shock. The numerical solution with a finite number of compatibility conditions is far more simple compared to a finite difference method and requires relatively very small computer time. Application of the theory to multidimensional problems is also possible.

REFERENCES

Anile, A.M. and Russo, G., 1986, Generalised wavefront expansion I : Higher order corrections for the propogation of weak shock waves, Wave Motion, 8 : 243.

Anile, A.M. and Russo, G., 1988, Generalised wavefront expansion II : The propagation of step shock, Wave Motion, 10 : 3.

Grinfel'd, M.A., 1978, Ray method for calculating the wavefront intensity in nonlinear elastic material, PMM J. Appl. Math. Mech., 42, 958.

Hugoniot, H., 1889, Sur la propagation du movement dans les corps et specialement dans les gaz parfaits, J. de l' e'cole polytechnique, 58 : 1.

Lazarev, M.P., Ravindran, R. and Prasad, P., 1992, Shock propagation in gas dynamics : explicit form of higher order compatibility conditions.

Maslov, V.P., 1978, English translation - Propagation of shock waves in an isentropic nonviscous gas, J. Soviet Math, 1980, 13:119.

Prasad, P., 1982, Kinematics of a multi-dimensional shock of arbitrary strength in an ideal gas, Acta Mechanica, 45 : 163.

Prasad, P., 1990, On shock dynamics, Proc. Ind. Acad. Sci. (Math.Sci.) 100 : 87.

Prasad, P., 1992, "A Theory of Shock Propagation", to appear as a Research Notes in Mathematics, A. Jeffrey, ed. Longman, London.

Prasad, P., and Ravindran, R., 1990, A new theory of shock dynamics, Part II : Numerical results, Appl. Math. Lett. 3 : 107.

Prasad, P., Ravindran, R., and Sau, A., 1991, The characteristic rule for shocks, Appl. Math. Lett. 4 : 5.

Rankine, W.J.M., 1870, On thermodynamic theory of waves of finite longitudinal disturbance, Trans. Roy. Soc. of London, 160, 277.

Ravindran, R., and Prasad, P., 1990, A new theory of shock dynamics, Part I : Analytic considerations, Appl. Math. Lett., Vol.3:77.

Ravindran, R. , and Prasad, P., 1992, On an infinite system of compatibility conditions along a shock ray.

Srinivasan, R., and Prasad, P., 1985, On the propagation of multi-dimensional shock of arbitrary strength, <u>Proc. Ind. Acad. Sci. (Math. Sci.)</u>, 94 : 27.

Stokes, E.E., 1848, On a difficulty in theory of sound, <u>Philisophical Magazine</u>, <u>33</u> : 349.

Thomas, T.Y., 1961, "Plastic Flow and Fracture in Solids", Academic Press, New-York and London.

Whitham, G.B., 1957, A new approach to shock dynamics, Part I, <u>J. Fluid Mech.</u>, 2 : 146.

Whitham, G.B., 1959, A new approach to problems of shock dynamics, Part II, Three dimensional problems, <u>J. Fluid Mech.</u>, 5 : 369.

INDEX

Abel hierarchy, 84–87
Abel transform, 78, 80, 107, 177
Abelian differential, 3, 8, 9, 16, 106–108, 115, 162
Abelian integral, 108, 109, 288
Action, 166, 173, 216, 241, 274, 278, 305–306, 311
Adams–Bashfort method, 333–335, 337, 341
Adiabatic, 105, 113, 216, 274
Angle-representation, 276, 279, 281, 283–288
Asymptotic ansatz, 113, 114, 179
Atomic measure, 251–253
Attractor, 39–40, 52
Autocorrelation function, 183–201
Averaged Whitham equation for Toda lattice, 5–10

Bäcklund transformation, 118, 146
Baker vector/Baker function, 121, 122
Baker–Akhiezer function, 9, 79–82, 109, 112, 118, 120, 177
Benney equations, 53, 55–59, 61, 65, 143–155, 168–170
Benney hierarchy, 61–62, 65
Bifurcation, 39–40, 48, 305, 306, 309, 311
Billiards, 219–234
Binary oscillation, *see* period-two oscillation, 331–332
Boltzmann type equation, 220, 225, 233
Boussinesq equation, 61, 64–65
Branch points of a Riemann surface, 3, 8, 10, 11, 16, 41, 56, 59, 65, 70, 71, 74, 76, 80, 84, 106–108, 110, 111, 113–115, 143, 179
Breather, 40–51, 283
Brownian motion, 233
Bunimovich–Sinai theorem, 219, 220, 233
Burger's equation, 28, 136–140, 157, 160, 161, 329, 330

Cantor set, 301, 304, 305, 310, 311
Cauchy's kernel, 65, 79–82
Caustic, 23–24, 34–35, 240–243, 279
Central limit theorem, 233
Chaos / chaotic, 40, 105, 331, 336, 339, 342
Characteristic speed/velocity, 4, 5, 14, 15, 64
Christophel symbols, 144
Chromatography, 144, 145, 153
C. Neumann problem, 279, 285, 286, 288
Compatibility conditions in shock dynamics, 357–359

Complex structure, deformation of, 70, 73, 77, 78
Complex tori, 75
Configuration space, 205
Conformal field theory, 70
Connection, 77, 274, 283
Conoidal waves, 203, 255
Conservation laws, 5ff, 37, 68, 90, 94, 145, 146, 149–153, 236, 240, 330, 331, 332, 345, 357, 360
Contact discontinuity, 329, 333, 336, 341, 342, 345
Contraction principle, 351
Contravariant derivative, 290
Convection, 315, 316
Couette flow, 315, 316
Covariant derivative, 290
Critical magnetic field, 183–201
Curvature, 144
Cusp / cuspidal, 119, 126, 133, 139–140
Cuts, 62–64
Cycle, 42–43, 71–72, 75, 106, 108, 216, 282

Darboux transformations, 117–134
Defocusing NLS equation, 21–27, 235–238, 242–245, 248, 249, 251, 276, 279, 283
Degree of mapping $R^1 \to S^1$, 315, 320
Diffusion coefficient, 219, 220, 226, 228, 233, 234
Diffusion equation, 219, 226, 228
Dirac operator, 246
Dispersionless Lax equations, 61–66
Dispersionless limit of integrable systems, 165–174
Dispersive hydrodynamics, 89–104
Divisor, 120, 122, 125, 177, 179, 297, 302, 304
Doppler frequency, 324, 325
Dressing, 174, 177
Driver amplitude, 44–45
Driver frequency, 43
Dubrovin equations, 107, 114–115
Duhamel integral, 212
Dyson formula, 260

Egorov class, 145–147
Eikonal equation, 22, 28, 34, 35, 36, 240, 241
Einstein–Kubo formula, 233–234
Electrophoresis, 144, 145, 153
Ellipsoid, 276, 277
Elliptic curve, 120, 122

Elliptic functions, 68, 265
Elliptic integrals, 68, 74, 94–96, 103, 135, 137, 143, 160, 204, 265, 267, 269
Energy density, 316
Energy-phase modulation equations, 43–44
Entropy solution of Hopf equation, 333
Euler hydrodynamics equations, 94
Euler–Poisson type equation, 90, 96, 136–139
Euler system, 237, 238, 254
Evolution equations, 37, 111

Finite-gap, 69–70, 78–80, 82–84, 175–180, 204, 205, 208, 209, 211, 215, 216, 289
Finite horizon property of a billiard, 222–223, 228, 230, 233–234
Flaschka's form of Toda lattice equations, 6
Focusing NLS equation, 27–37, 235, 236, 242–245, 249, 283
Fractional power, 62
Free energy, 316, 319
Fundamental frequency, 300, 304

Galilean invariance of NLS, 28
Galilei transformations, 145–147, 149, 151, 153
Gas dynamics, 22, 169, 173
Gas law, 22, 23
Gauge invariance / gauge transform, 168, 172
Gauss hypergeometric function, 100
Gelfand–Dikii "fractional power" ansatz, 62
Gelfand–Levitan integral equations, 177
Generalized hodograph method, 89, 90–94, 99, 109, 154
Genus, 2, 4, 5, 8, 13, 16, 17, 70, 71, 75, 81, 106, 118, 175, 254, 267, 269
Geodesic flow, ergodic property of, 219
Geometric phases, 273–295
GHM, *see* generalized hodograph method
Goursat problem, 90, 97, 101
GP problem, *see* Gurevich–Pitaevsky problem
Grassmanian, 118–120
Green's formula, 231
Green's function, 304
Gurevich–Pitaevsky problem, 89, 90–94, 96–99, 203

Hamiltonian, 61–62, 105–107, 112, 173, 174, 183, 275–280, 286, 287, 289, 290, 298, 299, 305
   Riemann invariant form of, 61
Hamiltonian system, 68, 105, 107, 143–146, 150–153, 167, 173, 216, 236, 257, 273, 278, 280, 281, 285–288, 298, 302
Hamiltonian flow, 105, 114, 276, 278, 279, 285, 286, 287
Hamilton–Jacobi equation, 63, 65, 173, 174, 213
Harmonic oscillator, 299–300
Heat transfer, 316
Helix, 320, 326
Hilbert–Schmidt operator, 230
Hilbert space, 303
Hilbert transform, 254, 350
Hodograph, 4, 16, 61, 63, 68–69, 89, 90–94, 99, 102, 103, 109, 136–140, 154

Holomorphic differential, 42, 72, 73, 75, 77, 106, 177
Homoclinic, 40, 273–275, 279, 285–288
Hopf equation, 249, 329, 331, 333–337, 341–344, 359
Hydrodynamic-type equations/systems, 53, 67, 68, 89, 143, 145, 149–154, 225
Hydrodynamic symmetries, 92, 149
Hyperelliptic, 70–78, 80, 106, 126, 175, 204, 209, 211, 215, 276
Hypergeometric equation, 99
Hypergeometric function, 100

Integrable / nonintegrable numerical scheme, 329, 331, 336, 341, 342, 345
Integrable system, 165–174
Invariant tori, 105, 216, 276, 279, 300, 302
Inverse scattering problem method, 53–59, 61, 135, 167, 170, 176, 178, 246, 248, 251, 259, 348
Inverse scattering transform, 67, 69, 79, 81, 249
Ising chain, 183–201
Isospectral symmetries, 79–82
ISP method, *see* inverse scattering problem method
Its–Matveev formula, 175, 204

Jacobi identity, 144
Jacobi *q*-function, 204
Jacobi transformation, 267
Jacobi problem, 286, 288
Jacobi variety, 273, 274, 279, 281, 283, 286, 289
Jacobian, 106
Jacobian matrix, 332
Jost functions, 54

Kac–van Moerbeke lattice, 331, 341, 342
KAM theory, *see* Kolmogorov–Arnold–Moser theory
Kadomtsev-Petviashvili equation, 67, 69, 78–84, 123, 132, 167–168
   dispersionless, 167–168
   finite-gap solutions, 69, 78–79, 82–84
   finite-gap inverse scattering transform for, 81
   isospectral symmetries, 79–82
   nonisospectral symmetries, 82–84
   Lax pair for, 81
KdV equation, *see* Korteveg–de-Vries equation
KdV functions, 212
KdV hierarchy, 84, 347–356
KdV hydrodynamics, 89, 91
KdV–Whitham problem, 2, 3
KdV–Whitham problem *vs.* Toda–Whitham problem, 2
Kinetic energy, 305, 306, 310
KN equation, *see* Krichever–Novikov equation
Kolmogorov–Arnold–Moser theory, 114, 180, 205, 302
Kolmogorov–Petrovsky–Piskunov problem, 322
Korteveg–de-Vries equation, 2, 4, 53–55, 61, 64, 67–88, 89, 90–93, 95, 96, 119, 126–128, 131–133, 143–155, 157, 162, 163, 175–181, 203, 205–209, 216, 238, 248, 249, 259, 275–280, 284–291, 298, 330, 345, 347–356
   averaged, 4, 68, 70, 143–155

Korteveg–de-Vries equation (*cont.*)
 dispersionless, 4, 53–55, 61, 64, 180
 finite-gap solutions, 69, 79–80, 82–84, 175–180,
  208, 209, 216
 isospectral symmetries, 79–82
 L-equation for, 53, 55
 L-A pair for, 79
 Lax pair for, 79
 linearized, 206
 moving waves, 67
 nonisospectral symmetries of, 67–88
 quasi-periodic solutions, 4
 reflecting coefficient for, 54
 shock problem for, 176–179
 symmetries of, 69
 Tsarev equations for, 90, 162
 wave breaking problem for, 89, 90–93
 weak dispersion limit of, 2, 4
KP, *see* Kadomtsev–Petviashvili equation
KP flow, 117, 118, 125
KP hierarchy, 119
KP vertex operator, 130
Krasnoselskii theorem, 306
Krichever–Novikov equation, 119, 127, 131–133
Krichever–Novikov's formalism, 120, 123
Krichever's scheme, 109
Kruskal symmetries, 149, 150, 152
Kuzmak–Whitham averaging ansatz, 105, 114

L-A pair, 79
Lagrangian, 173
Lagrangian submanifold, 278, 280, 281, 286, 289
Laminar state, 315, 316
Laplace–Beltrami type operator, 290
Laplace integral, 261
Lax entropy, 237
Lax equations, 61–66, 167
Lax equivalence theorem, 227, 233
Lax–Levermore minimizer, 157–164
Lax–Levermore problem, 203
Lax–Levermore–Venakides theory, 14, 135, 162
Lax pair, 6, 79, 81, 257
Legendre transform, 305
Lie algebras, 289
Lie derivative, 77
Line bundle, 118, 120–122, 126
Linear fractional transformation, 123, 127
Linear stability, 318–319
Liouville equation, 233–234
Liouville theorem, 186
Liouville tori, 105–116
Lorentz gas of hard spheres, 219–227, 229–230
Lorentz gas with accomodation reflection, 221–223,
  227–228, 230–233
Lyapounov functional, 316
Lyapounov–Schmidt decomposition, 302, 304–305
Lyapounov theorem, 316

Magnetic field, 183–201
Markov partition, 219, 233
Maslov method, 289

Modulation equations, 27–28, 37, 39–52, 89, 331–
  333, 342, 345
Moduli space, 76, 78, 273
Monodromy, 273–295
Moving frame, 258
Multifrequency averaging theory, 203–217
Multiphase asymptotics, 204, 215
Multi-phase averaged system, 135, 143, 149

Nash–Moser method, 297, 302, 304
C. Neumann problem, 279, 285, 286, 288
Neumann system, 105, 111, 112, 114
NLS equation, *see* non-linear Schr_dinger equation
Nondissipative shock waves, 89–92, 97, 98, 99, 102
Nonisospectral integrable equations, 69
Nonisospectral symmetries, 78–79, 82–87
Non-linear Schrödinger equation, 21–38, 55, 61, 93–
  95, 169, 180, 235–255, 276, 279–281,
  283, 291, 298, 328
 defocusing, 21–27, 89, 93–94, 235–238, 242–
  245, 248, 249, 251, 276, 279, 283
 focusing, 27–37, 235, 236, 242–245, 249, 283
 Galilean invariance of, 28
 semiclassical theory of, 21–37, 235–255
Nonlinear wave equation, 299, 303, 305
Non-linear waves, 297–313
NSW, *see* nondissipative shock waves
Number density, 220–221, 223–225, 228, 233
Numerical scheme, 329, 342

Optical fibers, 24–27
Optical pulses, 24
Optical shocks, 21–37
Oscillations arising in numerical experiments, 329–
  346

Painleve equation, 105
Parabolic structure, 121, 122
Pauli matrices, 183
Period matrix, 41–43, 78
Periodic solutions, 297, 299, 300–302, 304, 307,
  310
Period-two oscillations, 331–333, 336–345
Perturbed modulation equations, numerical integra-
  tion of, 44–51
Phase space, 220, 276, 279, 280, 297, 302
Poisson bracket, 62, 144, 174, 236, 247
Potential, 80, 81, 90, 94–96, 105, 107, 112, 150,
  177, 179, 211, 235, 236, 240, 299, 305–
  307, 311
Potential metric, 146, 151–153
Presoliton solutions, 291
Pre-chaotic, 40, 52
Principle of least action, 305–306, 311, 312

Quadric, 274, 276, 288
Quasiclassical limit, 53–59, 207
Quasimomentum, 72, 80, 211
Quasi-periodic geodesic, 277, 279
Quasi-periodic solutions, 286, 291, 297
Quasi-simple wave, 89, 90, 92, 99–103

Reflection laws, 219–222, 234
Regularization of integrals, 53, 57
Reynolds number, 315
Resonance, 105, 203–217, 297–313
Resonant group, 300–302, 304
RH problem, *see* Riemann-Hilbert problem
Riccati equation, 54, 124
Riemann bilinear relations, 72–74, 268
Riemann constants, 78, 81
Riemann (diagonal, characteristic) form, 68, 84, 89,
    254
Riemann–Hilbert problem, 176, 183–200, 249, 254
Riemann invariants, 22, 62, 68–70, 92, 143, 238,
    239, 249, 255
Riemann manifold, 290
Riemann metric, 290
Riemann problem, 75–78
Riemann surface, 3, 8, 10, 11, 16, 41–43, 56, 59, 65,
    70–78, 80, 81, 112, 143, 146, 216, 254,
    269, 274, 276, 280, 282–284, 286, 287,
    289
    branch points of, 3, 8, 10, 11, 16, 41, 56, 59, 65,
        70, 71, 74, 76, 80, 106–108, 110, 111,
        143
    cycles on, 42–43, 71–72, 106, 108, 282
    deformations of, 75–78
    differentials on, 71–75, 106
    genus of, 70, 71, 75, 81, 106, 254, 269
    local parameter, 71
    matrix of periods, 78
    moduli space of, 75–78
Riemann theta function, 41, 78, 106, 112, 113, 118,
    204, 211
Riemann wave, 92, 99
Riemannian metric, 144
Robust, 40, 50
Runge–Kutta method, 333

Saddle point method, 322, 324
Schrödinger equation / Schrödinger operator, 21–38,
    55, 61, 65, 93–95, 149, 169, 176, 180,
    235, 241, 242, 247, 249, 290, 318, 342,
    348
Schur polynomials, 131
Schwarzian derivative, 127
Semiclassical theory, 24–27, 235–255, 289, 291
Semihamiltonian system, 144–146, 149–154
SG, *see* sine-Gordon
Sheaf, 117, 118, 126
Shift operator, 6, 8
Shock discontinuities, 330
Shock dynamics, 357–363
Shock front speed, 259, 261, 270
Shock manifold / shock surface / shock curve, 357–
    359
Shock wave solutions / shock waves / shocks, 2, 4,
    11, 13, 17, 21–37, 176, 179, 329, 330,
    333, 357, 359–362
Shock waves, regularization of, 11–18
Short optical pulses, 24–25
Simplectic operator, 299

Simplectic structure, 274, 275, 298
Sine–Gordon, 39–41, 89, 149, 275, 283, 291
Sobolev space, 196, 350, 351
Soliton / solitonic, 23, 68, 76, 79, 99–103, 154, 167–
    169, 173, 177, 180, 208, 251, 273–291, 330
Sound speed / sound velocity, 15 17, 18, 22
Spectral equations, compatibility conditions for, 6
Spin, 183
Steepest descent, 249, 253
Strang-type convergence theorem, 330, 332
Strange gas law, 23
String theory, 70
Sturm–Liouville operator / Sturm–Liouville prob-
    lem, 105, 106 205, 211, 299, 307
Supercritical convection, 315, 316

t-function, 70, 275, 283
Taylor vortices, 315, 316
Temple class of systems, 145, 154
Toda chain, 289
Toda equations, 269
Toda lattice, 1–19, 173, 180, 257, 258, 259, 270,
    289, 331, 333
Toda shock problem, 257–271
Torus, 105, 216, 274, 276, 280, 300, 302, 310, 312
Transference, 117–133
Transport equation, 22, 35, 213, 219–234, 240, 241
Transverse Ising chain, 183–201
Tsarev equations, 55, 63, 90, 92, 161, 162

Umbilic, 273–279, 283–286, 288
Unit shift operator, 6

"Vacuum" eigenfunctions and eigenvalues, 208, 209
Virasoro algebra, 69–70
Vortex lattices, defects of, 315–328

Wave breaking problem, 89–104
Wave operator, 118
Wave packet, 203, 204
Wavenumber, 24, 28, 316, 325
Wavetrain, 2, 3, 7, 8, 39–40, 45, 49
Webs, 154
Weierstrass P-function, 67, 123
Whitham averaged system, 135–141
Whitham averaging method, 331
Whitham deformations, 105–116
Whitham equation, 1–19, 61, 65, 67–88, 94, 99, 108,
    115, 137, 143, 145, 146, 148–153, 157–
    162, 179, 204, 255, 270, 273
    for KdV, 3ff, 180
    for Toda lattice, 5–10, 17
    genus of solution, 2, 13, 16, 17
    hyperbolic structure of, 10–11
    nonisospectral symmetries, 84–87
    Riemann diagonal form, 68, 84
    symmetries of, 67–88
Whitham–Flaschka–Forest–McLaughlin equations,
    204, 212
Whitham hierarchy, 87
Whitham modulation equation, 89

Whitham system, 109, 110
WKB-asymptotics, 214
WKB expansion, 223

WKB method, 68, 240, 249–251
Zabolotskaya–Khokhlov equation, 65

The manufacturer's authorised representative in the EU is Springer
Nature Customer Service Centre GmbH, Europaplatz 3, 69115 Heidelberg,
Germany. If you have any concerns regarding our products, please
contact ProductSafety@springernature.com

Printed and bound by CPI Group (UK) Ltd, Croydon, CR0 4YY
23/04/2026
02095625-0015